Sustainable Crop Protection Strategies

Volume-1

Sustainable Crop Protection Strategies

Volume-1

H.R. Sardana
O.M. Bambawale
D. Prasad

2024

Daya Publishing House®
A Division of
Astral International Pvt. Ltd.
New Delhi – 110 002

First Published, 2010

Reprinted, 2024

ISBN: 978-81-70356-29-5 (HB)

Disclaimer:

Every possible effort has been made to ensure that the information contained in this book is accurate at the time of going to press, and the publisher and author cannot accept responsibility for any errors or omissions, however caused. No responsibility for loss or damage occasioned to any person acting, or refraining from action, as a result of the material in this publication can be accepted by the editor, the publisher or the author. The Publisher is not associated with any product or vendor mentioned in the book. The contents of this work are intended to further general scientific research, understanding and discussion only. Readers should consult with a specialist where appropriate.

Every effort has been made to trace the owners of copyright material used in this book, if any. The author and the publisher will be grateful for any omission brought to their notice for acknowledgment in the future editions of the book.

Published by : **Daya Publishing House®**
A Division of
Astral International Pvt. Ltd.
– ISO 9001:2015 Certified Company –
4736/23, Ansari Road, Darya Ganj,
New Delhi-110 002
Ph. 011-43549197, 23278134
E-mail: info@astralint.com
Website: www.astralint.com

Printed at : **Replika Press Pvt. Ltd.**

प्रो. स्वपन कुमार दत्ता
उपमहानिदेशक (फसल विज्ञान)
Prof. Swapan Kumar Datta
Deputy Director General
(Crop Science)

भारतीय कृषि अनुसंधान परिषद्
कृषि भवन, डा. राजेन्द्र प्रसाद मार्ग, नई दिल्ली - 110114
INDIAN COUNCIL OF AGRICULTURAL RESEARCH
KRISHI BHAWAN, DR. RAJENDRA PRASAD ROAD, NEW DELHI-110114

Foreword

Sustainable production of crops depends upon the several biotic and abiotic factors and particularly on the management practices. Among them, sound pest and disease management strategies contribute substantially to minimize losses caused by pests. Chemical method of control of pests is commonly adopted in many crops the world over. However, the indiscriminate use of pesticides has resulted in pest out breaks and pest resurgence on one hand and health hazards on the other hand. The importance of achieving sustainable food production through the use of eco-friendly sustainable pest management techniques is being realized more and more in the recent past. Thus, with the increasing importance of sustainable agriculture, the concept of integrated pest management (IPM) has emerged. To assess the present scenario in sustainable pest management, it becomes vital to systematically compile the recent advances in pest management to develop better crop protection programmes for future. The present book is an effort in this direction.

The editors of book '*Sustainable Crop Protection Strategies*' Drs. H.R. Sardana, O.M. Bambawale and D. Prasad have made a concerted effort and done a commendable job to compile the work done at various institutions under National Agricultural Research System (NARS) into 38 thought provoking chapters. Various aspects of pest management covered in the book on recent developments in the fields of pheromone technology, indigenous knowledge, biological control, biodiversity, simulation modeling and climate change, forecasting and geospatial technologies have been extensively and elegantly presented.

This book would not only serve as a reference source but also as a guide and inspiration for future research in the area of plant protection to achieve the most cherished goal of economically and ecologically sustainable agriculture. I hope the efforts of editors will certainly enriched the knowledge of the readers of crop protection. I cong ratulate the editors for this timely publication.

S. K. Datta

Swapan Kumar Datta

Preface

Over exploitation of natural resources and excessive chemicalisation of agriculture in the past few decades, have led to poor sustainability of farm production. Over dependency and indiscriminate use of agricultural chemicals have resulted in the problems of pest resurgence, secondary pest outbreaks, development of pest resistance, decimation of natural enemies on one hand and disturbances to environment and health hazards through residues in food chain on the other hand. These side effects have raised public concern about the safety of pesticides. Today we are confronted with the herculaean task of providing environmental and food security to our expanding population. This calls for reorientation of our strategies to minimize the use of chemical inputs in agriculture and depend more on eco-friendly sustainable crop protection approaches without causing disruption to fragile eco-system. Integrated pest management (IPM) is one such environmentally sensitive approach to pest management that relies on combination of common- sense approaches and fits into the concept of sustainable agriculture. IPM incorporates ecological and economic factors into agricultural ecosystem and decision making and addresses public concern about environmental quality and food safety. Today, IPM is considered to be the springboard to sustainable crop management and forms a focus of the present book.

The thirty eight chapters covered in these 2 volumes deal with various facets of pest management. The book begins with a chapter on maintenance of diversity in sugarcane ecosystem highlighting its importance from pest management perspective. The ramifications of introduction of new pests along with the movement of plant materials and germplasm are well known. The chapter 'Indian plant quarantine system' focuses on such sanitary and phytosanitory issues. The role of geospatial technologies, simulation modeling, entomopathogens and chemical pesticides in IPM programmes has been highlighted and discussed. There is need to integrate traditional knowledge of farmers with other advanced techniques in order to develop a profitable and sustainable approach to pest management. The chapter on 'ITK's' stresses upon the indigenous knowledge in cotton ecosystem.

Adequate focus is on pheromone technology which is a major tool for monitoring of pests for initiating pest control measures. The sustainable IPM technologies developed and practiced for major cropping systems *viz;* cotton, rice, vegetables, pulses, oilseeds, oilpalm and spices have been discussed in various chapters. In all, the book comprises of 38 chapters on various topics with recent technological advances such as maintenance of diversity in pest management, national plant quarantine system, geospatial technologies, simulation modeling and climate change, disease resistance, utilization of pheromones, diagnostics of viruses, eco-friendly pest management, mites and rodents pest management, bio-agents and entomopathogens, weed management and IPM in temperate horticultural crops.

The book '*Sustainable Crop Protection Strategies*' attempts to comprehend and consolidate the scattered knowledge on sustainable pest management strategies. The book would not only serve as a reference but also as a guide and inspiration for shaping future research on modern plant protection tactics. The book will also prove useful to all those interested in promoting the cause of IPM so that sustainability in agricultural system and environmental protection for future generations is achieved.

We received a tremendous response and support from all the contributors for preparing their chapters in tune with the theme of book, for which we express our sincere gratitude to them.

We are also grateful to Shri Anil Mittal of Daya Publishing House, New Delhi for timely taking up the task of publication of this book and providing us all technical help required during the preparation and finalization of the manuscript for the book.

H.R. Sardana

O.M. Bambawale

D. Prasad

Contents

— Volume 2 —

Chapter 1

Maintenance of Diversity in Sugarcane Crop System: A Pest Management Perspective

J. Srikanth
Section of Entomology, Sugarcane Breeding Institute,
Coimbatore – 641 007, Tamil Nadu

Introduction

Agricultural systems based on large numbers of different crop species and varieties traditionally bestowed a high degree of genetic heterogeneity on crop habitats. Farming systems with animals and plants further enhanced diversity of agro-ecosystems. Increased specialization of crop and variety practiced in the recent past, however, drastically reduced diversity at farm as well as crop level. Such a trend is often governed by economic significance of the crop to specific industry and as a cash crop sugarcane occupies a prominent position in this category. In India, the constant expansion of sugar industry accelerated the area under its cultivation concomitantly (Anonymous, 2008) leading to largely spatial and temporal monoculture sugarcane tracts. Expansion of area under cultivation fuelled by demand for byproducts in the coming years might further diminish diversity. Although the general negative effects of monoculture on soil fertility and crop productivity are well understood or conjectured, systematic attention or action is seldom extended to quantify such impact on crops like sugarcane. Besides the influence on abiotic environment of agro-ecosystems, narrow diversity affects biotic components such as pests and diseases by providing uninterrupted food over space and time. Socio-economic factors associated with the crop often disallow implementation of radical measures for macro-level manipulation of agro-ecosystem to alter the situation. Nevertheless, diversity that can not be adopted at the agro-ecosystem or individual farm level can be followed on a limited scale, *i.e.* at the individual plot level in a manner that balances the merits and demerits, and enhances the advantage of an inherent stable crop system.

Crop Diversity and Pest Abundance

It is well known that habitat diversity reduces herbivore abundance in complex vegetation systems and the phenomenon is as much applicable to crop systems as it is to natural systems. In a diversified agro-ecosystem, low pest abundance is explained by theories such as spatial dilution of the primary resource, chemical or structural interference with host location and use by herbivores, and enhanced natural enemy populations (Showler *et al.*, 1990). In other words, habitat diversity affects the ability of an insect, especially a specialist, to locate its spatially diluted host plant through visual and olfactory cues. From the parasitoid and predator point of view, properly planned plant diversity, even at the plot level, is expected to provide alternative prey and adult food sources like nectar and pollen. Diversity manipulation has been considered plausible and traditionally practiced with perennial orchards crop. However, the long duration of sugarcane confers similar advantages on the crop in terms of its amenability to polycultures combined with crop geometry.

Nature of Diversity

Two different approaches of diversity manipulation, namely crop-crop diversity and crop-weed diversity can be envisaged at the plot or farm level in agro-ecosystems. Crop-crop diversity can be maintained in the form of intercrops or border crops, the former with wider economic implications and the latter more as trap crops for direct attraction and destruction of insect pests. Overall, intercrops have been an accepted form of polyculture from crop management point of view in several crop systems despite the mixed results they produce. In sugarcane, the minimal utilization of soil, water and sunlight, and the consequent sluggish growth in the formative phase allow cultivation of suitable intercrops in tropical India (Sundara, 2003) where the mild climate imposes no restrictions on cropping pattern of main and intercrops. On the other hand, intercropping has greater relevance in the climatically severe sub-tropical India where the seasonality of autumn and spring sugarcane plantings limit or deny the cultivation prospects of *kharif* or *rabi* crops (Singh, 2007). Sugarcane based intercropping systems with cereals, oilseeds, pulses and other crops affect growth and yield of sugarcane in positive, neutral and negative manner (reviewed in Kailasam, 1994). Similarly, intercrops practiced in sugarcane influence sugarcane pests in positive, negative and neutral manner (reviewed in Srikanth *et al.*, 2000). While the negative impact on pests of main crop is the anticipated outcome of the practice of polyculture, positive effect on pest densities of main crop necessitates special consideration from their management point of view. On the other hand, crop-weed diversity approach through deliberate and selective maintenance of weeds in the crop habitat for natural enemy enhancement and pest reduction, though demonstrated in sugarcane (Ali *et al.*, 1984; Ali and Reagan, 1985; Srikanth *et al.*, 1997), needs strong evidence of economic gains to receive acceptance as a concept since weeds themselves are generally considered pests.

Pest Abundance vis-à-vis Crop-Crop Diversity

The major objective of intercropping is to manipulate pest abundance on the normally tall statured and long duration main crop. Several studies established the manipulative effect of intercrop on pests of sugarcane during the latter's formative phase, albeit in a variable manner. The various obvious and unknown reasons include varied levels of apparency due to structural and phytochemical complexity of the two crops, diverse dilution effects resulting from relative density of the crops, correspondence in the pest species on both crops, etc. However, it is not uncommon to encounter or expect variable pest abundance on the intercrop in polycultures, obviously due to similar reasons, and the damage the intercrop suffers contributes to the overall yield loss of the crop combination.

Effect on Main Crop Pests

The effect of sugarcane polyculture on pests of main crop varied from the desired 'reduction' to undesirable 'enhancement' through 'stable' densities. Such diverse response was recorded in both tropical and subtropical India, with monocot and dicot intercrops, and for limited groups of pests. Since intercrops are maintained in the field for a short period, their effect on those pests that remain active on the main crop during the corresponding period were examined in most of these studies. Thus, early season pests like shoot borer *Chilo infuscatellus* Snellen (Lepidoptera: Crambidae) and top borer *Scirpophaga excerptalis* Wlk. (Lepidoptera: Pyralidae) had been the subject of intense scrutiny in different investigations; internode borer *Chilo sacchariphagus indicus* (Kapur) (Lepidoptera: Crambidae) and root borer *Polyocha (=Emmalocera) depressella* Swinhoe (Lepidoptera: Pyralidae) had been examined in a few studies. Most studies investigated single species whereas some studies included additional species that appeared within the duration of the intercrop.

Density Reduction

In a low shoot borer incidence (1.6–8.9 per cent) study in subtropical India with spice intercrops like coriander, onion, garlic, etc. (Varun *et al.*, 1994), borer reduction amounted to about 80 per cent in intercropped plots. In a two year moderate shoot borer incidence (<13 per cent) study with green gram, cowpea, pigeonpea, sunflower, maize, sorghum, okra, mint, black gram and sunnhemp intercrops, higher borer incidence was recorded in sugarcane sole crop; only green gram significantly reduced incidence of the borer (Sardana, 2000). In another low shoot borer (5.1 per cent) and top borer (8.0 per cent) incidence study in the subtropics, levels of borer attack were the lowest in coriander intercropped sugarcane followed by the combination with potato intercrop; yield and net returns were highest in potato intercropped sugarcane followed by coriander (Giri, 2005).

In one of the earliest studies in tropical India on the impact of intercrops on shoot borer, onion, coriander and *palak* considerably reduced the pest density compared to sugarcane sole crop (Sikchi *et al.*, 1988). In a high shoot borer incidence (39 per cent) study with pulse intercrops like green gram, black gram and soybean, a 51 per cent maximum reduction in borer incidence was observed (Rajendran *et al.*, 1998). Intercropping and *in situ* incorporation of *dhaincha* integrated with other components reduced shoot borer damage by 68.23 per cent (Subramanian and Kannappan, 2003). Among coriander, soybean, French bean and okra intercrops, marked control of shoot borer was obtained in coriander intercropped sugarcane (Rachappa and Naik, 2004). In field experiments with green gram, black gram, cowpea, sunnhemp, *dhaincha* and soybean intercrops that were planted twice, *i.e.* along with sugarcane planting and 120 days after planting (DAP) after incorporating the first intercrop, shoot borer incidence, at a moderate level of 25.8 per cent in control, was significantly reduced in all intercrop combinations (Thirumurugan *et al.*, 2004 a). In sugarcane intercropped with cowpea at 1:1 ratio and incorporated at earthing-up, shoot borer showed 51.2 per cent reduction over control at the considerably high incidence of 37.3 per cent in control (Thirumurugan and Koodalingam, 2005).

In a study on internode borer, coriander, *palak* and *methi* planted one month after plating sugarcane significantly reduced borer intensity compared to sole crop at six months age of sugarcane (Sikchi *et al.*, 1988). In another interesting study with green gram and soybean intercrops planted twice on 90 and 180 DAP, soybean intercrop reduced internode borer incidence to a greater extent than green gram or sole crop (Thirumurugan *et al.*, 2004 b).

In studies on top borer, mainly in subtropical India where due to distinct broods it constitutes a more serious pest than in tropical India, coriander intercrop reduced its incidence (Tewari and Om Prakash, 1980); intercrops like wheat, potato, coriander or mustard reduced its incidence while coriander also reduced *Sesamia inferens* Walker (Lepidoptera: Noctuidae) (Misra and Hora, 1982); spice crops like garlic and fennel reduced its first three broods (Verma *et al.*, 1981); all the spice crops intercropped, *viz.* coriander, *ajavain*, onion, garlic, fenugreek, fennel and mangrail significantly reduced its incidence with the maximum reduction in *ajavain* intercropped cane (Varun *et al.*, 1990); field and vegetable crops like sunflower, *gobhi sarson*, radish and turnip reduced its first brood (Singla *et al.*, 1994). In spring-planted sugarcane, incidence of second brood was the lowest when sunflower was the intercrop during all the three years of experimentation; the variability in row spacing of sugarcane with intercrops did not affect the borer damage significantly (Singla *et al.*, 1996). Sugarcane intercropped with *ajavain* harbored the lowest incidence of the first (2.12 per cent), second (1.88 per cent) and third (2.88 per cent) broods; the crop also registered the lowest pest incidence at harvest (7.84 per cent) and percentage of egg laying (0.0 per cent) (Singh *et al.*, 2003).

In a two-year study on the effect of a variety of cereal, pulse and vegetable intercrops on root borer in summer sugarcane, green gram was consistently effective in reducing early incidence of the borer before the incorporation of intercrops; cumulative incidence of the borer at harvest did not show significant differences (Sardana, 2001).

Density Enhancement

A few studies reported enhancement of pests in sugarcane intercropped with monocots, particularly wheat, maize and jowar. In subtropical India, wheat grown as intercrop significantly increased the incidence of pink borer *S. inferens* (Kalra *et al.*, 1975; Misra and Hora, 1982) and armyworm *Mythimna separata* Walker (Lepidoptera: Noctuidae) (Kalra *et al.*, 1975). Wheat intercrop enhanced shoot borer incidence 14 fold over the sole crop and reduced yield also significantly over other intercrop combinations and sole crop (Giri, 2005). Maize intercrop more or less doubled shoot borer incidence in sugarcane at 1.5 and 2.5 months age. Internode borer intensity was higher by about three times in maize intercrop at six months age and five times in sorghum intercrop at eight months age than in sugarcane sole crop (Sikchi *et al.*, 1988).

Such enhancement was occasionally observed with dicot intercrops too: shoot borer incidence was higher in 25 day and 65 day old soybean intercropped sugarcane than pure sugarcane crop of corresponding age in an on-farm trial in tropical India (Srikanth *et al.*, 2000); among coriander, soybean, French bean and okra intercrops, higher shoot borer attack was noticed in okra combination (Rachappa and Naik, 2004).

Density Constancy

Intercrops, more often dicots, occasionally maintained *status quo* with regard to the abundance of main crop pests in both subtropical and tropical India. In replicated trials in tropical India, shoot borer incidence at 90-100 DAP did not differ significantly among sugarcane intercropped with pulses like green gram, black gram, cowpea and soybean; borer incidence did not differ in a 30 day old soybean intercropped sugarcane in a grower's plot (Srikanth *et al.*, 2000). Soybean, black gram, sunnhemp and *dhaincha* intercrops planted in combination with other factors like crop geometry and nutrient supply systems failed to influence shoot borer population positively or negatively (Srikanth *et al.*, 2002). In another trial combining row spacing and fertilizer treatments in tropical India, onion intercrop did not affect shoot borer incidence (Mahadevaswamy and Martin, 2002).

Internode borer intensity in several intercrop combinations was on par with sugarcane sole crop at eight months age (Sikchi *et al.*, 1988). In field experiments with pulse and green manure crops planted twice, internode and top borers incidence in different combinations did not differ significantly (Thirumurugan *et al.*, 2004 a). Companion crops such as radish, turnip, sunflower and *gobhi sarson* did not reduce damage due to termites (Singla *et al.*, 1994).

Effect on Natural Enemies

Parasitoids

In a study on the effect of spice intercrops on top borer, the extent of parasitism by *Telenomus beneficiens* Zehntner (Hymenoptera: Scelionidae) was 1-7 per cent while parasitism by *Trichogramma* sp. (Hymenoptera: Trichogrammatidae) was 1-4 per cent with no clear-cut differences among intercrops (Singh *et al.*, 2003). The egg parasitoid *Trichogramma chilonis* Ishii was most active in sugarcane-coriander combination while it was less active in okra intercropped sugarcane; shoot borer activity showed a negative trend with *T. chilonis* activity (Rachappa and Naik, 2004). Sequential releases of *T. chilonis* in green gram and soybean intercropped sugarcane produced enhanced parasitism rates to a greater extent in the former than the latter or in sugarcane sole crop (Thirumurugan *et al.*, 2004 b).

Predators

The more apparent general groups of predators were examined in an experimental plot with green gram, black gram, cowpea and soybean intercrops. While the abundance of predators, comprising spiders and coccinellids, was significantly lowest in cowpea intercrop plot, it was uniform in the other intercrops and sugarcane sole crop; similarly, predator numbers did not differ significantly between soybean intercrop and monocrop in three on-farm plots of different age (Srikanth *et al.*, 2000). In a similar study in South Africa, pitfall trap catches of arthropods–comprising beetles, ants and spiders–were five times more abundant in sugarcane intercropped with maize or sorghum than the sole crop (Conlong, 1995).

Effect on Intercrop Pests

Intercrops generally harbored low pest abundance in most polyculture studies analyzed (Risch, 1983) and the situation seems to be true for sugarcane too. For example, in a sugarcane-cotton diculture with different cotton varieties, the latter did not suffer from jassids or boll worms (Wankhede and Parashar, 1975); beans planted as intercrop with sugarcane suffered low levels of leaf hoppers and leaf beetles (Garcia *et al.*, 1979); population build-up of two noctuid pests in potato was lower when intercropped with sugarcane than with vegetable or maize (Parihar and Singh, 1992). Enhancement in intercrop and/or decrease in the main crop of pest incidence were, however, observed in other monocot-pulse combinations (Mahadevan and Chellaiah, 1985; Leroi *et al.*, 1990), though not in sugarcane.

Crop-Weed Diversity vis-à-vis Pest-Natural Enemy Complex

Several studies demonstrated the influence of deliberately maintained weedy habitats of sugarcane crop on activity of predators: predatory ants showed differential foraging ability in sugarcane selectively maintained as grass, broad-leaf or weed-free habitats, with broad-leaf plants providing more prey per amount of dry biomass (Ali *et al.*, 1984); greater abundance and diversity of both ground and foliage associated predators and prey were supported in weedy habitats which enhanced returns by 19 per cent (Ali and Reagan, 1985); activity of predators was enhanced and borers were reduced (Showler *et al.*, 1990); abundance of predatory spiders increased significantly (Srikanth *et al.*, 1997). Weedy habitats also negatively influenced pest populations occasionally: shoot borer incidence in sugarcane crop maintained with different weed regimes showed significant differences at 75 DAP; despite the overlapping and indistinct nature of the differences, a general decreasing trend in plots maintained weedy beyond 30 DAP and an opposite trend in plots maintained weed free beyond 30 DAP was observed (Srikanth *et al.*, 2002).

Overview

A close examination of polyculture studies conducted with sugarcane and an array of intercrops reveals interesting causes that apparently govern the diverse effects produced. In studies on shoot borer, the 80 per cent 'density reduction' with the spice intercrops coriander, onion, garlic, etc. in subtropical India (Varun *et al.*, 1994) was marked by low borer incidence (1.6–8.9 per cent), far below the general economic threshold of 15–22 per cent (David and Sithanantham, 1986). As opposed to this, the 51 per cent maximum reduction of borer with the pulse intercrops green gram, black gram and soybean (Rajendran *et al.*, 1998) in the high borer incidence (39 per cent) study still maintained the pest around the general economic threshold in tropical India. Other low to medium shoot borer intensity studies with onion, coriander (Sikchi *et al.*, 1988; Rachappa and Naik, 2004; Giri, 2005) and green gram (Sardana, 2000; Thirumurugan *et al.*, 2004 a) supported the suitability of both spices and pulses as intercrops. Reduction of top borer in sugarcane intercropped with coriander, *ajavain*, onion, garlic, fenugreek, fennel (Tewari and Om Prakash, 1980; Verma *et al.*, 1981; Misra and Hora, 1982; Varun *et al.*, 1990; Singh *et al.*, 2003), sunflower, *gobhi sarson*, radish and turnip (Singla *et al.*, 1994 and 1996) in subtropical India also proved the suitability of spice, field and vegetable crops alike. These studies not only indicated that intercrop effect is independent of borer density, agro-climatic conditions and type of crop but also suggested the role of other factors. In the more infrequent 'density enhancement', as much as 14 fold enhancement of borers with the monocot intercrops wheat, maize and jowar (Kalra *et al.*, 1975; Misra and Hora, 1982; Sikchi *et al.*, 1988; Giri, 2005) prompts the conclusion of taxonomic, structural or phytochemical similarity inducing allurement of pests involved as the obvious cause. However, enhancement of shoot borer with the dicot intercrops soybean (Srikanth *et al.*, 2000) and okra (Rachappa and Naik, 2004) suggested that enhancement too, like reduction, is independent of type of crop. In studies that showed 'density constancy' with a variety of intercrops, their evaluation in conjunction with other factors (Mahadevaswamy and Martin, 2002; Srikanth *et al.*, 2002) and documentation of pest intensity apparently after harvest of intercrop (Sikchi *et al.*, 1988) or late in the season after two cycles of intercrops (Thirumurugan *et al.*, 2004 a) may have partly confounded or overrated the intercrop effect. However, the distinct stability effect in experimental plots and growers' farms (Srikanth *et al.*, 2000) pointed to the validity of the phenomenon albeit with exclusion of type of crop as possible cause in view of the variety of the dicot crops evaluated. No effect on termite damage with radish, turnip, sunflower and *gobhi sarson* companion crops (Singla *et al.*, 1994) ruled out the role of spatial dilution and structural or chemical interference in host location by the subterranean pest. Since most intercrop studies used equal proportions of main and intercrops populations, the variation in results of various studies could partly be due to the variable numbers of contiguous rows of each crop alternating with the other. For example, main and intercrops planted in variable numbers of rows reduced shoot borer (Varun *et al.*, 1994) and top borer (Verma *et al.*, 1981) incidence. Besides, intercrops planted one month after planting sugarcane significantly reduced internode borer incidence (Sikchi *et al.*, 1988). Thus, crop geometry and temporal separation of main and intercrops could result in different levels of spatial dilution of the primary resource and chemical or structural interference with host location. Despite evidence that variability in row spacing of sugarcane with intercrops did not affect shoot borer (Mahadevaswamy and Martin, 2002) and top borer damage (Singla *et al.*, 1996) significantly, establishment of effects of crop geometry and sequential planting in dicultures necessitates simultaneous evaluation of several spatial and temporal combinations under identical agro-climatic conditions.

Increase in sheath moisture due to trash mulching is known to reduce shoot borer incidence (Avasthy and Tiwari, 1986). Significant reduction of shoot borer when *dhaincha* (Subramanian and Kannappan, 2003) and cowpea (Thirumurugan and Koodalingam, 2005) intercrops were incorporated in to the soil suggested such mulching effect. Uniform shoot borer incidence in nutrient combinations of organic form, biofertilizers and incorporation of *dhaincha* intercrop (Srikanth *et al.*, 2002) suggested lack of effect of organic matter content, despite the possibility of confounding effect of variables evaluated. The variable root borer incidence at harvest of sugarcane, long after incorporation of intercrops, was suggested to be due to increased organic matter (Sardana, 2001). However, such enhancement could be due to other nutritional factors and natural buildup of the pest and not due to the moderate levels of nutrients added by the incorporated intercrop. In general, the direct impact of an agronomic practice like intercropping on pests of main crop can be expected to last until the intercrop is harvested at about three months age of the main crop. The lack of effect of second intercrop incorporated at seventh month age on internode borer incidence in the 10th month (Thirumurugan *et al.*, 2004 a) obviously indicated that intercrops did not play any role in visual or chemical interference with host location by pests of the dense sugarcane four months after its planting.

Sugarcane crop habitat *per se* displays considerable diversity and abundance of natural enemies like spiders (Easwaramoorthy *et al.*, 1994), ground beetles (Easwaramoorthy *et al.*, 1997) and ants (Easwaramoorthy *et al.*, 1998). This could probably be due to the crop's undisturbed nature, natural weed cover in and around cropped area, and availability of alternative food for general predators. Crop-crop diversity can influence activity of these general predators more by providing suitable habitat than alternative food during the short life period of intercrop, despite the possibility of decreased predation rate and abundance of certain insect predators and parasitoids in high density polycultures (Risch *et al.*, 1982). The less significant effect of sugarcane-pulse diversity systems (Srikanth *et al.*, 2000) than sugarcane-cereal combinations (Conlong, 1995) on abundance of general predators like beetles, ants and spiders pointed to the non-generality of the suggestion that broad-leaf plants in crop habitats like sugarcane provide more prey per amount of dry biomass (Ali *et al.*, 1984) and consequently support greater natural enemy activity. Also, the limited role for spiders in the reduction of borers (Srikanth *et al.*, 2000) could be due to the general feeding habits of the former (Easwaramoorthy *et al.*, 1996) and tissue boring habits of the latter. Due to their oligophagous habit, predatory coccinellids are likely to be influenced to a greater extent than spiders by crop-crop phytochemical complexity. However, they would be beneficial only if they proliferate on some pest of the intercrop and, after its harvest, move over to sugarcane and feed on eggs of different early season borers or sucking pests. The intercrops attempted with sugarcane hitherto, with the exception of wheat and the pest *S. inferens*, neither harbor pests nor natural enemies common to both crops. This is more pertinent to predatory coccinellids that might buildup populations on the intercrop but may face lack of prey on the main crop after the harvest of intercrop, eventually leading to their dispersal from the habitat. This would have been applicable to the invasive woolly aphid *Ceratovacuna lanigera* Zehntner (Homoptera: Aphididae) for whose control intercropping with pulses has been suggested (Srikanth, 2004). The aphid generally occurs from fifth month after planting, a couple of months after harvest of any pulse intercrop, which would force the potential oligophagous predatory coccinellids out of the habitat. On the other hand, greater parasitism of the egg parasitoid *T. chilonis* on shoot borer in sugarcane-coriander diculture than sugarcane-okra combination (Rachappa and Naik, 2004) and on internode borer in sugarcane-pulse combination than sole crop (Thirumurugan *et al.*, 2004 b) suggested a definite role for allelochemicals in the tri-trophic interactions involving host-specific parasitoids.

Natural enemies generally play a more important role in crop-weed diversity systems than in annual crop-crop diversity systems in population regulation of herbivores (Baliddawa, 1985). Weeds within the plot could exert the same effect as intercrops in enhancing spatial dilution, and altering structural and phytochemical complexity. The luxuriant *Parthenium hysterophorus* L. that often outgrows sugarcane seedlings is more likely to interfere structurally or chemically with host location by shoot borer than enhance the abundance of natural enemies. Despite instances of positive effect on ground and foliage associated predators and prey in weedy habitats with enhanced crop returns (Ali and Reagan, 1985), the abundant general predators like spiders in weedy crop systems (Srikanth *et al.*, 1997) could have minimal impact on pest populations due to their general feeding habits (Easwaramoorthy *et al.*, 1996) and the reasons discussed above. Selective manipulation of weed flora within individual plots or in the habitat would unravel the abundance, foraging habits, relative predatory efficiency, and impact of different parasitoid and predatory groups on pest populations. Such studies might also reveal the possible interference among predatory groups, as was observed between ants and coccinellids (Srikanth *et al.*, 2001).

In a majority of polyculture studies involving several phytophagous species, more than 50 per cent of the pest species were less abundant in the intercrop (Risch, 1983). Such reduction in pests of cotton (Wankhede and Parashar, 1975) and potato (Parihar and Singh, 1992) in sugarcane polycultures indicated benefit to the intercrop. Monocot-pulse combinations unpredictably reduced borer incidence in the monocot crop, probably due to oviposition deterrence (Mahadevan and Chellaiah, 1985), but increased seed losses in pulse crop due to bruchids (Leroi *et al.*, 1990). Although no such enhanced pest incidence on intercrop was recorded in sugarcane polycultures, growers of soybean intercrop (Srikanth *et al.*; 2000) observed that sucking pests often warranted use of insecticides. This may enhance their load in the otherwise low insecticide consumption sugarcane crop (David, 1987) with possible impact on natural enemy complex. Nevertheless, low pest attack on beans sown 45 days after sugarcane (García *et al.*, 1979) and on late planted crop in both combinations in other crop systems like maize and bean diculture (Altieri *et al.*, 1978), obviously due to differential canopy growth, suggested that pest problems in intercrop, and main crop too, can be tackled by staggered planting of sugarcane and intercrops. Similarly, asynchrony with peak pest activity can be achieved by manipulating the planting date in the conventional manner.

Conclusion

Judicious vegetational diversification was suggested to lead to an ecologically stable plant protection system in sugarcane (Ali *et al.*, 1984; Ali and Reagan, 1985; Showler *et al.*, 1990). Crop-crop diversity is an acceptable agronomic practice relatively easy to adopt, more due to enhanced productivity than pest reduction that comes as an entomological value addition. Since intercrops by default are of short duration, pests that occur in the first 100 days of the crop such as shoot borer, top borer, early attack of woolly aphid, whitefly in endemic areas and termites are more likely to be influenced by the adoption of the practice. Selection of intercrop is indeed governed by a combination of regional, economic, agronomic and finally entomological considerations. Crops like wheat and potato are specifically suitable for subtropical India while a variety of pulses, spices and vegetabies can find use in tropical India too. Wheat, maize and sorghum should be scrupulously avoided in view of their taxonomic affiliation with sugarcane as well as their history as enhancers of pest densities on sugarcane (Kalra *et al.*, 1975; Misra and Hora, 1982; Sikchi *et al.*, 1988; Giri, 2005); so is the case with soybean (Srikanth *et al.*, 2000) and okra, the latter enhancing pest densities in tropical (Rachappa and Naik, 2004) but not subtropical (Sardana, 2000) India. Coriander, which was as good as potato in terms of net returns but suppressed shoot borer most, should be preferred in pest endemic areas (Giri, 2005) to obtain long term protection against the pest. The relative response of currently cultivated pest susceptible and resistant varieties should be determined under intercropping to utilize them efficiently in polyculture systems. Agronomically, intercropping requires adoption of wide-row and paired-row planting methods which, incidentally, are a pre-requisite for large-scale mechanization of sugarcane cultivation in the coming years. Micro-irrigation and fertigation through surface and sub-surface drip irrigation practices, which improve water use efficiency and yield (Ramesh *et al.*, 1994) and incidentally conserve predatory spiders (Srikanth *et al.*, 1997), may become incompatible with intercropping. Similarly, planting of intercrop twice, *i.e.* first at planting of sugarcane and second after harvest of first intercrop, though reduced pest incidence in sugarcane (Thirumurugan *et al.*, 2004 a and b), is not only unfeasible but also impedes regular agronomic operations. Despite evidence in favor of enhanced natural enemies (Srikanth *et al.*, 1997) and significantly lower borer population in weedy plots (Srikanth *et al.*, 2002), the possibility of maintaining such vegatative diversity may be limited by agronomic considerations as well as growers' acceptance of the concept. At best, maintenance of weedy habitat could be an incidental benefit arising out of minimum-effort sugarcane cultivation by resource-poor farmers. In the long run, crop-weed and/or crop-crop diversity would be the forerunner for a more diversity-oriented agro-ecological approach for sugarcane pest management.

Acknowledgements

I had my first exposure to sugarcane based intercropping through my association with late Dr. C. Kailasam, Principal Scientist (Agronomy). He was liberal in giving me access to his experimental plots for pest related investigations and frank in his discussions on the merits and demerits of the subject. I gained first hand knowledge of the prospects and problems when I visited intercropped fields of growers with the personnel of M/s Sakthi Sugars Ltd., Tamil Nadu. Dr. N. Vijayan Nair, Director, Sugrcane Breeding Institute, Coimbatore, encouraged by allowing me to crystallize my ideas in this compilation.

References

Ali, A.D. and Reagan, T.E., 1985. Vegetation manipulation impact on predator and prey populations in Louisiana sugarcane ecosystem. *J. econ. Ent.* 78(6): 1409-1414.

Ali, A.D., Reagan, T.E. and Flynn, J.L., 1984. Influence of selected weedy and weed-free sugarcane habitats on diet composition and foraging activity of imported fire ant (Hymenoptera: Formicidae). *Environ. Ent.* 13(4): 1037-1041.

Altieri, M.A., Francis, C.A., Schoonhoven, A. van and Doll, J.D., 1978. A review of insect prevalence in maize (*Zea mays* L.) and bean (*Phaseolus vulgaris* L.) polycultural systems. *Fld Crops Res.* 1(1): 33–49.

Anonymous 2008. Sugar statistics. *Coop. Sug.* 39(7): 45-79.

Avasthy, P.N. and Tiwari, N.K., 1986. The shoot borer, *Chilo infuscatellus* Snellen. In: H. David, S. Easwaramoorthy and R. Jayanthi (eds), *Sugarcane Entomology in India*. Sugarcane Breeding Institute, Coimbatore, India, pp. 69-92.

Baliddawa, C.W., 1985. Plant species diversity and crop pest control: an analytical review. *Insect Sci. Applic.* 6(4): 479-487.

Conlong, D.E., 1995. Results of preliminary pitfall trapping trials for potential arthropod predators of *Eldana saccharina* Walker (Lepidoptera: Pyralidae).. *Proc. Ann. Congr. S. Afr. Sug. Technol. Ass.* 69: 79-82.

David, B.V., 1987. Trends in production and use of pesticides in India. In : M. V. Rao and S. Sithanantham (eds), *Plant Protection in Field Crops, Natn. Semin. Pl. Prot. Fld Crops.* Central Plant Protection Training Institute, Hyderabad, India, pp. 1–14.

David, H. and Sithanantham, S., 1986. Integrated pest management. In: H. David, S. Easwaramoorthy and R. Jayanthi (eds), *Sugarcane Entomology in India.* Sugarcane Breeding Institute, Coimbatore, India, pp. 459-473.

Easwaramoorthy, S., David, H., Kurup, N.K. and Santhalakshmi, G., 1994. Studies on the spider fauna of sugarcane ecosystem in southern peninsular India. *J. Biol. Control* 8(2): 85-93.

Easwaramoorthy, S., Srikanth, J., Santhalakshmi, G. and Kurup, N.K., 1996. Life history and prey acceptance of commonly occurring spiders in sugarcane ecosystem. *J. Biol. Control* 10: 39-47.

Easwaramoorthy, S., Srikanth, J. and Kurup, N.K., 1997. Ground beetles in sugarcane ecosystem: new records and seasonal fluctuations. *J. Soil Biol. Ecol.* 17(1): 73-75.

Easwaramoorthy, S., Kurup, N.K. and Santhalakshmi, G., 1998. New records and population dynamics of minor arthropod predators in sugarcane ecosystem. *Insect Environ.* 4: 98-99.

Garcia, J., Cardona, C. and Raigosa, J., 1979. [Evaluation of populations of insect pests in mixed sugarcane and beans and their relation to yield]. Evaluación de poblaciones de insectos plagas en la associacion cana de azucar–frijol y su relacion con los rendimientos [Es] *Revta Colombiana de Entomologia* 5(1/2): 17–24.

Giri, S., 2005. Influence of different intercrops on the incidence of borer pests, productivity and profitability of autumn planted sugarcane in Gangetic alluvial zone of West Bengal. *Indian Sug.* 55(2): 105-108.

Kailasam, C., 1994. Evaluation of nitrogen levels and seed rates for short duration sugarcane (Co 8338) intercropped with three soybean cultivars of varying growth habits. *Ph.D. Thesis* Tamil Nadu Agricultural University, Coimbatore, India.

Kalra, A.N., Varma, A. and Srivastava, A.N., 1975. Companion cropping of sugarcane and wheat: pest problem and how to tackle it. *Sug. News* 7(7): 11–12.

Leroi, B., Alzouma, I. and Huignard, J., 1990. The influence of intercropping millet (*Pennisetum typhoides* Burm) with cowpea (*Vigna unguiculata* Walp) on the egg-laying and development of *Bruchidius atrolineatus* (Pic) (Coleoptera: Bruchidae). *Agric. Ecosystems Environ.* 31(1): 39–48.

Mahadevan, N.R. and Chelliah, S., 1985. Influence of intercropping legumes with sorghums on the stem borer, *Chilo partellus* (Swinhoe). In: A. Regupathy and S. Jayaraj (eds), *Behavioural and Physiological Approaches in Pest Management.* Tamil Nadu Agricultural University, Coimbatore, India, pp. 188–190.

Mahadevaswamy, M. and Martin, G.J., 2002. Weed infestation and shoot borer incidence in sugarcane intercropped with aggregatum onion (*Allium cepa* var. *aggregatum*) under garden land conditions. *Sug. Tech* 4(3 and 4): 165-167.

Misra, S.C. and Hora, B.S., 1982. Note on the incidence of top-borer (*Tryporyza nivella* Fabr.) and pink-borer (*Sesamia inference* Wlk.) on sugarcane grown in companion cropping pattern. *Indian J. agric. Sci.* 52(1): 45–46.

Parihar, S.B.S and Singh, O.P., 1992. Effect of cropping patterns on build up of *Agrotis ipsilon* and *Helicoverpa armigera* (Hub.) in potato. *Natn. Acad. Sci. Lett.* 15(2): 57–58.

Rachappa, V., Naik, L.K., 2004. Integrated management of early shoot borer, *Chilo infuscatellus* (Snellen) in sugarcane. *Ann. Pl. Prot. Sci.* 12(2): 248-253.

Rajendran, B., Haneefa, A.M., Umapathy, G. and Venkatakrishnan, J., 1998. Cultural measures to curb shoot borer menace in sugarcane. *Insect Environ.* 4(2): 48–49.

Ramesh, P., Kailasam, C. and Srinivasan, T.R., 1994. Performance of sugarcane (*Saccharum officinarun* L.) under surface drip, sub surface drip (biwall) and furrow methods of irrigation. *J. Agron. Crop Sci.* 172: 237–241.

Risch, S.J., 1983. Intercropping as cultural pest control: prospects and limitations. *Environ. Mgmt* 7(1): 9–14.

Risch, S.J., Wrubel, R. and Andow, D., 1982. Foraging by a predaceous beetle, *Coleomegilla maculata* (Coleoptera: Coccinellidae) in a polyculture: effects of plant density and diversity. *Environ. Ent.* 11(4): 949–950.

Sardana, H.R., 2000. Impact of companion cropping on the incidence of early shoot borer, *Chilo infuscatellus* of sugarcane. *Ann. Pl. Prot. Sci.* 8(2): 239-240.

Sardana, H.R., 2001. Influence of summer inter-cropping on the incidence of root borer, *Emmalocera depressella* of sugarcane. *Indian J. Ent.* 63(1): 49-51.

Showler, A.T., Reagan, T.E. and Knaus, R.M., 1990. Sugarcane-weed community interactions with arthropods and pathogens. *Insect Sci. Applic.* 11(1): 1-11.

Sikchi, B.R., Patil, A.S., Hapase, D.G., Kangane, J.B. and Jaiswal, P.M., 1988. Effect of intercropping on borer incidence in sugarcane. *Coop. Sug.*, 19(11): 797-801.

Singh, J.P., 2007. Studies on intercropping in sugarcane crop. *Coop. Sug.* 38(8): 41-44.

Singh, S., Dixit, A., Singh, S.B. and Yadav, S.S., 2003. Effect of spices intercropping on the incidence of top borer (*Scirpophaga excerptalis* Wlk.) and its biocontrol agents in sugarcane. *Proc. Ann. Conv. Sug. Technol. Ass. India* 65: A178-A185.

Singla, M.L., Duhra, M.S., Dhlaliwal, Z.S. and Sandhu, S.S., 1994. Effect of intercropping in autumn planted sugarcane on the incidence of *Scirpophaga excerptalis* Wlk. and termites. *J. Insect Sci.* 7(2): 199–201.

Singla, M.L., Dhlaliwal, Z.S., Duhra, M.S., and Sandhu, S.S., 1996. Effect of intercropping on the incidence of top borer, *Scirpophaga excerptalis* Wlk. in the spring crop of sugarcane. *J. Insect Sci.* 9(1): 86-88.

Srikanth, J., 2004. The epidemic of sugarcane woolly aphid *Ceratovacuna lanigera* Zehntner (Homoptera: Aphididae) in western India: an appraisal. In: N. Balasundaram (ed.), *Proceedings of National Seminar on Use of Appropriate Varieties and Management Practices for Improving Recovery of Sugarcane.* Sugarcane Breeding Institute, Coimbatore, January 19-20, 2004, pp. 169-180.

Srikanth, J., Easwaramoorthy, S., Kurup, N.K. and Santhalakshmi, G., 1997. Spider abundance in sugarcane: impact of cultural practices, irrigation and post-harvest trash burning. *Biol. Agric. Hort.* 14: 343-356.

Srikanth, J., Easwaramoorthy, S. and Kurup, N.K., 2000. Borer and predator incidence in sugarcane intercropped with pulses. *Sug. Tech* 2(1&2): 36-39.

Srikanth, J., Easwaramoorthy, S. and Kurup, N.K., 2001. *Camponotus compressus* F. interferes with *Cryptolaemus montrouzieri* Mulsant activity in sugarcane. *Insect Environ.* 7(2):51-52.

Srikanth, J., Salin, K.P., Easwaramoorthy, S. and Kailasam, C., 2002. Incidence of sugarcane shoot borer under different levels of weed competition, crop geometry, intercropping and nutrient supply systems. *Sug. Tech* 4(3 and 4): 149-152.

Subramanian, S. and Kannappan, K., 2003. Evaluation of IPM components in the management of shoot borer of sugarcane in Cauvery delta region, Tamil Nadu. *Coop. Sug.* 34(11): 879-882.

Sundara, B., 2003. Sugarcane based intercropping systems for tropical India. *Coop. Sug.* 34(12): 963-971.

Tewari, R.K. and Om Prakash 1980. Effect of intercrops on sugarcane top borer. *Fmr and Parliament* 15(7): 35–36.

Thirumurugan, A. and Koodalingam, K., 2005. Ecofriendly approaches for the management of sugarcane shoot borer. *Coop. Sug.* 36(7): 571-575.

Thirumurugan, A., Venkatachalam, S.R. and Shah, S.E., 2004 a. Management of borer complex in sugarcane through companion cropping under drought condition of Palar river basin area. In: N. Balasundaram (ed.), *Proceedings of National Seminar on Use of Appropriate Varieties and Management Practices for Improving Recovery of Sugarcane.* Sugarcane Breeding Institute, Coimbatore, January 19-20, 2004, pp. 187-190.

Thirumurugan, A., Shah, S.E. and Venkatachalam, S.R., 2004 b. Enhancement of *Trichogramma chilonis* activities against internode borer, *Chilo sacchariphagus indicus* through intercrops and botanicals. In: N. Balasundaram (ed.), *Proceedings of National Seminar on Use of Appropriate Varieties and Management Practices for Improving Recovery of Sugarcane.* Sugarcane Breeding Institute, Coimbatore, January 19-20, 2004, pp. 181-186.

Varun, C.L., Singh, S. and Singh, H.N., 1990. Effect of intercropping of spices on the incidence of top borer (*Scirpophaga excerptalis* Wlk.) in sugarcane under *bhat* soil condition. *Indian Sug.* 39(10): 751-752.

Varun, C.L., Singh, S., Pandey, K.P. and Singh, S.B., 1994. Influence of companion cropping of spices on the incidence of early shoot borer (*Chilo infuscatellus* Snell.) in sugarcane. *Indian Sug.* 44(1): 21-22.

Verma, R.S., Motiwale, M.P., Chauhan, R.S. and Tewari, R.K., 1981. Studies on intercropping of spices and tobacco with autumn sugarcane. *Indian Sug.* 31(7): 451-456.

Wankhede, N.P. and Parashar, K.S., 1975. Studies on intercropping of cotton (*Gossypium hirsutum* L.) in spring planted sugarcane (*Saccharum officinarum* L.). *Indian Sug.* 24(11): 951-954.

Chapter 2

Evolution of National Plant Quarantine System in India in Tune with International Phytosanitary Regime

B. Sarath Babu[1], K.S. Varaprasad[1], Kavita Gupta[2] and R.K. Khetarpal[2]

[1]National Bureau of Plant Genetic Resources–Regional Station, Rajendra Nagar, Hyderabad – 500 030, Andhra Pradesh

[2]National Bureau of Plant Genetic Resources, Pusa Campus, New Delhi – 110 002

Introduction

India has an agrarian economy which plays a leading role in national economic transformation. The increasing production demands on the limited agricultural resources however threaten the sustainability of its agricultural ecosystems. Within the available framework of resources and opportunities in agricultural production systems, pests, diseases and harmful pesticides in combination have threatened the natural ecosystems in India and continue to reduce crop yields and pose a major threat to life and environment. Amongst the natural disasters such as invasion by migratory pests like locusts and army worms cause severe damage to variety of food crops leading to total loss of production systems in Africa and Asia. Invasive alien pests or endemic pests also contribute towards the chronic losses on an annual or seasonal basis. We do have records of new pests causing destruction of crops that sometimes lead to devastation, hunger and famines. Therefore, it is not quite hard to understand the ramifications of introduction of new pests along with the movement of plant germplasm and agricultural produce.

The devastating effects resulting from diseases and pests introduced along with international movement of planting material, agricultural produce and products are well documented. (Ram Nath *et al.*, 1988). The Irish famine of 1845, which forced the people to migrate *en masse* from Europe, was the result of almost total failure of potato crop due to attack of late blight pathogen (*Phytophthora infestans*) introduced from Central America. Powdery mildew (*Uncinula necator*) root eating aphid (*Viteus vitifoliae*) and downy mildew (*Plasmopara viticola*) of grape was introduced in succession into France in mid nineteenth century from North America. These pests caused extensive damage to the grapevines and virtually annihilated the wine industry of the country. Coffee rust (*Hemileia vastatrix*) appeared in Sri Lanka in 1875. The disease was so severe that production went down from 400 million pounds in 1870 to just 5 million in 1889. The disease entered India in 1876 from Sri Lanka and within a decade, the coffee Industry of South India was badly affected. Germany put embargo in 1875 on the import of potatoes from USA to keep out the dreaded Colorado potato beetle (*Leptinotarsa decemlineata*). In 1898, most European countries put embargo on American import directed mainly to keep out San Jose´ scale (*Quadraspidiotus perniciosus*). Among the first few countries to promulgate comprehensive plant quarantine regulations were France, Germany, USA, UK and Australia.

In India, heavy losses in grain yield of *Cicer arietinum* crop in states of Haryana, Madhya Pradesh, Punjab and adjoining areas occurred during 1981-82 due to the introduction of virulent biotype of Aschochyta blight from the Middle East. Bunchy top of banana entered India from Sri Lanka and causes loss to banana of over Rs 4 Crores annually (Khetarpal and Ram Nath, 1998). The dreaded Golden nematode (*Heterodera rostochiensis*) introduced from Europe along with exotic seed material in 1960s has been causing severe infestation of potato tubers in the Nilgiris. A noxious weed *Lantana camara* was introduced as ornamental plant in 1809 from Central America, which spread to wastelands and pastures causing displacement of native species. All these examples clearly demonstrate that imported seed/planting material, especially bulk imports, without proper quarantine may result in introduction and establishment of quarantine pests into new areas which may severely damage the crop production and economy of a nation.

India being a signatory to the WTO Agreement quickly adopted itself to the international order and modified its domestic laws in order to ensure its due share in the international trade in general and to promote its overall growth in exports. Plant quarantine system in India has been continuously evolving strongly and positively in tune with the international developments. An account of these developments in sanitary and phytosanitary issues which eventually had a bearing on evolution of Indian plant quarantine system is discussed in detail in the present chapter.

International Developments

The World Trade Organization (WTO) established on January 1, 1995, is the legal and institutional foundation of the multilateral trading system. It provides the principal contractual obligations determining how countries frame and implement domestic trade legislation and regulations. It is the platform on which trade regulations among countries evolve through collective debate and negotiation. It has a broad scope in terms of commercial activity and trade policies to which it applies. Several provisions under the International Treaty on Plant Genetic Resources (ITPGRFA) gave foundation for developing and imparting safeguards in movement and exchange of plant material in connection with international programmes on food and agriculture (Murthy *et al.*, 2007). The WTO Agreement contains some 29 individual legal texts covering everything from services to government procurement, rules of origin and intellectual property. The WTO Agreement on Agriculture is designed to provide increased fairness in farm trade. The one on intellectual property, Agreement on Trade Related

Intellectual Property Rights will improve conditions of competition where ideas and inventions are involved. The Agreement on Technical Barriers to Trade ensures that technical regulations and certification procedures do not create obstacles to trade (http://www.wto.org).

It is the Agreement on application of Sanitary and Phytosanitary (SPS) Measures, which is directly related to the science of plant protection and has major implications on phytosanitation and trade. The SPS Agreement aims to overcome health-related impediments to market access by encouraging the "establishment, recognition and application of common sanitary and phytosanitary measures by different Members". Implementation of SPS measures at national and international level requires lot of scientific studies and data generation before it could benefit the stakeholder and promote safe trade (Khetarpal and Rajan, 1999, Khetarpal and Gupta, 2002).

Sanitary and Phytosanitary Measures

SPS measures are defined as any measure applied to;

1. Protect animal or plant life or health within the territory of the Member State from risks arising from the entry, establishment or spread of pests, diseases, disease- carrying organisms or disease causing organisms;
2. Protect human or animal life or health within the territory of the Member State from risks arising from additives, contaminants, toxins or disease causing organisms in food, beverages or foodstuffs;
3. Protect human life or health within the territory of the Member State from risks arising from diseases carried by animals, plants or products thereof, or from the entry, establishment or spread of pests; or
4. Prevent or limit other damage within the territory of the Member State from the entry, establishment or spread of pests.

Existing International Organizations for Developing Standards

In order to achieve the targets international standards need to be developed for which WTO has assigned the responsibility as follows:

For Food Safety

The Codex Alimentarius Commission (CAC) based in Rome is a subsidiary organ of the Food and Agricultural Organization of United Nations and the World Health Organization. The SPS Agreement designates Codex as the authority for all matters related to international food safety evaluation and harmonization. Codex also develops methodologies relating to food additives, veterinary drugs, pesticide residues, and contaminants, methods of analysis and sampling, and codes and guidelines of hygienic practice.

For Animal Health and Zoonosis

Zoonosis is animal diseases transmittable to humans. The standards, guidelines and recommendations are developed under the auspices of the World Organization for Animal Health (IOE), Paris. The IOE develops manuals on animal diseases, standards for diagnosis, vaccination, disease control, disinfection and certification.

For Plant Health

The International Plant Protection Convention (IPPC), Rome plays a vital role in trade as it is the organization recognized by the WTO under the SPS Agreement, as the source for International Standards for the Phytosanitary Measures (ISPMs) affecting trade. The guidelines and recommendations are developed under the auspices of the Secretariat, IPPC.

The above three organizations are often referred to as the "Three Sisters". They are observers and important contributors to SPS committee meetings. They are also called in as experts to give advice to WTO dispute settlement panels. For matters not covered by the above organizations, appropriate standards, guidelines and recommendations are promulgated by other relevant international organizations open for membership to all members of WTO, as identified by the SPS Committee.

The WTO agreements suggest that countries not only update their quarantine or plant health services to improve health status of their exports; they should also incorporate pest risk analysis (PRA) for making regulatory decisions during imports and exports. To this end the Food and Agriculture Organization (FAO) has been working in consultation with various regional plant protection organizations such that of Europe and North America. It has published ISPM Part 1 on Guidelines for pest risk analysis, dated February, 1996 under the provisions of SPS Measures. The standards, guidelines and recommendations for SPS have been developed under the auspices of the secretariat of the IPPC. The essence of the ISPM and SPS documents is regulatory action for International trade. Imports should be based on an acceptable level of risk through PRA and take transparent action having the least negative effect on the flow of international trade but still commensurate with the level of risk.

Quarantine services of most countries determine the entry status of imported agricultural commodities based on perceived risk. The entry status is the combination of a country's law, regulations, guidelines, policy and decisions of inspectors at ports or points of entry. A biologically based position about entry status is that if the risk (*i.e.* chances of pest entry) is perceived to be high, the entry status of imported articles should be conservative, or conversely if the risk is perceived to be low, the entry status should be liberal. The entry status is also influenced by the use of effective and practical safeguards. If the risk of introduction is low but the entry status is conservative, then it is not based on scientific facts but alleged to be a trade barrier. If the risk is high and the entry is liberal, again the status is not scientifically sound.

A PRA conducted according to international standards can provide the basis for our national planners to allocate resources to areas facing the highest risk and thereby, enabling them to better protect our agriculture and environment from the entry and establishment of exotic pests. PRA offers an improved method for assessing and managing risk by providing a format to consider and perhaps weigh the relative importance of the various factors leading to a more biologically sound approach to pest risk and mitigation. PRA, which consists of risk assessment and risk management, is a procedure by which quarantine services can technically justify new regulations, phytosanitary guidelines or safeguards, operational procedures, entry status of imported articles and resource allocations (Gupta *et al.*, 2002).

The IPPC requirement for a PRA consists of three stages (*viz.*, Initiation, Risk Assessment and Risk Management) whose compilation should be fully documented in the event a review or dispute arises. The PRA should clearly document information sources and the rationale used in reaching a management decision regarding mitigation measures that have been taken or are to be taken.

Plant Quarantine Scenario in India

Plant quarantine is a government endeavour enforced through legislative measures to regulate the introduction of planting materials, plant products, soil, living organisms, etc. in order to prevent inadvertent introduction of pests, pathogens and weeds harmful to the agriculture of a country/ state/region and if introduced, prevent their establishment and further spread (Kahn *et al.*, 1989).

The quarantine measures are of utmost relevance to a country like India whose economy is largely agriculture based. The awareness to quarantine measures in India started in early 20th century when the Indian Government in 1906 ordered compulsory fumigation of imported cotton bales to prevent introduction of Mexican cotton boll weevil (*Anthonomus grandis*). The policy of using government authority to prevent entry of dangerous exotic pests is based on the ground that it is preferable to undergo and tolerate some inconvenience in an effort to exclude pests than to submit to the expenses of controlling them for an indefinite period. Realizing the importance of plant quarantine, Government of India legislated the DIP Act in 1914 under which various notifications have been issued from time to time, restricting the entry of plants and plant material from other countries as well as from one state to another state within the country (domestic quarantine). The latest order promulgated by the government under the DIP Act is the Plant Quarantine (Regulation of Import into India) Order 2003 which superceded the 'Plants, Fruits and Seeds (Regulation of Import into India) Order, 1989' popularly known as PFS Order, 1989 (Sarath Babu *et al.*, 2002).

Three categories of materials are being imported under the PQ Order, 2003: a) bulk consignments for consumption, b) bulk consignments of seeds/planting materials for sowing/planting, and c) samples of germplasm in small quantities for research purposes. The 35 Plant Quarantine Stations under the Directorate of Plant Protection Quarantine and Storage undertake quarantine processing and clearance of consignments of the first two categories. The DIP Act is currently under review and revision. It has specified plants/planting materials that are i) prohibited to imported into India (Schedule IV) ii) permitted for import with additional declarations only through authorized institutions (Schedule V), and iii) permitted to import with additional declarations and special conditions (Schedule VI) under restricted conditions. Import permit and Phytosanitary Certificate are essential for importing plant material into the country. National Bureau of Plant Genetic Resources (NBPGR), being the nodal institution for genetic resources management, has been authorized to issue import permit and to handle quarantine processing of germplasm material including transgenic planting material under exchange for research by both public and private sectors.

There is a need to protect our agriculture against the losses from introduction of these pests without restricting the international movement of people or commodities. One of the primary missions of plant protection organizations/institutions is to identify exotic pests of crops that are important in Indian agricultural scenario and assessing the potential damage that those pests could cause if introduced in our country (Varaprasad *et al.*, 2008).

Out of the three different categories of material being imported, bulk imports for sowing/planting carry maximum risk as thorough examination and treatment becomes difficult and planting area required is also far too large. Quarantine processing is often restricted to smaller samples derived from them and based on results of these samples the whole consignment is rejected/detained or released after giving fumigation treatment or fungicidal/insecticidal treatments. The bulk consignments meant for consumption pose lesser hazards. However, small samples meant for research purposes are of immense quarantine importance. Since these samples usually comprise of germplasm material or wild relatives or landraces of a crop, they are more likely to carry diverse biotypes/races/strains of the pest/pathogen. In case of true seed generally risks are more with deep-seated infections than with surface-borne contamination. However, import of vegetative propagules present a much higher order of risk than true seeds.

Plant quarantine operations in India are carried out by the Directorate of Plant Protection, Quarantine and Storage, which functions under the aegis of the Ministry of Agriculture. The administrative structure of plant quarantine is shown in the organizational chart (Figure 2.1).

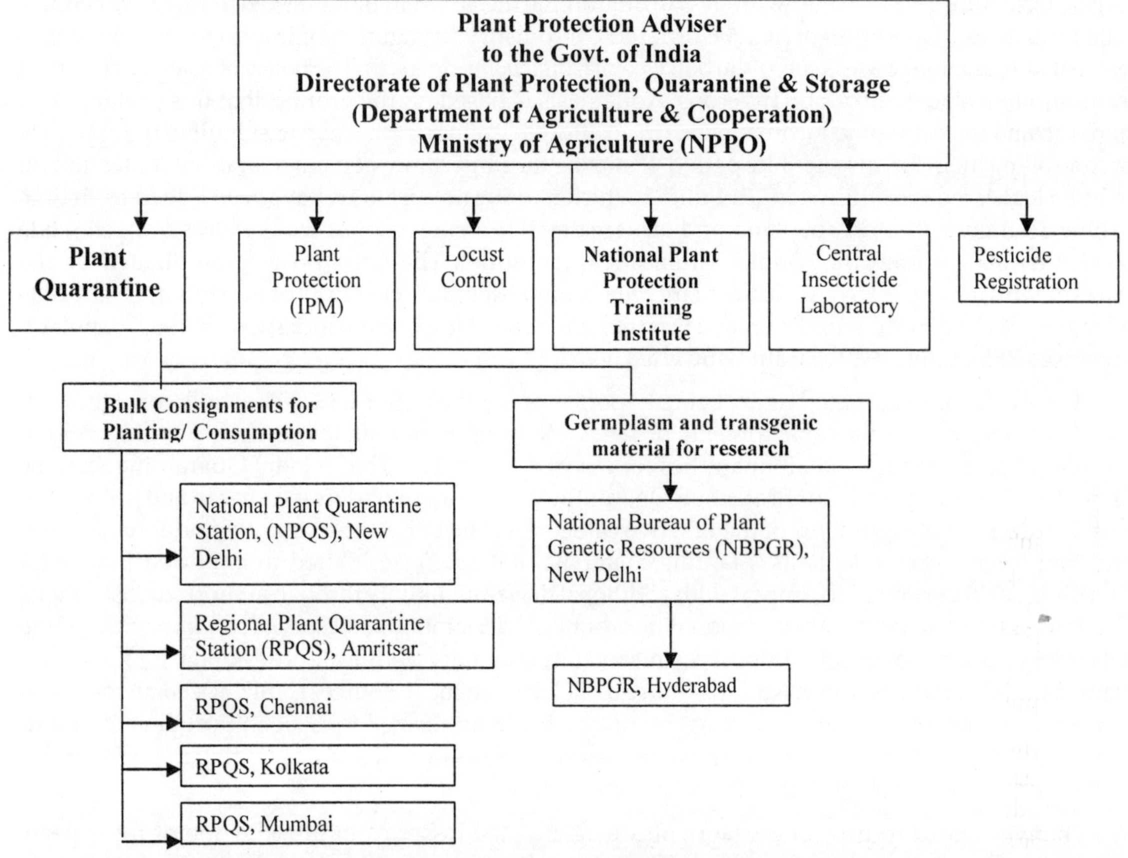

Figure 2.1: DPPQS Organizational Chart

The development of the Plant Quarantine (Regulation of Import into India) Order, 2003 (referred to hereafter as "the PQ Order") reflects the primary plant quarantine concerns of the Government of India. These are to:

1. Prevent the introduction and spread of exotic pests that are destructive to the country by regulating the import of plants and plant products through adequate policy and statutory measures
2. Support India's agricultural exports through credible export certification
3. Facilitate safe global trade in agriculture by assisting producers, exporters and importers and by providing technically comprehensive and credible phytosanitary certification.

Prior to the notification of the PQ Order in November 2003, the hitherto existing plant quarantine statute and regulations dated back as early as 1914. Until recently, regulatory measures in India operated on the basis of The Destructive Insects and Pests Act, 1914, which was promulgated to prevent introduction and spread of destructive pests affecting crops, and the Plant, Fruits and Seeds (Regulation of Import into India) Order, 1989. Some other Rules were promulgated for regulating import of live insects (1941), fungi (1943) and cotton (1972).

The New Seed Policy, 1988 was formulated to provide Indian farmers with access to the best available seeds and planting material, domestic and imported. The Plants, Fruits and Seeds (Regulation of Import into India) Order, 1989, prohibiting and regulating the import into India of plants, plant materials and the like, is based on post-entry quarantine checks. It has now been replaced by the PQ Order of 2003.

Role of NBPGR in Plant Quarantine

NBPGR is the nodal institution for exchange of plant genetic resources and has been empowered under Plant Quarantine (Regulation of Import into India) Order 2003 to handle quarantine processing of germplasm and transgenic planting material being imported for research purposes in the country by both public and private sectors. The NBPGR Regional Station at Hyderabad undertakes quarantine inspection and clearance of the germplasm meant for the International Crops Research Institute for Semi-Arid Tropics, Patancheru, Andhra Pradesh and other institutes of South India (Sarath Babu, 1997 and Varaprasad *et al.*, 1997). NBPGR has developed well-equipped laboratories and green house complex. A containment facility of CL-4 level has been established for processing transgenics. At NBPGR, adopting a workable strategy, a number of pests of great economic and quarantine importance have been intercepted on exotic material, many of which are yet not reported from India *viz.*, insects like *Acanthoscelides obtectus* in *Cajanus cajan*, *Anthonomus grandis* in *Gossypium* spp., *Ephestis elutella* in *Macadamia* nuts and *Vigna* spp., *Quadrastichodella eucalytii* in Eucalyptus, nematodes like *Heterodera schachtii*, *Ditylenchus dipsaci*, *D. destructor*, *Rhadinaphelenchus cocophilus*, etc. in soil clods and plant debris as contaminations with crop seeds of sugar beet, onion and garlic and coconut respectively, fungi like *Claviceps purpurea* in seeds of wheat and barley, *Peronospora manshurica* on soybean, *Fusarium nivale* on wheat, barley and *Aegilops*, *Uromyces betae* on sugarbeet, bacteria like *Xanthomonas campestris* pv. *campestris* on *Brassica* spp. and viruses like *Barley stripe mosaic virus* in barley, *Cherry leaf roll virus* on French bean, *Cowpea mottle virus* in cowpea, *Pea seed borne mosaic virus* in broad bean, *Tomato black ring virus* on cowpea and French bean, etc. (Khetarpal and Kavita Gupta, 2006). Rich information concerning various aspects has been published by NBPGR including a comprehensive publication on weed species(Anitha *et al.*, 2007) whose entry into the country is prohibited in the wake of promulgation of PQ Order 2003 and its enforcement.

Compliance to International Developments

With a view to modernize, upgrade and enhance the outdated plant quarantine system, its capacities and the related legal and administrative framework, the Government of India has recently approved the notification of a new Plant Quarantine Order for the country. This new Order is a step forward in harmonizing India's regulatory framework with the International Plant Protection Convention and internationally accepted standards and the tenets of the SPS Agreement of the WTO. Other supporting and managerial steps are also being taken to improve, to international standards, the entire gamut of the country's quarantine activity and phytosanitary border controls, including import and export inspections, on-field surveillance for pests and vectors, treatment standards and processes, and certification methodology. India is making imports of plants and plant materials subject to pest risk analysis to protect its crops from risk of introduction of alien pests. Efforts are also under way to improve the export certification process and standards to ensure that such phytosanitary certification gives an assurance of freedom from quarantine and regulated pests and vectors, including alien species for importing countries.

Development and Enforcement of New National Quarantine Legislation

There were many reasons prompting the development of new legislative provisions for plant quarantine in India. They include the following:

1. Liberalized trade in agriculture as a consequence of the WTO Agreements, although offering economic opportunities, also implied fresh challenges of amplified pest risks, a corollary to the increased volumes and array of international trade in agricultural commodities.
2. With diverse agroclimatic zones, varied agricultural produce and surplus food production, India finds itself in a position to exponentially expand its agricultural trade. However, what were required were accredited national standards for all critical phytosanitary activities. Relevant entities and other interested parties invariably felt the need for compliance with stringent international phytosanitary regulations.
3. There was an urgent need to fill the gaps in the existing plants, fruits and seeds order of 1989, namely by regulating the import of germplasm, genetically modified organisms and transgenic plant material, as well as live insects and fungi (including biocontrol agents etc).
4. India needed to facilitate safe conduct of global trade in agriculture and thereby fulfill its legal obligations under the relevant international agreements.
5. It was necessary to protect the interest of the country's farmers by preventing the entry, establishment and spread of destructive pests, vectors and alien species.
6. It was necessary to protect the national plant life and environment.
7. It was necessary to safeguard the national biodiversity from threats of invasive alien species.

The need was felt for incorporation in regulation of additional special declarations for the freedom of import commodities from quarantine and alien pests. These declarations would be based on standardized pest risk analysis, particularly for seeds and planting materials.

Salient Features of the PQ Order

1. The PQ Order 2003 has widened the scope of plant quarantine activities with the incorporation of additional definitions and made pest risk analysis a precondition for imports.
2. It places a prohibition on the import of commodities contaminated with weeds and/or alien species. Import of packaging material of plant origin is restricted unless the material has been treated.
3. The new Order includes provisions for regulating the import of:
   - ☆ Soil, peat and sphagnum moss
   - ☆ Germplasm, genetically modified organisms and transgenic material for research
   - ☆ Live insects, microbial cultures and biocontrol agents
   - ☆ Timber and wooden logs.
4. Agricultural imports are now classified as: (a) prohibited plant species; (b) restricted species where import is permitted only by authorized institutions; (c) restricted species permitted only with additional declarations of freedom from quarantine pests and subject to specified

treatment certifications; and (d) plant material imported for consumption or industrial processing permitted with normal phytosanitary certification.

5. A permit requirement is now enforced on imports of seeds, including flower seeds, propagating material and mushroom spawn cultures. Additional declarations are specified in the new Order for the import of >600 agricultural commodities, specifically listing as many as >300 quarantine pests and 31 weed species.
6. Notified points of entry have increased dramatically: there are now 130 such entry points, where previously there were 59. The new Order also rationalizes the structure of certification fees and inspection charges.

Till October 2008, 12 amendments to the PQ Order were notified revising definitions, clarifying specific queries raised by quarantine authorities of various countries, with revised lists of crops under the Schedules VI, VII and quarantine weed species under Schedule VIII. The revised list under Schedules VI and VII (including proposals made in four draft amendments) now include 641 and 288 crops/ commodities, respectively. The Schedule VIII now include 31 quarantine weed species. The PQ Order ensures the incorporation of "Additional/Special Declarations" for import commodities free from quarantine pests, on the basis of pest risk analysis (PRA) following international norms, particularly for seed/planting material (http://www.agricoop.nic.in/gazette.htm).

New Order in Harmony with the IPPC

India's new Order for plant quarantine aligns with the framework of the International Plant Protection Convention in several important ways:

1. The phytosanitary measures are designed to prevent global spread of noxious pests and are based on justified scientific principles with pest risk analysis as the cornerstone.
2. Provisions have been made applicable to packages and transportation.
3. Inspection and certification provisions accord with Article IV of the IPPC.
4. Phytosanitary certificates are in the format of Article V of the IPPC and according to the plant quarantine requirements of the importing country. They are to be issued after careful inspection and the required treatment.
5. There is emphasis on capacity enhancement and development and training of staff.
6. The new Order has been hosted on the Web site of the Ministry of Agriculture (and is available at www.agricoop.nic.in) and made accessible to everyone. It is transparent and applies uniformly to all exporting countries or parties.

Issue of Import Permits

Features of the new regulation of imports include the following:

1. Separate formats have been devised for applications for the issue of import permits and also for the permit letters issued for consumption purposes as opposed to those for propagative plant materials.
2. Commercial imports of seeds of coarse cereals, pulses, oil seed, fodder crops and planting materials of fruit plant species require prior clearance.
3. Applications for seeds and planting materials must be accompanied by (1) a registration certificate issued by the National Seeds Corporation or the Director of Agriculture or Director

of Horticulture of the state government and (2) a certificate of approval of post-entry quarantine facilities issued by the designated inspection authority.

4. Permits are to be issued within a maximum period of three working days of submission of an application.
5. Pest risk analysis has been made a precondition for import of new agricultural commodities.
6. Permits for import of soil or peat and for import of live insects, microbial cultures or biocontrol agents are to be issued only by the Plant Protection Adviser, the technical head of the plant quarantine service.
7. Permits for import of germplasm, genetically modified organisms and transgenic material are to be issued by the Director, National Bureau of Plant Genetic Resources, New Delhi.
8. Issued permits are valid for six months. This may be extended a further six months.
9. Permits are not transferable and no permits are to issued for landed consignments.
10. Relaxations from the conditions of the new Order, necessitated by emergency or unforeseen circumstances, are to rest with the Ministry of Agriculture.

Addressing Wider Ramifications–Biosecurity

Biosecurity is emerging as one of the most pressing issues before the international community. Essentially, it means the "management of all biological and environmental risks associated with food and agriculture, including forestry and fisheries", a sector that covers food safety, and the life and health of plants and animals. The risks include everything from genetically modified organisms, alien species and introduced plant and animal pests, to the erosion of biodiversity, the spread of transboundary cattle diseases, toxic weapons of war, and "mad cow" disease etc.

The need for biosecurity in food and agriculture has intensified with economic globalization, rapid improvements in communications, transport and trade, technological progress, and increased awareness of biodiversity and the environment. Internationally, the most comprehensive coverage of the issue is provided in the WTO Agreement on the Application of Sanitary and Phytosanitary Measures (or SPS Agreement). In environmental agreements–notably the Convention on Biological Diversity (CBD)–countries have agreed to take measures to protect biodiversity, and the CBD's Cartagena Protocol on Biosafety, adopted in January 2000, provides an international regulatory framework for the transboundary movement of living modified organisms (LMOs). Countries have agreed to cooperate, through Codex Alimentarius, the International Plant Protection Convention and the International Office of Epizootics, to protect human, plant and animal health, respectively.

The broad scope and complexity of biosecurity is reflected in the myriad of international agreements, laws and intergovernmental bodies created to address it. In fact, in an FAO report to the Committee on Agriculture, the international policy and regulatory framework for biosecurity have been described as "disjointed and incomplete". The report recommends a global approach to the issue, and calls for technical assistance to help developing countries establish, rationalize and optimize their national capacity for biosecurity.

The FAO report notes that "most international instruments deal with aspects of biosecurity from distinctly sectoral perspectives, whether this be disarmament, biological diversity, wetland protection, plant, animal or human health, marine resources management, nature conservation or concern with the introduction of genetically modified organisms or invasive alien species and genotypes." The SPS Agreement, which provides disciplines for the application of sanitary, phytosanitary and zoosanitary measures together, "does so predominantly from the point of view of their impact on trade, rather than their own intrinsic effectiveness". Needed is a coordinated, though not necessarily unified, approach to biosecurity, drawing on common components that have emerged so far. "In particular," FAO says, "there is a need to critically examine areas of potential overlap and perhaps areas of potential conflict, such as the interrelationship between sanitary and phytosanitary measures on the one hand, and the control of invasive alien species, biological diversity and the control of GMOs on the other."

The primary aim of a more coordinated international regulatory framework–taking into account agricultural sustainability, trade issues, food security, and environmental protection and conservation–would be to avoid duplication and inconsistencies among the various international and regional organizations and instruments. At national level, a similar rationalization is needed. Despite trends toward greater coordination, national authority for biosecurity matters is still scattered over a variety of ministries–typically agriculture, health, environment, forestry, fisheries, and trade and industry. There may be a particular need, FAO says, for technical assistance to developing countries, both in the implementation of national controls and negotiation of instruments at the international level.

Need for Modernizing National Plant Quarantine System

The establishment of WTO coupled with the liberalization policies of the government has provided unlimited opportunities for flow of materials. Quarantine set-up needs to be upgraded as this could result in globalization of pests. Further, the increase in import of transgenic planting material and the stipulation of WTO under its SPS Agreement has brought about additional elements of challenge to be faced by quarantine personnel.

Looking at our National Plant Protection Organization in totality there is a need to upgrade its manpower, infrastructure and capabilities to bring it up to the international standards. A National Plant Quarantine Information System needs to be established to develop a national database on PRA, Pest Free Areas, quarantine methodologies, policies and related issues. There is a need to reprioritize our research projects based on critical gaps identified to meet the requirement for export and import. A few that need immediate emphasis are:

- ✰ Survey and surveillance of disease/pests of national and international importance to have endemic pest data
- ✰ Generation of a comprehensive epidemiological data on important pests in order to be able to fix *tolerance limits*, to develop PRA and identify PFAs as per the WTO norms
- ✰ Development of diagnostic protocols using molecular techniques for detection of exotic pests and their variants not easily detectable otherwise but of grave economic concern
- ✰ Development of digitized keys for identification of exotic pests
- ✰ Evaluation of the biosafety risks in import of transgenics and beneficial biocontrol agents
- ✰ Development of methods to detect the presence of transgenic/GM contamination (deliberate or inadvertent mixing of GM seed with non-GM seed) in unknown sample during quarantine processing.

There is also a dire need to develop an accessible platform for getting information on biosecurity for the policy makers, administrators and the industry groups. A website, *viz.*, http://www.plantquarantineindia.org consisting of national database on legislation, quarantine procedures, methodologies, etc. designed by DPPQS is available. Compilations on potential quarantine pests of various crop groups are being taken up at NBPGR (Dev *et al.*, 2005). However, an internet-based portal mechanism for exchange of official information on food safety, and animal and plant health and the environment (like the International Portal for Food Safety and Animal and Plant Health), to facilitate communication among countries in these sectors needs to be developed for easy accessibility of information.

To address the issue of emerging pests in proper perspective recently, there was an endeavour by the government to establish a National Institute for Emerging and Exotic Plant Pests with a focused mandate of development of an early warning system for the pest outbreak, and models for pest/disease forecasting and risk analysis; development of novel diagnostic approaches for detecting pests, imparting multiple resistance to enhance the durability of resistance by pyramiding genes or developing management strategies by developing transgenics resistant to diseases. This institute could possibly work in network mode to facilitate research by those laboratories that are not well equipped and may boost research in case of many regional and chronic plant diseases (Khetarpal and Gupta, 2007).

Finally, awareness about the biosecurity among scientific fraternity and general public is greatly required to generate respect for the regulations in the interest of national security. Besides, the domestic quarantine regulations already promulgated for eight destructive pests (fluted scale, San José scale, coffee berry borer, codling moth, banana bunchy top and mosaic viruses, potato cyst nematode, potato wart and apple scab) seem to be a complete failure as there are reports of spread of these pests to neighboring states and also to different parts of the country. Extensive eradication measures and surveys are needed in regions of different states that would finally contribute in identifying the pest free areas, a task that is also required to be accomplished as per the international norms for fulfilling the commitments made under the SPS Agreement. There is also a dire need to check not only spread of above pests but also to promulgate domestic quarantine against certain important pests which have been introduced/detected in the country in the recent years and which are likely to spread fast.

1. American serpentine leaf miner (*Liriomyza trifolii*) from Karnataka in 1991
2. Spiraling white fly (*Aleurodiscus dispersus*) recorded from Tamil Nadu in 1993
3. Papaya ring spot virus in Maharashtra and Madhya Pradesh in 1994
4. Banana bract mosaic potyvirus and banana streak mosaic badnavirus in Trichy area of Tamil Nadu in 1995
5. Sunflower downy mildew (*Plasmopara halstedii*) recorded for the first time from Maharashtra in 1984
6. Peanut stripe potyvirus initially recorded at Raichur in Karnataka in 1987

These pests are now being reported from certain other States. Unlike in USA and Australia, there is a total lack of organized services of plant quarantine at State level in India. It is essential to set-up plant quarantine check-posts at inter-state border roads/inland container depots/rail roads manned by trained personnel to control the movement of planting material, transport carriers and containers so as to have effective check of the pests.

The importance of the subject of plant protection and intricacies involved in the risk analysis during trade is amply demonstrated by the following case study on wheat import.

Wheat Import: A Case Study

In the year 2005-06 there has been six million tonnes of wheat import from various countries such as Australia, Canada, Ukraine etc for which crop-based PRAs were conducted by the Government.

Wheat import conditions as per PQ Order 2003. The specifications for import of wheat for consumption/ processing are as follows:

Additional Declarations Required to be Incorporated into Phytosanitary Certificate

Freedom from:

1. Granary weevil (*Sitophilus granarius*)
2. Ergot (*Claviceps purpurea*)
3. Dwarf bunt (*Tilletia contraversa*)

Special Conditions of Import Include

Fumigation with Methyl bromide @ 32 gm/cu. m at @ 28 °C and above under NAP and the treatment shall be endorsed on phytosanitary certificate or by any other fumigant/substance in the manner approved by the Plant Protection Adviser for this purpose.

Besides, Import also Requires Freedom from the Following Quarantine Weed Species (Schedule-VIII of the PQ Order)

1. *Allium vineale*
2. *Ambrosia maritime*
3. *Ambrosia psilostachya*
4. *Ambrosia trifida*
5. *Apera-spica-venti*
6. *Bromus secalinus*
7. *Cenchrus tribuloides*
8. *Centaurea diffusa*
9. *Centaurea maculosa*
10. *Centaurea solstitialis*
11. *Cichorium pumilum*
12. *Cichorium spinosum*
13. *Cordia curassavica*
14. *Cuscuta australis*
15. *Cynoglossum officinale*
16. *Echinochloa crus-pavonis*
17. *Froelichia floridana*
18. *Helianthus californicus*
19. *Helianthus ciliaris*
20. *Heliotropium amplexicaule*
21. *Leersia japonica*
22. *Matricaria perforatum*
23. *Polygonum cuspidatum*
24. *Proboscidea lovisianica*

25. *Salsola vermiculata*
26. *Senecio jacobaea*
27. *Solanum carolinense*
28. *Striga hermonthica*
29. *Thesium australe*
30. *Thesium humiale*
31. *Viola arvensis*

The Ministry of Food, Govt. of India started importing wheat into India during the year 2006-07 mainly for Public Distribution System (PDS) throughout India. A total of 60,88,565 MTs wheat had been imported through various Ports such as Chennai, Cochin, Tuticorin, Mumbai, Vizag, Kakinada, Kandla, Mundra and Manalore from various countries viz; Australia, Russia, Hungary, Ukraine, Canada, France, Kazakhstan, Romania, Bulgaria, Czech. Republic, the Netherlands, Argentina.

Relaxed Phytosanitary Norms for Import of Wheat in to India Based on Recommendations made by Technical Committee of Experts (Valid from July, 2006–July 2007)

Wheat grains for consumption and processing to be allowed import by private importer as well as by STC on govt. account up to in accordance with the following conditions:

1. The wheat consignment would be required to be certified to be free from granary weevil (*Sitophilus granarius*).
2. The wheat consignment would be required to be certified to contain less than 0.01 per cent Ergot (*Claviceps purpurea*) w/w.
3. The wheat consignment would be required to be certified to contain less than 0.005 per cent Dwarf bunt (*Tilletia controversa*)
4. The wheat consignment would be required to be certified to contain less than average of 100 quarantine weed seeds, as per Schedule VIII of this Order, per 200 kg of samples drawn as per the prescribed procedure.
5. The wheat consignment would be required to be fumigated with Methyl Bromide @32 g/cu m at 21°C and above for 24 hours under Normal Atmospheric Pressure and the treatment endorsed on the Phytosanitary Certificate OR the wheat consignment would be required to be fumigated with Aluminium Phosphide @ 40g/1000 cu ft for a period of at least 21 days prior to arrival into India provided the ship hold has provision for post re-circulation of phosphine gas during the period of transit.

Ever since the revised phytosanitary conditions came into force, 47 out of total 145 consignments imported from Australia, Russia, Canada, Ukraine, France, Romania, Czech. Republic, the Netherlands and Argentina were found to be infested/infected with quarantine pests of consignments imported from various countries. Besides, a large number of non-quarantine pests especially weeds were also intercepted from the wheat consignments. The PRA conducted revealed that the spread of serious exotic pests was unavoidable in case the imported material is distributed by FCI under Public Distribution System (PDS) or sold by private traders as:

1. Sclerotia of Ergot (*Claviceps purpurea*) intercepted from 39 of consignments could survive under adverse conditions and remain viable in soil for upto 10 years and can survive on wild hosts/weeds prevalent in the wheat growing areas.

2. Unrestricted distribution throughout the country via PDS with multiple rural outlets might also pose serious threat to the wheat growing areas by spreading infected grain and weed seeds.
3. Majority of weed seeds intercepted could easily spread to wheat growing areas through air, water, transport, planting materials, etc. and in addition to agriculture may also pose serious health problems as evidenced in case of introduction of Parthenium seeds along with imported wheat from USA in early 1950s.

(*Source:* DAC, Ministry of Agriculture)

Wheat is a staple food for a large population of India and invasion through grains by any of the quarantine pests might result in loss of food for a large population. Hence, using PRA as a decision making tool, India could technically justify its decision of importing wheat from specific countries based on a rational risk analysis prior to import and avoid the potential threat posed by the exotic pests to its biosecurity.

Task Force for Plant Protection

The Government of India began a process of consultation on how to improve the capacity of the Plant Protection/Quarantine activities in the country in June, 2007. This was undertaken with the aim to introduce flexibility in the public sector; improve information sharing and analysis afforded by new Information and Communications Technology; and a general desire to improve transparency and provide more efficient customer service in the area of plant protection. A Coordination Committee on Plant Protection (CCPP) was constituted to ensure the validity of this plan and key public sector stakeholders were involves in reviewing the plan. Five major areas were identified as most significant based on the experience of problems in the areas plant protection and quarantine, one of them being Strengthening Phytosanitary Capacity in India. The need to upgrade the technical phytosanitary capacity in India was driven by a number of factors:

- ✰ Heightened risks associated with changing agricultural practices and/or human-induced environmental changes
- ✰ Projected increases in India's trade in agricultural commodities implying the need for strengthening and better coordination in routine tasks of sampling and testing shipments along with adequate testing facilities and staff
- ✰ Structural change in the Indian public sector to introduce more flexibility and transparency and a willingness to outsource services
- ✰ Emergence of new technologies for information sharing and access to data

Although there has been considerable support for investment in the area, by both FAO and UNDP, the Plant Quarantine Network in India still provides a rather uneven coverage when viewed across the whole country. The better-equipped, resourced and run Plant Quarantine Stations (PQS), and certain of the Regional Plant Quarantine Stations, are able to run a very professional regime of inspection and certification for exports, and of inspection with a view to making pest or disease interceptions for imports. However, others lack trained staff, infrastructure, and management system to implement such measures fully. As there is free movement within the country once products have been imported, the overall system of plant quarantine is only as good as its weaker links.

The CCPP short-listed five action points for the Task Force based on the discussions held in July 2007 New Delhi as follows:

1. Update and develop SOPs which are needed across the PQ system to ensure consistency and effectiveness in the system as a whole. This needs to include emergency response system which is presently missing in Indian PQ.
2. Develop information sharing tools to improve consistency and efficiency of the PQ system, internally within PQ stations on interceptions, endemic pest information, exotic pest information, develop a national database of phytosanitary certificates (PSCs), to eventually create a one single system issue of of PSCs in order to allow users to apply online and develop import inspection system and release of consignments compatible with the Customs department
3. Undertake capacity building in terms of human resources by imparting trainings on SOPs. Training need assessment required based on which training modules, frequency of training to be finalized.
4. Develop a strategy for public awareness on plant biosecurity / PQ matters
5. Creation of Plant Quarantine authority- conceptualizing, preparation of legislation, preparation of EFC note and draft cabinet note

The Task Force met and developed an action plan and classified the issues into short term and medium term for ease of addressing them (Khetarpal and Gupta, 2008)

Indian plant quarantine system has certainly come a long way starting with the legislation *viz.*, Destructive Insects and Pests Act way back in 1914. However modernizing quarantine mechanism is much more crucial today in the light of globalization not only to protect the crops from invasive foreign pest species but to safeguard the economy by ensuring enhanced productivity and farm incomes. It is in this direction that we can draw plenty of useful information from the better evolved biosecurity and quarantine systems in several other countries like the United States, Europe, Australia and New Zealand, which can be modified to suit our own requirements.

References

Anitha, K., Sivaraj, N., Chakrabarty, S.K., Sarath Babu, B., Prasada Rao, R.D.V.J., Varaprasad, K.S. and Khetarpal, R.K., 2007. Weeds of Quarantine significance to India. National Bureau of Plant Genetic Resources, New Delhi. India 71P.

Gupta, Kavita, Pandey, B.M. and Khetarpal, R.K., 2002. Pest Risk Analysis and its significance to India under WTO Regime. Seed Tech News 32: 266- 269. http: / / www.fao.org http: / / www.wto.org

Kahn Robert P., 1989. Plant Protection and Quarantine: Selected Pests and Pathogens of Quarantine Significance Vol II. CRC Press Inc. Florida, 265 p.

Khetarpal, R.K. and Gupta, Kavita, 2006. Policy Issues in Plant Protection. In: *One hundred years of Plant Pathology in India: An Overview* (eds. S S Chahal, R K Khetarpal and T S Thind) Indian Society of Mycology and Plant Pathology, Scientific Publishers, Jodhpur, 381- 401.

Khetarpal, R.K. and Gupta, Kavita, 2008. Discussion Paper INDIA Open-ended Working Group on Building National Phytosanitary Capacity FAO Headquarters, Rome, Italy 8- 12 December, 2008 available at: https: / / www.ippc.int / servlet / BinaryDownloaderServlet 208303_Paper. do.doc?filename=1228321600516_2008_OEWG_Dec_14_India_Discussion_Paper.doc &refID=208303

Khetarpal, R.K. and Gupta, Kavita, 2007. Plant Biosecurity in India- Preparedness. *Indian Journal of Plant Protection*. 35 (2): 168-178.

Khetarpal, R.K. and Gupta, Kavita, 2002. Status of Quarantine as a Phytosanitory Measure under WTO; National Scenario. In: National Seminar on Resources Management in Plant Protection during twenty First Century, B. Sarath Babu, Varaprasad, K.S., Anitha,K., Prasada Rao, R.D.V.J., Chakrabarty,S.K. and Chandurkar,P.S. (Eds), Plant Protection Association of India, Hyderabad, pp.104-113.

Khetarpal, R.K. and Rajan, 1999. Sanitary and Phytosanitary Agreement of WTO. Seed Tech News 29: 51-53.

Khetarpal, R.K. and Nath, Ram, 1998. Evolving national plant quarantine system: harmonization of current quarantine regulations with sanitary and phytosanitary (SPS) agreement. In: National Dialogue (ISPGR Dialogue 1998): Issues on plant genetic resources, December 1-2, 1998, NBPGR, New Delhi.

Murthi, N., Anishetty, K.S., Varaprasad and Sivaraj, N., 2007. International treaty on Plant Genetic Resources for food and Agriculture (ITPGRFA): Scope and Provision. In Biodiversity in India: Issue and concern Kannaiya-, S. and Gopalan.A (eds). Associated Publishing Company, New Delhi.

Nath, Ram, Mathur, V.K. and Verma, B.R., 1988. Plant quarantine activities in the Indian context. In: Plant Genetic Resources–Indian Perspective. R.S. Paroda, R.K. Arora and K.P.S. Chandel (ed). NBPGR, New Delhi, pp 470-480.

Sarat Babu, B., K.S.Varaprasad, K.Anitha, R.D.V.J.Prasada Rao, S.K.Chakrabarty and Chandurkar,P.S. (Eds.) 2002. Resources Management in Plant Protection, Vol. II, Plant Protection Association of India, Hyderabad, India. 201p.

Sarath Babu, B., Varaprasad, K.S., Anitha, K., Prasada Rao, R.D.V.J., Chakrabarty, S.K. and Chandurkar, P.S. (Eds) 2002. Resources Management in Plant Protection, Volume I, Plant Protection Association of India, Hyderabad, India. 303pp.

Sarath Babu. B., 1997. Detection of insect pests of quarantine significance in screening of germplasm under exchange programme during 1989-1995. *Indian. J. Ent.Res.*, 21(3): 295-297.

Task force for Plant Protection, http://dacnet.nic.in/ipmweb/taskforce/national_plan.htm

Varaprasad, K.S., Anitha, K., Chakrabarty, S.K. and Khetarpal, R.K., 2008. Role of Phytosanitary policies in rice trade. *Current Science* 94 (3): 303-304.

Varaprasad, K.S., K. P. S. Chandel, A.M. Ghanekar, and Y. L. Nene 1997. Safe germplasm export of ICRISAT mandate crops, NBPGR Regional Station, Hyderabad Plant Quarantine Unit, ICRISAT, Patancheru 36 pp.

Chapter 3

Disease Resistance in Rice: Progress and Perspectives

Ram Singh[1], S. Sunder[1], C.S. Reddy[2] and D.S. Dodan[1]
[1]**CCS HAU Rice Research Station, Kaul – 136 021, Haryana**
[2]**Directorate of Rice Research, Rajendranagar – 500 030, Andhra Pradesh**

Introduction

Rice (*Oryza sativa* L.) is a major cereal crop, which provides nutritional and livelihood security for half of the human race. In India, it is cultivated over 44.6 million hectares with a total production of 93.1 million tonnes and provides 43 per cent of calorie requirement for more than 70 per cent of Indian population. The rice yield is most often adversely affected by the diseases and insect-pests, which are known to cause 18.6 per cent losses. Of about 75 diseases reported to occur on rice, bacterial leaf blight, blast and sheath blight continue to be the major threat in realizing yield potential of cultivars. Several relatively minor diseases like false smut, sheath rot, stem rot, bakanae and brown spot have also emerged as potential diseases in several states.

Disease management technology has mainly relied on use of pesticides, which has led to a number of environmental problems and development of resistance in the pathogens. Planting of resistant varieties is an eco-friendly strategy of disease control and is an integral component of integrated disease management system. Moreover, no effective chemical control measures are available for certain diseases like bacterial blight. Therefore, the most effective and economical way to control rice diseases is by developing resistant cultivars. The present chapter deals with the concepts and levels of resistance, identification of sources of resistance to single and multiple diseases, mechanism and inheritance of resistance of important rice diseases.

Concepts and Levels of Resistance

Van der Plank (1968) recognized two basic types of resistance on the basis of genetic and epidemiological concepts namely, Vertical resistance, also known as monogenic, complete or race-specific resistance and Horizontal resistance, also called polygenic, incomplete or race non-specific resistance. Race-specific resistance is qualitative in nature and often characterized by hypersensitivity in a host genotype.It can either be immunity or high level of resistance and does not allow the pathogen to sporulate, but it is short-lived in nature and is often known to break down, with serious economic consequences. In Korea, the resistance of the 'Tongil' variety for blast was effective for a 5-year period. In Japan, the longevity of resistant blast varieties is 3 years (Marchetti and Bonman, 1989). In Colombia, resistant varieties to blast released during 1969-1986 lasted only for a year or two before being overcome by previously unidentified virulent races (Ahn and Mulekar, 1986). Although race-specific resistance conferred by a single gene or a combination of few genes is generally short-lived, however, there are instances where such genes (*Pi-zt*, *Pi-b*, *Pi*-ta 2, etc.) proving effective for long periods in many countries.

Race non-specific resistance is a quantitative resistance, which has also been called field resistance, generalized resistance, race-reducing resistance, dilatory resistance (Lee *et al.*, 1989). Later, a few other terms like partial and durable resistance have also come into vogue for this type of resistance. Varieties that lack complete resistance but sustain only minor losses are referred to as partially resistant. This kind of resistance is race-specific (Toriyama, 1975) or race non-specific (Ezuka, 1979; Yeh and Bonman, 1986). In some rice-growing environments highly conducive to diseases like blast, partial resistance may not be sufficient. In such situations, breeders exploit more complete resistance through gene pyramiding or super imposing complete resistance onto varieties possessing partial resistance (Bonman and Mackill, 1988). Durable resistance is retrospective rather than prospective. It is based on the longevity of resistance in a given ecosystem irrespective of genetic control and magnitude. It essentially describes the commercial utility and longevity of a particular resistant variety. It remained effective for a long period while a cultivar possessing it has been widely cultivated in an environment favouring the disease (Johnson, 1981). The definite longevity required for durable resistance is not predetermined. Resistance is referred to as durable if resistance of one variety remains effective even though another succumbs to an epidemic.

Breeding for disease resistance in rice was initiated as early as 1930 in Indonesia against rice tungro virus and during 1940's in Japan against bacterial blight. However, it gained momentum after mid-sixties when reduced genetic variability, continuous cultivation of input-responsive high yielding varieties made them genetically more vulnerable to diseases. The research developments regarding host plant resistance to various rice diseases are described below.

Bacterial Blight (*Xanthomonas oryzae* pv. *oryzae*)

It is one of the major constraints to rice productivity in South East Asia causing 6-60 per cent loss in grain yield. The pathogen is known to produce three types of symptoms, kresek or wilt, leaf blight and pale yellow leaf. Of these leaf blight is the most common phase, which is characterized by the presence of water soaked stripes with wavy margins on leaves. Yellow droplets of bacterial ooze are often seen on leaf blade in humid weather.

Host Resistance

Breeding and deployment of resistant cultivars carrying major resistance genes (R) has been the most effective approach for controlling bacterial blight. Most of the workers have evaluated rice genotypes for resistance to bacterial blight using local bacterial isolate(s) with unknown virulence. However, some researchers have identified bacterial blight donors against known virulence(s) prevalent in the region. The genotypes found resistant to bacterial blight in different agro-climatic regions of the world, reviewed by Shukla *et al.* (2003), are presented in Table 3.1.

Table 3.1: Rice Genotypes Found Resistant to Bacterial Blight in Different Countries

Country	*Resistant Genotypes*
Japan	Aikoku-soto-sango-kei, Aka-Shinriki, Hatsugasumi, Hen 2-1-24, Hokubu 2, Hokubu 2-1014, Kameji 3, Kogyoku 1, Kogyoku-asahi, Lead rice, Magatama, Mii, Nakashin 120, Nigeria 5, Sengoku 4, Shigasekitori 11, Shinseki 2, 4, 13, Sugaippon, TKM 6, Zensho 26, 70 X-38, 70 X-42, 70 X-43, 70 X-46
Philippines	3 Baifufugoya, Balilla, BJ 1, DD-96, DV-29, DV-52, DZ-60, DZ-78, Early Prolific, Giza 38, Keng Chi-ju, PI 162319, M. Sungsong, Semora Mangga, Sigadis, Tainan 9, TKM 6, Wase-Aikoku 3, Zenith
China	Aus 373, Aus 69232-2, DP 5148, DP 5165, Dular, DV 85, IRBB 5, IRBB 8, Jia 45, Jian 12, Jian 029, 95-Jian-25, 96-Jian-35, Zao Sheng Ai Guo 3-97-G 104
Indonesia	Arimbi, Batara, Bengawan, Dewi Tara, Djelita, Pelita1/1, Pelita 1/2, Ramadja, Sigadis, Syntha
India	AD 9246, Ajaya, BAM 10, BAU 4090-1, BG 90-2, BJ 1, BK 670, BPT 1235, BR Nos. 50-3-4, 51, 51-282-8, 171-2B-8, C 1321-9, CB II, CR Nos. 10-52, 129-118, 167-7, 200-788-3, 243-43-150, 294-548, 318-548-7, 319-644, 837, DV 85, DV 86, DZ 78, Govind, HAU 118-104, HAUK Nos. 9-102-1-4, 10-37-2, 10-211-1-5, 12-52-4-2, HKR Nos. 91-119, 95-128, 95-131, 95-138, HKR 120, HKR 122, HPU 804, IET Nos. 3262, 4140, 4141, 7172, 8326, 9910, 9912, 10312, 10313, 10413, 10418, 12174, Indira, IR 20, IR 20/Leaung 152-737, IR 29, IR 36, IR 54, IR Nos. 442-46-3-3-3, 878-13-45, 1712-166-1, 2055-481-2-6-2, 2798-3, 4432-1-644, 5657-33-2-1-2, 6808-189-2-2-1-3, 9763-11-2-2-3, 9801-9-3, 13429-299-2-1-3, 22082-412-2, IRON 83-105, KJT 24, Kogyoku, KR 10-47, MR 1550/Ratna-576, MRC 603-303, MRC 603-383, MSS, MSS/Zenith-548, Nam Sagui, Nam Sakouy, NDR 301, NDR 302, NRL 326-3, OR 45-61-23, OR 173-1-1, *Oryza barthii, O. latifolia, O. longistaminata, O. malampuzhaensis, O. meridionalis, O. nivara, O. punctata, O. rufipogon*, P 50746-55-13-3-1-1, Pant Dhan 4, Patong 32, PAU Nos. 122-73-1-4-1, 164-102-1-12-1-1-1, 1966-77-22-1, 2991-40-2, 2991-48-2, 2991-102-1, PR 103, Prasad, R 8-2535, R 35-27-51, Ramakrishna, Rasi, RAU 49, RAU 4056-53-5, RP Nos. 632-94-1-2-1-7, 633-C, 1033-43-2, 1036-35-2-5-1-1, 1064-84-6-2-4, 1575-636-6-1, 1670-1418-2205-1582, 2151-4-21-1, 2151-21-1, 2151-33-2, 2151-33-4, 2151-40-1, 2151-64-3-1-1, 2151-173-1, 2151-175-1, 2151-192-1, 2151-200, 2151-224-4, Sigadis, SKL 6, Tainan 3, TKM 6, UPR Nos. 79-164, 81-44, 82-1-1, 82-1-7, 103-44-2, 174-14, 221-124-1, 254-21-1-1, 1536-4-1, W 1263, Wase Aikoku 3

In addition, OR 142-99, RP 1854-111-4-3-1, Savitri (Panda *et al.*, 1994); AS 330, Kola Ahu, IR 64 (Kalita *et al.*, 1995); NPT 63K-2-109, NPT 57K-11-104 (Kumar *et al.*, 1997); RP 2151-21-22, RP 2151-33-2-11-21, RP 2151-40-1-71, Ruchi (Agrawal *et al.*, 1997); HKR 95-129, PAU 1061-19-2-2, PAU 1920-100-2-1-3-3, PAU 1973-121-1-2-1-1, PAU 2212-25-1-2 (Rekha *et al.*, 2002); BC 4F 1-111-120-2735-12, CR 837, HKR 95-75, HKR 99-57, HKR 99-58, HKR 99-60, HKR 2000-57, HKR 2000-115, IR 17494-32-1-1-3-2, IRBB 53, PAU 2020-10-3-1, PAU 3075-20-3, PR 106-P_2 (xa 5, xa 13, Xa 21), PR 106-P_9 (xa 13, Xa 21), PR 116, PR 118 (Sunder *et al.*, 2008) were also observed to exhibit bacterial blight resistance in different parts of India. Lore *et al.* (2004) reported that IR 71604-4-4-3-7-2-2-3-3, IR71693-197-4-4, IR 72101-4-159-1-3-3, IR 72102-4-159-1-3-3, PAU 2338-12-3-3-1, PAU 2881-22-1-3-1, PAU 2881-22-1-3-2, PAU 2909-2-2-2-3, PAU 2962-89-1-3-3, PAU 2963-42-1-3-2, PAU 2964-26-1-4-3, PAU 2964-30-1-2-3, PAU 2964-30-1-6-3, PR 106-P_2 and PR 106-P_9 were resistant to all the pathotypes of *X. oryzae* pv. *oryzae* prevalent in Punjab, India while three lines, PAU 2884-61-1-2-3, PAU 2888-3-1-2-3, PAU 3075-35-1 (PR 118) were resistant to pathotypes I, II, III, IV and moderately resistant to pathotypes V. High degree of horizontal resistance has been reported in Boro II, Chinsurah, Dora, IR 20, IR 22, IR 29, Jelita, Nakashim 120, Palita 1/1, Remadja in Indonasia; Asominori, Gomashirazu Koguoku, Kunihikari, Niphonbare, Satominori, Shimotsuki, and Yaeho in Japan (Shukla *et al.*, 2003). Recently, Laha *et al.* (2008) identified BC4F-1-111-120-2735-12, HKR 95-138, HKR 2001-115, HKR 2001-116, PAU 1973-121-1-2-1-1, PAU 3042-3-3-4, PR 106-P_2 and PR 106-P_9 possessing broad-spectrum resistance to multiple isolates of the pathogen. In China, D 601, D 602 and D 603 (Xu Jian Long *et al.*, 1995); Yang Fu Xian, Fu 4 Xiao Shiu 664 and Wei You 64 (Qian Haan Liang *et al.*, 1997) have also been reported to have bacterial blight resistance.

A varied response of rice plants to *X. oryzae* pv. *oryzae* at different growth stages has been reported. Resistance to the kresek was reported in BJ 1, CO 13, Kanto 51, Lacrosse, Nagarbhog, Raminad Strain 3, Sachikazi, Zenith/Nira, Zukkoku (Reddy and Mohanty, 1977); CR 294-54-8, OR 173-1-1, TCM 2, TCM 3 (Chand *et al.*, 1985); RP 2151-175-1, RP 2151-192-1 (Ram Singh and Chand, 1986). Kaku and Kimura (1978) observed that Dewirath, Sigadis and TKM 6 had seedling resistance while Chugoku 45, Kogyoku and Wase Aikoku 3 possessed adult plant resistance (Mew, 1987). Horino (1981) demonstrated that IR 28 and Tetep had high level of resistance at both seedling and adult plant stage to bacterial group 1.

Mechanism of Resistance

Scanty information is available on the mechanism of bacterial blight resistance. Kiryu and Mizuta (1955) found that cultivars with few, short, narrow and erect leaves have low infection than those having luxuriant growth and spreading leaves. Cultivars with hairy leaves showed maximum disease while the disease was very low in cultivars with glabrous leaves due to retention of more inoculum by the hairy cultivars. Resistant varieties possess lower stomatal index than susceptible ones (Shukla *et al.*, 2003). A negative correlation has been observed between disease development and frequency of distribution of silicated cells in coastal region (Kaul and Sharma, 1987). However, some workers did not find any correlation between posture, colour, length and thickness of leaves, number of tillers/hill and duration of cultivars with bacterial blight incidence (Singh and Rao, 1971; Authors, unpublished data).

Resistant cultivars have a higher ratio of reducing sugar to total nitrogen, higher contents of polyphenols, lower contents of some free amino acids and production of a non-specific phytoalexin (Ou, 1985; Mahto *et al.*, 1987). A high phenyl alanine ammonia lyase activity (Purushothaman, 1974), increased activity of polyphenol oxidase and peroxidase along with increase in total phenol content (Minni Sandhu *et al.*, 1995) has been noticed in resistant cultivars due to synthesis of more phenyls than susceptible ones. Rao and Nayudo (1979) reported that increased phenolic content favours a high peroxidase activity in diseased tissues of resistant cultivar providing resistance through the oxidation of phenols into toxic quinones, which are inhibitory to the pathogen. In IR 28 or Tetep, the bacterial cells became irregular in shape and were immobilized by fibrillar material induced from the cell wall of host. The development of such fibrillar material has been reported to be a defence mechanism associated with incompatible reaction (Horino, 1981).

Inheritance of Resistance

Considerable information is available on genetics of resistance in rice against bacterial blight pathogen. Resistance has been found to be dominant or recessive and monogenic or polygenic (Shukla *et al.*, 2003). Currently 30 genes for bacterial blight resistance (Nino-Liu *et al.*, 2006; Kuldeep Singh, R.M. Sundaram-personal commuication) are presented in Table 3.2.

Most of the known bacterial blight resistance genes have been tagged with DNA markers (He *et al.*, 2006) and six genes *viz.* Xa 1, Xa 3, xa 5, xa 13, Xa 21, Xa 26 and Xa 27 have been cloned (Dai *et al.*, 2007). Deployment of few genes has been highly successful. Xa 4 has been most widely used in the tropics and in many parts of China (Sun *et al.*, 2003; Mew *et al.*, 2004; Vikal *et al.*, 2004) while Xa 1 and Xa 3 are widely used in Japan and Korea (Mew *et al.*, 2004). In general, genes that condition resistance at seedling stage (xa 5, Xa 10) remain effective in adult stage also. However, genes for adult plant resistance such as Xa 3, Xa 4, Xa 6, xa 8, xa 9, Xa 21 do not necessarily provide resistance at the seedling stage indicating that the two types of resistance are distinct and distinguishable. Kaur *et al.* (2005) reported that bacterial blight resistance in *O. nivara* Acc. 81825 was governed by a single dominant gene. The gene was subsequently designated as Xa 30(t) and found effective at both seedling and adult stage (Kuldeep Singh-personal communication).

Table 3.2: Genes for Resistance to Bacterial Blight in Different Sources

Gene	*Chromosome Location*	*Linked Marker*	*Donor*	*Features*
Xa 1	4	Npb 235	IRBB 1, Kogyoku	Effective against Japanese race I; ineffective against all Philippine races
Xa 2	4	V 61, E 10736	IRBB 2, Wase Aikoku 3	Linked to Xa 1; effective against Japanese race II; susceptible to all Philippine races
Xa 3/Xa 26/ Xa 6/xa 9	11	C 481 S, Y 6855 RA	IRBB 3, IRBB 26, Wase Aikoku 3, Chugoku 45, Zhachang long	Resistance to Philippine races 1, 2, 4, 5 and all Japanese races at booting stage only, resistance to Philippine race 3 at all growth stages
Xa 4	11	Nbp 181	IRBB 4, TKM 6	Linked to Xa 26; resistance to Philippine races 1, 4, 5, 7, 8 and 10
xa 5	5	RG 556	IRBB 5, IR 1545-248, DV 85, BJ-1, IR 1291-7	Resistance to Philippine races 1, 2, 3, 5, 7, 8, 9 and 10
Xa 7	6	G 1091	IRBB 7, DV 85, DV 87	Resistance to Philippine race 1 at booting stage and to 2, 3, 5, 7, 8 and 10 at all growth stages
xa 8	7	RM 500, RM 533	IRBB 8, P 1231129	Resistance to Philippine races 5 and 8
Xa 10	11	M 419, M 491	IRBB 10, Cas 209	Linked to Xa 4, resistance to Philippine races 2, 5 and 7
Xa 11	–	–	IRBB 11, IR 8, IR 944, RP 9-3	Resistance to japanese races IB, II, III A and V; ineffective against all Philippine races
Xa 12	4	–	IRBB 12, Kogyoku, Java 14	Resistance to Japanese race V
xa 13	8	RG 136, xa 13-prom	IRBB 13, BJ 1 (Aus Boro)	Resistance to Philippine race 6
Xa 14	4	VAZ 190 B, RG 163	IRBB 14, TN 1	Resistance to Philippine races 5 and 8
xa 15	–	–	IRBB 15, M 41	Resistance to Japanese races
Xa 16	–	–	IRBB 16, IR 24, Tetep	Resistance to Japanese isolates H 8581 and H 8584
Xa 17	–	–	IRBB 17, Asminori	Resistance to Japanese isolates H 8513
Xa 18	–	–	IRBB 18, Toyonishiki	Resistance to Burmese isolates BM 8417 and BM 8429; ineffective against all Philippine races
xa 19	–	–	IRBB 19, XM 5	Resistance to Philippine races 1, 2, 3, 4, 5 and 6
xa 20	–	–	IRBB 20, XM 6	Resistance to Philippine races 1, 2, 3, 4, 5 and 6
Xa 21	11	PTA 248	IRBB 21, *Oryza longistaminata*	Resistance to Philippine races 1, 2, 3, 4, 5, 6, 7, 8 and 9

Contd...

Table 3.2–Contd...

Gene	Chromosome Location	Linked Marker	Donor	Features
Xa 22	11	L 363 B	IRBB 22, Zhachang long	Linked to xa 26, broad-spectrum resistance
Xa 23	11	RG 1109, RM 254	IRBB 23, *O. rufipogon*	Strong resistance at all growth stages to all Philippine races and most japanese and chinese races
xa 24	–	–	IRBB 24, IR 28, DV 85, DV 86, Aus 295, Dungano	Resistance to Philippine race 6
Xa 25(a)	–	–	IRBB 25, HX-3, a somaclonal mutant of Minghui 63	Resistance to Philippine races 1, 3 and 4 and to several Chinese races
Xa 25(b)	–	–	IRBB 25, Minghui 63	Resistance to Philippine race 9
Xa 27	6	M 1081, M 1095	IRBB 27, *O. minuta*	Resistance to Philippine races 2 and 5
xa 28	–	–	Lota Sail	Resistance to Philippine races 2 and 5
Xa 29(t)	–	–	*O. officinalis*	Xa 29 is a tentative designation, not fully characterized
Xa 30(t)	–	–	*O. nivara* Acc 81825	Xa 30 is a tentative designation, effective against all the pathotypes prevalent in northern India

Sridhar *et al.* (1999) observed that all the R-genes are not equally effective. Lines possessing Xa 1, Xa 3, Xa 4, Xa 7, Xa 10, Xa 11 and Xa 14 individually were susceptible to the local population of the pathogen at Cuttack, India; however, genes xa 8 and Xa 21 were singly effective. All the combinations of 2-, 3- or 4-genes were resistant, presumably because of quantitative complementation between resistance genes (Yoshimura *et al.*, 1995; Huang *et al.*, 1997). Goel *et al.* (1998) also made similar observations and reported that genotypes carrying Xa 1, Xa 3, Xa 10, Xa 11, Xa 14 and Xa 18 were susceptible or moderately susceptible whereas, IRBB 8 (xa 8), DV 85 (xa 5 + Xa 7) and BJ 1 (xa 5 + xa 13) were resistant to all the isolates analysed in Punjab, India. Shanti *et al.* (2001) showed that each near-isogenic line with known single resistance gene was susceptible to one or more of the 10 strains representing eight pathotypes existing in Orissa, Madhya Pradesh and Uttar Pradesh, India. No single strain could overcome all resistance genes. The gene combinations Xa 4 + xa 5, xa 5 + Xa 21 and Xa 4 + xa 5 + Xa 21 conferred a broad-spectrum of resistance to all the strains evaluated, supporting the strategy of pyramiding appropriate resistance genes. The genes xa 5, xa 13 and Xa 21 together have shown resistance to all known races of the pathogen (Huang *et al.*, 1997). In Haryana, India, a new virulence of the pathogen was encountered during *Kharif* 2001, which produced susceptible reaction in genotypes possessing Xa 1, Xa 3, Xa 4, xa 5, Xa 7, xa 8, Xa 10, Xa 11, xa 13, Xa 14, Xa 18, Xa 21, Xa 4 + xa 5, Xa 4 + xa 13 and Xa 4 + Xa 21. Only the near isogenic lines having combinations of xa 13 + Xa 21, Xa 4 + xa 5 + Xa 21 and Xa 4 + xa 5 + xa 13 + Xa 21 were resistant (Authors' unpublished data). Shanti and Shenoy (2005) also observed broad-spectrum resistance in genotypes having combinations of Xa 4 + xa 5 + xa 13 + Xa 21.Using marker assisted selection, Gopalakrishnan *et al.* (2008) developed a bacterial blight resistant line, Pusa 1460-01-32-6-67 (released as improved Pusa Basmati-1/Pusa 1460) by transferring two bacterial blight resistant genes (xa 13, Xa 21) from the non-basmati resistance donor line IRBB 55 into a popular, bacterial blight susceptible, basmati rice variety Pusa Basmati-1. Likewise, three bacterial blght resistance genes, Xa 21, xa 13 and xa 5 have been introgressed into Samba Mahsuri which exhibited high level of resistance to *X. oryzae* pv. *oryzae* (Sundaram, *et al.*, 2008).

Pathogenic Variability

Following the break down of resistance in cultivar Asakabe, the studies on host-pathogen interaction were carried out in different countries. The pathogen is highly variable and over 30 races of *X. o.* pv. *oryzae* have been reported world wide (Nino-Liu *et al.*, 2006; Gonzalez *et al.*, 2007). Mew (1987) complied the work done in different countries on host-pathogen relationship and reported the reaction of existing seven races (IA, IB, II, IIIA, IIIB, IV, V) in Japan, six (1-6) in Philippines, nine (A-I) in Indonesia, two (1, 2) in Thailand, three (Ia, Ib, II) in India, two (A, B) in Korea and two (N-1, N-2) in Nepal on a set of 10 differential cultivars *viz.* Kinmaze, Kogyoku, Rantaj-emas, Wase Aikoku 3, Java 14, IR 8, IR 20, IR 1545-339-2-2, Cas 209 and DV 85. Buddenhagen and Reddy (1972) found that virulence was low in Japan and Australia but high in India, Indonesia and Sri Lanka.

Now-a-days IRRI differential system comprising of 22 near-isogenic lines, possessing single or multiple resistance genes is commonly used to determine virulence in *X. oryzae* pv. *oryzae*. Rehman *et al.* (1993) identified five races *viz.* Ia, Ib, II, III and IV from India and Nepal. Virulence of Sri Lankan isolates resembled to that of the Indian pathotype Ia (Dissanayake and Ponnawila, 1992). The isolates of *X. oryzae* pv. *oryzae* from Nepal and Bangladesh were categorized into five virulence groups and 12 races, respectively (Shukla *et al.*, 2003). The population of the pathogen prevalent in Punjab, India was classified into 8 pathotypes (Goel *et al.*, 1998) while that of Orissa, Madhya Pradesh and Uttar Pradesh, India into 17 haplotypes and 9 pathotypes (XH-1, XE-2, XC-3, XA-12, XA-13, XA-25, XB-26, XA-59 and XA-63) employing repetitive PCR DNA finger printing and virulence analysis, respectively (Shanti *et al.* 2001).

Blast (*Pyricularia grisea*)

It is one of the most widespread and serious constraints to rice production in both tropical and temperate regions. The disease appears on leave as spindle shaped spots with greyish centre and dark brown margins. Nodes turn black and get weakened due to tissue disintegration, resulting in breakage of stem at the nodal region. Neck infection results in production of dirty white earheads bearing half filled or chaffy grains. The disease has caused several epiphytotics in different regions of the world.

Host Resistance

Massive screening programmes operated at various levels have identified several blast resistant rice genotypes namely, MTUEP 1243-8, VL93-6052 (Anonymous, 1999a); CRMS 30, HKR 90-403, HKR 90-404, HKR 90-413, HKR 90-421, HKR 91-406, HKR 95-435, HKR 95-458, HKR 95-505, HKR 95-511, RP 3228-44-12-9-7, RP 3238-96-14-9-5, UPR 1071-21-1-2 (Ram Singh *et al.*, 1999); Basmati Gola, Basmati Mehtrah, Bor Chokura, Chandan, G-65, HKR 97-401, HKR 97-416, HKR 97-418, HKR 97-422, Haryana Basmati No. 1, IR 841-85-1-1-2, Kalijira 8-1, MTU 4407, N 50, NIRRI 140-6, Pusa 743, RP 2144-19-6-2, RP 3120-15-12-9-4, RP 3135-17-12-8-8, RP 3238-33-15-7-1, RPU 41-62-1-1-3, Sonale (Ram Singh *et al.*, 2004); Fukuton, Fu She 94, IAC 25, IR 20, IR 42, Kurok and Xiang Zaoa (Gupta and Kapoor, 2002); Bharani, IET 2508, IET 3262, Vargan (Rangaswami and Mahadevan, 2004); HPR 1057, HPR 1171, RP 2421 (Sharma and Kapoor, 2002); China 972, China 1007, K 23-98, K 78-13/Barkat, K 681-9-1, SKAU 105, SKAU 338, SKAU 681-3-3-1 (Anwar *et al.*, 2003); Akashi, Arroz, Dos Indos (HRC 1010), CAN 108-8-42, CAN 4136, IRAT 267 and Pallavi (Chauhan *et al.*, 2000); Raminad Strain 3, Zenith (Mohd. Najeeb *et al.*, 2006); Binam, BR 24, IR 64 and Tequing (Srinivas Prasad *et al.*, 2008) in diffrent parts of the country. Some genotypes namely, Camponi SML, Columbia 1 and 2, Dissi hatif 73127, IR 9292-21-1, Tetep, VL 8 (Tandon and Bhatt, 1982; Bhatt *et al.*, 1984); HKR 90-403 and HKR 91-405 (Ram Singh and Dodan, 1996) have been reported to possess resistance to both leaf and neck blast while genotypes like

BJ 1, IET 8892, IR 20, IR 36, Rasi (Satyanarayana and Reddy, 1990); VL 91-5834 and VL 1044 (Anonymous, 1999a); VHC Nos. 85, 218, 976, 1080, 1098, 1296, 1404, 1487, 1488, 1493, 1494, 1508, 1583, 1759, 1760, 1761 and 1824 (Bhatt and Chauhan, 1984) displayed slow blasting. The primary donors for resistant cultivars in Indian conditions are Tetep, Tadukan, Zenith, Co 4 and MTU 5. Beside these, ADT 25, ARC 7098, Beganbisi 2, Carreon, Ceysron, Co 24, Co 25, Co 26, Dawn, Dular, Gajepsali, Gampai, Kalamdani, Kolimekuri, Madrisali, Milyang 83, Moroberekan, Mugisali, *O. minuta*, *O. rufipogon*, *O. glaberrima*, PTB 10, Rikhojoi 2, Rongaahu, Rongaguitia, Sigadis, Sollpona, Tarabali and TKM-1 have also been used to incorporate blast resistance in improved rice varieties.

Paul *et al.* (1993) evaluated 377 breeding lines for blast resistance and found that 16 lines namely, CT 8008-3-5-2P-M, CT 8008-3-5-5P-M, CT 8008-3-5-6P-M, CT 8008-3-5-7P-M, CT 8008-3-5-8P-M, CT 8222-4-1-8P-M, CT 8238-6-14P-M, CT 8285-13-1-3P-M, CT 8285-13-5-2P-M, IR 47761-27-1-36, IR 53306-36-1-3-1, IR 53649-AC 5-2, IR 54791-19-2-3, IR 56446-94-3-1-2 NAR 210-1-4 and NAR 210-1-1-4 were resistant in Guyana. In, France, Silue and Notteghem (1991) studied blast resistance in 99 *glaberrima* varieties and found that BG 155, BG 162, BG 178, MG 26, TG 10, YG 12, YG 15, YG 247 and YG 304 were resistant to all strains used. Similarly in China, 25 varieties namely, A 36 line 16, C 102, Chente 232, Dangen 1, Dushouxi 1, Erjiufeng, Guizaoshen (japonica type), Ji 86A-48, Jiahu 5, Jiangxinu, Jiannong zao 9, Ming 119, Mingnu 706, Sanyangai 1, Shuanzao 25, Suweon 290, Xiangya 1, Xiangzhou 5, Zhe 852, Zhefu 802, Zhi 20-5, Zhuke 2, 528-4, 1344, 81005 showed a broad-spectrum of resistance to 50 isolates tested (Sun Guochang and Sun Shuyuan, 1992).

Of several blast resistant varieties developed, Govind, Heera, IR 36, IR 64, Jaya, Jyothi, Pant Dhan 4, Rasi, Swarnadhan, Sona Mahsuri and Tulasi have been widely cultivated to combat blast in different parts of the country. Out of 836 varieties released in India (Rangaswami and Mahadevan, 2004; Shobha Rani *et al.*, 2008), 247 varieties were found resistant and 176 were moderately resistant to blast. Durable blast resistance has been reported in IR 36 in tropical Asia (Yeh and Bonman, 1986); Milyang 30, Milyang 42 in Korea (Lee *et al.*, 1989); IAC 25, IAC 47, Moroberekan from West Africa (Nottenghem, 1985); Zhen Shan 97 and Zhen Luon 13 in China (Lee *et al.*, 1989). However, blast resistance in improved rice varieties is short-lived due to dynamic changes in race composition of the pathogen.

Mechanism of Resistance

Different substances in plants such as epicuticular waxes, free phenols, cell wall bound phenol oxidases and phytoalexins were reported to operate against the blast pathogen. Cuticular waxes of varieties susceptible to blast favour development of a large number of appressoria when compared with those of resistant varieties. Intense tissue browning, a characteristic feature of resistant varieties, limits the growth and sporulation of the pathogen. The biochemical processes associated with tissue browning affect pathogen development. Cell walls of rice leaf blades contain cinnamate derivatives and the toxic activities of p-coumarate, ferulate and their oxidized products are associated with resistance. A few compounds such as probenazole were developed for controlling blast disease, which augment development of phytoalexin that reduced fungal growth. However, none of these mechanisms are universal in nature (Sridhar *et al.*, 1990).

Inheritance of Resistance

Identifying and using broad-spectrum blast resistant genes in rice breeding have been considered as the most effective and economical strategy to control the disease. Resistance in rice cultivars to blast has been reported to be governed by genes that were either monogenic dominant, monogenic recessive, two dominant independent genes, two dominant complementary genes, two recessive duplicate genes or three dominant genes with any two conferring resistance. The resistance to leaf blast may be governed by three pairs of dominant genes. The involvement of an inhibitory gene or cytoplasmic factor in the expression of blast resistance was also reported in several rice cultivars (Muralidharan, 2006). Blast resistance has been observed to be conferred by independent dominant genes designated as Pi-1(t), Pi-2(t) and Pi-3(t) in the near isogenic lines C 101 LAC, C 101 A 51 and C 104 PKT, respectively. Blast resistance in C 101 PKT and C 105 TTP-4 were conferred by dominant alleles at an additional locus, designated as Pi-4^a(t) and Pi-4^b(t), respectively (Mackill and Bonman, 1992). In many instances of investigation complete resistance for blast was monogenic (Kiyosawa, 1981; Marchetti *et al.*, 1987).

Gene pyramiding seems promising to provide broad-spectrum and durable resistance. Combining several genes and monitoring their presence is difficult by conventional methods because of their epistatic interactions. Mapping blast resistance genes and locating closely linked markers has made it possible to confirm the presence of given gene in a variety or genotype. Thus, even when the phenotype is ambiguous, the presence of a gene can be monitored over successive segregating generations. To date, at least 50 R genes have been identified and mapped using molecular markers. These genes are distributed on 10 or 12 rice linkage groups, with the exceptions of chromosome 3 and 10, and many of them are clustered on chromosome 6, 11 and 12. Among the genes identified, the gene Pik^h, derived from the variety Tetep and located on chromosome 11, confers good level of resistance against many races of *M. grisea.* The blast resistant genes mapped in rice along with linked molecular markers, relevant to India, are given in Table 3.3.

Table 3.3: Blast Resistance Genes (*Pi*) Mapped in Rice

Gene	Chromosome	Markers
Pi-1	11	RM 144, RM 224, RG 303, R 10, RZ 536
Pi-2	11	RG 64, RM 527, MSM-1, MSM-4
Pi-k^h	11	RM 206, TRS 26, TRS 33
Pi 4	8	RM19211, RM19212, RM19213
Pi-9	6	C 26348, S 47656, RM 12705

Only few attempts have been made to identify blast resistance genes in Indian cultivars. They include *Pi-k^p* in HR 22, *Pi-z^t*, *Pi*-a in Co 25, and *Pi-z^t* in TKM 1 and Co 4 and Pi-k^a in Charnack. However, at IRRI, Philippines, thirteen dominant resistance genes as *Pi-a*, Pi-*i*, *Pi-k*, *Pi-k^s*, *Pi-k^p*, *Pi-k^h*, *Pi-m*, *Pi-ta*, *Pi-ta^2*, *Pi-z*, *Pi-z^t*, *Pi-b* and *Pi-t* were identified at eight loci and near-isogenic lines in the background of Co 39 comprising single gene or multiple genes in various combinations were developed (Mackill and Bonman, 1992; Khush and Brar, 2001). Subsequently, more than 43 major resistance genes have been identified (Zhu *et al.*, 2004) and twelve blast resistance genes *viz. Pi-1, Pi-2, Pi-4, Pita, Pi-5(t), Pi-6, Pi-7, Pi-9, Pi-10(t), Pi-11(t),Pi-b* and *Pi-20* have been mapped with molecular markers (Khush and Brar, 2001). Muralidharan *et al.* (2003, 2004) demonstrated that a high degree of resistance was imparted by single gene *Pi-1, Pi-4b* or *Pi-2*; or in combination of two genes with *Pi-2* and *Pi-4*; and three genes with *Pi-1, Pi-2* and *Pi-4*. The genotype A 57 carrying three resistance genes *Pi-1, Pi-2* and *Pi-4* exhibited resistance to blast across the country in all rice growing environments. The performance of BL 245 with two resistant genes *Pi-2* and *Pi-4* and C 101 LAC (*Pi-1*) was comparable to A 57 and was marginally superior to resistant checks.

Efforts have also been made for molecular tagging of a few blast resistance genes. PCR based DNA markers linked to blast resistance gene Pi-2(t) at 2.8 cM has been used as a useful marker in rice breeding programme (Hittalmani *et al.*, 1995). Naqvi *et al.* (1995) reported a RAPD marker to be linked to Pi-10(t) gene on chromosome 5. In a program on gene pyramiding at Directorate of Rice Research, Hyderabad, Pik^h is being introgressed into the genetic background of an elite fine-grained variety Samba Mahsuri with the help of the marker RM 206. At NRCPB-IARI, New Delhi, a systematic approach to map and clone *Pi-k*^{h} gene present in rice line Tetep was undertaken (Sharma *et al.*, 2005b). Tetep has been used as a resistant parent to develop an F_2 mapping population of more than 200 plants and developed a molecular linkage map of $Pi\text{-}k^h$ locus with SSR markers at 0.5 and 0.7 cM distances on chromosome 11 and the gene was finally cloned by using map based cloning approach (Sharma *et al.*, 2005a).

Pathogenic variability

The existence of strains of *P. grisea* differing in pathogenicity was first noticed by Sasaki (1922). However, intensive studies on pathogenic races of fungus were initiated in Japan only after the breakdown of resistance of rice cultivars such as Futaba after cultivation for about a decade. *P. grisea* exhibited an extreme degree of variability in pathogenicity and several physiological races have been recognized (Table 3.4) in different regions of the world based on infection type developing on differential hosts.

However, the races reported by different researchers were not comparable as the differentials used by them were not identical. To overcome this problem, use of common set of differentials comprising of international differentials, recombinant inbred lines (RIL), donors and commercials cultivars (A 57, BL 122, BL 245, C 101 A 51, C 101 LAC, C 101 PKT, C 104 PKT, C 105 TTP 4 L 23, Caloro, Co 39, Dular, HR 12, IR 64, Kanto 51, NP 125, *O. minuta*, Raminad Strain 3, Rasi, RIL 10, RIL 29, Shia tia-tsao, Tadukan, Tetap, Usen and Zenith) and designation of races as international races (IA, IB— IH) has been suggested. Recently, the isolates of *P. oryzae* representing 17 locations of the country have been divided into two major groups based on their virulence pattern on above set of differentials. Group 1 isolates from DRR, Hyderabad, Diphu, Almora, Nawagam, Barapani, Coimbatore, Gudalore, Faizabad, Ghagharghat and Titabar were more pathogenic at all the locations than group 2 isolates representing Nellore, Rajendranagar, Hazaribagh, Mugad, Raipur and Navasari (Anonymous, 2008). Beside virulence pattern, the molecular marker technology like RFLP and DNA finger printing have been used widely for understanding the population structure of *P. grisea* occurring in Philippines, Korea, USA, Columbia, China, Thialand, Vietnam and in India. In southern United States, 79 isolates were grouped into eight pathotypes which were partitioned into eight discrete, distinct lineages using MGR DNA finger printing while in Columbia, 50 pathotypes were grouped into seven lineages. Likewise, 151 isolates of the blast fungus were differentiated into six genetic lineages while virulence analysis revealed the presence of 39 pathotypes among these isolates (Sridhar and Singh, 2001). In India, 36 lineages of the fungus occurring in southern India were identified by Gnanamanickam *et al.* (1993) and Sivaraj *et al.* (1995, 1996) employing RFLP technique using MGR 586 as a hybridization probe. The population of Jagadalpur, Madhya Pradesh was diverse as it consisted of six lineages of the seven detected by Sridhar *et al.* (1996). Two lineages were detected among the isolates from Sikkim which were also detected in Madhya Pradesh and Orissa (Sridhar and Singh, 2001). Sharma *et al.* (2002) categorized 250 isolates of *M. grisea* collected from North-western Himalayan regions into 25 DNA fingerprint groups or lineages using RAPD analysis. Similarly, the finger printing patterns of 29, 31 and 43 isolates of *P. grisea* from Banki, Dhenkanal and Berhampur revealed the presence of 23, 24 and 43 haplotypes, which were differentiated into three, four and twelve lineages, respectively (Mishra *et al.*, 2006). However, 119 isolates of the pathogen were grouped into 52 pathotypes on the basis of disease reaction on international differential rice lines.

Table 3.4: Physiological Races of *P. oryzae* Identified in Different Countries

Country	Races	Dominant Race	Differentials Used	Reference
Japan	13	–	T-group: Tetep, Tadukan C-group: Usen, Chokotou, Kanto 51, Yakeiko N-group: Ishikarishiroke, Ginga, Homarenishiki, Norin 22, Aichi-asahi, Norin 20	Goto, 1965; Sridhar and Singh, 2001
	34	–	–	Matsumoto *et al.*, 1969
USA	20	–	Zenith, Lacrose, Caloro, CI 5309, CI 8970-P, CI 8970-S, PI 180061 (Dular), PI 201902 (NP 125), PI 231128 (Raminad Strain 3), PI 231129 (Wag-wag), Roxoro (Sup), Taichung 65 (Sup)	Muralidharan, 2006
China	19	Race 12	Chianung 242, Chianung 280, Cutsugulcul, Kaao-chio-liu-chou, Kanto 51, Kaohsiung-ta-li-chen-yu, Kung-shan-wu-shen-ken, Kwang-ful, Natala, Norin 21, Pia-kantao, Sensho, Taichung 33, Taichung 65, Taichung 171, Taichung-ti-chio-wu-chien	Muralidharan, 2006
Taiwan	27	–	–	Chien, 1967
India	31	IC 17	Caloro, CI 8970, Dular, Kanto 51, NP 125, Raminad Strain 3, Usen, Zenith	Padmanabhan *et al.*, 1970
	17	–	–	Venkatarao and Apparao,1976
	At least 54	–	–	Rangaswami and Mahadevan, 2004
	IB 17, IB 18, IB 49	–	–	Padhi and Chakrabarti, 1972
	IA, IC, ID, IE	–	–	Mariappan, 1984; Mohd. Najeeb *et al.*, 2006
Korea	18	–	Ayanishiki, Ishikari-Shiroke, Kanto 51, Norin 1, Norin 17, Norin 22, Pi 1, Sensho, Tonewase, Zenith	Lee and Matsumoto, 1966
Philippines	252	IA 110	Chokoto, CI 5309, Co 25, Kataktara DA 2, Khao-tah-haeng 17, Lacrose, Peta, Pia-kan-tao, Raminad Strain 3, Shia-tiao-tsao (S), Taichung-t-c-w-c, Wag-wag	IRRI, 1975
Columbia	14	ID 8 (US 6)	Aichi-asahi, Bluebonnet 50 (Sup), Caloro, Dular, Ishikari-shiroke, Kanto 51, Napal (Sup), NP 125, Raminad Strain 3, Sha-tiao-tsao (S), Usen, Zenith	Galvez and Lozano, 1968
	> 45		International and local differentials	Correa-Victoria and Zeigler, 1993
Nigeria	11	–	11 differentials	Awoderu, 1970

Sheath Blight (*Rhizoctonia solani*)

The pathogen produces elliptical or oval to irregular greenish gray spots on leaf sheath and leaves with white centre and brown or purplish margin. Brown silky mycelium and brown to dark sclerotia are produced on the surface of the lesions. It is one of the most widespread and damaging diseases of rice in India and known to cause up to 69 per cent loss in grain yield.

Host Resistance

Despite an extensive search, high level of sheath blight resistance has rarely been found among the cultivars, breeding lines or related wild species of *Oryza*. Cultivars from India, Thialand, Burma, Europe and North America have been found more resistant than local *Japonica* cultivars (Ou, 1985). Moderately resistant sources identified in different countries included Zenith, Kataktara DA-2, Kuang fu 1, PI 160641 and Mamoriaka in Philippines (IRRI, 1972; Gangopadhyay and Chakrabarti, 1982), 12814/SI 26 in Indonesia (James, 1974), Jasmine 85 (Marchetti *et al.*, 1998) and Dixiebelle rice in USA derived from RU 8303181 X CB 801 (Mc Clung *et al.*, 1998). Rice genotypes may be resistant or susceptible to sheath blight at either or both stages of growth *i.e.*, seedling and adult stage. Prasanna Kumari Pillai (1990) screened more than 150 rice accessions using cut stalk laboratory assay technique and found that 12 entries namely, P 578-1-157-129-18, P 598-163, P 598-163-6-102-2, P 615-1-1-36, P 615-1-1-37, P 615-1-2-39, P 615-1-2-40, P 745-1, P 1037-3-13, P 1037-7-4, P 1037-88-1 and P 1037-88-6 were free from infection of *R. solani*. The results matched very well with the reaction in field tests. Amongst 182 genotypes evaluated for sheath blight resistance under artificial inoculation conditions, five genotypes *viz.* HRI 126, HKR 99-103, HKR 2000-26, HKRH 1059 and IR 64683-87-2-2-3-3 were moderately resistant to sheath blight in nursery (seedling stage) while in field (adult stage), only two genotypes–HKR 2000-514 and IR 64683-87-2-2-3-3 proved resistant and three genotypes HKR 99-103, HKR 2001-537 and HKRH 1059 were moderately resistant. Three genotypes namely, HKR 99-103, HKRH-1059 and IR 64683-87-2-2-3-3 showed resistant/moderate resistant reaction both under nursery and field conditions. Regression analysis showed a significant but low correlation (r=0.473) between nursery and field tests indicating that the nursery screening is not an absolute indication of resistance in main field. None of the genotypes showing susceptible to highly susceptible (disease score 7-9) reaction in nursery was found to be resistant/moderately resistant in field indicating that the nursery screening can be utilized in short listing of rice genotypes for screening against sheath blight in main field (Authors' unpublished data).

Premlatha Dath (1990) had compiled the rice genotypes possessing good level of resistance to sheath blight, which is presented in Table 3.5.

Pathogenic Variability

In Taiwan, 300 isolates of *R. solani* were classified into seven cultural types and six physiologic races using 16 differential rice cultivars (Chien and Chung, 1963). In contrast, Haque (1975) and Tsai (1973) did not observe any distinct differential reaction among 40 single basidiospore cultures and 25 field isolates of the pathogen in Bangladesh and Taiwan, respectively. Tu (1967) found that the Taiwan isolates with less aerial growth were more aggressive. The hyphal anastomosis revealed at least 12 anastomosis groups (AGs) of *R. solani*, which are further divided into intraspecific groups. The sheath blight fungus has been placed in AG-1 IA whose population differed considerably in cultural characters, morphology and virulence pattern (Sunder *et al.*, 2003). In China, Yi *et al.* (2002a,b) classified *R. solani* isolates into four and five groups using UPGMA-SAHN clustering analysis and RAPD techniques, respectively. The virulence variation had no significant correlation with geographical region and RAPD groups.

Mechanism of Resistance

Sheath blight is usually severe on fertilizer responsive dwarf cultivars vis-a-vis tall cultivars. In Japan late maturing cultivars have been observed to escape sheath blight than early cultivars due to lower temperature during autumn (Ou, 1985). Various features of the host imparting resistance were more pyrophosphatase activity, tight encirclement of stem by sheath, amount of wax deposits on outer leaf sheath and failure of the fungus to produce infection cushion (Premlatha Dath, 1990). Reduction in total soluble sugars and starch could be correlated with increasing disease intensity (Ou, 1985). Lee and Rush (1983) reported that in southern U.S.A. highest level of resistance was found in short and medium grain types which were more like japonica rice. Most of the early senescent tall types showed more field resistance than the late senescent short ones in Bangladesh (Gangopadhyay and Chakrabarti, 1982).

Table 3.5: Rice Genotypes Possessing Resistance to Sheath Blight in Different Countries

Cultivars	*Breeding Lines*	*Reference*
India	A 15-100-1-3-1, ADT 5, ADT 22, ARC Nos. 5925, 5943, 10572, 10618, 10836, 14342, 14529, 15762, 18119, 18275, 18545, ASD 1, B 1757 d-Sm-6-1, B 1757 d-Sm-47-1, B 1757 d-Sm-47-3, B 1757 d-Sm-47-6, B 1757 d-Sm-50-6, B 2012 d-Sm-52-1-2, Bhasamanik, BK 105, BR Nos. 3, 4, 4-30-51-2, 7, 8, 51-26-10, 51-49-6, 51-95-2, 51-196-2, 52-85-7, Co 20, Co 22, CR Nos. 1-6, 44-11, 44-121-1, 57-11-2, 117-1, 215-SS 54-1, 280-5, 1014, Cult. Nos. 55-100, 2100, 2210-2, 3868, 3872, 3875, 3879, 3885, 3896, 3900, 3904, 3948, 3964, 3965, 4076, 4295, 5143, 5996, 6001, DA 29, DGWG FH 423, Dudshar, Dular, HKR 109, HPU Nos. 804, 2203, 2205, 2206, IET Nos. 1991, 2812, 2815, 4699, 5656, 5891, 6234, 6235, 8134, 8252, 8340, 8748, IR Nos. 8, 20, 32, 42, 62, 64, 532-1-18(S), 1156-501-P, 1544-340, 2035-244-3-2-3, 2053-200-4, 2070-132-2-2-5, 2071-588-4-4, 2071-588-5-1, 2071-648-4-4-6, 2071-803-5-2-6, 2071-843-1-1-6, 2098-115-2-3, 2328-198-1-1-1, 2793-97-3-2, 2796-44-2, 3464-149-3-2, 4422-98-3-6, 4531-10057-5-1-1, 4531-10057-5-1-2, 4641-143-1, 21818-138-2-3, 21820-154-3-2-2-3, 28150-84-3-3-2, Java 14, JBS 40, JC Nos. 5, 18, 41, 56, 66, 75, 118, 126, 220, JR 49, JW 55, Kam Ban Ngan, Kattichamba, KAU 796, KRC Nos. 346, 355, 356, M 11-57-5-1, M 14-59-2, M 15-36-2 (released as MO 5), M 15-37-1, M 22-67-3-2, M 22-184-1-1, M 23-16-1-1, M 23-17-1-1, M 23-69-1-1, M 23-74-1-1, M 23-83-1-1, M 23-100-2-2, M 24-211-1, Mahsuri, MGL 14, Milyang 58, MR 44, MTU 17, MTU 5783, NC 678, Nermar, Nizersail, OR 42-26-2, *Oryza rufipogon*, *O. barthii*, OS 4, OYT 6241, OYT 6245, Pankaj, PAU 41-10-1-3-262-1-5, PAU 317-B-2, Ponni, PTB 33, Pusa 245-1-17-1, R 245-30-41, R 2357, R 2386, R 2393, R 2401, R 2423, R 2433, R 15561-6, Ramedija, RNR 74822, Rojasail, RP Nos. 4-14, 5-32, 661-12-4-4-2, 967-11-1-3-9, 1057-393-1, 1699-174-9-1, 1855-99-1-1B, 2151-40-1, S 22, Saket 1, Shangsanba, Saibham, Sinna Sivappu, Suduwee, Suweon 294, T 141, T 1185, Tabend, Ta-poo-cho-Z, Tadukan, Tetep, Tillainayagam (SAR), TKM 6, Toga, Tumai Angouba, V 42, V 228, W 449, Wase Aikoku, WGL 28712, Zinya 31	Premlatha Dath, 1990
	IR Nos. 54742-1-11-17-26-2, 54742-6-1-14-15-3, 54742-6-20-3-9-3, 54742-11-1-9-15-1, 54742-50-19-19-1-1, 54742-50-19-19-1-3, 54745-2-23-19-8-2, 54751-3-38-10-15-3	Lakshmanan and Vėlusamy, 1991
	ARC 10531, ARC 10532, ARC 10635	Reddy, 1993
	IR 40, KK 2	Ram Singh and Dodan, 1995a
	RNR 15336, RNR 82096	Reddy *et al.*, 1997
	NPT 57K-11-104, NPT 63 K-7-108, NPT 63 K-8-12, NPT 63 K-12-51, NPT 63 K-2-109	Kumar *et al.*, 1999; Tiwari and Kumar, 2000
	ACM-56, AVT-M (H) 1808, AVT-M (H) 1810, AVT-M (H) 1814, IR 51673-50-2, IVT-J 2503, IVT-J 2507	Meena *et al.*, 2000

Contd...

Table 3.5–Contd...

Cultivars	*Breeding Lines*	*Reference*
	Andrewsali, Monoharsali, Chingdar	Kalita *et al.*, 2000; Singha and Borah, 2000
	Bhadra, MTU 2400	Sudhakar *et al.*, 2000
	Dee-Geo-Woo-Gen, Dinesh, IET 16307, IR 54, Matangini, Malta, Palman 579, Sabita, Seetabhog	Biswas, 2001b
	CRK 26-1-2-1, CRM 03-1, HKRH 2, IR 48725-B-B-86-2-2, PAU 3075-20-3	Authors' observations
Venezuela	Araure-1, R 2386	Premlatha Dath, 1990
Philippines	Kuang-fu 1, Mamoriaka, PI 160641, Zenith	IRRI, 1979
	ARC Nos. 12208, 12248, 12256, 12417, 12547, 12622, 12637, 12655, Bahagia, BG 34-11, Carreon, Colombia, CPSLO, CS 2, IR ATOM 20, IR 8/4 X Zenith, IR 442-2-58, IR 533-1-89, IR 1093-148-3, IR 1330-3-2, IR 1360-87-1, IR 1917-3-17, Kataktara DA-2, Mehran, OS 4, PTB 18, Ratna, Sinaloa A 68, T 442-148-29, Ta-poo-cho-Z	Premlatha Dath, 1990
	Oryza minuta Acc. 101089, *O. rufipogon* Acc. 100907	Amante *et al.*, 1990
USA	PI 7701052, P! 7801020, PI 7803097, Tadukan, Tetep	Premlatha Dath, 1990
China	IET 4699, Java 14, Tetep	Guo *et al.*, 1985
	Guyanal, Ta-poo-cho-z, Tetep	Sha and Zhu, 1990
	HB 14, HB 15, Jiayu 948, Zhefu 910, Zhonfu 955, Zhongzu 1, Zhomgfu 9818	Cheng *et al.*, 2000

Inheritance of Resistance

The inheritance of resistance to sheath blight is not yet thoroughly understood due to lack of an international set of differentials. Marshall and Rush (1980) observed that resistance to sheath blight is inherited as a dominant character while Goita (1985) suggested that two pairs of complimentary genes controlled resistance which is modified by epistasis. Sha and Zhu (1989) reported that there are no major genes for sheath blight resistance and it is a typical quantitative trait controlled by polygenes. However, some researchers proposed that sheath blight resistance in some rice cultivars is controlled by major genes (Hashioka, 1951; Masajo, 1976; Xie *et al.*, 1990). Zou *et al.* (2000) identified six quantitative trait loci (QTLs) namely, QSB-2, QSB-3, QSB-7, QSB-9-1, QSB-9-2, QSB-11 located on chromosomes 2, 3, 7, 9 and 11, respectively contributing to sheath blight resistance. The respective alleles of QSB-2, QSB-3, QSB-7 and QSB-9-2 from Jasmine 85 could explain 21.2, 26.5, 22.2 and 10.1 per cent of the total phenotypic variation, respectively while the alleles of QSB-9-1 and QSB-11 from Lemont could explain 9.8 and 31.2 per cent of the total phenotypic variation. Pan *et al.* (1999a) reported that only partial resistance is available for sheath blight and that has been considered to be polygenic. They further reported that resistant cultivars, Jasmine 85 and Teqing, possessed a non-allelic dominant major resistance gene. Pan *et al.* (1999b) identified three major QTLs in an F_2 population of Jasmine 85/ Lemont by bulked segregant analysis. The dominant resistance gene in transgenic rice '4011' was named as Rsb 1 and mapped on rice chromosome 5 (Che *et al.*, 2003). The Xa 21 gene, Bt gene and chitinase gene were combined in a single rice line, which showed high tolerance of sheath blight (Datta *et al.*, 2003).

Foot Rot and Bakanae (*Gibberella fujikuroi = Fusarium moniliforme*)

The pathogen produces lanky, pale yellow and taller seedlings, which die before or after transplanting. The presence of white or pink mycelial growth and adventitious roots from lower internodes of the infected plants are other diagnostic features of the disease. It is widely distributed through out the world including India causing 3.0-95.4 per cent yield loss.

Host Resistance

The nature of resistance to bakanae is not clear due to evaluation of limited germplasm against this pathogen. The rice genotypes, ADT 8, Aryan, Agami Monthakab, C 4-64, Co 18, Co 22, CR 4–1048, CR 83-23, GEB 24, HR 98, HR 101, HR 105, IR 20, IR 26, IR 32, IR 38, IR 44, IR 45, Norin 18, Norin 20, P 84-283, PTB 7, RP 5-3, RP 5-32, RP 5-48, RP 176-6, Wateribune from India; Apostal, Guinongang Strain 1, Macan Bino, Macan Binan from Philippines; Akage No. 3, Kairyo-Mochi No. 1, Shiroka from Japan; Charnock, Dharial, IR 8, Panbira, Surjamukhi from Bangladesh; MN 62M, MR 7 from Thialand and Basmati 370 and its dwarf mutant DM 107-4, IR 6, IR 8, IR 9, KS 282, PK 1339-12-1-1-0-6, PK 3303-15-1, PK 3713-9 and Shadab from Pakistan have been found to exhibit resistance (Ram Singh and Sunder, 1997; Cheema *et al.*, 1998; Tasleem *et al.*, 1999a). In Haryana, India, Sunder *et al.* (1998) found that resistance against bakanae is more common in non-scented rice cultivars. Amongst 124 scented and 97 non-scented rice genotypes evaluated under artificial epiphytotic conditions, only two scented genotypes namely, C 4-64 and Karjat X 13-21 were resistant where as amongst non-scented cultivars, 10 genotypes *viz.* AS 34011, BG 936, BR 1067-84-1-3-2-1, BR 1257-31-1-1, BR 4363-8-11-4-9, CN 3886, HKR 86-104, IR 39464-54-1-3-2-1-3, IR 58109-109-1-1-3 and PR 106 showed resistance to bakanae. Later, Bagga and Kumar (2000) reported that 13 *basmati* lines *viz.* IR 58755-158-1-3-2, IR 64668-54-3-1, IR 66229-45-3-2, IR 67409-98-1-3-3-1, IR 67418-131-2-3-3, IR 67418-204-4-4-2, IR 67423-208-2-1-1, IR 67423-234-3-3-3, PAU 2343-60-2-3, PAU 2383-2-3-1-1, Double haploids 1, Double haploids 7 and Double haploids 19 were free from bakanae in Punjab, India under artificial inoculation conditions.

Mechanism of Resistance

Kanjanasoon (1965) observed that unlike susceptible varieties, the fungus cannot grow well on the surface of crown or the surrounding leaf sheath of resistant cultivars and can not form a mycelial mass dense enough to penetrate the host tissues. Thakur and Subramanian (1966) observed low nitrate reductase activity in roots and a high enzyme activity in the shoots in susceptible rice variety MTU 9 with infection while no such differences were noticed in resistant cultivar GEB 24. Thakur and Ghosal (1976) found that total phenols were more in the resistant cultivar GEB-24 than susceptible variety MTU-9. The phenol content increased sharply after infection in resistant but slightly in susceptible cultivar.

Inheritance of Resistance

The inheritance pattern of bakanae resistance has not been studied thoroughly. However, Tasleem *et al.* (1999b) found that it is monogenic, recessive in rice variety IR 8 and dominant in KS 282.

Pathogenic Variability

Heterokaryosis (vegetative compatibility) is responsible for generating and maintaining genetic variability in *F. moniliforme*. Puhalla and Spieth (1985) found that vegetative incompatibility is controlled by nuclear genes (*vic* genes). There are at least 4 *vic* loci in group C strains attacking rice. A significant variability in strains of *F. moniliforme* has been observed with respect to vegetative compatibility, production of gibberellins, fusaric acid, pectic enzymes and their ability to cause disease (Ram Singh and Sunder, 1997; Sunder and Satyavir, 1998). In Haryana, India, Sunder and Satyavir (1998) categorized 28 isolates of the pathogen into five groups based on GA_3 production in culture and virulence pattern on five paddy cultivars. Vegetative compatability was independent of pathogenicity and of ability to produce GA_3. However, a positive correlation existed between GA_3 production in culture and bakanae incidence in trays (Sunder and Satyavir, 1998) and pectic enzymes and foot rot development (Thakur, 1974).

False Smut (*Ustilaginoidea virens = Claviceps oryzae sativae*)

The pathogen attacks the florets producing large, velvety, green smut balls (pseudomorphs) causing considerable quantitative and qualitative losses. The disease, usually regarded as a minor disease, has attained economic significance in several states particularly due to cultivation of input responsive high yielding cultivars and hybrids.

Host Resistance

Screening of rice germplasm to identify false smut donors has mainly been done under natural disease pressure. The rice genotypes found resistant to *U. virens* under natural epiphytotics included IET Nos. 954, 1985, 2005, 2610, 2671, 2685-2, 2721, 3145, 3181, 48173 in Andhra Pradesh (Rao and Rao, 1979); CN 758-1-1-1, CR 155-5029-216, RP 1854-566-1-1-1, TNAU-BSR 16 (Ansari *et al.*, 1988); IET 9710 and Savithri in the Andaman Islands (Ram and Ansari, 1989); IET Nos. 7511, 7720, 7978, 8350, 8364, 8511, 8588, 8653, 8958, 9313, 9799, 9843, IR 50, IR 29652-65-2-3, RP 16691-2897-5626 (Singh and Khan, 1988); FH Nos. 259, 264, 355, 538, 556, 561, 591, 612, 621 and NDU Nos. 82, 83, 84, 89, 97, 98, 99, 100, 101, 102, 103, 106, 107, 110, 111, 112, 118, 119, 120, 121, 124, 125, 126 in Uttar Pradesh (Singh, 1989); RP 4-14, SKL 6, SKL 6-1-23, SYE 75 in Maharastra, India (Patil and Moghe, 1990). Bharadwaj (1990) found that B 3719 C-TB 8-1-4, China 988, HPU 2202, HPU 5101, Nag 1-38, UL 501 and VRS-1 were free from false smut infection under upland condition. Out of 160 accessions of Assam rice collection, ARC 11868 was found to be the promising donor for resistance breeding against false smut (Urmila Dhua and Dhua, 1999). Likewise, 15 selections from IR 8, one each from IR Nos. 66, 109, 271, 534, two each from IR Nos. 126, 279, 280, 281, 298, 305, 407, 423 and four from IR 224 in Bangladesh (Ashrafuzzaman, 1974), Heukjinjubyeo in Korea (Shim *et al.*, 2001) and Sakha 102 in Egypt (Atia, 2004) have also shown resistance to this disease. Some other genotypes including Ananga, Anupama, Badami, Bala, Barathi, Bhuban, Cauvery, Ch 45, China 988, derivatives of the cross W 1263/W 12787, Jagannath, HPU 870, IET Nos. 1782, 1789, IR Nos. 30, 90, 9729-67-3, 10198-66-2, 18482-plp 3-1-2-1-1-1, 18482-plp 3-2-3-1-1, 53426-Nag-2-1, 53426-Nag-5-2, 53426-Nag-6-1, 53451-Nag-8-1, 53451-Nag-10-1, Lalat, Parijat, Pathara, Pusa NR 57-24, Pusa NR 57-26, R 68-1, Rambha, Sabarmati, Savitri, Vijaya, VR Nos. 116-1-2, 379, 381, 395, 412, 524, 1000-1, Zenith have also been reported to possess resistance against false smut (Dodan and Ram Singh, 1996). Eight mid-early to medium duration rice hybrids namely, HKRH-1005, IAHR-200-006, IAHS-200-005, MTUHR-2048, NDRH-11, PERH-1031, PERH-1037 and UPHR-1107 remained free from both false smut and kernel smut diseases under natural field condition in West Bengal, India (Biswas, 2001a). Screening tests carried out at CCS HAU Rice Research Station, Kaul, Haryana under natural infection during *Kharif* 2004-2007 have indicated that HKR 47, HKR 98-418, HKR 99-491, HKR 2000-603, HKR 2000-645, HKR 2001-482, HKR 2002-6, IR 64, IR 65907-91-1-B, MTU 1067, Govind and hybrid HKRH 21 possessed false smut resistance (Authors' observations).

Only limited germplasm has been evaluated under artificial epiphytotics owing to lack of reliable and reproducible inoculation technique for inciting false smut. Singh and Gangopadhyay (1981) observed least infection in Zenith while Singh (1984) reported that IR 22, IR 26, IR 28, Vijaya were completely free from infection and Anupama, Cauvery, IR 30, CH 45, Jayanty, Prasad, Sabarmati, Saket 4, Satya and Sona had trace infection under artificial inoculation condition in Uttar Pradesh,

India. Paicos-1 and HKR 47 have been developed as false smut resistant/moderately resistant varieties from Manipur and Haryana, respectively (Singh, 1987, Authors' unpublished data).

Mechanism of Resistance

The mechanism of resistance has not been studied well. The scented tall genotypes have been observed to contract significantly less false smut than high yielding dwarf varieties and hybrids (Bhardwaj, 1990; Authors' observations). The resistant genotypes have been found to possess higher amount of phenols and ortho dihydroxy phenol compared to susceptible genotypes.

Pathogenic Variability

Very little information is available on variability of the pathogen. Singh (1984) and Verma and Singh (1988) differentiated 8 isolates of *Claviceps oryzae sativae,* collected from different parts of India, into four groups based on differences in symptoms and morphological and cultural characteristics. However, Zhou *et al.* (2004) did not find significant genetic diversity among 56 strains of *U. virens* in China.

Brown Spot (*Cochliobolus miyabeanus=Drechslera oryzae*)

The pathogen causes small, round to oval, brown spots surrounded by a typical yellow halo on leaves in addition to grain discolouration and rotting of panicle stalk. The disease is famous for 'The Great Bengal Famine' when as high as 90 per cent loss in grain yield of paddy was recorded during the year 1942.

Host Resistance

The genotypes namely, BAM 10, CH 13, CH 45, Co 20, Dakar Nagra 273-32, Kalma 219, Nagra 41-14, Patnai 549-33, T-141, T-498-2A, T-960, T-998m, T-2112, T-2118 from India; Hainan No. 217, Chiu Tiu Chiu, Hukubozu, Diakokuwase, Yamahaku, Hinomaru, Norin 17 from Japan; Leuang Yai 34, Leuang Thong 82, Muey Nawng 62 M, PJP 54 x IRH 8-BKN 56-7-106, TN-1 from Thailand; IET 13238, JB 781 from Philippines; CI 9515 and Dawn from USA have been reported to possess resistance against brown spot by earlier researchers (Ou, 1985). Subsequently several genotypes including C 5309, Chainan 8, Chainung 242, Fa Yiu Tsai, Hoyoku, IRRI Nos. 29, 31, 44, 47, 48, 49, 50, Japan 11, Kinmaze, Kusaful, Rexoro, Tainan 1, Tainan 3, TN-1 (Srivastava and Maheshwari, 1971); C 1837, C 5409, C 8188, C 8291, C 10652, C 10754, CH 45, CR 1-6, CR 42-38-173, CR 63-168, CR 64-109, CR 64-167, IR 24, MTU 5152, MTU 5314, Padma (Singh and Sharma, 1975); BKNFR 2606-3-2-1-2, CR 98-7216, CR 1002, CRRP 34, IR 42 (Prasad and Singh, 1985); IET 11459, MTU 7633, Suraksha (Saifulla, 1993); BD-47, Cauvery, Poorva, Sattary (Rajput and Bartaria, 1995); a breeding line IR 54742-11-17-26-1 derived from *O. officinalis* (Saini and Goel, 1998); ACM 60 P, KAUM 59-29-7-1-2, OR 165-86-12, OR 165-97-15, OR 764-39, OR 930-1-1-2, Rewa 6-85-3, Rewa 14-174, RNRM 70, RP 2479-18715-1197, RP 2526-14767-1143, TTB 151-51-3-5, TTB 151-91-3, UPRI 201-1-5-3 (Singh *et al.*, 2000); Aditya, Areba, Dular, Kalamkata, Lohana 1, Pallavi, Rangi (Chauhan *et al.*, 2000); OR 165-97-15, Rewa 14-174 (Jha *et al.*, 2004) have been reported to possess brown spot resistance by various workers in India. Recently, in multilocation testing under All India Coordinated Rice Improvement Programme, none of the test genotypes was found to be resistant at all the test locations, however, eight genotypes *viz.* CN 1919, HPR 2177, HUR 18-6, NLR 3041, UPR 2856-10-1-3, VL 30918, VL 30919 and VL 30920 exhibited an average severity index < 4.0 (Anon. 2008). In Haryana, India, resistance to stalk rot phase has been recorded in BPT 1788, MTU 1067 and Swarnadhan by Sunder *et al.* (2005).

Mechanism of Resistance

Anatomical features of rice plant like thicker epidermal cells and more silicated cells have been found associated with brown spot resistance (Ou, 1985). However, Purkayastha and Chattopadhyay (1975) did not find any correlation between anatomical differences and silica content with resistance. Incidence of brown spot increased with leaf angle, infection being lesser on the top most leaves (Gangopadhyay and Chattopadhyay, 1974). Misra and Prasad (1964) noticed that in resistant cultivars the stomata opened for a short period. Number of total tillers/hill and percentage of non-effective tillers showed a negative relationship with disease incidence while length and breadth of flag leaf had a positive effect. Panicle length and grain density was positively correlated with disease but spikelet sterility and grain weight had no effect. Synthesis of polyphenols and their oxidation products, dry weight, total N, reducing sugars like ascorbic acid, free amino acids and total proteins, presence of a pathogen inhibitory alkane in waxy layer have been correlated with host resistance (Ou, 1985). Hegazi *et al.* (1993) observed a high correlation between the levels of phenols, lignin, flavonoids, silicon and oxidative enzymes in rice leaf tissue and the susceptibility of rice cv. Giza 180 and Arabi to brown spot.

Inheritance of Resistance

Nagai and Hara (1930) working with a Korean strain of the pathogen found that resistance was a dominant character, whereas Adair (1941) indicated it as recessive, several genes possibly being involved. Subsequently, Bedair *et al.* (1976) found that leaf and kernel resistance to *C. miyabeanus* was inherited independently by a single dominant and recessive gene, respectively. However, Prasad *et al.* (1998) indicated that resistance was controlled by two pairs of recessive genes. Harahap (1976) observed a significant correlation between seedling and mature plant reaction. The seedling stage resistance was found to be recessive. Omar *et al.* (1979) have also observed a highly positive correlation (r= 0.93) between seedling and adult plant reactions but both had low correlation with grain reaction. However, Prabhu *et al.* (1996) reported a highly +ve correlation between mean severities of brown leaf spot and grain discolouration

Eruotor (1986) observed that a minimum of 3 major genes controlled lesion size and at least one gene controlled lesion number. Balal *et al.* (1979) reported that seedling and adult plant leaf reaction were similar and governed by 3 genes (Her 1, Her 2, Hes 1); the first two were dominant for resistance and third for susceptibility. Kernel resistance was dominant over susceptibility and controlled by 2 genes (Hekr 1 and Hekr 2).

Pathogenic Variability

Eruotor (1980) studied 29 isolates of *C. miyabeanus* on 12 varieties of rice and indicated no physiological specialization in the fungus, however, Misra (1985) and Harish *et al.* (2007) observed considerable differences in virulence of fungal isolates.

Stem Rot (*Sclerotium oryzae*)

The disease appears as small, blackish, irregular lesions on leaf sheath and stem resulting in rotting, collapsing and premature lodging of the crop. Dark gray mycelium and small, round black sclerotia are found on splitting the infected stem. The disease is particularly severe in water-logged areas and known to cause substantial quantitative and qualitative losses due to increased lodging, smaller panicles, production of light chalky grains and poor milling quality.

Host Resistance

Several workers have reported that resistance to stem rot is either not available or rarely found (Raina *et al.*, 1980; Chand *et al.*, 1985; Jiten Konthoujam and Chhetry, 2007). However, the sources of resistance to stem rot identified by various researchers in different parts of the world have been compiled by Ram Singh *et al.* (2002), which is mentioned in Table 3.6.

Table 3.6: Rice Genotypes Showing Resistance to Stem Rot in Different Countries

Country	*Genotypes*
Sri Lanka	Hondurawela
Philippines	Araban, Elonelon, Kinaturay, Ramay, Raminad Strain 3
USA	Aikoku, Akage, Arkrose, Asahi, Asahi Mochi, Bozu, Butte, Caloro, Colusa, D 16, Fukui Shioda, Iyo Benkei, Kameji, L 202, M 101, M 201, Magnolia, Mizuno, Mubo Aikoku, Mubo Aikoku 22, Nakate Shinriki, Nishiki, Onsen, Wild species of *Oryza rufipogon* and *O. nivara*, S 201, Spain Jap., Zenith
Japan	Gimbozu, Strains of Shinriki and Sembon
Bangladesh	AU-30-49, AU-41-15-18/E, AU-41-15-18/L, B 5441-6-Kn-7-1-2-3, BG 375-1, BR 8, Chandrasail 1, Chainung-Sen-Yu-6, DJ 684D, IR 32, IR 944-85-1-2-11-2-1, IR 2071-586-5-6-4, IR 2451-90-4-3, IR 2793-80-1, IR 3464-75-1-1
Iraq	IR 8, IR 22
Argentina	Colombiano, El Paso 144, Fortuna, IRGA 410, P 796 L 386
India	AC 1773, ADT 4, ARC 12751, ARC 12753, Basmati 217, Basmati 370, BJ 1, Bock, BR 51-331-4, C 5342, C 6664, C 6730, C 722355, Calaro, CAMOR, Ch 49, China 27, China 972, Chugoku 45, CR 57-49-5, CR 137-36-30, CR 318-548-7, Dudshar, Early Prolific, EC 2506, Hansraj 353, HAU 118-87, HAU 118-110, HAU 118-111, HAUK 12-52-4-2, HKR 86-197, HKR 120, HKR 213, HKR 216, HKR 219, HKR 220, IET Nos. 125, 4141, 5633, 9700, 9701, 10294, 10413, 10418, 10478, Improved Sona, IR Nos. 4-67-23, 6-53-2-3, 8-24-2-1, 8-64-3-1, 11-166-3-2, 22, 42, 262A-43-8-11, 2588-34-5-6, 29723-143-3-2-1, 32809-26-3-3, 33238-25-2-3-2, 48725-B-B-38-1, 48725-B-B-103-2-3, 49455-20-2-2-3-3-1, 54751-3-38-10-17-3, IRON 89-59, Jalmagna, Latisail, MNP 234, MNP 236, MTU 9997, NP 125, OR 149-99, Pakistani Basmati, Pankaj, Pankhari 203, PTB 31, Pusa 2-21, Rasi, RP Nos. 4-41, 6-516-31-1-1, 270-48-4, 1057-393-1, 2151-166-4-1, 2151-166-4-4-1, 2151-175-1, 2151-192-1, 2151-192-2-5, 2151-221-4-2-4, 2151-224-4, RPSC 136, Sabarmati improved, Sigadis, Tadukan, Taichung 65, Taiwan 3, TCM 3, TKM 6, TN 1, Type 1132, V 18, Wag-wag, Ziri Karnal

Besides, the genotypes Gobind, IR 35546-17-3-1-3, KJT 72-1-173-44 and RYT 3/13 have been identified as promising donors for resistance breeding programme against stem rot as these genotypes showed consistent resistance amongst 71 rice cultures evaluated in a sick plot following three scoring systems based on the disease incidence, disease severity and coefficient of infection (Ram Singh and Dodan, 1997). Resistance to stem rot has also been reported in Bam 9, Bozu, Co 4, Co 25, IR 71, IR 272, Raminad Strain 3 (Gupta and Kapoor, 2002); BR 51-282-8-HR 45, IR 48725-B-B-103-2-3 (Kumar *et al.*, 2003); Bara 62, Basmati 3, Muskan 7, Muskan 41 (Rangaswami and Mahadevan, 2004). Recently, Sunder *et al.* (2008) reported that of 788 genotypes evaluated in a sick plot, HKR 2001-456, Narmada, Prasanna, RP 3895 and OR 1914-8 were resistant in Haryana, India.

Mechanism of Resistance

A little information is available on nature of resistance to stem rot. Nonaka and Yoshi (1958) observed less carbohydrate and more nitrogen compounds in infected culms resulting in low C:N ratio which favoured the disease development. A high correlation was observed between the lowering of the physiological functions of the culm base and damage caused by the disease (Watanabe, 1952).

Inheritance of Resistance

Ferreira and Webster (1975a,b, 1976) reported that host resistance and pathogen virulence are quantitatively inherited while virulence and fitness (production of conidia and sclerotia, growth rate, etc.) are inherited independently.

Pathogenic Variability

Meagre information is available on pathogenic variability in *S. oryzae*. Variable reaction of fungal isolates on a set of rice cultivars has been reported at different locations (Jain, 1973; Ahuja and Srivastava, 1990; Mazzanti de Castanon *et al.*, 1997). Contrary to this, Ferreira and Webster (1975a) failed to notice significant differences in virulence of different isolates. They further reported that virulence of individual isolate can be determined by measuring their ability to incite lesions (Ferreira and Webster, 1976). A set of nine rice genotypes *viz.*, CR 155-88, M Sung Sang, MNP 236, TKM 1, Tainan 3, Bhawani, Dudshar or Nagheri Baō, Chinan 8 and ARC 13287 or Bharati had been proposed as differentials for strain variation in *S. oryzae* isolates (Gangopadhyay and Das, 1983). However, Ahuja (1987) failed to identify variability among *S. oryzae* isolates using above set of differentials.

Sheath Rot (*Sarocladium oryzae*)

The disease is characterized by the production of greyish brown lesions/blotches on the upper most leaf sheaths enclosing the young panicles which either remain unemerged or partially emerged and distorted. Such panicles have partially filled/unfilled discoloured grains. Abundant whitish powdery growth is usually found inside affected sheaths and young panicles may rot completely. The disease is widely distributed in different parts of the world and is known to cause 6-57 per cent loss in grain yield depending upon the disease incidence.

Host Resistance

The research work on host plant resistance to sheath rot conducted at various institutes has been reviewed by Ram Singh and Dodan (1995b) and mentioned in Table 10.7.

Table 10.7: Rice Lines/Cultivars Resistant to Sheath Rot in Different Countries

Country	*Cultivars/Lines*
Cameroon	IR 7167-33-2-3, TOX 3145- TOC 34-2-3, TOX 3344-TOC 3-4
Korea, Taiwan	Japonica type varieties
India	ADT 36, ADT 38, ARC 7717, AS 2992, AS 5821, AS 6975, ASD Nos. 5, 7, 8, 11, 17, 18699, 19703, Ballia and Ghazipur selection of Mahsuri and Hansraj, Cult. Nos. 1287, 3226, 6403, CR 44-120-1, CR 93-4-2, D 25-4, Dissi hatif, Homthong, IET Nos. 7511, 7720, 7978, 8022, 8023, 8340, 8588, 9965, Intan, IR Nos. 8, 20, 24, 26, 50, 54, 60, 1529, 2071-176-1-2-1, 2071-685-3-5-4, 9179-209-2-2-3-3, 13423-10-2-3, 13423-17-1-2-1, 13927-40-2-3-3-3-3, 17494-32-1-1-3-2, 18349-53-1-3-1-3, 28143-60-3-3-2, 29652-65-2-3, 29725-3-1-3-2, 31802-48-2-2-2, 31851-63-1-2-3-2, 31868-64-2-3-3-3-3, 32429-45-3-2-6, 32429-47-3-2, 54742-6-34-17-1-1, 54742-18-3-8-22-1, 54742-22-14-3-7-2, 54745-2-21-12-7-1, Kalanamak, Kanto, M 23-17-1-1, M 23-69-1-1, Manoharsali, Pankaj, PAU 122-73-1-4-1, Pusa 37-6-2-3, R 2410, Raminad Strain 3, Ramtulasi, RP 974-112-1-6, RP 1825-71-4-1, RP 6801, RP 8884-8-1, RP 16691-2897-5526, Sigadis, SR 26-B, Tadukan, TCM 6012, Tellahamsa, Tetep, TKM 6, TN 1, Vishnubhog, Zenith

In addition, Taichung Sen Yu 229 (Lin and Tsai, 1977); Gajgor, Mahsuri (Mukherjee and Singh, 1980); Asha, Dubraj, Mathuri, Safri 17, Usha (Upadhyay and Diwakar, 1984); ARC 14529, CR 333-1-1-3, IET 8584, IR 9202-21-1, RB 2069-2-6-1, RP 1057-35-1-1, RP 2071-18-1-1, VRS 286-4-1 (Lewin and

Vidhyasekaran, 1987); ACM 1, AD 85002, AU 42/1, DP 11091/5, P 837, Ponni, TKM 1, TM 4300, TM 8381, TM 10232, TNAU 80030, TNAU 80042, White Ponni (Saroja *et al.*, 1987); BPMS 40A, BPMS 69, KSR White (Singh, 1988) have also been reported to possess sheath rot resistance in different parts of the world. Recently, in multilocations testing, several rice cultures namely, CB 04-15, CB 04-20, CB 04-27, CN 1448-10-1-1-3, CN 1450-14-12-2, CR Nos. 2071-247-1, 2076-180-6-1-1, 2251-3-5-3-1, 2286-6-4-2-1, 2458-99, IET Nos. 19766, 20421, 20453, 20457, 20460, NDR 40114, NDR 9930013, NLR 20022, OR 1898-16, PAU 3111-8-5-3-4, PAU 3419-1-3-2, RGL 20545, RGL 21066, RGL 21349, RNR 855, RPHR 650-2-5, TTB 103-3-1, TTB 302-1-2, TTB 303-1-2-5, TTB 303-3-2-1, VL 7318, VL 30336 and VOHP 3102 were found resistant to this disease (Anonymous, 2008).

Mechanism of Resistance

Mohan and Subramanian (1977) described sheath rot as a high sugar disease. A considerable reduction in total sugars and phenolic contents of leaf sheaths has been recorded in infected leaf sheath (Balakrishnan and Nair, 1983). Purkayastha *et al.* (1983a,b) detected momilactone in infected, dark brown coleoptiles of resistant tall cultivars. Significant differences in momilactone contents of infected leaf sheath of tall and semi-dwarf cultivars were also noticed. The abiotic elicitors GA_3, NAN_3 and penicillin induced the production of momilactone A in rice that reduced the sheath rot incidence. Cross reactivity tests indicated that the common antigen relationship between the cultivars and the pathogen could be altered by the application of GA_3 and NAN_3 (Anima Ghosal and Purkayastha, 1987). Velazhahan (1987) and Velazhahan and Ramabadran (1992) reported that the nitrogen fertilization increased the susceptibility of the host by decreasing the phenol content while potassium application reduced the disease by augmenting the synthesis of phenol. The peroxidase and polyphenol oxidase activities increased in rice plants inoculated with *S. oryzae* (Biswas and Purkayastha, 1988). Phenyl alanine ammonia lyase (PAL) activity decreased gradually upto 7 days in healthy and inoculated plants but level was higher in inoculated plants. All enzyme levels increased after GA_3 treatment but decreased in plants treated with Aureofungin and carbendazim. They suggested that accumulation of phenols might inhibit pathogen growth or act as precursor of lignin biosynthesis.

Inheritance of Resistance

Information available on inheritance of resistance to sheath rot of rice is limited. Srinivasachary *et al.* (2002) identified nine QTLs for sheath rot on seven chromosomes. Of these, three QTLs flanked by RS 109-RG 236, RG 612-RG 140 and RG 192-RG 172 recorded the LOD value of >3.5. The QTL flanked by RG 109-RG 236 recorded 18.8 per cent variation. The QTL identified for sheath rot ressistance flanked by RG 612-RG 140, RZ 276-CD 920 markers on chromosome 1 have also been reported to confer resistance to leaf blast.

Pathogenic Variability

Phookan and Hazarika (1994) did not observe significant morphological differences among eight isolates of *S. oryzae* collected from various rice growing areas of North-eastern India. However, the isolates were categorized into three distinct groups based on cultural, physiological and pathological tests on eight rice genotypes. Subsequently, Joe and Manibhushan Rao (1995) categorized the isolates of S. *oryzae* into five groups as SO 1 to SO 5 and those of *S. attenuatum* as SA 1 to SA 5 in order of virulence based on pathogenic tests on rice. The relative capacity of isolates to secrete CMCase, PTE, invertase and protease *in vitro* showed a positive correlation with degree of pathogenicity to rice.

Table 3.8: Rice Genotypes Possessing Resistance to Two or More Diseases

Location	Resistant Genotypes	Description	Reference
Kerala, India	M 23-17-1-1, M 23-69-1-1, TCM 6012	R to ShB, ShR	Balakrishnan *et al.,* 1981
Uttar Pradesh, India	Dissi hatif	R to LBl, NBl, ShR, MR to LSc	Bhatt *et al.,* 1984
	Tetep	R to LBl, NBl, MR to ShR, LSc	
	VL-8	R to NBl, ShB, MR to LBl, ShR, LSc	
	IR 9202-21-1	R to LBl, NBl, MR to ShR, LSc	
	Camponia SML	R to LBl, MR to NBl, LSc	
	Ta-poo-cho-z	R to LBl, NBl, MR to ShR, LSc	
	IR 4547-6-2-5	R to LBl, NBl, MR to ShR	
	IR 3237-339-2-5	R to LBl, ShR, MR to NBl	
Haryana, India	HAU Nos. 118-87, 118-110, 118-111, BR 51-282-8, IR 42, IR 2588-34-5-6	R to BB, SR	Chand *et al.,* 1984
	TCM 3, HBC 19, Pakistani Basmati HAUK 12-52-4-2	R to SR, kresek phase of BB R to SR, leaf blight phase of BB	Chand *et al.,* 1985; Chand and Ram Singh, 1986
	RP 2151-175-1, RP 2151-192-1, HKR 95-75	R to BB, SR	Ram Singh and Chand, 1986; Sunder *et al.,* 2007
	HKR 120, HKR 122, IR 54, PR 4141, RP 2151-21-1	R to BB, MR to SR	Panwar *et al.,* 1988
	HAU 118-104 CAMOR, IR 29723-143-2-1 IR 40	R to BB, MR to SR R to SR, MR to ShB R to ShB, MR to BB	Ram Singh and Dodan, 1995a
	HKR 05-60, IR 48725-B-B-86-2-2	R to ShB, SR	Authors' observations
Tamil Nadu,	ASD 18699, ASD 19703	R to BB, BS	Manuel *et al.,* 1985a
India	ASD 5, ASD 11	R to BB, Bl, ShR, BS, RTV	Manuel *et al.,* 1985b
	IET 9370, 9424, 9490, 9585	R to BB, BS	Natrajan *et al.,* 1988
	IR Nos. 54742-1-11-17-12-1, 54742-1-11-17-12-3, 54742-33-9-14-26-4, 54742-33-18-20-3-2, 54745-2-10-17-8-3, 54745-2-21-12-17-4, 54745-2-34-3-10-2	R to ShB, BS	Lakshmanan and Velusamy, 1991
Andhra Pradesh,	RNR Nos. 12088, 29692, 52118,	R to BB, ShB	Raj *et al.,* 1987
India	52124, 52156, 57979		
Bihar, India	Bhognabhari, Bhojana, Bibisall, Bhimsall, Chapasair, HRC 702, Kalamdani, Kalamkadhi, Katki, Lakhansall, Nanhiya, Satalchini, Sikkinalhiyi	R to ShR, BS	Shukla *et al.,* 1995
	Areba, Aditya, Dular, Pallavi	R to LBl, MR to BS	Chauhan *et al.,* 2000

Contd...

Table 3.8–Contd...

Location	Resistant Genotypes	Description	Reference
Assam, India	Badsabhog, Borjoha	R to Bl and BB	Singha and Pathak, 1998
	PBCMS 2A X Ranjit, PBCMS 3A X Pankaj	R to BB, MR to ShB, GD	Singha *et al.*, 2000
Madhya Pradesh, India	NPT 57K-11-104, NPT 63K-8-12	R to BB, MR to ShB	Kumar *et al.*, 1999
Punjab, India	CR 695-1, CR 703-8, HKR 28, R 257-63-3-1	R to BB, BS, MR to ShB	Raina *et al.*, 1992
	NDR 8002, PAU 2013-44-3-2, PAU 2036-40-2-2	R to BB, MR to ShB	Raina *et al.*, 1999
	IET 14252	R to BB, ShR, MR to BS	Goel and Lore, 2004
	IET Nos. 14277, 17048, 17159	R to BB, MR to ShB, BS	
	IET14777	R to BB, ShR, MR to BS	
	IET Nos. 15639, 15839, 15941, 16822, 17135, 17159, 17246	R to BB, ShR, MR to BS	
	IRGC Nos. 81940, 81941, 103303, 105165, 105268, 105270, 105272	R/MR to ShB, ShR	Ritu Bala and Goel, 2007
	IRGC 105272	R to BB, ShB, ShR	
Philippines	Pyramided lines from IR 72	R to BB, MR to ShB	Datta *et al.*, 2003
DRR, Hyderabad, India	CR 294-54-8, HKR 117, IR 32429-45-3-2-6, RP 2151-116-4-1, RP 2151-178-1-4-2, RP 2071-18-1-1, VL-8	R to BB, Bl	Siddiq, 1990
	RP 2076-46-4-2	R to BB, Bl, ShB, BS	
	RP 1125-1548-1-48	R to BB, Bl, RTV	
	RP 2151-173-1-8	R to BB, ShB, RTV	
	RP 1057-35-1-1	R to BB, ShB	
	HKR 120	R to BB, Bl, ShR	
	KD 14-1-39	R to Bl, ShB, BS, RTV	
	IET 19542, 20066	R to BB, Bl	Anonymous, 2007
	IET 19739	R to ShR, BS	
	CB 02-012	R to ShB, ShR, BS	
	CB 03-008	R to Bl, ShB, ShR	
	IET 19144	R to BB, Bl	
Anonymous, 2008			
	IET 19958, IET 20421	R to Bl, ShR	
	IET 19163, IET 19402	R to BB, NBl	
	IET Nos. 19931, 20228, 20247, 20388	R to NBl, ShR	
	IET 20604	R to NBl, ShB	
	IET 20441	R to Bl, BS	
	IET 20433, IET 20442	R to Bl, ShB	
	IET 20455	R to NBl, BS	
	IET 19766	R to ShR, BS	
	IET 20448	R to BB, NBl, ShB	
	IET 20453	R to NBl, ShB, ShR, BS	

BB: Bacterial blight, Bl: Blast, LBl: Leaf blast, NBl: Neck blast, ShB: Sheath blight, ShR: Sheath rot, SR: Stem rot, BS: Brown spot, GD: Grain discolouration, RTV: Rice tungro virus; LSc: Leaf sclad.

Multiple Disease Resistance

A rice crop in a given area is attacked by different pathogens at different growth stages. Moreover, certain diseases are known to predispose the crop to other diseases. Rice tungro virus or bacterial blight infections in the early growth stages predispose the plants to sheath rot infection. The simultaneous occurrence of two or more diseases causes extensive crop losses, indicating the need to develop resistance for more than one disease (Reddy *et al.*, 1986). Evaluation of germplasm and advanced breeding lines for multiple stresses under various environments in the country led to identification of several rice genotypes possessing resistance to two or more diseases (Table 3.8). Similarly, several multiple disease resistant rice cultivars have been released at IRRI, Philippines (Siddiq, 1990; Table 3.9) and in India for commercial cultivation in different agro-ecosystems (Shobha Rani *et al.*, 2008; Table 3.10). Out of the later cultivars, IR 36, Pant Dhan 4, Sahyadari 2, Swarnadhan and Tulasi are popular in the country.

Table 3.9: Selected IRRI Rice Varieties with Multiple Resistance to Major Diseases

Sl.No.	*Variety*	*Resistance to*					
		Blast	*Bacterial Blight*	*Rice Tungro Virus*	*Grassy Stunt*	*Narrow Brown Leaf Spot*	*Brown Spot*
1.	IR 20	R	R	MR	–	–	MR
2.	IR 28	R	R	R	R	MR	–
3.	IR 34	R	R	R	R	–	–
4.	IR 36	R	R	R	R	MR	MR
5.	IR 40	R	R	R	R	–	–
6.	IR 42	R	R	R	R	MR	MR
7.	IR 44	R	R	R	–	–	–
8.	IR 48	R	R	R	R	R	MR
9.	IR 50	–	R	R	R	R	R
10.	IR 52	–	R	R	R	–	–
11.	IR 54	R	R	R	R	R	MR
12.	IR 58	R	R	R	R	–	R
13.	IR 60	R	R	R	R	–	–

Table 3.10: Varieties Resistant/Moderately Resistant to more than Two Diseases in India

Sl.No.	*Variety*	*Released by/in*	*Resistance to*				
			Blast	*Bacterial Blight*	*Rice Tungro Virus*	*Sheath Blight*	*Brown Spot*
1.	Bhuthnath	CVRC	R	–	R	MR	R
2.	Dhanrasi	CVRC	R	MR	MR	–	–
3.	IR 36	CVRC	MR	R	MR	–	MR
4.	Manasarovar	CVRC	R	MR	MR	MR	–
5.	Richa	CVRC	R	–	–	MR	MR
6.	Salivahana	CVRC	R	MR	–	MR	MR

Contd...

Table 3.10–Contd...

Sl.No.	Variety	Released by/in	Resistance to				
			Blast	Bacterial Blight	Rice Tungro Virus	Sheath Blight	Brown Spot
7.	Swarnadhan	CVRC	R	–	MR	MR	MR
8.	Tulasi	CVRC	R	MR	MR	–	–
9.	Apurva	Andhra Pradesh	R	R	–	–	R
10.	Chandan	Andhra Pradesh	R	MR	–	MR	–
11.	Chandrama	Assam	R	MR	MR	MR	–
12.	Rajendradhan 201	Bihar	R	R	MR	–	–
13.	Bamleshwari	Chhattisgarh	R	R	–	MR	MR
14.	Birsadhan 103	Jharkhand	R	R	–	–	MR
15.	Neeraja	Kerala	R	MR	–	–	MR
16.	Ratnagiri 3	Maharashtra	R	MR	–	–	MR
17.	Sahyadri 2	Maharashtra	R	R	MR	–	–
18.	Madhuri	Madhya Pradesh	R	R	–	–	MR
19.	Bhanja	Orissa	R	MR	–	–	MR
20.	Birupa	Orissa	R	R	R	–	–
21.	Ghanteswari	Orissa	R	MR	–	MR	–
22.	Jagabandhu	Orissa	R	MR	R	–	–
23.	Jogesh	Orissa	R	MR	–	MR	R
24.	Kanchan	Orissa	R	R	–	MR	–
25.	Manika	Orissa	MR	MR	R	–	–
26.	Mehar	Orissa	R	MR	M	–	–
27.	Nilagiri	Orissa	R	MR	MR	–	–
28.	Pratikshya	Orissa	MR	–	–	MR	R
29.	Rajeshwari	Orissa	MR	MR	R	–	–
30.	Sidhanta	Orissa	R	MR	–	R	R
31.	Uphar	Orissa	R	MR	–	MR	–
32.	Annalakshmi	Pondicherry	MR	–	R	–	MR
33.	ADT 39	Tamil Nadu	R	R	–	–	MR
34.	Co 45	Tamil Nadu	R	R	R	–	–
35.	Pant Dhan 4	Uttarakhand	R	MR	–	–	R
36.	Pant Dhan 6	Uttarakhand	R	R	–	–	R

*CVRC: Central variety release committee.

Conclusion

Utilization of commercial rice varieties possessing resistance for effective and economic management of diseases has helped to stabilize rice production. Satisfactory progress with respect to development of bacterial blight and blast resistant varieties has been made in different countries. Bacterial blight and blast resistance traits are well defined and studied both at genetic and molecular levels. Hence, the technology of marker assisted selection (MAS) offers great promise for development of rice lines resistant to these diseases. Many of the bacterial blight and blast resistance genes have already been mapped and cloned, and the markers tightly linked to these genes are available in the public domain. Marker-assisted breeding and gene pyramiding would be an ideal strategy to be used for getting targeted results. However, systematic resistance breeding programme for other diseases like sheath blight, false smut, sheath rot, stem rot and brown spot etc. is lagging behind. International donors resistant to different strains of these pathogens need to be developed so that work on this aspect may be initiated/strengthened. There is an urgent need to evolve an efficient and reliable inoculation technique for emerging diseases like false smut to locate stable resistant sources for their utilization in resistance breeding programme. Continuous emergence of new pathogens or races/ pathotypes has been posing new challenges to the ongoing crop improvement efforts. In India, the resistance of majority of the bacterial blight resistant varieties like HKR 120, PR 109, PR 110, PR 111, PR 113, PR 114, PR 115, PR 116, PR 118, PR 4141, Pant Dhan-6, Madhuri, Ruchi, Sita, Padmini, Pallavi, Pathara, Tapaswani, Lunglinaphou, Sahyadri-2, Birsa Dhan 105, Birsa 107, Ajaya, IR 36, Narendra Dhan 359 etc. has been overcomed due to change in virulence of the pathogen in different parts of the country. Further efforts are needed to breed multiple disease resistant varieties with polygenic background to confer wide-range resistance. Ecological understanding of the pathogen population structure helps in developing suitable deployment strategies for achieving durable resistance. A large proportion of the identified sources of resistance still remain unutilized. Untapped resistant genes available from land races and wild accessions are yet to be explored. Cultivated germplasm, land races and wild rices should be evaluated to identify resistance genes effective against various races of the pathogen in different agro-climatic zones, which may be pyramided in the same genetic background through use of MAS.

References

Adair, C.R., 1941. Inheritance in rice of reaction to *Helminthosporium oryzae* and *Cercospora oryzae*. *Tech. Bull.* USDA No. 772, 18 pp.

Agrawal, K.C., Thrimurty, V.S., Lakpale, N. and Kumar, R., 1997. Varieties for effective management of bacterial blight of rice. In: *Indian Phytopathol. Soc. Golden Jubilee Intern. Conf.*, N. Delhi, India, Nov. 10-15, 1997 (Abstr.) p. 3-013.

Ahn, S.W. and Mulekar, A., 1986. Rice blast management under upland conditions. In: *Progress in Upland Rice Research*, IRRI, Philippines, pp. 363-374.

Ahuja, S.C., 1987. Variability and pathogenesis of *Sclerotium* spp. causing stem rot of rice. *Final Technical Report*, Directorate of Research, HAU, Hisar: 67 pp.

Ahuja, S.C. and Srivastava, M.P., 1990. Stem rot of paddy-A review. *Agric. Rev.* 11(2): 61-78.

Amante, A.D., Pena, R. de la, Sitch, L.A., Leung, H. and Mew, T.W. (1990). Sheath blight resistance in wild rices. *Intern. Rice Res. Newsl.* 15(3): 5.

Anima Ghosal and Purkayastha, R.P., 1987. Biochemical response of rice (*Oryza sativa* L.) leaves to some abiotic elicitors of phytoalexins. *Indian J. Exptl. Biol.* 25 (6): 395-399.

Anonymous 1999a. *Draft Proceedings-XXXIV, Annual Rice Group Meeting,* DRR, Rajendranagar, Hyderabad, April 11-14, Appendix I, IV and pp. 50-51.

Anonymous 2007. *Progress Report 2006, Vol. 2, Crop Protection (Entomology, Plant Pathology)* AICRIP, ICAR, DRR, Rajendranagar, Hyderabad-500 030 (A.P.), India.

Anonymous, 2008. *Progress Report 2007, Vol. 2., Crop Protection (Entomology, Plant Pathology)* AICRIP, ICAR, DRR, Rajendranagar, Hyderabad-500 030 (A.P.), India.

Ansari, M.M., Ram, T. and Sharma, T.V.R.S., 1988. Yield loss assessment in promising rice cultures due to false smut. *Oryza* 25: 207-209.

Anwar, A., Ahmad, N., Zargar, M.A., Singhara, G.S., Ashraf Rather, M., Bhat, G.N., Parray, G.A. and Salroo, M.Y., 2003. Evaluation of elite rice genotypes against blast disease (*Magnaporthe grisea*) under epiphytotic conditions in Kashmir. *Pl. Dis. Res.* 18: 77-79.

Ashrafuzzaman, M.H., 1974. Varietal screening under natural field conditions for disease resistance in rice. II. Testing of rice strains for resistance to bacterial leaf blight, leaf yellowing and false smut. *Indian Phytopath.* 27: 263-264.

Atia, M.M.M., 2004. Rice false smut (*U. virens*) in Egypt. *Z. Pflkrankh. Pflschutz.* 111(1): 71-82.

Awoderu, V.A., 1970. Identification of races of *Pyricularia oryzae* in Nigeria. *Pl. Dis. Reptr.* 54: 520-523.

Bagga, P.S. and Kumar, V., 2000. Resistance to bakanae or foot rot disease in basmati rice. *Indian Phytopath.* 53: 321-322.

Balakrishnan, B. and Nair, M.C., 1983. Observations on post-infectional changes of rice plants infected by sheath rot fungus, *Acrocylindrium oryzae* Sawada. *Agric. Res. J. Kerala* 21: 73-74.

Balakrishnan, B., Rajan, K.M., Mathai, G. and Remabai, N., 1981. Disease reaction of advanced rice cultures. *Intern. Rice Res. Newsl.* 6(4): 5-6.

Balal, M.S., Omar, R.A., El-Khadem, M.M. and Aidy, I.R., 1979. Inheritance of resistance to the brown spot disease of rice, *Cochliobolus miyabeanus. Agril. Res. Rev.* 57(8): 119-133.

Bedair, F.A., Shaalan, M.I., El-Khishen, A.A. and Gomma, M.E., 1976. Inheritance of resistance to the brown spot disease of rice, *Cochliobolus miyabeanus. Egyptian J. Genet. Cytol.* 5(2): 433-449.

Bhardwaj, C.L., 1990. False smut incidence on rice relative to plant characters and environmental factors. *Intern. Rice Res Newsl.* 15(3): 29-30.

Bhatt, J.C. and Chauhan, V.S., 1984 Evaluation of hill rice germplasm for leaf blast resistance for hilly regions in India. *Intern. Rice Res. Newsl.* 9(5): 4.

Bhatt, J.C., Garg, D.K. and Tandon, J.P., 1984. Rice with multiple disease and insect resistance in hilly regions in Uttar Pradesh. *Intern. Rice Res. Newsl.* 9(1): 11.

Biswas, A., 2001a. Field reaction of hybrid rice varieties to false smut and kernel smut diseases in West Bengal, India. *Environ. Ecol.*, 19: 229-230.

Biswas, A., 2001b. Evaluation of rice germplasm for sheath blight disease resistance in West Bengal, India. *Environ. Ecol.*, 19: 247-248.

Biswas, S. and Purkayastha, R.P., 1988. Induction of resistance in rice plants against sheath rot disease. *Indian Phytopath.* 41: 51-56.

Bonman, J.M. and Mackill, D.J., 1988. Durable resistance in rice. *Oryza* 25: 103-110.

Buddenhagen, I.W. and Reddy, A.P.K., 1972. The host, the environment, *Xanthomonas oryzae* and the researcher. In: *Rice Breeding*, IRRI, Los Banos, Philippines, pp. 289-295.

Chand, H. and Ram Singh 1986. Evaluation of scented rice germplasm against stem rot and bacterial blight. *Indian Phytopath.* 39: 488-489.

Chand, H. and Ram Singh and Panwar, D.V.S., 1984. Varietal resistance to stem rot and bacterial blight. *Intern. Rice Res. Newsl.* 9(5): 4.

Chand, H., Ram Singh and Panwar, D.V.S., 1985. Search for multiple disease resistance in rice. *Indian Phytopath.* 38: 452-455.

Chauhan, J.S., Variar, M., Shukla, V.D., Maiti, D., Bhattacharya, N. and Lodh, S.B., 2000. Screening rice genetic resources for major diseases of uplands and quality traits of resistant donors. *Indian Phytopath.* 53: 80-82.

Che, K.P., Zhan, Q.C., Xing, Q.H., Wang, Z.P., Jin, D.M., He, D.J. and Wang, B., 2003. Tagging and mapping of rice sheath blight resistant gene. *Theor. Appl. Genet.* 106: 293-297.

Cheema, A.A., Awan, M.A. and Ali, Y., 1998. Screening of basmati rice mutant against prevalent diseases in the Punjab province. *Pakistan J. Phytopathol.* 10(1): 39-41.

Cheng, G.Y., Zhang, G.A., Zhang, Z.D., Huang, H.M., Nie, X.G., Li, Y.S., Mao, H.P. and Yu, L.F., 2000. Comparison of resistance to rice sheath blight caused by *Rhizoctonia solani* in early rice cultivar. *J. Huazhong agril. Univ.*, 19: 547-549.

Chien, C.C., 1967. Studies on the physiologic races of the rice blast fungus, *Pyricularia oryzae* Cav. *Bull. Taiwan agric. Res. Inst.* 26: 1-63.

Chien, C.C. and Chung, S.C., 1963. Physiologic races of *Pellicularia sasakii* in Taiwan. *Agril. Res. Taiwan* 12(2): 1-6.

Correa-Victoria, F.J. and Zeigler, R.S., 1993. Pathogenic variability in *Pyricularia grisea* at a rice blast hot spot breeding site in eastern Columbia. *Plant Dis.* 77: 1029-1035.

Dai, L., Liu, X., Xiao, Y. and Wang, G., 2007. Recent advances in cloning and characterization of disease resistance genes in rice. *J. Integrative Pl. Biol.* 49: 112-119.

Datta, K., Baisakh, N., Thet, K.M., Tu, J. and Datta, S.K., 2003. Pyramiding transgenes for multiple resistance in rice against bacterial blight, yellow stem borer and sheath blight. *Theor. Appl. Genet.* 106: 1-8.

Dissanayake, N. and Ponnawila, W.J., 1992. Identification of bacterial blight (BB) pathotype of *Xanthomonas oryzae* pv. *oryzae* in Batlagoda, Sri Lanka. *Intern. Rice Res. Newsl.* 17(1): 25.

Dodan, D.S. and Ram Singh 1996. False smut of rice: Present status. *Agric. Rev.* 17: 227-240.

Eruotor, P.G., 1980. Inheritance of brown leaf spot of rice caused by *Cochliobolus miyabeanus* and varietal reaction to isolates of this fungus. *Dissertation Abstracts International, B.* 41(4): 1192B.

Eruotor, P.G., 1986. Inheritance of sheath colouration and of brown leaf spot of rice. *Z. Pflschutz.* 96: 47-52.

Ezuka, A., 1979. Breeding for genetics of resistance in Japan. In: *Proc. Rice Blast Workshop*, IRRI, Philippines, pp. 27-40.

Ferreira, S.A. and Webster, R.K., 1975a. Genetics of stem rot resistance in rice and virulence in *Sclerotium oryzae. Phytopathology* 65: 968-971.

Ferreira, S.A. and Webster, R.K., 1975b. The relationship of sporulation, sclerotia production and growth rate to virulence and fitness of *Sclerotium oryzae. Phytopathology* 65: 972-976.

Ferreira, S.A. and Webster, R.K., 1976. Evaluation of virulence in isolates of *Sclerotium oryzae*. *Phytopathology* 66: 1151-1154.

Galvez, G.E. and Lozano, J.C., 1968. Identification of races of *Pyricularia oryzae* in Columbia. *Phytopathology* 58: 294-296.

Gangopadhyay, S. and Chakrabarti, N.K., 1982. Sheath blight of rice. *Rev. Pl. Pathol.* 61: 451-460.

Gangopadhyay, S. and Chattopadhyay, S.B., 1974. Correlation between leaf angle and incidence of brown spot disease of rice incited by *Helminthosporium oryzae*. *Indian J. Mycol. Pl. Pathol.* 4: 34-39.

Gangopadhyay, S. and Das, K.M., 1983. Reaction of rice to eight isolates of *Sclerotium oryzae* in India. *Indian Phytopath.* 36: 219-222.

Gnanamanickam, R., Sivaraj, R., Levy, M., Ranganathan, N. and Hamer, J.E., 1993. Genetic organization of the rice blast fungus in India. In: *6th Annual Meeting International Programme on Rice Biotechnology*, China National Rice Research Institute and Purdue University, West Lafayette, IN, USA, p. 232 (Abstr.).

Goel, R.K., Kaur, L. and Saini, R.G., 1998. Effectiveness of different *Xa* genes against *Xanthomonas campestris* pv. *oryzae* population causing bacterial blight of rice in Punjab (India). *Rice Genet. Newsl.* 15: 131-133.

Goel, R.K. and Lore, J.S., 2004. Some sources of multiple disease resistance in rice in the irrigated agro-ecosystem in North-western India. *Pl. Dis. Res.*, 19:167-168.

Goita, M., 1985. Inheritance of resistance to sheath blight in long grain rice. *Dissertation Abstracts International Sci. Eng.* 46: 1808B.

Gonzalez, C., Szurek, B., Manceau, C., Mathieu, T., Sere, Y. and Verdier, V., 2007. Molecular and pathogenic characterization of new *Xanthomonas oryzae* strains from West Africa. *Molecular Plant-Microbe Interaction* 20: 534-536.

Gopalakrishnan, S., Sharma, R.K., Rajkumar, K.A., Joseph, M., Singh, V.P., Singh, A.X., Bhat, K.V., Singh, N.K. and Mohapatra, T., 2008. Integrating marker assisted background analysis with foreground selection for identification of superior bacterial blight resistant recombinants in Basmati rice. *Plant Breeding* 127: 1-9.

Goto, K., 1965. Physiologic races of *Pyricularia oryzae* in Japan. In: *The Rice Blast Disease*, John Hopkins Press, Baltimore, USA, pp. 237-242.

Guo, C.J., Chen, Z.Y. and Wang F.M., 1985. Pathogenic variability in *Thanatephorus cucumeris* (Frank) Donk and techniques for identifying varietal resistance. *Sci. agric. Sin.* 5: 50-57.

Gupta, V.K. and Kapoor, A.S., 2002. Fungal diseases of rice. In: *Disease of Field Crops* (eds. V.K. Gupta and Y.S. Paul), Indus Publishing Co., New Delhi, pp. 92-94.

Harahap, Z., 1976. Inheritance of resistance to brown spot disease of rice caused by *Helminthosporium oryzae* Breda de Haan. *Dissertation Abstracts International, B.* 37(5): 2004B-2005B.

Harish, S., Saravanakumar, D., Kamalakannan, A., Vivekananthan, R., Ebenezar, E.G. and Seetharaman, K., 2007. Phylloplane microorganisms as a potential biocontrol agent against *Helminthosporium oryzae* Breda de Hann, the incitant of rice brown spot. *Archiv. Phytopathol. Pl. Prot.* 40 (2): 148-157.

Hashioka, Y., 1951. Inheritance of resistance to sheath blight in rice varieties. *Annals Phytopathol. Soc. Japan* 15: 98-99.

Haque, M.A., 1975. *Variation in cultural characteristics, physiology and pathogenicity of field isolates of Thanatephorus cucumaris (Frank) Donk, the causal organism of rice sheath blight disease.* M.S. Thesis, Univ. Philippines, Los Banos, Laguna, Philippines, 79 pp.

He, Q., Li, D., Zhu, Y., Tan, M., Zhang, D. and Lin, X., 2006. Fine mapping of Xa 2, a bacterial blight resistance gene in rice. *Mole. Breed.* 17: 1-6.

Hegazi, M.F., Dorria, I.H., Mostafa, M.H. and Ibrahim, I.K., 1993. Changes in some metabolites and oxidative enzymes associated with brown leaf spot of rice. *Annals agril. Sci. Cairo* 38(1): 291-299.

Hittalmani, S., Foolad, M.R., Mew, T.W., Rodriguez, R.L. and Huang, N., 1995. Development of a PCR-based marker to identify rice blast resistance gene, *Pi-2(t)*, in a segregation population. *Theor. Appl. Genet.* 91: 9-14.

Horino, O., 1981. Ultra structural histopathology of rice leaves infected with *Xanthomonas campestris* pv. *oryzae* on Kogyoku group rice varieties with different levels of resistance at the seeding stage. *Annals Phytopathol. Soc. Japan* 47: 501-509.

Huang, N., Angeles, E.R., Domingo, J., Magpantay, G., Singh, S., Zhang, G., Kumaradivel, N., Bennett, J. and Khush, G.S., 1997. Pyramiding of bacterial blight resistance genes in rice: marker assisted selection using RFLP and PCR. *Theor. Appl. Genet.* 95: 313-320.

IRRI 1972. *Annual Report, 1971,* IRRI, Los Banos, Philippines: 105-107.

IRRI 1975. *Annual Report, 1974,* IRRI, Los Banos, Philippines.

IRRI 1979. *Annual Report, 1978,* IRRI, Los Banos, Philippines: 164-165.

Jain, S.S., 1973. Specialization in pathogenicity of *Sclerotium oryzae* Catt., the causal organism of the stem rot disease of rice. *Oryza* 10: 59-66.

James, W.C., 1974. Assessment of plant diseases and losses. *Annu. Rev. Phytopathol.* 12: 27-49.

Jha, A.C., Rai, B. and Jha, M.M., 2004. Screening of resistant sources against brown spot disease. *Annals Biol.* 20(2): 189-193.

Jiten Konthoujam and Chhetry, G.K.N., 2007. Reaction of some rice cultivars to stem rot caused by *Sclerotium oryzae* Catt. in Manipur. *Indian Phytopath.* 60: 218-221.

Joe, Y. and Manibhushan Rao, K., 1995. Biochemical constituents of differentially virulent isolates of *Sarocladium* spp. causing sheath rot disease of rice. *Z. Pflkrankh. Pflschutz.* 102 (3): 291-298.

Johnson, R., 1981. Durable resistance: Definition of genetic control and attainment in plant breeding. *Phytopathology* 71: 567.

Kaku, H. and Kimura, T., 1978. Reaction types of rice cultivars to strains of *Xanthomonas oryzae. Bull. Chugoku Natn. agric. Exptl. Stn. Ser. E* 13:17-43.

Kalita, M.K., Pathak, K. and Saikia, M.K., 2000. Reaction of some rice cultivars against leaf and sheath blight of rice (*Rhizoctonia solani* Kuhn) in southern Assam. *Annals Agri-Bio Res.* 5(2): 163-164.

Kalita, U.C., Baruah, D.K. and Ali, M.S., 1995. Evaluation of bacterial blight resistance in rice germplasm. *Intern. J. Trop. agric.* 13(1/4): 157-160.

Kanjanasoon, P., 1965. *Studies on bakanae disease of rice in Thailand.* Doc. Agric. Thesis, Tokyo Univ., Japan.

Kaul, M.L.H. and Sharma, K.K., 1987. Bacterial blight in rice–A review. *Biol. Zent. Bl.* 106: 141-167.

Kaur, R., Grewal, N., Das, A., Vikal, Y., Singh, J., Bharaj, T.S., Sidhu, J.S. and Singh, K., 2005. Inheritance of bacterial blight resistance in two accessions of wild rice, *Oryza nivara*. *Rice Genet. Newsl.* 22: 78-82.

Khush, G.S. and Brar, D.S., 2001. Rice genetics from Mendel to functional genomics. In: *Rice Genetics IV. Proc. 4th Rice Genetics International Symp.* (eds. G.S. Khush, D.S. Brar and B. Hardy), Oct. 22-27, 2000, pp. 3-26, IRRI, Philippines.

Kiryu, T. and Mizuta, H., 1955. On the relation between habits of rice and varietal resistance of bacterial leaf blight. *Kyushu agril. Res.* 15: 54-56.

Kiyosawa, S., 1981. Gene analysis for blast resistance. *Oryza* 18: 196-203.

Kumar, A., Ram Singh and Jalali, B.L., 2003. Evaluation of resistance to stem rot and yield losses caused by the disease in rice. *Indian Phytopath.* 56: 403-407.

Kumar, A., Tiwari, R.K.S. and Janoria, M.P., 1997. Host-pathogen interaction of new plant type breeding lines. In: *Indian Phytopathol. Soc. Golden Jubilee Intern. Conf.*, N. Delhi, India, Nov. 10-15, 1997 (abstr.).

Kumar, A., Tiwari, R.K.S., Parihar, S.S., Pandya, K.S. and Janoria, M.P., 1999. Performance of prototype rice lines from ideotype breeding. *Oryza* 36: 146-148.

Laha, G.S., Ram Singh, Sunder, S., Krishnaveni, D., Dodan, D.S. and Muralidharan, K., 2008. Evaluation of selected cultures for broad-spectrum resistance to bacterial leaf blight. *J. Mycol. Pl. Pathol.*: In press.

Lakshmanan, P. and Velusamy, R., 1991. Resistance to sheath blight and brown spot in lines derived from *Oryza officinalis*. *Intern. Rice Res. Newsl.* 16(6): 8.

Lee, E.H., Zhang, Q. and Mew, T.W., 1989. Durable resistance to rice diseases in irrigated environments. In: *Progress in Irrigated Rice Research*, IRRI, Philippines, pp. 93-110.

Lee, F.N. and Rush, N.C., 1983. Rice sheath blight: a major rice disease. *Pl. Dis.* 67: 829-832.

Lee, S.C. and Matsumoto, S., 1966. Studies on the physiologic races of rice blast fungus in Korea during the period of 1962-63. *Annals Phytopathol. Soc. Japan.* 32: 40-45.

Lewin, H.D. and Vidhyasekaran, P., 1987. Sources of resistance to sheath rot. *Intern. Rice Res. Newsl.* 12(1): 17.

Lin, T.F. and Tsai, T.T., 1977. Indica rice varietal improvement for major diseases and insect resistance. *J. agric. Assoc. China Newsl. Ser.* 102: 86-95.

Lore, J.S., Singh, N. and Bharaj, T.S., 2004. Improved sources of genetic resistance against bacterial blight of rice. *Pl. Dis. Res.*, 19: 169-170.

Mackill, D.J. and Bonman, J.M., 1992. Inheritance of blast resistance in near isogenic lines of rice. *Phytopathology* 82: 746-749.

Mahto, B.N., Singh, R.N., Awasthi, C.P. and Abidi, A.B., 1987. Sugars and phenolic compounds in rice leaves in relation to varietal resistance to bacterial blight (BB) pathogen. *Intern. Rice Res. Newsl.* 12(4): 12-13.

Manuel, W.W., Ganesan, K., Rajagopalan, C.K. and Mariappan, A., 1985a. New sources of resistance to major rice diseases. *Intern. Rice Res. Newsl.* 10(1): 7.

Manuel, W.W., Ganesan, K., Rajagopalan, C.K. and Mariappan, A., 1985b. Reaction of ASD varieties to various rice diseases. *Intern. Rice Res. Newsl.* 10(1): 7.

Marchetti, M.A., Bollich, C.N., Webb, B.D., Jackson, B.R., Mc Clung, A.M., Scott, J.E. and Hung, H.H., 1998. Registration of 'Jasmine 85' rice. *Crop. Sci.* 38: 896.

Marchetti, M.A. and Bonman, J.M., 1989. Rice blast management. In: *Rice Farming Systems: New Directions*, IRRI, Philippines, pp. 175-183.

Marchetti, M.A., Lai, X.H. and Bollach, C.N., 1987. Inheritance of resistance to *Pyricularia oryzae* in rice cultivars grown in United States. *Phytopathology* 77: 799-804.

Mariappan, V., 1984. Rice blast pathogen *Pyricularia oryzae* Cav. in Tamil Nadu, India. *Intern. Rice Res. Newsl.* 9(5): 12.

Marshall, D.S. and Rush, M.C., 1980. Relation between infection by *Rhizoctonia solani* and *Rhizoctonia oryzae* and disease severity in rice. *Phytopathology* 70: 941-946.

Masajo, T.M., 1976. *Varietal resistance in rice to sheath blight caused by Thanatephorus cucumeris (Frank) Donk Rhizoctonia solani Kuhn*. Ph.D. Dissert. Louisiana State Univ., Baton Rogue.

Matsumoto, S., Kozaka, T. and Yamada, M., 1969. Pathogenic races of *Pyricularia oryzae* Cav. in Asia and some other countries. *Bull. Natn. Inst. agril. Sci. Tokyo* C 23: 1-36.

Mazzanti de Castanon, M.A., Mazza, de Gaiad, S. and Gutierrez de Arriola, S.A., 1997. Pathogenic variability in *Sclerotium oryzae* in Argentina. *Fitopatologia* 32: 58-63.

Mc Clung, A.M., Marchetti, M.A., Webb, B.D. and Bollich, C.N., 1998. Registration of 'Dixiebelle' rice. *Crop Sci.* 38: 898.

Meena, B., Ramamoorthy, V., Banu, J.G., Thangavelu, R. and Muthusamy, M., 2000. Screening of rice genotypes against sheath blight disease. *J. Ecobiol.* 12(2): 103-109.

Mew, T.W., 1987. Current status and future prospects of research on bacterial blight of rice. *Annu. Rev. Phytopathol.* 25: 359-382.

Mew, T.W., Leung, H., Savary, S., Vera Cruz, C. M. and Leach, A., 2004. Looking ahead in rice disease research and management. *Critical Rev. Plant Sci.* 23: 103-127.

Minni Sandhu, Parmil Singh and Sharma, S.C., 1995. Effect of *Xanthomonas oryzae* infection in rice on polyphenol oxidase and peroxidase activity and total phenol content. In: *Integrated Disease Management and Plant Health* (eds. V.K. Gupta and R.C. Sharma), Scientific Publisher, Jodhpur, pp. 49-53.

Misra, A.K., 1985. Variability in *Drechslera oryzae*–the causal organism of brown spot disease of rice. *Indian Phytopath.* 38: 168-169.

Misra, A.P. and Prasad, Y., 1964. The nature of resistance of paddy to *Helminthosporium oryzae* Breda de Hann. *Indian Phytopath.* 17: 287-295.

Mishra, D., Singh, U.D., Dash, A.B., Reddy, J.N., Sridhar, R., George, M.L.C., Vera Cruz, C.M., Bernardo, M.A. and Leung, H., 2006. Analysis of *Pyricularia grisea* populations from three different blast epidemics. *Intern. Rice Res. Notes* 31(1): 22-24.

Mohan, R. and Subramanian, C.L., 1977. Influence of nitrogen nutrition and sheath rot disease on sugar contents of two paddy varieties. *Curr. Sci.* 46(21): 753-755

Mohd. Najeeb, Sabiya Bashir, Kalha, C.S. and Ahmed, M., 2006. Identification of races in rice blast pathogen *Pyricularia grisea* with differential hosts. *Pl. Dis. Res.* 21: 178-179.

Mukherjee, P. and Singh, B.P., 1980. Testing of indigenous rice germplasm against sheath rot by artificial inoculation. *Indian Phytopath.* 33: 149.

Muralidharan, K., 2006. Variability, the weapon in the combat between plant and pathogen as illustrated with rice and blast. In: *Plant Protection in New Millennium, Vol. 1* (eds. Ashok V. Gadewar and B. P. Singh), Satish Serial Publishing House, New Delhi, pp. 69-108.

Muralidharan, K., Krishnaveni, D., Laha, G.S., Reddy, C.S., Srinivasprasad, M. and Sridhar, R., 2003. Evaluation of blast resistance genes in India. *Rice Genet. Newsl.* 20: 94-96.

Muralidharan, K., Krishnaveni, D., Laha, G.S., Reddy, C.S., Srinivasprasad, M. and Sridhar, R., 2004. Performance of rice blast resistance genes in multi-environment test in India. *Indian Phytopath.* 57: 260-266.

Nagai, I. And Hara, S., 1930. On the inheritance of variegation disease in a strain of rice plant. *Japanese J. Genet.* 5: 140-145.

Natrajan, S., Jeyarajan, R., Narayanasamy, P. and Ranganathan, T.B., 1988. Varietal screening for bacterial leaf blight of rice. In: *Adv. Res. Pl. Pathol.: Bacteria Based on Proc. Natn. Symp. Phytobacteriol.*, Univ. of Madras, Madras, India, March 14-15, 1986 (eds. S.S. Gnanamanickam and A. Mahadevan), Today and Tomorrows Printers and Publishers, New Delhi, India, pp. 95-96.

Naqvi, N.I., Bonman, J.M., Mackill, D.J., Nelson, R.J. and Chatto, B.B., 1995. Identification of RAPD markers linked to a major blast resistance gene in rice. *Molecular Breeding* 1: 341-348.

Nino-Liu, D.O., Pamela, C. Ronald and Adam, J. Bogdanove 2006. *Xanthomonas oryzae* pathovars: model pathogens of a model crop. *Mol. Pl. Pathol.* 7(5): 303-324.

Nonaka, F. and Yoshi, H., 1958. Carbohydrate and nitrogen contents of lower parts of rice culms affected with stem rot fungus, *Leptosphaeria salvinii. Sci. Bull. Fac. agric. Kyushu Univ.* 16: 447-454; 459-463.

Nottenghem, J.L., 1985. Blast resistance methodologies in the Ivory Coast, 1972-1985. In: *Progress in Upland Rice Research*, IRRI, Philippines, pp. 305-316.

Omar, R. A., Bala, M.S., El-Kazzaz, M.K. and Aidy, I.R., 1979. Reaction of rice varieties to the brown spot disease of rice, *Cochliobolus miyabeanus*, at different stages of growth. *Agril. Res. Rev.* 57(8): 103-118.

Ou, S.H., 1985. *Rice Diseases*, 2nd edn., CMI, Kew, England, 370 pp.

Padhi, B. and Chakrabarti, N.K., 1972. Occurrence of new pathogenic races of *Pyricularia oryzae* in nature. *Proc. Indian Sci. Congr. IV*. 59: 32 (Abstr.).

Padmanabhan, S.Y., Chakrabarti, N.K., Mathur, S.C. and Veeraraghavan, J., 1970. Identification of pathogenic races of *Pyricularia oryzae* in India. *Phytopathology* 60: 1574-1577.

Pan, X.B., Rush, M.C., Sha, X.Y., Xie, Q.J., Linscombe, S.D., Stetina, S.R. and Oard, J.H., 1999a. Major gene, non-allelic sheath blight resistance from the rice cultivars Jasmine 85 and Teqing. *Crop Sci.* 39: 338-346.

Pan, X.B., Zou, J.H., Chen, Z.X., Lu, J.F. and Zhu, L.H., 1999b. Tagging major quantitative traits loci for sheath blight resistance in a rice variety, Jasmine 85. *Chinese Sci. Bull.* 44: 1783-1789.

Panda, S.K., Karmakar, S.K. and Samal, T., 1994. Evaluation of long duration rice cultivars for stem borer and bacterial leaf blight resistance. *Orissa J. agril. Res.* 7 (Suppl.): 80-81.

Panwar, D.V.S., Ram Singh and Chand, H., 1988. Stable resistance to bacterial leaf blight and stem rot in rice. *Oryza* 25: 94-96.

Patil, B.G. and Moghe, P.G., 1990. Reaction of promising rice cultivars to major diseases in eastern Vidarbha zone, India. *Intern. Rice Res Newsl.* 15(5): 11.

Paul, C.R., Mon Repos and Nanda, J.S., 1993. Reaction of some rice germplasm to leaf blast in Guyana. *Intern. Rice Res. Newsl.* 18(4): 11.

Phookan, A.K. and Hazarika, D.K., 1994. Variability in *Sarocladium oryzae*, incitant of sheath rot of rice (*Oryza sativa*). *Indian J. agric. Sci.* 64(1): 64-66.

Prabhu, A.S., Soave, J., Zimmermann, F.J.P., Filippi, M.C., Souza, N.R.G., Curvo, R.C.V., Lopes, A.M., Sobral, C.A.M., Ferreira, R.P., Kobayashi, T. and Galvao, E.U.P., 1996. Genetic variability for disease resistance in Brazilian upland rice native germplasm. *Pesquisa Agropecuaria Brasileira* 31(6): 413-424.

Prasad, K., Shukla, V.D. and Sinha, P.K., 1998. Varietal screening and inheritance of resistance to brown spot disease in rice (*Oryza sativa*). *Indian J. agril. Res.* 68(5): 258-260.

Prasad, Y. and Singh, R.S., 1985. Screening for brown spot resistance. *Intern. Rice Res. Newsl.* 10(3): 7.

Prasanna Kumari Pillai 1990. Laboratory screening of rice cultivars for sheath blight resistance. *Extended Summary, Proc. Intern. Symp. Rice Res.: New Frontiers*, Nov. 15-18, 1990, DRR, Hyderabad: 237-239.

Premlatha Dath, A. (1990). *Sheath Blight of Rice and its Management.* Associated Publishing Co., Shidipura, Karol Bagh, New Delhi: 129 pp.

Puhalla, J.E. and Spieth, P.T., 1985. A comparison of heterokaryosis and vegetative compatibility among varieties of *Gibberella fujikuroi* (*Fusarium moniliforme*). *Exptl. Mycol.* 7: 39-47.

Purkayastha, R.P. and Chattopadhyay, M., 1975. Studies on the resistance of rice cultivars to helminthosporiose. *Indian Phytopath.* 28: 33-37.

Purkayastha, R.P., Ghosal, A. and Biswas, S., 1983a. Momilactone synthesis associated with differential resistance of tall and semi-dwarf rice cultivars to sheath rot disease. *Indian Phytopath.* 36: 322-327.

Purkayastha, R.P., Ghosal A. and Biswas, S., 1983b. Production of momilactone associated with resistance of rice cultivars to sheath rot disease. *Curr. Sci.* 52(3): 131-132.

Purushothaman, D., 1974. Phenyl alanine ammonia lyase and aromatic amino acids in rice varieties infected with *Xanthomonas oryzae*. *Phytopathol. Z.* 80: 171-175.

Qian Hann Liang, Chen Qi Zhi, Zhang, J., Cheng Yong Jian, Yang Ding Bing, Xiao Yan Nong, Peng Guo Xiong 1997. On variability for pathotypes of *Xanthomonas campestris* pv. *oryzae* in Jiang Han Plain. *Acta Phytopath. Sin.* 27(4): 333-337.

Raina, G.L., Saini, R.K. and Dhindsa, H.S., 1992. Multiple disease resistance in rice in the Punjab. *Pl. Dis. Res.* 7: 272-274.

Raina, G.L., Singh, G. and Sidhu, G.S., 1980. Sources of resistance to major rice diseases in Punjab, *IRRI Newsl.* 5: 6-7.

Raina, G.L., Sidhu, J.S., Bharaj, T.S., Gautam, R.K. and Omvir 1999. Multiple disease resistance in rice. *Crop Improv.* 26: 63-65.

Raj, R.B., Tayaba Wahab, Rao, G.V., Rao, A.S. and Reddy, T.C.V., 1987. Evaluation of rice cultures against bacterial leaf blight and sheath blight disease. *Indian Phytopath.* 40: 397-399.

Rajput, R.L. and Bartaria, A.M., 1995. Reaction of rice cultivars to brown spot. *Agric. Sci. Digest* 15(4): 205-206.

Ram Singh and Chand, H., 1986. Reaction of rice genotypes to stem rot fungi and bacterial blight pathogen. *Intern. Rice Res. Newsl.* 11(4): 15.

Ram Singh and Dodan, D.S., 1995a. Reaction of rice genotypes to bacterial leaf blight, stem rot and sheath blight in Haryana. *Indian J. Mycol. Pl. Pathol.* 25: 224-227.

Ram Singh and Dodan, D.S., 1995b. Sheath rot of rice. *Intern. J. Trop. Pl. Dis.* 13: 139-152.

Ram Singh and Dodan, D.S., 1996. Screening of basmati rice genotypes against blast. *Intern. Rice Res. Newsl.* 21(2-3): 45.

Ram Singh and Dodan, D.S., 1997. Comparison of two scoring systems for evaluating stem rot resistance in rice and a new proposed rating scale. *Intern. Rice Res. Newsl.* 22(1): 48-50.

Ram Singh, Dodan, D.S. and Sunder, S., 2000. Stable resistant sources of bacterial leaf blight and stem rot of paddy. *HAU J. Res.* 30: 35-36.

Ram Singh, Kumar, A. and Jalali, B.L., 2002. Variability, predisposing factors and management of stem rot of rice caused by *Sclerotium oryzae* Catt.: An overview. *Annu. Rev. Pl. Pathol.* 1: 275-289.

Ram Singh and Sunder S., 1997. Foot rot and bakanae of rice: Retrospects and prospects. *Intern. J. Trop. Pl. Dis.* 15: 153-176.

Ram Singh, Sunder, S. and Dodan, D.S., 2004. Evaluation of scented rice genotypes for resistance to blast and its management by fungicides. *J. Mycol. Pl. Pathol.* 34: 280-281.

Ram Singh, Sunder, S., Nandal, D.P. and Ram, M., 1999. Management of paddy blast through cultural practices, host resistance and fungicides. *HAU J. Res.* 29: 109-114.

Ram, T. and Ansari, M.M., 1989. Multiple resistance in rice cultivar to bacterial leaf blight and fungal diseases. *J. Andaman Sci. Assoc.* 5(2): 158-159.

Rangaswamy, G. and Mahadevan, A., 2004. *Diseases of Crop Plants in India*, 4th edn., Prentice Hall of India Pvt. Ltd., New Delhi-110 001, pp. 171-175.

Rao, K.C. and Rao, K.N., 1979. Varietal reaction of rice to false smut due to *Ustilaginoidea virens* (Cke.) Tak. *The Andhra agric. J.* 26 (1& 2): 74.

Rao, N.S.R.K. and Nayudu, M.V., 1979. Enzymological and permeability changes involved in bacterial leaf blight of rice. *Phytopath. Z.* 96: 77-82.

Reddy, A.P.K., Miah, S.A. and Mew, T.W., 1986. Disease resistance in rainfed lowland rice. In: *Progress in Rainfed Lowland Rice*, IRRI, Philippines, pp. 253-262.

Reddy, C.S., 1993. Virulence in *Rhizoctonia solani* and sources of sheath blight resistance in rice. *Indian Phytopath.* 45(Suppl.) : 334 (Abstr.).

Reddy, M.M., Madhusudan, T., Kulkarni, N. and Kashikar, M., 1997. Sources of resistance to sheath blight. *Intern. Rice Res. Notes* 22(1): 25.

Reddy, P.R. and Mohanty, S.K., 1977. Varietal resistance to the kresek phase of bacterial blight. *Intern. Rice Res. Newsl.* 4(2): 5.

Rehman, F.U., Gnanamanickam, S.S., Adhikari, T.B. and Alvarez, A.M., 1993. New races of *Xanthomonas oryzae* pv. *oryzae* among strains representing major rice growing areas of India and Nepal. *Intern. Rice Res. Notes* 18(3): 33-34.

Rekha, Ram, L., Singh, R. and Ram Singh 2002. Screening of rice genotypes for leaf folder, *Cnaphalocrocis medinalis* (Guenee) and bacterial leaf blight, *Xanthomonas oryzae* pv. *oryzae* (Ishiyama) Dye. *Indian J. Pl. Genet. Resources* 15(2): 100-104.

Ritu Bala and Goel, R.K., 2007. Detection of sources of multiple disease resistance in different *Oryza* spp. *J. Res. PAU* 44(2): 110-112.

Saifulla, M., 1993. Screening of rice culture against blast, brown spot and leaf scald diseases. *Agric. Sci. Digest* 13(3-4): 171-172.

Saini, R.G. and Goel, R.K., 1998. Disease resistance in rice (*Oryza sativa* L.)–An appraisal of the Indian work. *Crop Improv.* 25(2): 141-154.

Saroja, R., Suriachandraselvan, M., Venkatakrishanan, J., Nilakantapillai, K. and Ranganathan, T.B., 1987. Field evaluation of rice culture resistant to leaf folder and sheath rot. *Intern. Rice Res. Newsl.* 12(2): 18.

Sasaki, R., 1922. Existance of strains in rice blast fungus. *J. Pl. Prot., Tokyo* 9: 631-644.

Satyanarayana, K. and Reddy, A.P.K., 1990. Identification of slow blasting genotypes in rice. *Extended Summary, Proc. Intern. Symp. Rice Res.: New Frontiers*, DRR, Hyderabad, Nov. 15-18, 216 pp.

Sha, X.Y. and Zhu, L.H., 1989. Resistance to sheath blight in China. *Intern. Rice Res. Newsl.* 14(2): 14.

Sha, X.Y. and Zhu L.H., 1990. Resistance of some rice varieties to sheath blight. *Intern. Rice Res. Newsl.* 15(6): 17.

Shanti, M.L., George, M.L.C., Vera Cruz, C.M., Bernardo, M.A., Nelson, R.J., Leung, H., Reddy, J.N. and Sridhar, R., 2001. Identification of resistance genes effective against rice bacterial blight pathogen in eastern India. *Pl. Dis.* 85: 506-512.

Shanti, M.L. and Shenoy, V.V., 2005. Evaluation of resistance genes and their pyramids against rice bacterial leaf blight pathogen *Xanthomonas oryzae* pv. *oryzae*. *Oryza* 42: 169-173.

Sharma, B.R. and Kapoor, A.S., 2002. Management of rice blast (*Pyricularia grisea*) through host resistance and chemicals in hills. *Pl. Dis. Res.* 17: 283-290.

Sharma, T.R., Chauhan, R.S., Singh, B.M., Paul, R., Sagar, V. and Rathour, R., 2002. RAPD and pathotypes analyses of *Magnaporthe grisea* populations from the North-western Himalayan region of India. *J. Phtopathol.* 150: 649-656.

Sharma, T.R., Madhav, M.S., Singh, B.K., Shanker, P., Jana, T.K., Dalal, V., Pandit, A., Singh, A., Gaikwad, K., Upreti, H.C. and Singh, N.K., 2005a. High-resolution mapping, cloning and molecular characterization of the gene of rice, which confers resistance to rice blast. *Mol. Genet. Genomics* 274: 569-578.

Sharma, T.R., Shanker, P., Singh, B.K., Jana, T.K., Madhav, M.S., Gaikwad, K., Singh, N.K., Plaha, P. and Rathour, R., 2005b. Molecular mapping of rice blast resistance gene Pik^h in the rice variety Tetep. *J. Plant Biochem. Biotech.* 14: 127-133.

Shim, H.S., Ryu, J.D. and Han, S.S., 2001. Incidence of rice false smut caused by *Ustilaginoidea virens* in different geographic regions and cultivars and its chemical control. *Res. Pl. Dis.* 7(2): 102-106.

Shobha Rani, N., Prasad, G.S.V., Rao, L.V.S., Sudharshan, I., Pandey, M.K., Babu, V.R., Ram, T., Padmavati, G., Muralidharan, K., Pasalu, I.C. and Virakatamath, B.C., 2008. High yielding rice varieties of India. *Tech. Bull. 33*, DRR, Rajendranagar, Hyderabad, India, 188 pp.

Shukla, S.N., Sunder, S., Ram Singh, Sharma, S.K. and Sinha, A.P., 2003. Management of bacterial blight of rice through conventional and non-conventional methods. *Annu. Rev. Pl. Pathol.* 2: 351-378.

Shukla, V.D., Chauhan, J.S., Variar, M., Maiti, D., Chauhan, V.S. and Tomar, J.B., 1995. Reaction of traditional rainfed rice accessions to brown spot, blast and sheath rot disease. *Indian Phytopath.* 48: 433-435.

Siddiq, E.A., 1990. Modern approaches to breeding for disease resistance. In: *Basic Research for Crop Disease Management* (ed. P.Vidhyasekaran), Daya Publishing House, Delhi–110 006, pp. 277-297.

Silue, D. and Notteghem, J.L., 1991. Resistance of 99 *Oryza glaberrima* Steud. varieties to blast. *Intern. Rice Res. Newsl.* 16(2): 13.

Singh, A.J., 1987. Paicos 1, a promising new rice for Manipur. *Intern. Rice Res Newsl.* 12(2): 6.

Singh, C.B. and Rao, Y.P., 1971. Association between resistance to *Xanthomonas oryzae* and morphological and quality characters in induced mutants of *indica* and *japonica* varieties. *Indian J. Genet. Pl. Breed.* 31: 369-373.

Singh, M. and Gangopadhyay, S., 1981. Artificial culture of *Ustilaginoidea virens* and screening of rice varieties. *Trans. Br. Mycol. Soc.* 77: 660-661.

Singh, R.A., 1984. False smut of rice. *Exptl. Stn. Res.Bull. III.* GBPUAT, Pantnagar: 108 pp.

Singh, R.A. and Sharma, V.V., 1975. Evaluation of rice germplasm and varieties for resistance to brown spot. *Il-RISO* 24: 383-386.

Singh, R. N., 1988. Status of sheath rot in eastern Uttar Pradesh, India. *Intern. Rice Res. Newsl.* 13(6): 33.

Singh, R. N., 1989. Field resistance of Faizabad rice lines against bacterial leaf blight, brown spot, sheath rot and false smut. *Narendra Deva J. agric. Res.* 4(1): 111-113.

Singh, R. N. and Khan, A.T., 1988. Multiple resistance to bacterial and four fungal diseases. *Intern. Rice Res Newsl.* 13(2): 9.

Singh, R.S., Singh, S.N. and Yadav, B.P., 2000. Management of brown spot disease of rice caused by *Drechslera oryzae. Madras agric. J.* 87 (7-9): 372-375.

Singha, K.D. and Borah, P., 2000. Screening of local upland rice cultivars of Assam against sheath blight. *Annals Biol.* 16(2): 161-162.

Singha, K.D., Borkakoti, R.P. and Pathak, A.K., 2000. Evaluation of hybrid rice for disease resistance against BLB, sheath blight and grain discolouration. *Annals Biol.* 16(2): 163-166.

Singha, K.D. and Pathak, P.K., 1998. Evaluation of aromatic and semi-glutinous rice germplasm of Assam for resistance against major diseases. *Oryza* 35: 87-88.

Sivaraj, R., Gnanamanickam, S.S. and Levy, M., 1995. Genetic and pathotype diversity in *Pyricularia grisea* populations in India. In: Proc. 4th *Annual Meeting National Rice Biotechnology Network, July 4-6, 1995*, National Chemical Laboratory, Pune and DRR, Hyderabad, India pp. 50-52 (Abstr.).

Sivaraj, R., Krishnamurthi, K., Viji, G., Dayakar, B.V., Gnanamanickam, S.S. and Levy, M., 1996. Approaches for the management of rice blast. Biological control with bacterial antagonists and pyramiding of blast resistance genes. In: Proc. 5th *Annual Meeting National Rice Biotechnology Network, Nov. 13-16, 1996*, IARI, New Delhi, pp. 101-104.

Sridhar, R., Mandakini Nayak and Kumar, S., 1990. Physiology of disease resistance in rice blast fungus. In: *Extended summary, Proc. Intern. Symp. Rice Res.: New Frontiers*, DRR, Hyderabad, India, 221 pp.

Sridhar, R., Reddy, J.N. and Agarwal, P.K., 1999. Usefulness of combinations of bacterial blight resistance genes at Cuttack, Orissa, India. *Intern. Rice Res. Notes* 24(2): 24-25.

Sridhar, R., Shanti, M.L., Singh, U.D., Variar, M., Maiti, D., Reddy, J.N., Sinha, P.K., George, M.L.C., Bernardo, M.A., Baraoidan, M.R. and Nelson, R.J., 1996. Molecular characterization of bacterial blight and blast pathogen population of India and breeding for cultivars with bacterial blight and blast resistance genes. In: *Proc. 5th Annual Meeting National Rice Biotechnology Network, Nov. 13-16, 1996*, IARI, New Delhi, pp. 95-97.

Sridhar, R. and Singh, U.D., 2001. Genetic and pathogenic diversity of the rice blast pathogen. In: *Major Fungal Diseases of Rice-Recent Advances* (eds. S. Sreenivasaprasad and R. Johnson), Prof. K. Manibhushan Rao Festschrift, Kluwer Academic Publisher, Netherland.

Srinivasachary, Shailaja Hittalmani, Girish Kumar, K., Shashidhar, H.E. and Vaishali, M.G., 2002. Identification of quantitative trait loci associated with sheath rot resistance (*Sarocladium oryzae*) and panicle exertion in rice (*Oryza sativa* L.). *Curr. Sci.* 82(2): 133-135.

Srinivas Prasad, M., Ramesha, M.S., Voleti, S.R., Pasalu, I.C. and Vijaya Kumar, C.H.M., 2008. Resistance to blast disease and tolerance to drought in rice genotypes. *J. Mycol. Pl. Pathol.* 38: 108-109.

Srivastava, M.P. and Maheshwari, S.K., 1971. Varietal resistance of rice to brown leaf spot of rice. *Il-RISO* 20: 271-273.

Sudhakar, R., Rao, K.C. and Reddy, C.S., 2000. Evaluation of medium duration rice varieties against sheath blight disease. *Indian J. Pl. Prot.* 28: 215-217.

Sun Guochang and Sun Shuyuan 1992. Characterisitics of some rice varieties resistant to *Pyricularia oryzae* Cavara. *Intern. Rice Res. Newsl.* 17(5): 9-10.

Sun, X., Yang, Z., Wang, S. and Zhang, Q., 2003. Identification of a 47-kb DNA fragment containing Xa 4, a locus for bacterial blight resistance in rice. *Theor. Appl. Genet.* 106: 683-687.

Sundaram, R.M., Vishnupriya, M.R., Biradar, S.K., Laha, G.S., Reddy, G.A., Shobha Rani, N., Sharma, N.P. and Sonti, R.V., 2008. Marker assisted introgession of bacterial blight resistance in Samba Mahsuri, an elight *indica* rice variety. *Euphytica* 160: 411-422.

Sunder, S., Kataria, H.R., Satyavir and Sheoran, O.P., 2003. Characterization of *Rhizoctonia solani* associated with root/collar rots and blights. *Indian Phytopath.* 56: 27-33.

Sunder, S., Ram Singh, Dodan, D.S. and Mehla, D.S., 2005. Effect of different nitrogen levels on brown spot (*Drechslera oryzae*) of rice and its management through host resistance and fungicides. *Pl. Dis. Res.* 20: 111-114.

Sunder, S., Ram Singh, Dodan, D.S. and Ram, L., 2007. Multiple resistance to major diseases and insect-pest in rice. *Pl. Dis. Res.* 22: 81-83.

Sunder, S. and Satyavir 1998. Vegetative compatibility, biosynthesis of GA_3 and virulence of *Fusarium moniliforme* isolates from bakanae disease of rice. *Plant Pathol.* 47: 767-772.

Sunder, S., Satyavir and Singh, A., 1998. Screening of rice genotypes for resistance to bakanae disease. *Indian Phytopath.* 51: 299-300.

Tandon, J.P. and Bhatt, J.C., 1982. Potential sources of blast resistance for hilly regions in India. *Intern. Rice Res. Newsl.* 7(3): 7.

Tasleem-uz-Zaman-Khan, Gill, M.A., Bajwa, K.M. and Khan, M.G., 1999a. Screening of rice varieties/ lines for resistance against bakanae disease. *Pakistan J. Phytopathol.* 11: 166-168.

Tasleem-uz-Zaman-Khan, Gill, M.A. and Khan, M.G., 1999b. Inheritance studies for bakanae disease of rice. *Pakistan J. Phytopathol.* 11: 181-182.

Thakur, K.S.S., 1974. Role of gibberellic acid, fusaric acid and pectic enzyme in the foot-rot disease of rice. *Il-RISO* 23: 191-207.

Thakur, K.S.S. and Ghosal, M., 1976. Study of phenolic contents of resistant and susceptible varieties of rice in relation to foot rot disease. *Il-RISO* 25: 87-91.

Thakur, K.S.S. and Subramanian, S., 1966. Nitrate reductase of rice (*Oryza sativa* L.) under foot-rot disease. *Planta* 71: 125-129.

Tiwari, R.K.S. and Kumar, A., 2000. Evaluation of new type rice breeding lines against biotic stresses. *J. Mycol. Pl. Pathol.* 30: 57-59.

Toriyama, K., 1975. Recent progress of studies on horizontal resistance in rice breeding for blast resistance in Japan. In: *Horizontal Resistance to Blast Disease of Rice,* CIAT, Cai, Colombia, pp. 65-110.

Tsai, W.H., 1973. *Variation in cultural characteristics and pathogenicity of single basidiospores of Thanatephorus cucumaris (Frank) Donk, causing sheath blight of rice.* M.S. Thesis Univ. Philippines, Los Banos, Laguna, Philippines, 79 pp.

Tu, J.C., 1967. Strain of *Pellicularia sasakii* isolated from rice in Taiwan. *Pl. Dis. Rep.* 51: 682-684.

Upadhyay, R.K. and Diwakar, M.C., 1984. Sheath rot in Chhatisgarh, Madhya Pradesh, India, *Intern. Rice Res. Newsl.* 9(5): 6.

Urmila Dhua and Dhua S.R., 1999. Evaluation of rice germplasm against false smut disease of rice. *Oryza* 36: 190-191.

Van der Plank, J.E., 1968. *Disease Resistance in Plants.* Academic Press, New York.

Velazhahan, R., 1987. *Studies on Plant Growth and Biochemical Changes in Rice (Oryza sativa L.) as Influenced by N and K Fertilization and Acrocylindrium oryzae Sawada Inoculation.* M.Sc. (Ag.) Thesis, Annamalai Univ., India.

Velazhahan, R. and Ramabadran, R., 1992. Influence of potassium fertilization on sheath rot incidence and phenolic contents of rice. *Madras agric. J.* 79(5): 294-298.

Venkatarao, G. and Apparao, G., 1976. Pathogenic races of *Pyricularia oryzae* in Andhra Pradesh. *Oryza* 11: 53-60.

Verma, R.K. and Singh, R.A., 1988. Variation in *Claviceps oryzae-sativae,* the incitant of false smut of rice. *Indian Phytopath.* 41: 48-50.

Vikal, Y., Goel, R.K., Singh, R. and Singh, K., 2004. Molecular characterization of some bacterial blight resistant germplasm lines using STS markers linked to Xa 4, xa 5, xa 13 and Xa 21. *Pl. Dis. Res.,* 19: 25-30.

Watanabe, B., 1952. The development of stem rot disease of rice plant and the physiological properties of the basal part of stem. *Bull. Kyushu agric. Exptl. Stn.* 1: 271-276.

Xie, Q.J., Rush, M.C. and Cao, J., 1990. Somaclonal variation for disease resistance in rice (*Oryza sativa*). In: *Pest Management in Rice.* (eds. B.T. Grayson, M.B. Green and L.G. Copping), Elsevier Appl. Sci., New York: 491-509.

Xu Jian Long, Jin Yi Zi, Weng Jin Ping and Zhao Xin Li 1995. Technique for transfer of the recessive gene xa 5 for resistance to bacterial blight in rice. *Acta Agriculturae Zhejiangensis* 7(2): 76-80.

Yeh, W.J. and Bonman, J.M., 1986. Assessment of partial resistance to *Pyricularia oryzae* in six rice cultivars. *Plant Pathol.* 35: 319-323.

Yi, R.H., Liang, C.Y., Zhu, X.R. and Zhou, E.X., 2002a. Genetic diversity of rice sheath blight isolates (*Rhizoctonia solani* AG-1 IA) from different rice cultivars. *Chinese Rice Res. Newsl.* 10(2): 7-8.

Yi, R.H., Liang, C.Y., Zhu, X.R. and Zhou, E.X., 2002b. Genetic diversity and virulence variation of rice sheath blight strains (*Rhizoctonia solani* AG-1 IA) from Guangdong province. *J. Trop. Subtrop. Bot.* 10(2): 161-170.

Yoshimura, S., Yoshimura, A., Iwata, N., Mc Couch, S.R., Abenes, L.L., Baraoidan, M.R., Mew, T.W. and Nelson, R.J., 1995. Tagging and combining bacterial blight resistance genes in rice using RAPD and RFLP markers. *Mol. Breed.* 1: 375-387.

Zhou, Y.L., Fan, J.J., Zeng, C.Z., Liu, X.Z., Wang, S. and Zhao, K.J., 2004. Preliminary analysis of genetic diversity and population structure of *Ustilaginoidea virens. Acta Phytopath. Sin.* 34(5): 442-448.

Zhu, M., Wang, L. and Pan, Q., 2004. Identification and characterization of new blast resistance gene located on rice chromosome 1 through linkage and differential analyses. *Phytopathology* 94: 515-519.

Zou, J.H., Pan, X.B., Chen, Z.X., Xu, J.Y., Lu, J.F., Zhai, W.X. and Zhu, L.H., 2000. Mapping quantitative trait loci controlling sheath blight resistance in two rice cultivars (*Oryza sativa* L.). *Theor. Appl. Genet.* 101: 569-573.

Chapter 4

Geo-spatial Technologies in Pest Management

A.K. Kanojia
National Centre for Integrated Pest Management (NCIPM)
LBS Bhawan, IARI, Pusa Campus, New Delhi – 110 012

Introduction

The factors which have sound bearing on crop production especially crop protection can be monitored in real time situation and be seen and visualized for specific location and region using Geographic Information System(GIS), Global Positioning System(GPS) and Remote Sensing(RS). The GPS with personal digital assistant (PDA) facility is quite useful and handy as survey device to provide spatial as well as non-spatial associated parameters, which decide pest build up, from the survey field itself and the collected information can then be uploaded to the system having GIS. The ability of GIS to depict the real-time concerned attribute scenario based on the data through GPS makes it an integral part for monitoring and reporting of changes in the entity in the question and devise accordingly to plan and manage at field as well as large scale level for a big region like our country. In this chapter emphasis has been laid upon GIS and GPS and their potential use in the concerned area of pest management.

The collection of data about the spatial distribution of significant properties of the earth's surface has long been an important part of the activities of organized societies. From the earliest civilizations to modern times, spatial data have been collected by navigators, geographers, surveyors and rendered into pictorial form by the cartographers. As scientific study of the earth advanced, new material needed to be mapped. The development in the assessment and understanding of natural resources: geology, geomorphology, soil science, ecology that began in the 19th century and have continued to

Figure 4.1: GPS Receiver

Figure 4.2: Scare Resources must be Used Intelligently

this day, provided new themes to be mapped. Resources management requires to analyse phenomena, processes and their interaction throughout space and time. The resource information system for agricultural management has to be considered as multidimensional *i.e.*, attribute dimension, spatial dimension and temporal dimension (Singh, 1998). Geographic Information System (GIS) offers such capabilities as they integrate multi-sector, multilevel and multi-period databases. New information systems have to be used by the scientific community to have overall idea and understanding of the existing pattern of the phenomena as they are associated with and to provide real time information for executing the necessary strategies by the administrators, the planners, the researchers etc. both effectively and timely. Since our earth is not a small piece of land of similar type of climate, soil, flora and fauna, water body etc. but it is full of diversities in every sphere of life and it is clear from these facts also that it is having total surface area of 509,700,000 sq. km., The area of land is 148,400,000 sq. km., while water bodies occupy 361,400,000 sq. kms. The variations in surface features, ocean depth, temperature, atmosphere etc. vary from one region to other over the earth. Consequently varied agro ecological regions, cropping patterns, plant pests, natural enemies are distributed on our earth. The diversities are also found in colours in our country and it is clear from the facts that it is 12 times of UK, 8 times of Japan, having 16 per cent of world population, it is 7th largest and 2nd most populous country in the world with geographical area of 32,87,263 sq. km, while its north-south extent is almost 3214 km and east-west extent: 2933 km., its coast line is around 7516 km (Kalpana, 2000) with different kinds of soils, vegetation, socio-ethnic groups, farming systems, patterns and modes of indigenous wisdom of local communities. Therefore, it is imperative to understand intricacies and complexities involved in the development of a phenomena over earth and that is why the use of such system like GIS helps decision making effectively as it is a powerful integrating technology which needs to be utilized by the community of natural resource scientists and the concerned planning people.

Spatial Technologies: GIS and GPS

The rapid development of spatial technologies in recent years has made available new tools and capabilities in crop production for management of spatial data. In particular, the evolution of geographic information systems (GIS), the global positioning system (GPS), and remote sensing (RS) technologies has enabled the collection and analysis of field data in ways that were not possible before the advent of the computer. Remote sensing part is not being taken as it may not be covered under one lecture and only GIS and GPS are being presented and discussed.

Geographic Information Systems (GIS)

GIS applications enable the storage, management, and analysis of large quantities of spatially distributed data. These data are associated with their respective geographic features. For example, water quality data would be associated with a sampling site, represented by a point. Data on crop yields might be associated with fields or experimental plots, represented on a map by polygons

There are several definitions to define GIS: "A system which uses a spatial database to provide answers to queries of a geographical nature." (Good Child, 1985). "A powerful set of tools for collecting, storing, retrieving at will, transferring and displaying spatial data from the real world." (Burrough, 1987). "An organized collection of computer hardware, software, geographical data and personnel designed to efficiently capture, store, update, manipulate, analyze and display all forms of geographically referenced information" (Environmental System Research Institute, 1990).

Thus, GIS is a combination of hardware, software, database and personnel and it needs to have geographical area over and for which pattern and trends of the selected phenomena has to be shown and queries are to be made. Once geographic information of any kind is translated into the digital form in a GIS, it becomes easy to copy, edit, analyze, manipulate and transmit it.

A complete GIS must answer:

What exists at a particular location?

Where can a specific feature be found?

What has changed over time?

What spatial pattern exists?

What if..?

Activities involved in building a GIS includes (a) getting spatial data into GIS by digitization, (b) making spatial data usable, (c) getting attribute data into GIS, (d)managing the data-base, (e) performing geographic analysis and (f) presenting the results of the analysis.

The data for a GIS may be from manually surveyed data, remote sensing data and secondary data. A Global Positioning System (GPS), which is a handholding device, may also be used for capturing the spatial data. It is used to provide accurate position of the captured data.

GIS acts as a decision making tool and therefore its application is found extensively in research organizations, government concerns, business–for a wide range of analysis including environmental monitoring, land use planning, locational analysis, water resources management, agronomical practices, tax appraisal, utility and infrastructure planning, demographic analysis, pest distribution, pest modelling, pest forecasting, habitat studies, archaeological analysis etc.

Table 4.1: Basic Functions of GIS

Functions	*Sub-functions*
☆ Data acquisition and preprocessing	☆ Digitizing
	☆ Editing
	☆ Topology Building
	☆ Projection transformation
	☆ Format conversion
	☆ Attribute assignment
☆ Database management and retrieval	☆ Data Archival
	☆ Hierarchical Modeling
	☆ Network modeling
	☆ Relational Modeling
	☆ Attribute query
☆ Spatial measurement and analysis	☆ Measurement operations
	☆ Buffering
	☆ Overlay operations
	☆ Connectivity Operations
☆ Graphic output and visualization	☆ Topographic maps
	☆ Statistical maps
	☆ 3D Bird's eye view

Why GIS?

- Digital World
- Accessibility of data
- Compatibility
- Accuracy
- Spatial Analysis
- Multiple Themes

It is an essential and user friendly tool for providing site-specific information in digital form making use of viable inputs from GPS and remotely sensed data.

Requirements of a GIS

Basic requirements for GIS work are the following:

- *Digitizer*: to convert map or survey-recorded geographic features like rivers, ponds, wells district/village/site boundary.
- *Work station*: to act as a desktop device like a PC.
- *GIS software*: to undertake and accomplish GIS task like popular GIS from ESRI.
- *Plotter/printer*: to take printout in the way you want.
- *IMS*: to publish useful GIs information over web in the same way as seen on desktop GIS and user may query the thing from anywhere in the world.
- *Data*: to act as viable input both spatial and attribute data of the study.
- *Man power*: to act as a controller of GIS and data and presenter of the themes.

There has always been advancement in spatial technologies too and it helps working the same task with less number of devices as these days digitizers are becoming out of fashion since desktop digitizing is quite easy and becoming popular and after scanning any image on-monitor digitizing is done in spite of pasting the image/map over the table and tracing the concerned features manually.

Inputs to GIS

- Manually surveyed data
- Aerial photos
- Remotely sensed satellites data
- Census data
- Published data
- Global Positioning System

The Global Positioning System (GPS)

GPS technology has provided an indispensable tool for management of agricultural and natural resources. GPS is a satellite- and ground-based radio navigation and locational system that enables the user to determine very accurate locations on the surface of the Earth. Although GPS is a complex and sophisticated technology, user interfaces have evolved to become very accessible to the non-

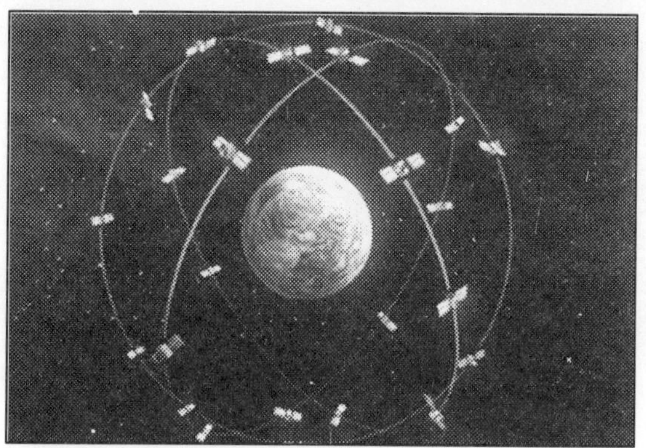

Figure 4.3: Concept of the GPS Satellite Constellation (Photo courtesy U.S. Department of Space)

Figure 4.3: A GPS Satellite (Photo courtesy U.S. Army)

technical user. Simple and inexpensive GPS units are available with accuracies of 5 to 3 meters, and more sophisticated precision agriculture systems can obtain centimeter level accuracies.

The Global Positioning System (GPS) is actually a constellation of 27 Earth-orbiting Satellites (24 in operation and three extras in case one fails), controlled by the US Department of Defense and paid for by the US taxpayer but provides free commercial service to everyone. A GPS receiver's job is to locate four or more of these satellites, figure out the distance to each, and use this information to deduce its own location.

GPS receiver calculates its position on earth based on the information it receives from four located satellites.

Initialization and selecting map datums and coordinate systems

A GPS receiver calculates the distance to GPS satellites by timing a signal's journey from satellite to receiver.

At a particular time (let's say midnight), the satellite begins transmitting a long, digital pattern called a pseudo-random code. The receiver begins running the same digital pattern also exactly at midnight. When the satellite's signal reaches the receiver, its transmission of the pattern will lag a bit behind the receiver's playing of the pattern.

GPS in PDA may be used easily for the following as far as pest management in agriculture is concerned for this chapter:

Geo-referencing of sampling points,

Acreage of crop,

Multi-locational point data,

Pest data,

Natural enemies data,

Weather data

Obtaining reliable data: The attribute data like crop, stage, variety, pest number, timing, sampled in the study field.

Integrated databases: Spatial and statistical data integration is possible through GPS using this GIS.

A standard GPS receiver will not only place you on a map at any particular location, but will also trace your path across a map as you move. If you leave your receiver on, it can stay in constant communication with GPS satellites to see how your location is changing. With this information and its built-in clock, the receiver can give you several pieces of valuable information:

- How far you've traveled (odometer)
- How long you've been traveling
- Your current speed (speedometer)
- Your average speed
- A "bread crumb" trail showing you exactly where you have traveled on the map The estimated time of arrival at your destination if you maintain your current speed.

Costs of GIS 60 per cent of the costs attributed to data, 30 per cent to personnel and 10 per cent to system hardware and software. It is easier to get hardware and software donated but no one can give you your data.

Data in GIS

1. Spatial data–describes locations of entity (where) *i.e.* map of points, lines and polygons
2. Attribute data–characteristics of spatial answers (what)
3. Image data- gives feel of entity use camera or video (how it looks)
4. Identifiers–link mechanism for spatial attribute an image data, very important must be unique, used to control data management process.

Two Major Data Models in GIS: Raster and Vector

Raster

- Area divided into grid of cells, each cell contains a value
- Cells numbered, one set of cells and values forms a layer
- Model tells "what occurs everywhere"

  Vector maps: each localization is stored in terms of map co-ordinates, and information will be represented on the map only for those points;

Vector

- Data represented by points, lines, areas/Polygons
- Model tells "where everything occurs"

  Raster maps: data will be represented by a regular grid, each point of the grid representing a given value.

Remote Sensing

The acquisition of analog (photo) and digital (image) data from platforms ranging from handheld devices to space-borne systems. Remotely sensed data may be interpreted manually, in analog or digital form, or analyzed digitally via image processing and classification techniques.

Continuously collected remote sensing data from space provides major benefits to scientists studying and understanding human impacts on the global environment, managing the earth's natural resources, and planning and conducting many other activities of scientific and social importance

In agriculture, color infrared (CIR) aerial photography or satellite images are used to determine the health of vegetation.

Remote sensing technologies are used to gather information about the surface of the earth from a distant platform, usually a satellite or airborne sensor. Most remotely sensed data used for mapping and spatial analysis is collected as reflected electromagnetic radiation, which is processed into a digital image that can be overlaid with other spatial data.

Figure 4.4: Infrared vs. Regular Color

Reflected radiation in the infrared part of the electromagnetic spectrum, which is invisible to the human eye, is of particular importance for vegetation studies. For example, chlorophyll strongly absorbs blue (0.48 mm) and red (0.68 mm) wavelength radiation and reflects near-infrared radiation (0.75–1.35 mm). Leaf vacuole water absorbs radiation in the infrared region from 1.35–2.5 mm (Samson, 2000). The spectral properties of vegetation in different parts of the spectrum can be interpreted to reveal information about the health and status of crops, rangelands, forests and other types of vegetation.

While reading the colors from imageries following things must be understood:

Chlorophyll, the green pigment in plants, reflects nearly all near infrared light (NIR) and absorbs nearly all visible red light; The more red the vegetation appears the healthier it is; Disease or stressed plants can be observed from aerial or satellite imagery. Deep red shows healthy vegetation, Pink: Stressed or immature vegetation, Dark Green: Black organic loam or higher soil moisture, Light Green: Lighter soil or shallow topsoil, Black: Clear standing water, White: Sand, clay or canola in bloom.

GPS/GIS/RS Applications

Guidance

It acts as a guiding force and tool for

- *Point Guidance*: Pont to point information is provided for attaining precise application of viable inputs in the field.

- *Swath Guidance*: Whole band of land is taken like scanning and the homogenous or heterogeneous patches are identified in the feild.

Control

Its application helps to control variable rate application of critical inputs, variable depth tillage, variable irrigation etc.

Mapping

Its application helps taking basic and applied decision for the field / region under study:

- Soil properties
- Chemical Application
- Chemical Prescriptions
- Tillage Maps
- Irrigation potential and actual usage
- Yield Mapping
- Pest Mapping
- Topographic Maps
- Planting Maps

GIS-GPS-RS technologies are used in combination for site-specific crop management as in pest management practices are required. Pest management techniques are to adopt and devise such tactics and options which may manage pests below economic threshold level (ETL), increase yield with minimum ecological imbalance giving more sustainable way of crop production with farmer participatory approach. Since precision farming techniques are employed to increase yield, reduce production costs, and minimize negative impacts to the environment (Zhang *et al.*, 1999), the importance. Using GIS analytical capabilities, variable parameters that can affect agricultural production can be evaluated. These parameters include yield variability, physical parameters of the field, soil chemical and physical properties, crop variability (*e.g.*, density, height, nutrient stress, water stress, chlorophyll content), anomalous factors (*e.g.*, weed, insect, and disease infestation, wind damage), and variations in management practices (*e.g.*, tillage practices, crop seeding rate, fertilizer and pesticide application, irrigation patterns and frequency) (Zhang, Wang, and Wang, 2002).

The modeling capabilities of GIS can be combined with remotely sensed landscape imagery to evaluate the effects of management practices and to assist resource managers and public decision makers in making informed decisions.

GIS Acts as an Integrating Technology

Figure 4.5, schematic diagramed themes may be generated and viewed integrating many layers of concerned information over one map to arrive at a concrete decision for the region of study.

The ability to interface GIS with relational databases enables integration of large data sets and many variables to support management decisions (*e.g.*, Arvanitis, Ramachandran, Brackett, Rasoul, and Du, 2000). Some progressive farmers are now beginning to use GPS for recording observations. Such as weed growth, unusual plant stress, coloring and growth conditions, which can then be mapped with a GIS programmes. In the years to come, GPS system role in precision agriculture may

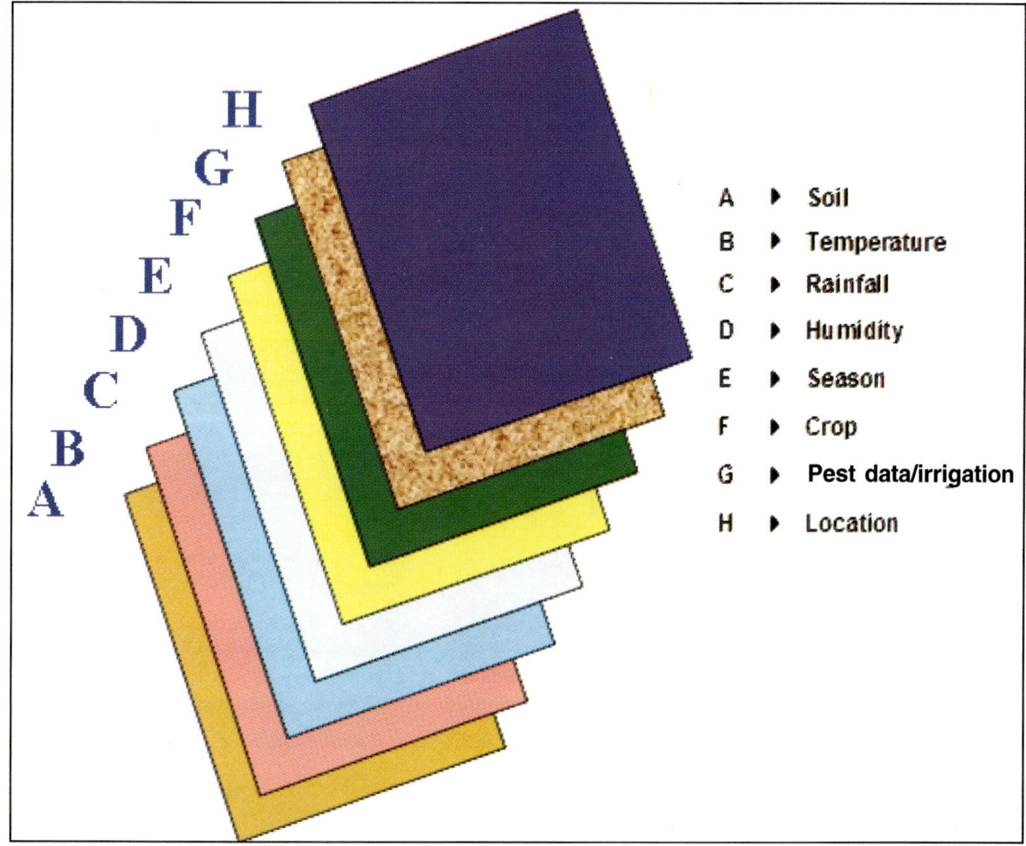

Figure 4.5: This is called a Visual Correlation of the Selected Themes

help the Indian farmers to harvest the fruits of frontier technologies without compromising the quality of land and produce.

Sending the Results to the End/Potential Users through GIS

Figure 4.6 throws light on how result may be sent to potential users over web so that the generated information may be used timely and efficiently by the concerned ones anywhere in the world and advisory may be given accordingly:

Automated Pest Mapping

Automated pest mapping is required to generate real time pest scenario and the generated pest scenario may be viewed from nay where in the world through world wide web using internet connected PCs even in remote villages by the concerned user and the information seeking end users or practitioners. Figure 4.7 shows the flow of information from each component to final view over web. This task has been initiated by the Centre to accomplish the task of getting automated pest mapping for major crops in India and this is being tried on pilot regions on the country using GPS enabled PDAs and GIS.

Pest/Irrigation
Spatial
Internet
Display Unit
(end user)
Internet map server
Leased lines
GIS
Data server

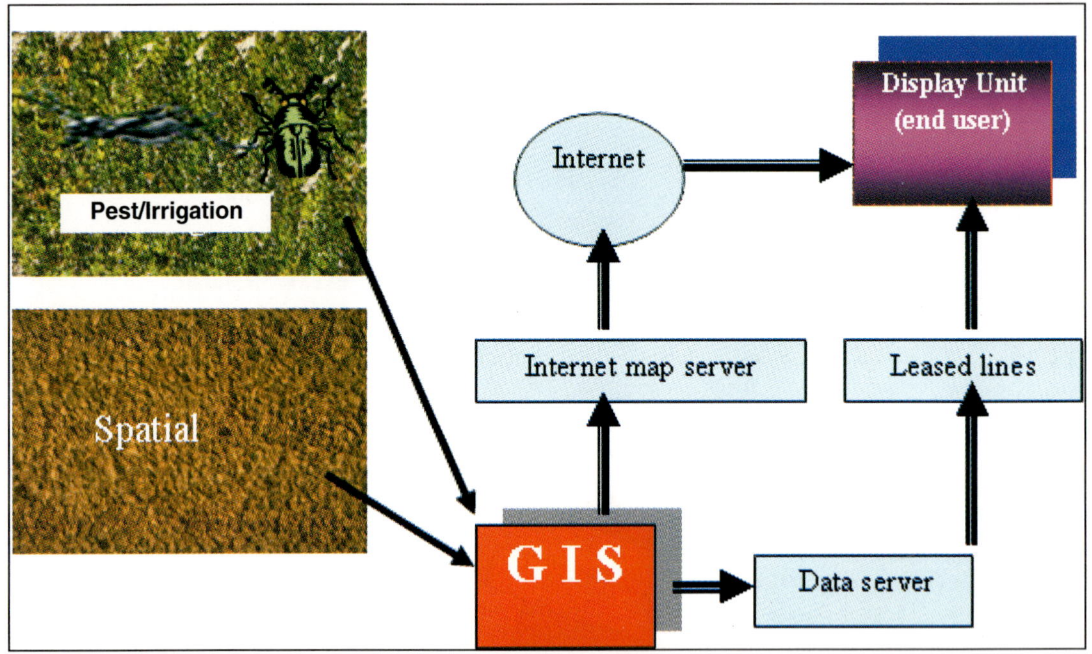

Figure 4.6

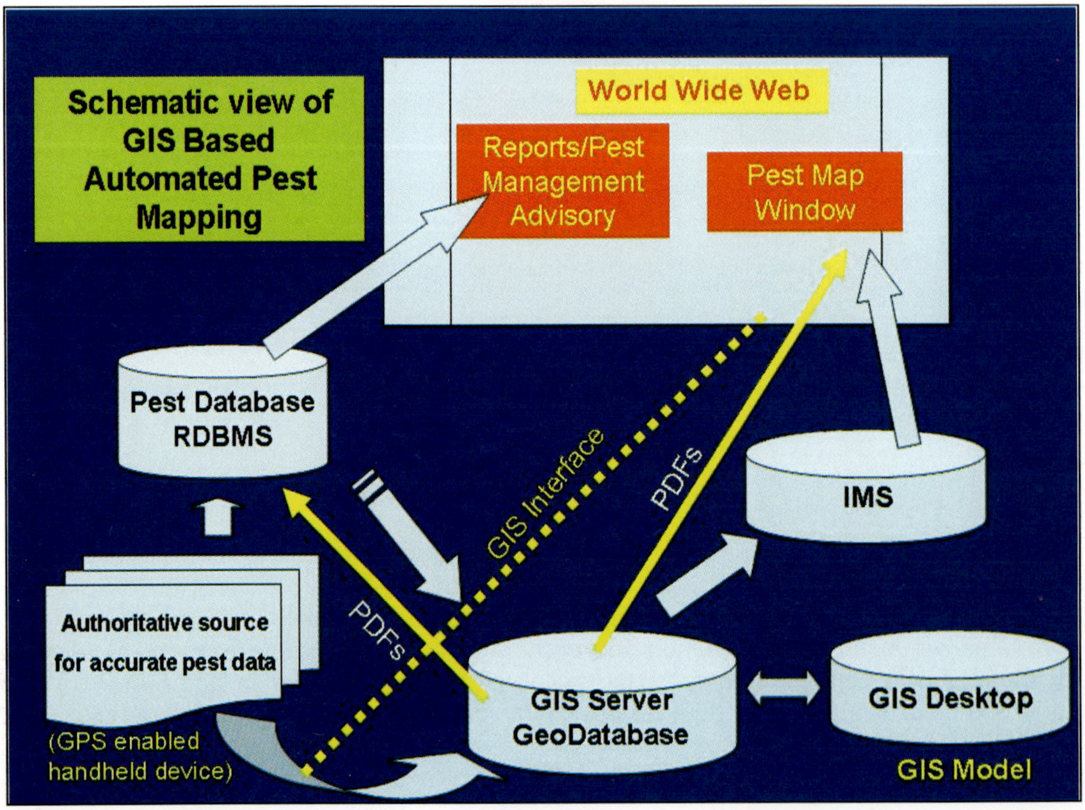

Schematic view of
GIS Based
Automated Pest
Mapping
World Wide Web
Reports/Pest
Management
Advisory
Pest Map
Window
Pest Database
RDBMS
GIS Interface
PDFs
IMS
Authoritative source
for accurate pest data
(GPS enabled
handheld device)
PDFs
GIS Server
GeoDatabase
GIS Desktop
GIS Model

Figure 4.7

Pest Management in Crop Production through Spatial Technologies

A GIS can manage different data types occupying the same geographic space. For example, a biological control agent and its prey may be distributed in different abundances across a variety of plant types in an experimental plot. Although predator, prey, and plants occupy the same geographic region, they can be mapped as distinct and separate features. Once patterns and correlations are elucidated, management practices can be modified to optimize yield and production costs, and minimize environmental impacts caused by excessive applications of fertilizers and pesticides. Site-specific applications of fertilizers, pesticides and other applications can be implemented by dividing a field into smaller management zones that are more homogeneous in properties of interest than the field as a whole (Zhang *et al.*, 2002).

Importance of Pest Management in Crop Production

Pest life cycle, stages, damage nature and parts of plant affected, ETL for specific pest, predator, prey, Soil moisture, temperature, rainfall, humidity, leaf area index, wind speed, sun shine hours, crop stage, variety, utilization of minerals/nutrients etc.

Efficient and Economic assessment: Proper appreciation of the relationship of Pest-Soil-Crops-Climate is essential.

Soil-moisture Measurement

This kind of data may be generated by taking geo-referenced soil samples using GPS at field and may be analyzed at field using experience or at laboratory level by following the established procedures. The generated data may becomes decision making parameter and may be used in GIS for irrigation application on the basis of the location specific requirement.

Soil Moisture for Important Crops

Irrigation practices for important crops must be taken into consideration for designing the expected model for pest status in the study area. Interval between two irrigations depends on the rate of soil-moisture for knowing the soil-moisture depletion and the depletion, the following formula may be applied in your model development exercise for your area:

Moisture deficit can be calculated by the following formula:

$$D = \frac{FC - Mc}{100} \times Bd \times DR$$

Deficit (D), Field Capacity (FC), Bulk density of soil (Bd), Depth of root-zone(DR)

Factors Affecting Pest Distribution

Climatic: Rainfall, humidity, temperature, sunshine, wind speed, snow as well as evapo-transpiration.

Physiographic: Crop, crop area, stage of crop, soil moisture, soil type, storage capacity of channel.

Geological: Lithologic composition and texture.

Assessment of Water Availability

Monthly Runoff= monthly Rainfall–Monthly Evaporation losses

Monthly Evaporation losses = 0.481 x Mean Monthly Temp.

Crop Field Characterization and Pest Assessment using GIS and GPS

The power of a GIS lies in its ability to analyze relationships between features and their associated data (Samson, 1995). This analytical ability results in the generation of new information, as patterns and spatial relationships are revealed.

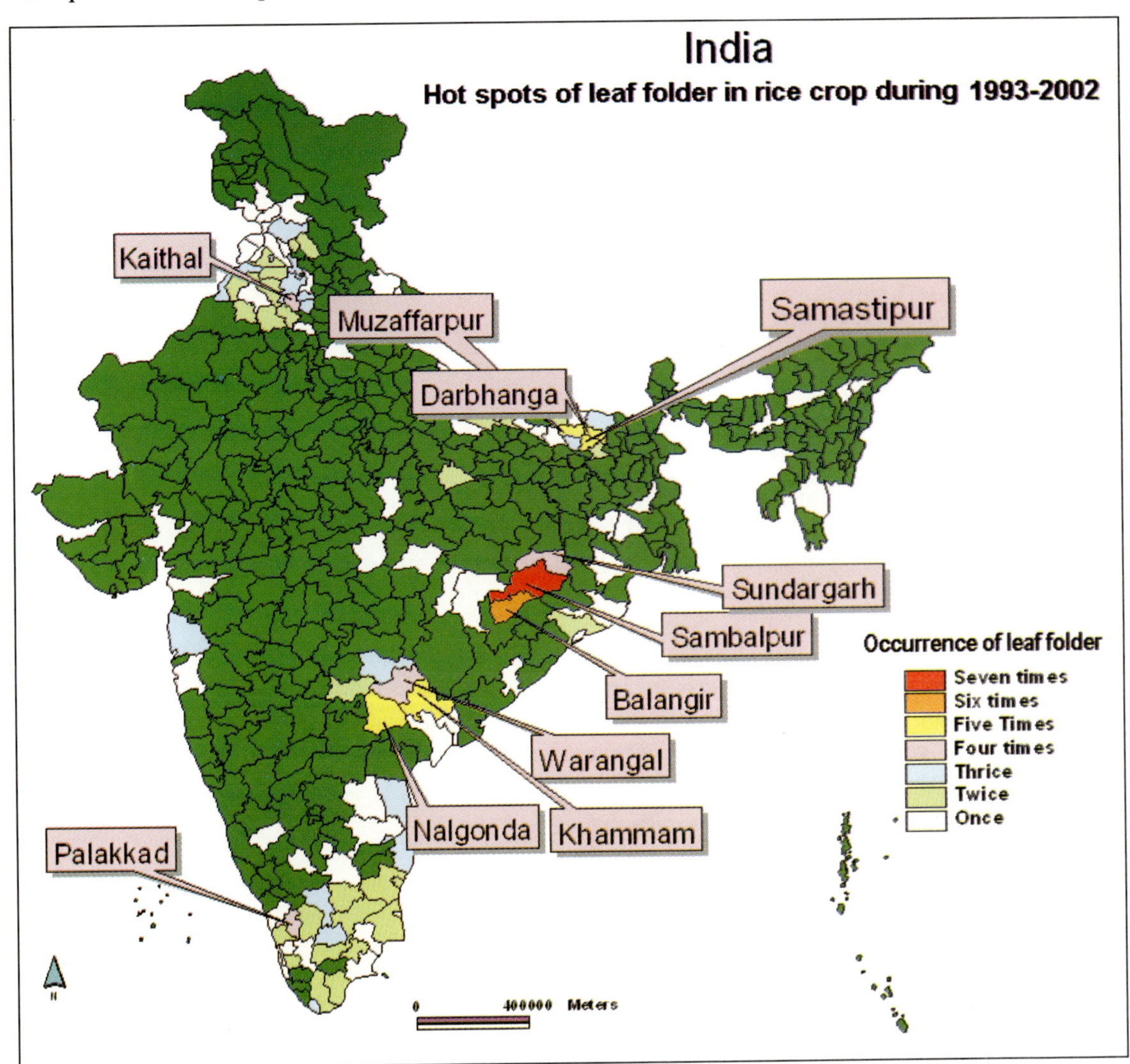

Figure 4.8

A GIS can manage different data types occupying the same geographic space. For example, a biological control agent and its prey may be distributed in different abundances across a variety of plant types in an experimental plot. Although predator, prey, and plants occupy the same geographic region, they can be mapped as distinct and separate features.

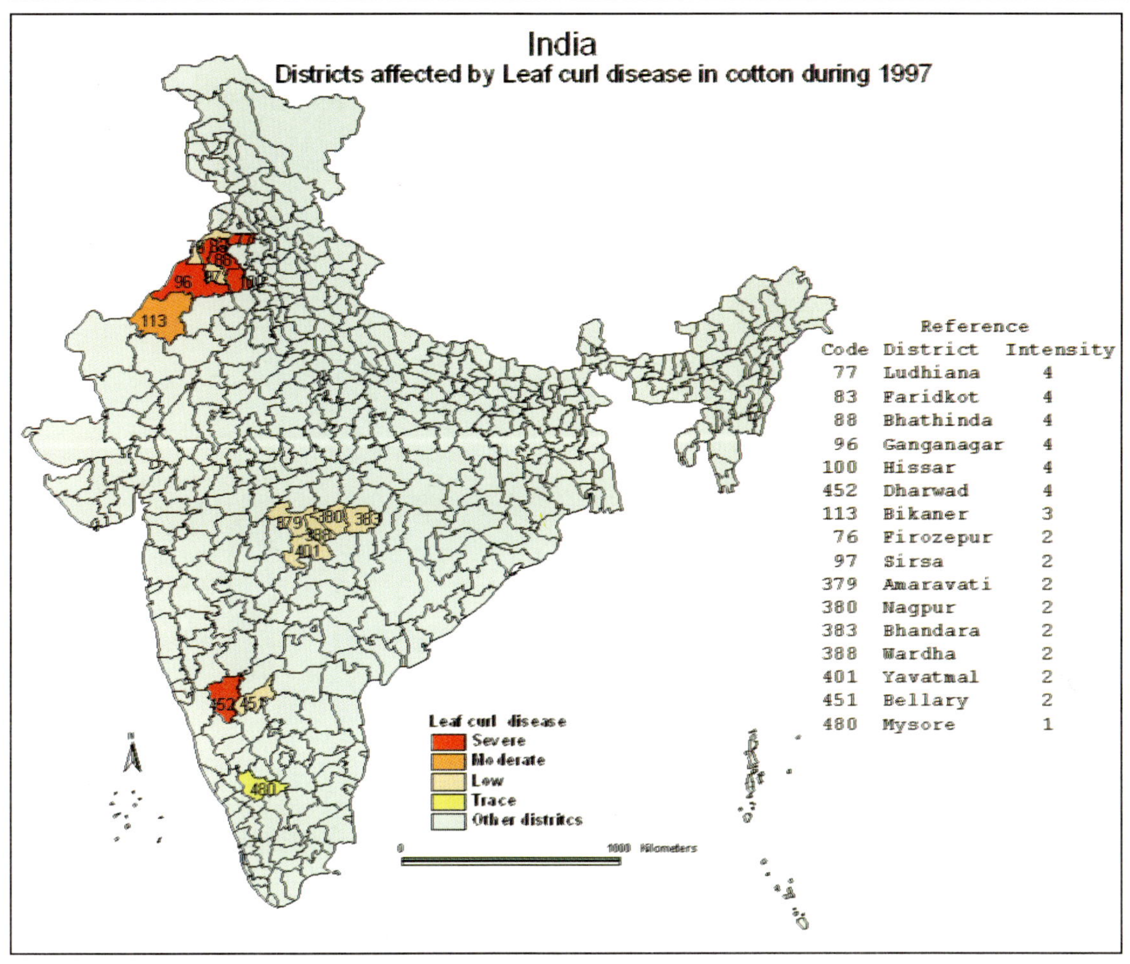

Figure 4.9

Creation of Digital database of Crop and Pest needs digitizing of crop field including its boundary, trap location, sample locations. Using GIS and survey data pest distribution maps have been generated by NCIPM, New Delhi. The prepared maps shows the locations affected by pests in rice crop in India and the area affected by Sugarcane Wooly Aphid in Maharshtra Region.

Wind speed, Contours, Soil-temperature, air-temperature, rainfall, humidity, sunshine hours, crop, crop stage, pest number, disease scale, plant parts affected, date and time, geo-location. Digital Pest status zone model may be generated using interpolation of geo-referenced trap locations as contours using GIS and then pest distribution and highly concentrated area Grid, Pest delineation zone, Delineated sub pest zone can be generated. Finally crop-pest units are obtained and thematic mapping by overlaying all the gathered information. This has been depicted in the given schematic diagram where flow of information and step by step procedure is given.

Forewarning of pests in agriculture is also an important component which has strong bearing on taking timely and appropriate decision to adopt mitigating measures for pests control. This task

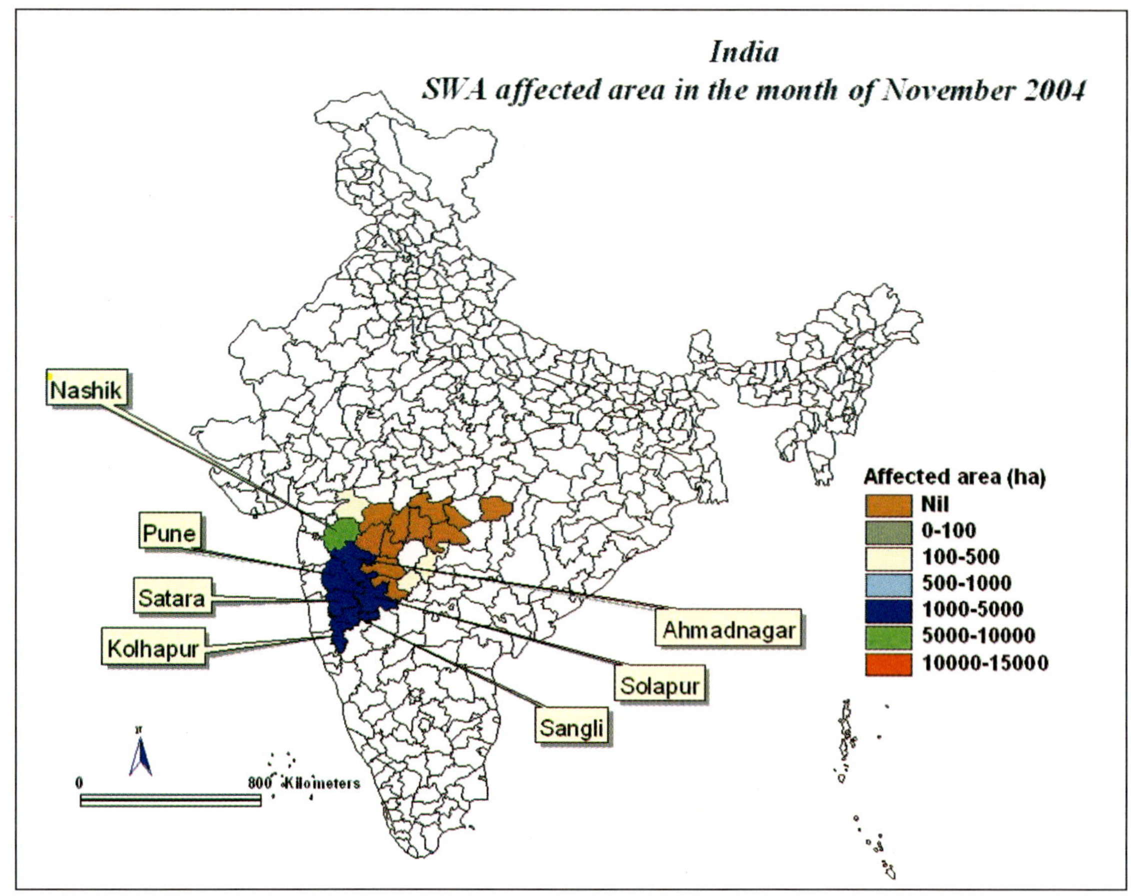

Figure 4.10

becomes more useful when coupled with geo-spatial technologies like GIS where forecasting is done on visual form through thematic maps for the prone area and inputs are obtained from the established mathematical models developed for the region. It is gaining momentum and popularity world over and works related to this aspect have been done at many places. Development of Geographic Information System based forecasting of *Helicoverpa armigera* attack on pigeonpea at Gulbarga district of Karnataka was accomplished by NCIPm, New Delhi in which an operationally feasible, working forecasting system operating at Taluk level using GIS tools initiated. In this task Crop information (variety, date of sowing, phenology, acrage), weather data–weekly collection and pest data–fortnightly collection (larval counts and pod damage), presence/absence of natural enemies were taken. Digitization of Gulbarga district map (vector) incorporating taluk boundaries, locations of met. observatories using GIS tools was done. On the basis of the collected thematic map for *Helicoverpa armigera* attack on pigeonpea at Gulbarga district of Karnataka were developed.

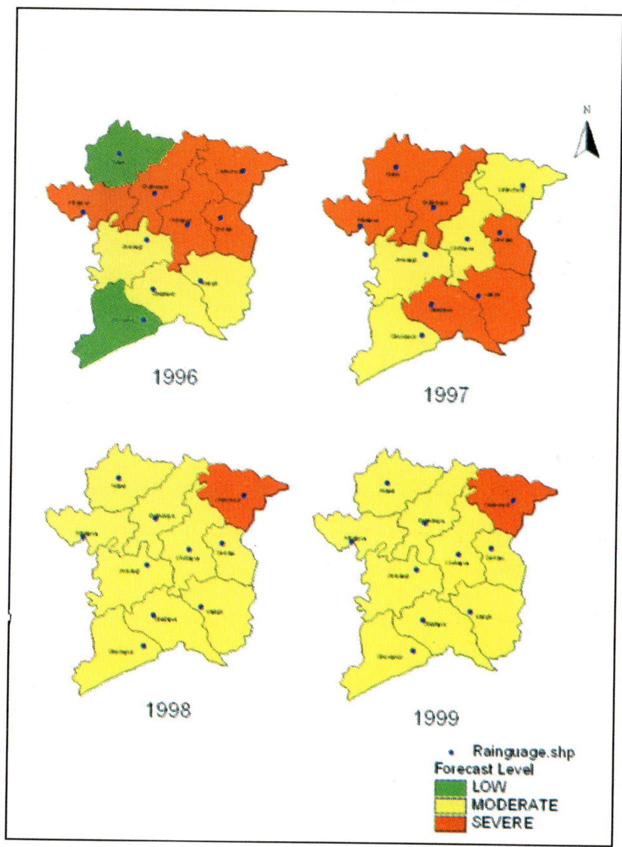

Figure 4.11: Forewarning of *Helicoverpa armigera* at Gulbarga District during 1996–1999

Conclusion

Geo-spatial technologies are need of the hour and are becoming important components for providing crucial information on pest distribution along with weather data and these technologies when coupled with real time information on crop and associated pests can definitely help timely management strategies for pests. The pest database created using spatial technologies can be used by plant protectionists, planners and researchers for further analysis and planning purposes, and to make overlays with other relevant spatial data. There is a high potential to use these technologies.

References

Ajai, 2002. *GPS and its applications*, Training course on Remote Sensing and GIS Applications in Agriculture, 27th May–7th June, 2002, RRSSC- Bangalore, pp: 89-99.

Anonymous, 1997. *The Maryland Forest Pest Management Report*, Cooperative Forest Health Annual Summary.

Arvanitis, L., Ramachandran, B., Brackett, D., Rasoul, H., and Du, X., 2000. Multiresource inventories incorporating GIS, GPS and database management systems: A conceptual model. *Computers and Electronics in Agriculture*, 28: 89-100.

Colvin, T.S. and Kerkman, E.W., 1999, Precision Agriculture Requires precise tuning. *Precision Agriculture-Proceedings of the 4th International Conference*. PP. 1083-1088

Duale, A.H. and Nwanze, K.F., 1999. Incidence and distribution in sorghum of the spotted stem bores *Chilo partellus* and associated natural enemies in farmers' fields in Andhra Pradesh and Maharashtra States. *Int. J. Pest Mgt.* 45 (1) : 3-7.

Kalpana, R., 2000. *Handbook of General Studies*. Spectrum India Pub., New Delhi, pp. 241.

Kanojia, A.K., Devi, T. Prameela, Kumar, Surender, Trivedi, T.P., Singh, Amerika, Sundrasarma, K.S., Natrajan, K. and Pawar, A.D., 2001. Distribution maps of insect pests and diseases of cotton crop in India, Tech. Bulletin, National Centre for Integrated Pest Management, New Delhi, India (in press).

Kenneth, E.F. and Margaret Lynch, 2000. *Geographic Information Systems as an Integrating Technology: Context, Concepts and Definitions*. The Geographer's Craft Project Report, Department of Geography, The University of Texas at Austin,U.S.A.

Kolk, A., Starzyk, J.R., Kinelski, S. and Dzwonkowski, R., 1996. Atlas of Forest Insect Pests. 705 pp.

Leibhold, A. M., Rossi, R. E. and Kemp, W.P., 1996. Geostatistics and geographic information systems in applied insect ecology. *Ann. Rev. Entomology* 38: 303-327.

Li, H., Gartner, D., Mou, P. and Trettin, C., 2000. A landscape model (LEEMATH) to evaluate effects of management impacts on timber and wildlife habitat. Computers and Electronics in Agriculture, 27, 263-292

Mishra, H.C., 1996. *A Handbook on GIS*. GIS India Pub., Hyderabad, A.P., India.

Ortigosa, R. G., De Leo, G. A., and Gatto, M., 2000. VVF: Integrating modeling and GIS in a software tool for habitat suitability assessment. *Environmental Modelling and Software*, 15: 1-12.

Parker, W.E. and Turner, S.T.D., 1996. Application of GIS modelling to pest forecasting and pest distribution studies at different spatial scales. *Aspects of Applied Biology*. 46: 223-230.

Peng, R.K., Christian, K. and Gibb, K., 1997. Distribution of the green ant, *Oecophylla Smaragdina* (F.) (Hymenoptera: Formicidae) in relation to native vegetation and the insect pests in cashewnut plantation in Australia. *Int. J. Pest Mgt.* 43(3): 203-212.

Puri, S.N, Tapati Banerjee, Barnwal, V.K., Surender Kumar and Sunderasarma, K.S., 1998. Distribution of insect pests and diseases of rice crop in India. Tech. Bultn. No. 7, National Centre for Integrated Pest Management, New Delhi, India.

Randall, K.L., 1999. Detection, documentation, and spatial analysis of resistance in Colorado potato beetle using bioassay and GIS technology. *Masters Thesis*, University of Guelph, Guelph, Ontario, Canada.

Reddy, M.A., 2000. *Remote Sensing and Geographical Information System*. Bookionics Pub., Hyderabad, Andhra Pradesh, India, pp. 126.

Samson, S.A., 1995. Determining need for a Geographic Information System (GIS). *Journal of Extension [On-line]*, 33(5).

Samson, S.A., 2000. Optical properties of vegetation in the visible, near- and shortwave infrared wavelengths. Mississippi State University Geospatial Technologies Extension Program, Geospatial Resource Paper #2.

Schell, S.P. and Lockwood, J.A., 1997. Spatial analysis of ecological factors related to rangeland grasshopper (*Orthoptera: Acrididae*) outbreaks in Wyoming. *Environmental Entomology*, 26:1343-1353.

Singh, R.L. and Singh, R.P.B., 1997. *Elements of Practical Geography*. Kalyani Publ., New Delhi, pp.1-7, 186-200.

Suthers, R.W., 1991. Predicting the survival of immigrant pests in new environments. *Crop Protection*, 10: 331-333.

Tiilikkala, K., Ojanen, H., Finch, S., Hartfield, C. and Brunel, E., 1999. Use of a geographical information system for forecasting the activities of carrot fly and cabbage root fly. *Proceedings on Integrated Control in Field Vegetable Crops*, held at China, Crete, pp.15-24.

Varma, R.V., 1991. Spatial and temporal distribution of *Ailanthus* pests, *Eligma narcissus* and *Atteva fabriciella* (Final report of the Research Project, Kerala Forest Research Institute, India), 78 : 39.

Zeki, C., 1996. Organization and operation of forecasting and warning systems against pests in Turkey. OEPP's Bulletin No. 26, pp. 3-4, 587-593.

Zhang, N., Runquist, E., Schrock, M., Havlin, J., Kluitenburg, G. and Redulla, C., 1999. Making GIS a versatile analytical tool for research in precision farming. *Computers and Electronics in Agriculture*, 22, 221-231.

Zhang, N., Wang, M., and Wang, N., 2002. Precision agriculture–A worldwide overview. *Computers and Electronics in Agriculture*, 36, 113-132

Chapter 5

Characterization, Diagnostics and Management of Viruses Infecting Onion in India

Meenakshi Arya and V.K. Baranwal
Virology Unit, Division of Plant Pathology, IARI, New Delhi – 110 012

Introduction

Onion (*Allium cepa*) is an important vegetable crop consumed by almost every community as food flavouring. According to the 2001 statistics of the FAO, onions are second only to tomatoes in value of vegetable crops cultivated world wide. Its productivity in India is much lower than the productivity in other countries. This reduced productivity is due to several biotic and abiotic stresses. One of these major constraints in increasing production of onion has been insect damage and fungal and viral plant diseases. Here we have discussed about viruses which infect onion. A number of viruses have been reported to infect onion (Hoa *et al.*, 2003) but many of them has not been fully characterized and their effect on yield and quality of onion has not been quantified. There are three important viruses which have been reported to infect onion in India *viz.*, an aphid borne *Potyvirus* such as *Onion yellow dwarf virus* (OYDV) (Ahlawat and Varma, 1997; Hoa *et al.*, 2003), a thrip borne *Tospovirus*, *Iris yellow spot virus* (Ravi *et al.*, 2005) and a mite borne *Allexivirus* (Sandeep Kumar, 2008). Another important potyvirus which infects onion is *Leek yellow stripe virus* (LYSV) but has not yet been reported from India. *Potyviruses* and *Allexiviruses* infect other *Allium* species such as garlic and shallot and have been extensively studied. This paper deals with characterization, diagnostics and mode of spread of these viruses infecting onion in India and their management.

Onion Yellow Dwarf Virus

Giddings (1916) was the first to report a mosaic disease of onion in West Virginia (USA), but he could not find the organism associated with the disease. Later Melhus *et al.* (1929) reported a disease on onion with stunting, crinkling and yellowing of the leaves and named it as "onion yellow dwarf". Epidemic due to OYDV in the Pleasant Valley onion district in the Mississippi river valley led to Henderson's classical paper on 'yellow dwarf, a virus disease of onions, and its control' (Henderson, 1935). He also showed the disorder to be readily transmissible through sap. In India, a disease similar to onion yellow dwarf was observed in seed crop (Dhingra and Naraini, 1963). Studies on characterization, detection and transmission of OYDV have been mainly done using garlic as material host plant.

The particles of OYDV are filamentous; non-enveloped; usually flexuous, 770-823 nm long, axial canal obscure, basic helix obscure (Bos, 1982). Particle size may vary depending upon a number of variables in case of flexuous rod-shaped viruses such as age of infection, a possible host influence, and the precise calibration of the electron microscope magnification, accurate measurement and method of preparation. The nucleic acid of OYDV particles contains single stranded RNA like that of other potyviruses (Bos, 1982).

The incidence of disease in onion was reported up to 95 per cent (Melhus *et al.*, 1929) from USA. In Argentina, 56-92 per cent of onion plants were infected in the field (Conci *et al.*, 1992). In 1993, at Nazafgarh in Haryana state, 65 per cent plants of onion cv. Agrifound Light Red in a three-acre seed crop were found infected with onion yellow dwarf disease. Severely affected plants did not develop flower resulting in a total loss in seed production (Ahlawat and Varma, 1997). The yield of bulb and seed was greatly reduced in virus infected onion plants (Bos, 1982).

Symptoms

Stunting is the main symptom of OYDV in onion. Symptoms on the leaves and flower stalks of the affected plant appear as mosaic mottling at the early stage of disease, short chlorotic streaks and general yellowing, yellow striping in flower stalks of affected plants appear later. Finally the plants become much stunted as compared to healthy plants. Crinkling and drooping of older leaves is also common (Ahlawat and Varma, 1997). Severely affected plants do not develop flower and bud resulting in a heavy loss to seed production. According to FAO 1997, detoriation during storage and premature sprouting of bulbs may also occur.

Host Range

Experimental host range of OYDV and its strains is narrow. Plants both in dicotyledonous and monocotyledonous are susceptible and they belong to three families: Amaryllidaceae, Chenopodiaceae and Leguminaceae (Havranek, 1973). The susceptible *Allium* host species are *A. cepa* var. *cepa*, *A. cepa* var. *ascalonicum* (Bos *et al.*, 1978); *A. sativum*, *A. longicuspis* (wild garlic) and great-headed garlic (OYDV-GhG) (van Dijk, 1993). OYDV also infects Chinese sacred lily (*Narcissus tazetla*) and Jonquil (*Narcissus jonquilla*) (Anonymous, 1932; Henderson, 1935). But these reports of *Onion yellow dwarf virus* infecting Chinese sacred lily and Jonquil have not been confirmed (Bos, 1982). Among dicots, *Chenopodium amaranticolor*, *C. quinoa*. *C. murale* (Graichen *et al.*, 1985; Diekmann, 1997), *Nicotiana clevelandii*, *N. megalosiphum* (Bos, 1982) are also susceptible to OYDV (Bos *et al.*, 1978; Ghosh and Ahlawat, 1997). In combination with other viruses, symptoms may be aggrevated (Diekmann, 1997).

Onion (*Allium cepa*) is used as diagnostic host species. It shows systematic leaf striping, curling and plant stunting (Bos, 1982). Most plants become readily infected after inoculation to obtain 100 per cent disease. *Chenopodium amaranticolor*, *C. quinoa*, *C. murale* have been used by many workers for diagnosis of virus but their infection depends on virus isolates/strains (Havranek, 1973). However, OYDV can be easily maintained and propagated on *Allium cepa* which has higher concentration of virus particles than in any other hosts (Bos, 1982). Bos *et al.* (1978) studied that *Allium porrum*, *Chenopodium amaranticolor*, *Chenopodium quinoa*, *Nicotiana clevelandii* are diagnostically insusceptible hosts.

Transmission

OYDV has been shown to be transmitted by over 50 aphid species in USA (Drake *et al.*, 1933) and in a non-persistent manner and by mechanical inoculation (Bos, 1976; FAO, 1997). Hardtl (1965) reported 6-29 per cent seed transmission in onion cv. "Stuttgarter Riesen" on the basis of field observation. Although the virus was detected in pollen from infected plants by Louie and Lorbeer (1966) seed transmission could not be obtained by many workers (Bos, 1976, 1982; Delecolle and Lot, 1981; van Dijk and Sutarya, 1992a). The main route of natural spread is by vector transmission and through vegetative propagation of infected bulbs. (FAO, 1997).

Henderson (1935) suggested *Myzus ascalonicus* to be the major vector of shallot yellow virus (OYDV) because other aphid species did not readily feed on shallot. Vuittenez (1957) reported that *Aphis fabae* mainly spread the virus in onion crop. Although *M. ascalonicus* may spread OYDV from virus infected plants, aphid infested bulbs after planting and possibly during storage, the main spread of OYDV and its strain will occur during spring and summer by the large number of migrating aphids of different species (Drake *et al.*, 1933). *Myzus persicae* (18 per cent) and *Rhopalosiphum maidis* (9 per cent) were the most important vector of WoYSV in Japan (Yoshino and Yasu, 1965). Although onion thrips (*Thrips tabaci* Lind.) has been reported to be the possible vector of OYDV in onion seed field, transmission of the potyvirus of Welsh onion (WoYSV) by onion thrips in Japan failed (Yoshino and Yasu, 1965).

Genome Organization of OYDV

The complete sequence of an isolate of *Onion yellow dwarf virus* (OYDV) infecting garlic from Yuhang, Zhejiang province, China, was determined by Chen *et al.*, 2003. The complete genome map of OYDV has been shown in the Figure 5.1. It is ~ 10538 nts in length and is predicted to encode a polyprotein of 3403 amino acids (aa) with a calculated Mr of 385.1 KDa. In their study they showed that OYDV isolates from China contain a P3 region much larger than other potyviruses. The predicted P3 protein was of 530 amino acids, larger than that of any of the potyviruses sequenced to date (344-378 aa). The additional sequence occurs at the N-terminus of the protein, does not represent duplication from elsewhere in the OYDV genome and could not be matched to any other sequences in the databases. Similar sequences were found in 4 other Chinese OYDV isolates. Phylogenetic analysis of the amino acid sequences of the polyprotein showed that OYDV is distantly related to Pea seed-borne mosaic virus and the potyviruses of grasses and cereals. Molecular weight of coat protein of OYDV was shown to be 30,000 Daltons (Bos, 1982; Kobayashi *et al.*, 1996). Based on coat protein gene sequences OYDV has been categorized as OYDV "garlic type", OYDV "wakegi-type" and LYSV group (Tsuneyoshi, *et al.*, 1997). But in India only garlic type of OYDV has been found infecting onion (Sandeep Kumar 2008) and garlic (Arya *et al.*, 2006).

Diagnostics

Electron Microscopy and Serological Detection

Polyclonal antibodies for OYDV have been prepared by several workers and used for detection of the virus in ELISA and ISEM (Delecolle *et al.*, 1985; Bos *et al.*, 1978). In India, OYDV antiserum was prepared by Hoa (1999) and used for the detection of OYDV by ISEM and ELISA.

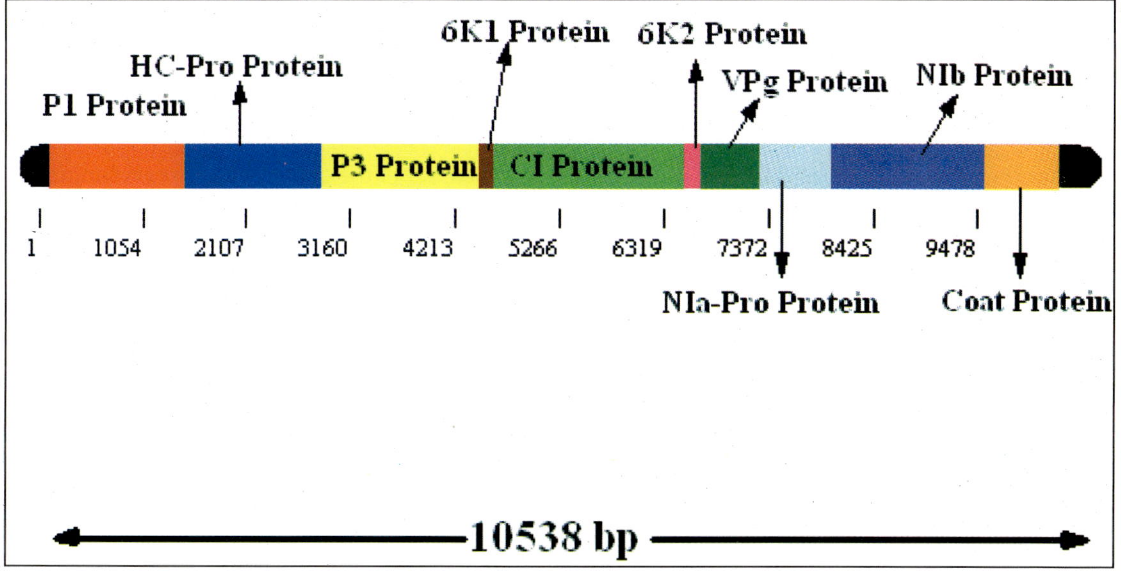

Figure 5.1: Schematic Diagram of Complete Genome of OYDV

Molecular Detection

The molecular detection of viruses includes the techniques such as Nucleic acid spot hybridization (NASH) and Reverse transcriptase- polymerase chain reaction (RT-PCR). The NASH is time consuming, labour intensive and includes the use of radioactive substances. RT- PCR on the other hand is simpler and quicker than NASH. It is 10-100 times more sensitive than ELISA (Dovas *et al.*, 2001). It would be also useful to detect virus in very small tissue samples such as meristem tip employed to obtain virus free propagation material.

Dovas *et al.*, 2001 developed and optimized a RT-PCR and Immunocapture RT-PCR (IC-RT-PCR) protocols for the sensitive detection of *Onion yellow dwarf virus*, *Leek yellow stripe virus* and allexiviruses infecting *Allium* species including onion. They demonstrated that RT-PCR was 10^2-10^4 times more sensitive than double-antibody sandwich-enzyme-linked immunosorbent assay (ELISA). Further, a 'one step' IC-RT-PCR assay was developed using plant leaf extract as template source, which proved to be 10^2 times more sensitive than ELISA and convenient for testing large numbers of leaf samples.

In India, Arya *et al.* (2006) designed primers from the coat protein (CP) region, RNA dependent RNA polymerase gene and 3′ UTR region of OYDV for its detection and characterization. In their study, they characterized the coat protein of one garlic isolate of OYDV from Delhi and one isolate from an unknown cultivar of onion from Nasik. Bulb tissues are better source of RNA template preparation than leaves for detection of OYDV in onion (Sandeep Kumar, 2008).

Lunello *et al.* (2004) developed a method for the detection of *Onion yellow dwarf virus* (OYDV) and *Leek yellow stripe virus* (LYSV), the two most prevalent garlic potyviruses, that combines IC-RT-PCR/ RT-PCR with the use of TaqMan® probes. Comparisons with ELISA results obtained with identical OYDV and LYSV infected samples showed sensitivity in detecting these viruses increased up to 10^6-fold. OYDV and LYSV were detected using different fluorochromes in the probe, thus allowing unequivocal diagnosis for each of them. The polyvalence of the designed virus-specific primers and probes was shown through their application to the detection of three isolates from very different geographical areas and from different hosts. This technique can be used for OYDV detection in onion.

Mite-Borne Filamentous Viruses (MbFV)/Allexiviruses

Another important virus infecting onion is MbFV, a member of the genus *Allexivirus* belonging to family *Flexiviridae. Allexivirus* are unusually flexuous cross banded particles of 700-800 nm in length. Mite-borne viruses were first reported in 1970 by Razvjakina (Dovas *et al.*, 2001) infecting onions and they were first named *onion mosaic virus* (OMV). These new viruses were first described by Van Dijk *et al.* 1991, who partially characterized *Onion mite-borne latent virus* (OMbLV) and *Shallot mite-borne latent virus* (SMbLV). Phylogenetically they are between poty- and carlaviruses. They are transmitted by the eriophyid mite *Aceria tulipae* (van Dijk *et al.*, 1991). Their genome organization differs from carlaviruses and potyviruses by the presence of an extra gene (ORF 4 = 42K) with unknown function, absent in these two virus genera (Van Regenmortel *et al.*, 2000). Allexiviruses have genomes of approximately 8 kb that contain six ORFs (Figure 5.2). At 3′ end they have a poly (A) tail (http://www.ncbi.nlm.nih.gov/). *Onion mite-borne latent virus* (OMbLV) in different cultivars of onion has been reported from Russia (van Dijk *et al.*, 1991) but this virus has only been tentatively kept under the genus *Allexivirus* (Adams *et al.*, 2004).

Figure 5.2: Schematic Diagram of Allexivirus Genome

The incidence of yield loss caused by Allexiviruses has not yet been reported in onion whereas in garlic, Carfune *et al.* (2006) studied the effect of two allexiviruses on garlic yield in Argentina. In their study, they found that GarV-A caused 14 to 32 per cent reduction in bulb weight and 6 to 11 per cent reduction in diameter compared with the negative control in the two cultivars, namely, Morado-INTA and Blanco-IFFIVE. GarV-C caused less damage than GarV-A (15 per cent in weight and 5 per cent in diameter) with respect to the negative control in cultivar Blanco-IFFIVE, and did not produce significant yield losses in cultivar Morado-INTA.

Symptoms

Generally MbFV induce no or only very mild symptoms (*e.g.* faint short stripes) in *Allium* spp., hence the name mite-borne latent viruses (FAO, 1997).

Host Range

The viruses have been reported from garlic, onion, shallot, sand leek, rakkyo and several wild leek species (*A. ampeloprasum* complex). Isolates seem restricted to particular *Allium* spp. Experimental host range seems to be restricted to *Allium* spp. Crow garlic appears to be a good experimental, although symptomless host for all MbFV isolates reported so far (FAO, 1997). Local lesions are produced only on *Chenopodium murale, C. amaranticolor, C. quinoa* or *Atriplex hortensis,* with not all viruses being able to infect all these test plants (van Dijk *et al.* 1991; van Dijk and van der Vlugt 1994).

Transmission

The eriophyid wheat curl mite *Aceria tulipae* is a very efficient vector on crops in the field as well as on stored bulbs (van Dijk *et al.* 1991). The transmission is also through mechanical inoculation while no data are available on seed transmission. The high incidence in many vegetatively propagated *Allium* spp. in Europe and Asia, especially garlic, onion and shallot, is because the viruses spread easily with infected planting material and viruliferous mites on harvested bulbs and cloves (FAO, 1997).

Diagnostics

Electron Microscopy and Serological Detection

Owing to high coat-protein variability, MbFV are difficult to be detected by serological methods. Typical highly flexuous morphology and cross-banding of particles distinguishes MbFV from poty- and carlaviruses in the electron microscope (FAO, 1997) but recently, Shahraeen *et al.*, 2008 has showed the presence of an *Allexivirus Garlic virus C* infecting onion in Iran by ELISA.

Molecular Detection

Most of the studies on molecular detection of Allexiviruses in *Allium* species have been undertaken using garlic as host material. However Dovas *et al.*, 2001 has developed a reliable an efficient RT-PCR method of detection of Allexivirus infecting onion. In India, Sandeep Kumar, 2008 have observed filamentous virus particles which were left undecorated in Immunosorbent electron microscopic detection using antisera of different Poty and Carla viruses infecting onion leaves. Therefore the virus particles were suspected to be of *Allexivirus*. A genus specific *Allexivirus* primer set was used for its detection in onion. RT-PCR could detect it in leaf, bulb and anther tissues in onion but consistent detection of the virus was observed only in bulb tissues. The sequencing of RT-PCR amplified products confirmed the *Allexivirus* showing closer identity to *Garlic virus C* (GVC). GVC has been recorded for the first time in onion crop in India.

Iris Yellow Spot Virus

In addition to the genera *Potyvirus* and *Allexivirus*, onion plants have often been infected with thrips borne *Tospovirus Iris yellow spot virus* (IYSV) which was first discovered in the United States in 1989 infecting onions grown for seed in Idaho. This disease is considered an emerging or re-emerging pathogen and the virus has been reported as a potentially devastating pathogen of onion in the United States. Since 1989, the disease has spread to nine states of USA *i.e.* Washington, Oregon, California, Nevada, Colorado, Utah, Arizona, Georgia and New Mexico. It has been also reported in Spain, Israel, Iran, Brazil, Australia, Netherlands, New Zealand and India. (Gera *et al.*, 1998; Pozzer *et al.*, 1999; Schwartz *et al.*, 2002; Coutts *et al.*, 2003; Creamer *et al.*, 2004; du Toit *et al.*, 2004; Gent *et al.*, 2004; Crowe and Pappu, 2005, Ward *et al.*, 2008). The disease is primarily a problem on onions (*Allium* spp.), garlic (*A. sativum*), chives (*A. schoenoprasum*), leeks (*A. porrum*), iris (Iris spp.) and lisianthus (*Eustoma grandiflorum*) (Coutts *et al.*, 2003). Weed hosts include jimsonweed (*Datura stramonium*), tobacco (*Nicotiana* spp.) and redroot pigweed (*Amaranthus retroflexus*) (http://cahe.nmsu.edu/pubs/_h/H-255.pdf). The genus Tospovirus consists of three genomic RNAs, with Large RNA in negative sense and Medium and Small (S) RNAs in ambisense. The S RNA codes for the nucleoprotein (N) gene which is being used to study the molecular variability among IYSV isolates.

IYSV typically doesn't kill plants; however, the virus reduces plant vigor and bulb size. Onions grown for seed will have reduced seed yield and quality. Additionally, the virus weakens plants, making them more susceptible to other diseases, pests and environmental stresses. The severity of the disease depends on the plant's overall health at the time of infection. Otherwise healthy plants may show few symptoms and maintain decent growth. Plants under environmental or cultural stress may show severe symptoms resulting in significant economic losses (http://cahe.nmsu.edu/pubs/_h/H-255.pdf). Losses of up to 100 per cent have been reported in Brazil and the Pacific North-west (Crowe and Pappu, 2005).

Symptoms

Plants infected with IYSV have characteristic yellow- to straw-colored lesions. Lesions may be more or less round with or without a necrotic center or may be diamond shaped. Lesions will appear on both the seed stalk and the leaves. Seed stalks may swell at the point of infection. Late in the season, infected seed stalks and leaves will lodge (fall over) (http://cahe.nmsu.edu/pubs/_h/H-255.pdf).

Transmission

The virus is transmitted by onion thrips (*Thrips tabaci*) but is not transmitted by other common *Thrips* species, such as Western flower thrips (*Frankliniella occidentalis*) (Crowe and Pappu, 2005). The virus is transmitted by both larvae and adults, but only larvae can acquire the virus from infected plants. Virus transmission is persistent: Once a thrips has acquired the virus, it can transmit the virus for the remainder of its lifetime. The disease has a potential to spread rapidly in fields with large numbers of viruliferous thrips. Distribution of infected plants in the field will be associated with feeding activity by the vector. In many cases, the damage is first noticed on field edges, in areas of stressed plants, or in locations with thin plant stands. The virus is not seed-borne and does not appear to be found in the bulbs. (http://cahe.nmsu.edu/pubs/_h/H-255.pdf).

Serological and Molecular Detection

There are several reports serological and molecular detection of IYSV infecting onion in different parts of world (Robène-Soustrade *et al.*, 2005; Gera *et al.*, 1998; Coutts, *et al.*, 2003; Crowe and Pappu, 2005, Miller *et al.*, 2006). IYSV was confirmed by ELISA using a commercially available antiserum (Agdia Inc., Elkhart, IN) and RT-PCR. RT-PCR gave DNA amplicons of the expected size. The DNA amplicons were cloned and sequenced. Nucleotide sequence comparisons with known IYSV N gene sequences confirmed virus identity (Crowe and Pappu, 2005).

In India, Ravi *et al.*, 2005 observed during 2002-03, field-infected onion (*Allium cepa*) plants exhibiting characteristic symptoms of chlorotic spindle or diamond shaped lesions on the leaves and scapes, with twisting or bending flower bearing stalks, were observed in the Jalna and Nasik regions of Maharashtra. Field-infected onion samples and greenhouse-inoculated plants were found positive for IYSV by ELISA. RT-PCR using primers designed to the capsid gene and flanking sequences of IYSV (GenBank Acc. No. AF001387) produced the expected 925 bp amplicon from infected but not healthy onions. The results of symptomatology, host range, ELISA and RT-PCR indicated that the causal agent was a strain of *Iris yellow spot virus*, which was a new report from an important onion growing regions of Maharashtra state, India.

Management of the Disease Caused by OYDV, IYSV and Allexiviruses

Among viruses infecting *Allium* species, potyviruses are most important because they induce disease which may result in large yield reductions. Their transmission takes place efficiently by various aphid species when briefly probing on plants. OYDV is extremely harmful in onion (van Dijk,

1993). In onion crops, the virus can effectively be controlled by the removal of sources of infection (Henderson, 1935; Vuittenez, 1957)

Dutch onion crops are free of OYDV because of the absence of seed onions (van Dijk, 1994). Stocks free of OYDV were obtained by meristem culture in France (Lafitte, 1986), England (Walkey *et al.*, 1987) and Switzerland (Darbellay *et al.*, 1988). By growing with true seed, shallot and onion cultivars may be freed from OYDV, but this may be readily reinfected. Factors favouring reinfection are vicinity of virus sources (shallot plants), large aphid population, and vegetative propagation when producing seed (van Dijk, 1994).

Under Indian condition, to minimize losses due to OYDV, Ahlawat and Varma (1997) suggested indexing for the presence of viruses in mother bulbs. In England, cross protection of garlic by mild potyvirus isolates against virulent isolates have been used for managing the disease (Walkey and Antill, 1989).

Allexiviruses are spread by the eriophyid mite during storage of bulbs where mites move freely. Spread of these viruses rarely occurs in the field. Therefore the management of Allexiviruses may include raising of onion plants from true seeds in a controlled conditions.

Good thrips and weed management are recommended for IYSV management. Onion thrips are best managed with insecticides. Although no cultivars are known to be resistant to the virus, research has shown that cultivars vary in their susceptibility to both the virus and the thrips vector (http://cahe.nmsu.edu/pubs/_h/H-255.pdf).

References

Adams, M. J., Antoniw, J. F., Bar-Joseph, M., Brunt, A. A., Candresse, T., Foster, G. D., Martelli, G. P., Milne, R.G. and Fauquet, C.M., 2004. The new plant family flexiviridae and assessment of molecular criteria for species demarcation. *Archives of Virology* 149: 1045-1060.

Ahlawat, Y. S. and Varma, A., 1997. Serological detection of mixed viral infection in onion seed crop and possible measure for its management. *Indian Phytopathology.*, 50: 137-140.

Anonymous., 1932. Botany and Plant Pathology Section. *Annual Report of Iowa Agriculture Experimental Sation* pp- 32-46.

Anonymous. http://www.ncbi.nlm.nih.gov/

Anonymous. http://cahe.nmsu.edu/pubs/_h/H-255.pdf.

Arya, M., Baranwal, V. K., Ahlawat, Y. S. and Singh, L., 2006. RT–PCR detection and molecular characterization of *Onion yellow dwarf virus* associated with garlic and onion. *Current Science* 91: 1230-1234.

Bos, L., 1976. Onion yellow dwarf virus. CMI/AAB *Descriptions of Plant Viruses* No. 158: 4.

Bos, L., 1982. Viruses and virus diseases of Allium species. *Acta Horticulture*. 127: 11-29.

Bos, L., N. Huijberts, H. Huttingaand Maat, D.Z., 1978. Leek yellow stripe virus and its relationships to onion yellow dwarf virus: characterization, ecology and possible control. *Netherland J. Plant Pathol.* 84: 185-204.

Cafrune, E. E., Perotto, M. C. and Conci, V. C., 2006. Effect of two Allexivirus isolates on garlic yield. *Plant Disease* 90: 898-904.

Chen, J., Adams, M. J., Zheng, H. Y. and Chen, J. P., 2003. Sequence analysis demonstrates that Onion yellow dwarf virus isolates from China contain a P3 region much larger than other potyviruses. *Archives of Virology.* 148: 1165-1173.

Conci, V. C., Nome, S. F. and Milne, R. G., 1992. Filamentous viruses of garlic in Argentina. *Plant Disease.* 76 (6): 594-596.

Coutts, B.A., McMichael, L. A., Tesoriero, L., Rodoni, B. C., Wilson, C. R., Wilson, A. J., Persley, D. M. and Jones, R. A. C., 2003. *Iris yellow spot virus* found infecting onions in three Australian states. *Australian Plant Pathology.* 32: 555-557.

Creamer, R., Sanogo, S., Moya, A., Romero, J., Molina-Bravo, R. and Cramer, C., 2004. Iris yellow spot virus on onion in New Mexico. *Plant Disease.* 88: 1049.

Crowe, F. J. and Pappu, H. R., 2005. Iris yellow spot virus in onion seed crops in Central Oregon. *Plant Disease.* 89: 105.

Darbellay, Ch., Granges, A. and Perko, J., 1988. La Milrac, une nouvelle variete d'cleatote. Revue Suisse de Viticulture, *Arboriculture et. Horticulture.* 20: 219-226.

Delcolle, B. and Lot, H., 1981. Viroses det'ail: I–Misc eneyidena at essays de carterisation par immunoelectromicroscopic d'un complexi de trios virus chez differentes population de'ail attaintes de mosaique. *Agronomie.* 1: 763-769.

Delecolle, B., Lot, H. and Michel, M. J., 1985. Application of ELISA for detecting onion yellow dwarf virus in garlic and shallot seeds and plants. *Phytoparasitica.* 13: 3-4.

Dhingra, K. L. and Nariani, T. K., 1963. A virus disease of onion. *Indian Phytopathology.* 16: 311-312.

Diekmann, M., 1997. Onion yellow dwarf potyvirus (OYDV). In: *Allium* species. FAO/IPGRI. *Technical Guidelines for the Safe Movement of Germplasm.* pp 28-29.

Dovas, C. I., Hatziloukas, E., Salomon, R., Barg, E., Shiboleth, Y.and Katis, N. I., 2001. Comparison of methods for virus detection in *Allium* spp. *Journal of Phytopathology.* 149: 731–737.

Drake, C. J., Tate, H. D. and Harris, H. M., 1933. The relationship of aphids to transmission of yellow dwarf of onions. *Journal of Economic Entomology.* 26: 841-846.

du Toit, L. J., Pappu, H. R., Druffel, K. L., Pelter, G. Q., 2004. Iris yellow spot virus in onion bulb and seed crops in Washington. *Plant Disease.* 88: 222.

FAO/IPGRI., 1997. *Allium* spp. Technical Guidelines for the Safe Movement of Germplasm (Ed) Diekmann, M. No. 18.

Gera, A., Cohen, J., Salomon, R., and Raccah, B., 1998. Iris Yellow Spot tospovirus detected in onion (*Allium cepa*) in Israel. *Plant Disease.* 82:127.

Gent, D. H., Schwartz, H. F. and Khosla, R., 2004. Distribution and incidence of *Iris yellow spot virus* in Colorado and its relation to onion plant population and yield. *Plant Disease.* 88, 446-452.

Giddings, N. J., 1916. Report of the Department of Pathology West Virginia Agriculture Experimental Station. pp 44.

Graichen, K., Raichen Bacher, D., Leistner, H. U., 1985. Nechweis des knoblaunchmosaik- virus (Garlic mosaic virs): *Archiv fur phytopathologie and Pflanzenschutz.* 21: 323-325.

Havranek, P., 1973. Occurrence of viruses in the genus *Allium* and virus free clones of common garlic (*A. sativum*) Proc. of the 7th Conference of Czechoslovakia Plant Virology, High Tatras., 1971. *Public House of the Slovakia Academy of Science,* Bratislava. 133-136.

Henderson, W. J., 1935. Yellow dwarf, a virus disease of onion, and its control. *Research Bulletin of Iowa agriculture Experimental Station*. 188: 209-255.

Hoa, N. V., 1999. Characterization of Onion yellow dwarf virus isolated from onion crop. M. Sc. Thesis. Indian Agricultural Research Institute. New Delhi, India.

Hoa, N.V., Ahlawat, Y. S., and Pant, R. P., 2003. Partial characterization of Onion yellow dwarf virus from onion in India. *Indian Phytopathology*. 56 : 276-282.

Kobayashi, K., Rabinowicz, P., Bravo-Almonacid, F., Helguera, M., Conci, V., Lot, H., and Mentaberry, A., 1996. Coat protein gene sequences of garlic and onion isolates of the onion yellow dwarf potyvirus (OYDV). *Archives of Virology*. 141: 2277-2287.

Lafitte, O., 1986. Du nonveau en echalote. Des vaieties certifiees plus productives. *Semences et progres*. 48: 26-37.

Louie, R., and Lorbeer, J.W., 1966. Mechanical transmission of OYDV. *Phytopathology* 56: 1020-1023.

Lunello, P., Carmen, M., Vilma, C. and Fernando, P., 2004. Ultrasensitive detection of two Potyviruses using a real-time fluorescent (Taqman®) RT- PCR assay. *Journal of Virological Methods*. 118: 15-21.

Melhus, I. E., Reddy, C.S., Henderson W.J. and Vestal, E., 1929. A new virus disease epidemic on onions. *Phytopathology*., 19: 73-77.

Miller, M. E., Saldana, R. R., Black, M. C. and Pappu, H. R., 2006. First Report of *Iris yellow spot virus* on Onion (*Allium cepa*) in Texas. *Plant Disease*. 90:1359.

Pozzer, L., Bezerra, I. C., Kormelink, R., Prins, M., Peters, D., de O. Resende, R. and de Ávila, A. C., 1999. Characterization of a tospovirus isolate of Iris yellow spot virus associated with a disease in onion fields in Brazil. *Plant Disease*. 83: 345-350.

Ravi, K. S., Kitkaru, A.S. and Winter, S., 2005. *Iris yellow spot virus* in onions: a new tospovirus record from India. *Plant Pathology*. 55: 288.

Robène-Soustrade, Hostachy, B., Roux-Cuvelier, M., Minatchy, J., Hédont, M., Pallas, R., Couteau, A., Cassam N. and Wuster G., 2005. First report of *Iris yellow spot virus* in onion bulb- and seed-production fields in Réunion Island. *Plant Pathology*. 55 (2): 288-288.

Sandeep Kumar., 2008. Molecular characterization and detection of *Onion yellow dwarf virus* and an *Allexivirus* associated with onion (*Allium cepa* var. *cepa* L.). M. Sc. Thesis. Indian Agricultural Research Institute, New Delhi, India.

Schwartz, H. F.,Brown, W. M., Blunt, Jr., T. and Gent, D. H., 2002. Iris yellow spot virus on onion in Colorado. *Plant Disease*. 86: 560.

Shahraeen N, Lesemann DE. And Ghotbi T., 2008. Survey for viruses infecting onion, garlic and leek crops in Iran. Bulletin OEPP/EPPO *Bulletin*. 38, 131-135

Tsuneyoshi, T., Ikeda, Y. and Sumi, S., 1997. Nucleotide sequences of the 3' terminal region of onion yellow dwarf virus isolates from *Allium* plants in Japan. *Virus Genes*. 15: 73-77.

van Dijk, P., 1994. Virus diseases of *Allium* species and prospects for their control. *Acta Horticultare*. 358: 299-306.

van Dijk, P. and Sutarya, R., 1992 a. Virus survey of garlic, shallot, and welsh onion in Java, Indonesia. Internal communication of the DLO center for plant breeding and reproduction research, Wageninger, the Netherlands. 70 pp.

van Dijk, P., 1993. Carlavirus isolates from cultivated *Allium* species represent three viruses. *Netherlands Journal of Plant Pathology*. 99: 233-257.

van Dijk, P. and R.A.A. van der Vlugt., 1994. New miteborne virus isolates from rakkyo, shallot and wild leek species. *European J. Plant Pathol.* 100: 269-277.

van Dijk, P., Verbeek, M. and Bos, L., 1991. Mite-borne virus isolates from cultivated *Allium* species, and their classification into two new rymoviruses in the family Potyviridae *Netherlands Journal of Plant Pathology*. 97: 381-399.

Van Regenmortel, M.H.V., Fauquet, C.M., Bishop, D.H.L., Carstens, E.B., Estes, M.K., Lemon, S.M., Maniloff, J., Mayo, M.A., McGeoch, D.J., Pringle, C.R., Wickner, R.B. (Ed.)., 2000. Virus taxonomy: The seventh report of the International Committee on Taxonomy of Viruses. San Diego: Academic Press,1162p.

Vuittenez, A., 1957. Methods de lutte contre la bigorrune de loignon. Les journees frutieres de Maraicheres d' Avignon. 25-26. *Janvier.*, 1957: 57-58.

Walkey, D. G. A. and Antill, D. N., 1989. Agronomic evaluation of virus-free and virus infected garlic (*Allium sativum* L.). *J. Horti. Sci.* 64: 53-60.

Walkey, D. G. A., Webb, M. J. W., Bolland, C. J. and Miller, A., 1987. Production of virus-free garlic (Allium sativum L.) and shallot (A. ascalonium L.) by meristem-tip culture. *J. Horti. Sci.* 62: 211-20.

Ward, L.I., Perez-Egusquiza, Z.,Fletcher, J.D., Ochoa Corona, F.M., Tang, J.Z., Liefting, L.W., Martin, E.J., Quinn, B.D., Pappu H.R. and Clover G.R.G., 2008. First report of *Iris yellow spot virus* on *Allium cepa* in New Zealand. New disease report. *Plant Pathology*. (Available on net).

Yoshino, M. and Yasu, M., 1965. studies on ecology of yellow dwarf of stone leek (*Allium fistulosum* L.) (In Japanese) *Bulletin of Saitama Prefecture Agricultural Experimental*. 26: 1-67.

Chapter 6

Pheromone Based Strategy for the Management of Red Palm Weevil *Rhynchophorus ferrugineus* (Olivier) in Palm Agro-ecosystems

J.R. Faleiro
ICAR Research Complex for Goa
Ela, Old Goa – 403 402

Introduction

Research on insect pheromones is rapidly accelerating in India and also at a global level (Yadav *et al.*, 2003; Boo and Park, 2005).There are over 7000 semiochemicals and pheromones identified and synthesized for insect control (El-Sayed,2005) and have been used to successfully monitor and control a wide range of Lepidopteran and Coleopteran pests with sex and aggregation pheromones, respectively. Synthesis of male produced aggregation pheromone (4-methyl-5-nonanol) by Hallett *et al.*, 1993, for the Red Palm Weevil (RPW) *Rhynchophorus ferrugineus* (Olivier) (Coleoptera: Curculionidae/ Rhynchophoridae/Dryophthoridae) has led to area-wide pheromone based IPM strategies to combat RPW in coconut, *Cocos nucifera* plantations of India and South–East Asia and also in and date palm, *Phoenix dactylifera* plantations in several Middle-Eastern countries.

Globally, RPW is reported from 15 per cent of the coconut growing countries, while in date palm 50 per cent of the date growing countries are currently reporting the pest, where the spread has been rapid during the last two decades after it was reported from the United Arab Emirates in the Middle–East in 1985 where approximately 5-6 per cent of the palms are infested (Faleiro 2006a; Zaid *et al.*, 2002). Although RPW occurs on 17 palms species (Estaban-Duran *et al.*, 1998), it is a key pest of coconut and date palm besides causing wide spread damage to the Canary island palm, *Phoenix canariensis* in Europe which is much sought after for its ornamental value. At present only the American continent is free of this pest.

RPW mostly attacks young palms below the age of 20 years and females prefer to lay eggs in palm tissue on cracks, crevices and wounds (Nirula, 1956). Newly hatched grubs bore into the surrounding plant tissue and infested palms exhibit one of the following symptoms (Abraham *et al.*, 1998) *viz.* (i) presence of tunnels on the trunk and base of frond petiole, (ii) gnawing sound due to feeding by grubs, (iii) oozing out of thick brown fluid from the tunnels, (iv) appearance of chewed plant tissues in and around opening of tunnels with a typical fermented odour, (v) fallen empty pupal cases and dead adults around a heavily infested palm and (vi) breaking of the trunk or toppling of the crown in case of severe and prolonged infestation. In date palm, infested offshoots become dry. These symptoms can be perceived either by sight, sound (gnawing sound due to feeding by grubs) or smell (typical fermented odour of infested palm tissue). Early detection of infestation in the field is the key to the success of any RPW-IPM strategy as palms in the early stage of infestation usually recover with insecticide treatment, while palms in the late stage of attack often harbour several overlapping generations of this pest, are difficult to treat and usually die. Pheromone based RPW-IPM programmes have been successfully used to combat this pest in several countries in coconut and date palm (Rajapakse *et al.*, 1998; El-Ezaby *et al.*, 1998; Abraham *et al.*, 2000; Vidyasagar *et al.*, 2000a, b; Al-Khatri, 2004; Faleiro, 2005; Soroker *et al.*, 2005; Oehlschlager, 2006; Sujata *et al.*, 2006; Jayanth *et al.*, 2007). Due the lethal and hidden nature of this pest and also the high value of the crops involve area-wide control of RPW is taken up at a low action threshold on just one per cent infested palms (Faleiro, 2006a,b). Sound decision making is crucial for initiating new and also validating on-going area-wide RPW-IPM programmes (Faleiro and Kumar,2008).

RPW-IPM

The components of the RPW- IPM strategy include

1. Surveillance based on periodic field surveys for detecting infestations and also through pheromone traps (monitors).
2. Palms and field sanitation.
3. Mass trapping adults using food baited pheromone traps, in endemic pockets and detection of infestations.
4. Preventive chemical treatments of wound and palms (treating palms infected with bud rot disease/infested with *Oryctes* attack in coconut to prevent attraction of adults).
5. Filling frond axil of young palms with mixture of insecticide and sand.
6. Curative treatment of infested palms in the early stage of attack.
7. Eradicating severely infested palms, treating hidden breeding sites including beheaded palms, neglected and closed gardens.
8. Cutting fronds in coconut if required at a distance of one meter from the frond base.
9. Educating and training farmers and agricultural officers.
10. implementing quarantine measures in date palm to ensure movement of pest free planting material (Abraham and Kurian, 1975; Abraham *et al.*, 1989; Abraham *et.al.*, 1998).

Following is an overview of the pheromone based RPW- IPM to manage this pest in palm agro-ecosystems including lure synthesis and efficiency, pheromone trap design,bait-lure synergy including trapping protocols and impact RPW- IPM programmes.

Lure Synthesis and Efficiency

Chemical attractants and food baits are known to attract the *Rhynchophorus* palm weevils. Both sexes of *R. palmarum* were reported to be attracted to a chemical mixture by Hagley,1965 while coconut log traps smeared with toddy mixed with yeast and acetic acid were found to be superior in attracting *R. ferrugineus* as compared to other baits tested (Kurian *et al.*, 1984). Subsequently, with the synthesis of the male produce aggregation pheromone "ferrugineol" (4-methyl-5-nonanol) by Hallett *et al.*, 1993 pheromone based area-wide RPW-IPM programmes have been widely used to manage this pest in coconut and date plantations. Addition of the second component (4-methyl-5-nonanone) in a small amount increased capture rates by nearly 65 per cent (Abozuhairah *et al.*, 1996). 4-methyl-5-nonanol is produced by both *R. ferrugineus* and *R. vulneratus* suggesting species synonomy, which was confirmed through morphological, molecular- genetic and breeding data (Perez *et al.*, 1996; Hallett *et al.*, 2004). Oehlschlager, 2005 reported that trapping is most efficient for all palm weevils if the aggregation pheromone is combined with the food bait and ethyl acetate. Ethyl acetate when added to RPW pheromone traps enhanced captures in date plantations of Oman (Abdallah and Khatri, 2005) while in coconut plantations of Goa weevil captures in RPW-pheromonetraps increased by 2.33 times as compared to traps loaded only with pheromone + food bait (Figure 6.1).

From Sri Lanka, ferrugineol was synthesized by Grignard reaction with butyl magnesium bromide and 2-methyl-1-pentanal (Gunawardena, 1994).Further, the same laboratory reported that *R. ferrugineus* was sensitive to the size and position of oxygen function, degree of olefinic bonds in the molecules and degree of unsaturation. Decanol elicited the lowest Electroantennogram response to ferrugineol while the activity could be enhanced when combined with n-pentanol (Gunawardena and Bandarage, 1995).For using pheromone technology in the field it is essential to commercially produce the lure with appropriate dispensing technology.

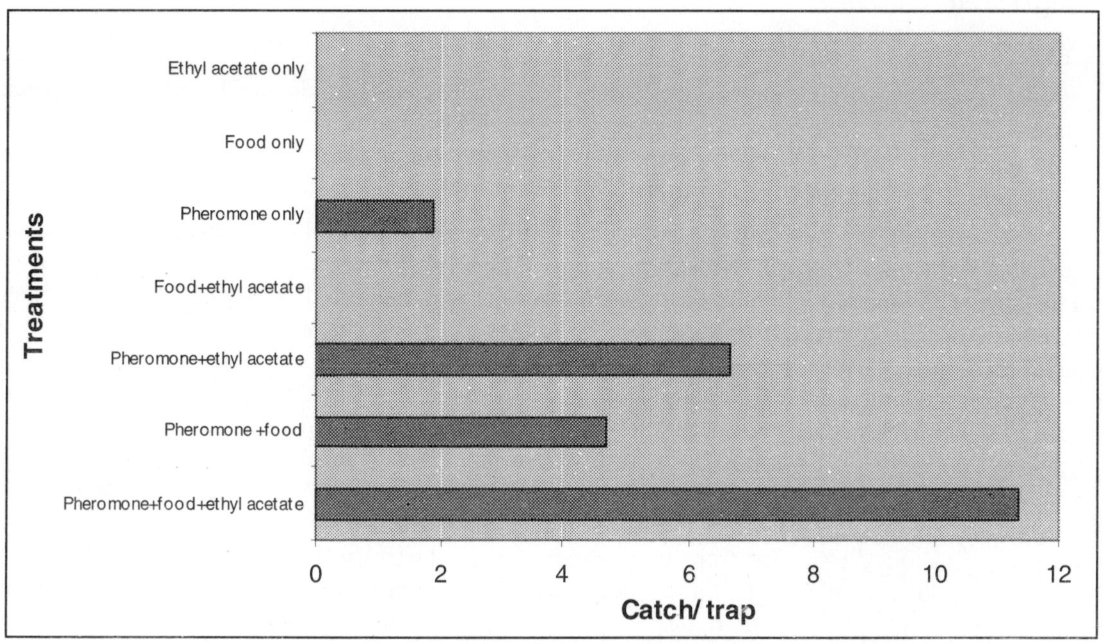

Figure 6.1: Influence of Ethyl Acetate on *R. ferrugineus* Captures in Food Baited Pheromone Traps

Chem Tica International, Costa Rica was the first to commercially synthesize and formulate pheromone lures (Ferrolure) for RPW during 1994, which are widely used in RPW-pheromone based control programmes in several countries. Ferrolure is composed of 4-methyl-5-nonanol, while Ferrolure+ has 10 per cent 4-methyl-5-nonanone in addition to 4-methyl-5-nonanol. Based on Sri Lankan technology the Central Plantation Crop Research Institute (CPCRI), Kerala, India synthesized RPW pheromone during 2000. Currently, Pest Control India (PCI), Bangalore is the leading manufacturer of RPW pheromone in India. RPW pheromone lures have also been commercially produced by Agrisense, UK; Plant Research International (Pherobank), Netherland and ISCA Technologies, USA (Faleiro, 2006b).Russell IPM,UK also commercially produces RPW pheromone lures. Field trials from India indicate that Tripheron+ was at par with Ferrolure+ (Krishnakumar *et al.*, 2004). Trials conducted on the trapping efficiency of RPW pheromone lures since 1999 in the Saudi Arabia and India have shown that mostly the Costa Rican lure from Chem Tica International sustained the trapping efficiency and was in some trials at par with the Pherobanklure (400mg) from the Netherlands and ISCA technologies, USA. However, the CPCRI lure from India was cost effective but needs to improve the dispensing technology which was done through glass ampoules (Faleiro *et al.*, 2000; Faleiro and Chellappan, 1999; Faleiro and Satarkar, 2003a; Faleiro *et al.*, 2004; Kalleshwaraswami *et al.*, 2004; Abraham *et al.*, 1999a). It is important to periodically evaluate the lures available in the market so that the best can be recommended.

Besides using an efficient lure, the dispenser used to discharge the pheromone in the field is important, especially in area-wide IPM operations. RPW pheromone lure has been dispensed in the field through thermoplastic spatulas, sachets, vials, glass ampoules and plastic cans (Faleiro and Satarkar, 2003a; Faleiro *et al.*, 2000).To sustain the trapping efficiency it is essential to identify lures that are exhausted and replace with new lures. For this it would be desirable to have a dispenser in which the lure can be seen.

RPW-Pheromone Trap Design

The type of trap used to capture any insect using pheromone lures is vital to sustain the trapping efficiency of a pheromone based IPM programme. The RPW pheromone trap was designed based on the experience of trapping *R. palmarum* in oil palm plantations of Latin America, where bucket traps were found to be most effective in trapping *R. palmarum* (Oehlschlager, *et al.*, 1993).

Trials with inverted and upright bucket traps in date plantations of Saudi Arabia in 1994 showed that the upright bucket trap (5L) with four windows (1.5 x 5 cm^2) cut equidistantly below the upper rim of the bucket having a rough outer surface was superior to the inverted model and was named as the reusable Saudi bucket trap (Anonymous,1998; Faleiro *et al.*, 1998).Besides the lure hung from inside the lid of the bucket with a piece of wire, this trap is baited with a kairomone releasing food bait (dates, green coconut petiole, sugarcane etc.) to orient entry of the adults in to the trap by generating bait-lure synergy which is discussed further in this chapter.In addition, the food bait is mixed in one litre non repellant insecticide (0.05 per cent carbofuran 3G, carbaryl 50WP) solution so as to kill and retain the trapped weevils (Anonymous, 1998; Oehlschlager, 1998). Hallett *et al.*, 1999, evaluated the trapping efficiency of the standard bucket trap (upright and inverted), funnel trap and vane trap with funnel in Indonesia and found that vane traps were superior to bucket traps. However, the ease in servicing (change of food bait and insecticide solution) has led to the preference of the upright bucket model, which is essential especially when mass trapping the pest in area-wide programmes. Higher weevil captures were recorded in black bucket traps as compared to white traps (Hallett *et al.*, 1999) while Ajlan and Abdulsalam, 2000 captured more weevils in green reusable bucket traps as compared to

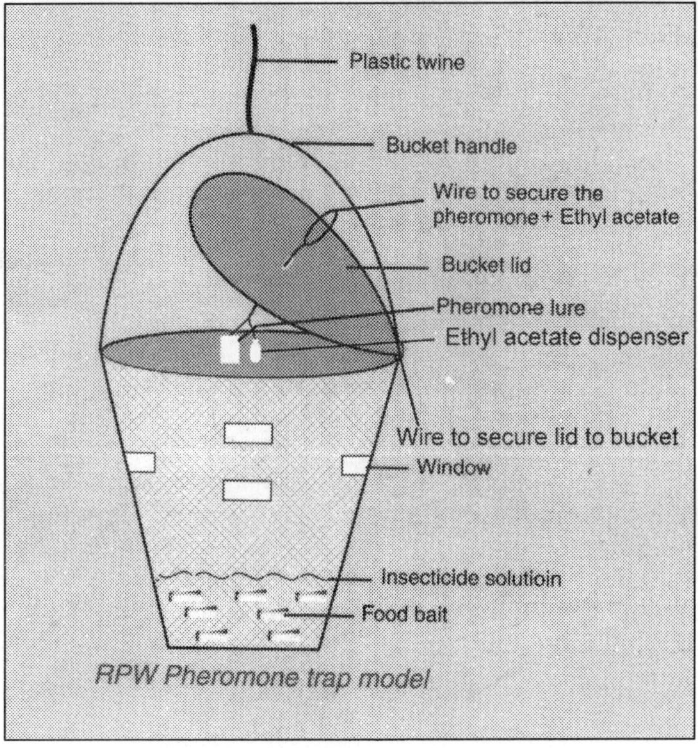

Figure 6.2: *R. ferrugineus* Pheromone Trap

white and yellow. Trials conducted in coconut plantations of Goa have shown that although trap colour did not significantly influence the trapping efficiency while, rough outer trap surface enhanced weevil captures (Faleiro, 2005). In UAE and India, specially fabricated plastic traps with a rough exterior surface have now been designed and are available commercially.

Bait-lure Synergy and Trapping Protocols

Prior to the synthesis of ferrugineol, food baits were used to attract and kill RPW adults and formed an important component of the RPW-IPM strategy in coconut plantations of India. Reports from Kerala in South India reveal that coconut log traps were superior to metal traps baited with pieces of the coconut trunk in attracting adults of RPW (Kurian, *et al.*, 1979). Later, coconut logs treated with coconut toddy, yeast and acetic acid were found to be significantly superior to other food bait combinations tested (Kurian, *et al.*, 1984).

Palm volatiles released from the food bait act synergistically with ferrugineol released from the pheromone dispenser to attract RPW adults in to the traps (Hallett *et al.*, 1999). Food baits added to RPW pheromone traps orient the attracted weevils into the trap. The synergy that goes on between the lure and the food bait is vital in enhancing and sustaining trapping efficiency of food baited RPW pheromone traps (Faleiro, 2006a). Weak bait–lure synergy would result in the attracted weevils orienting themselves towards the palm trees instead of the trap as according to Borden, 1985 host plants and insect-produced chemicals generate synergy for attraction, especially in the order Coleoptera. Weak bait-lure synergy could result in pheromone trapping being counter productive as the attracted weevils that fail to enter the trap may infest palms around RPW- pheromone traps. Rochat, 2006 reported increased infestations around RPW-pheromone traps in date plantations of Iran which could be attributed to weak-bait lure synergy that manifests when dry baits and inappropriate trapping protocols are used. The importance of adding kairomone releasing food baits in RWP pheromone traps has been emphasized by many. From the literature cited on this aspect sugarcane, dates (*khajur*), date palm tissue, coconut petiole, plantains, sugarcane molasses and coconut shavings are the potential food baits that are recommended for use in RPW food baited pheromone traps (Oehlschlager, 1994; Annonymous, 1998;Faleiro and chellapan, 1999; Hallett *et al.*, 1999; Faleiro and Satarkar, 2005; Nair, *et al.*, 2000; Muthiah *et al.*, 2005; Jayanth *et al.*, 2007). While selecting a food bait one would have to consider its cost and availability. Although, Faleiro and satarkar, 2005 reported that among the food baits evaluated (Figure 6.3), dates (*khajur*) gave the best weevil captures, in coconut plantations it would be advisable to use green coconut petiole bits which is readily available in coconut grooves. Dates are widely used as food bait in RPW pheromone traps in the Middle-East where it is found in plenty. Oehlschlager, 2006 reported that for efficient trapping of this pest it is essential to have a pheromone lure with the components (ferrugineol + ferruginone), ethyl acetate and moist food bait containing a good level of sugar. Rochat, 2006 suggested that more work needs to be done to understand the mechanism of synergistic responses to the combined odour of pheromone + palm. From Figure 2 it is evident that for most of the baits weevil captures reduced in the second week as compared to the captures in the first week (oilpalm fruit, pineapple, cashew apple, dates etc),while only with sugarcane weevil captures increased in the second week. This emphasizes the need to change the food baits periodically.

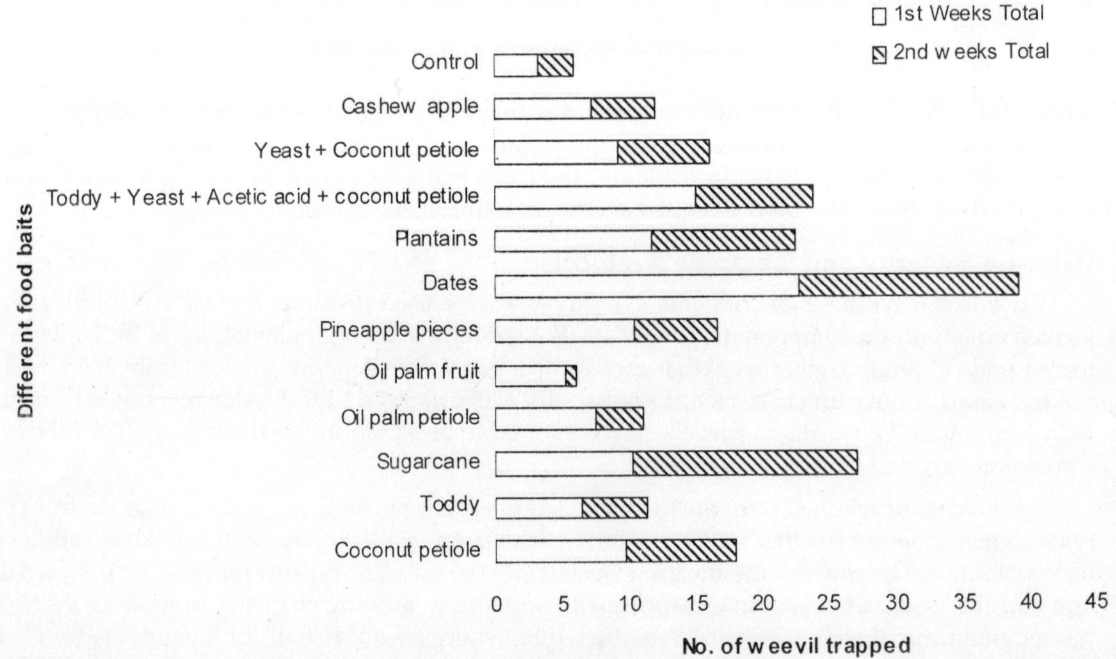

Figure 6.3: ***R. ferrugineus*** **Captures in Ferrugineol Based Pheromone Traps Using Different Food Baits**

To sustain the right amount of bait-lure synergy it is also important to adopt the optimum trapping protocols with regard to trap servicing (periodic replacement of food bait and insecticide solution), trap placement and trapping density density.

Trap Servicing

Periodic replacement of the food bait and insecticide solution is essential to sustain the trapping efficiency.Faleiro 2005, suggested that 200g of coconut petiole was sufficient for one trap and found that weevil captures progressively declined when traps were serviced (change of food bait + water) at 10, 20 and 30 days. However, addition of water in traps sustained the trapping efficiency when traps were not serviced beyond 15 days. This highlights the need to use water along with food baits, as baits that go dry may minimize bait–lure synergy and consequently restrict trapping efficiency, which could result in the attracted weevils landing on palms in the vicinity of such traps that generate low bait–lure synergy.

Retaining adult weevils that enter the RPW pheromone trap is essential from the management point of view. This could be achieved by killing the captured weevil with insecticide mixed with the bait or mechanically preventing escape of adults (using funnel traps Hallett *et al.*, 1999). In the commonly used bucket traps insecticides are used to kill and retain captured adult weevils. It is important that the insecticide used does not counter the odours produced by the pheromone lure and the food bait in the trap. Trials from Saudi Arabia have shown that among the insecticides evaluated deltamethrin had the least repellency as compared to quinalphos which exhibited the highest repellency to adult weevils, when used in RPW pheromone traps. (Abozuhairah *et al.*, 1996). Reports from India suggest that carbofuran and carbaryl (1L) were suitable for use in RPW pheromone traps to kill and retain trapped weevils (Faleiro and Satarkar, 2002; Abraham and Nair, 2001). Soap solution/detergent was used to kill and retain captured weevils bucket traps in Iran and India (Rochat, 2006, Jayanth *et al.*, 2007).

Trap Placement

Correct Trap placement in the field will ensure prolonged lure longevity and also facilitate easy servicing of traps which is crucial in pheromone based area-wide RPW-IPM operations. Care should be taken to avoid setting traps on palms in the susceptible age group to RPW attack of less than 20 years. Highest weevil captures were recorded when traps were placed at a height of 1.0m from the ground in coconut plantations of Goa (Faleiro, 2005). In date plantations Israel, UAE and Spain RPW pheromone traps are set at ground level. However, traps placed at waist height on tree trunks are convenient to service as compared to traps set at ground level (Faleiro, 2006b). In Kerala RPW was effectively mass trapped in area-wide pheromone based IPM programmes by setting traps 5 feet from the ground. In this trial traps set in the middle of the coconut plantation caught fewer weevils as compared to traps on the border.Due to fragmentation of the farms large scale community based mass trapping is likely to be more effective as compared to single farm trapping (Jayanth *et al.*, 2007). Community based area-wide RPW-IPM programmes have also been recommended by Nair and Nair, 2003 and Faleiro, 2005

Trap placement in the field influences lure longevity, with traps set in the shade of palm canopy retaining the chemical lure for a longer period as compared to traps that are exposed to sunlight (Faleiro *et al.*, 1999). Hence, it is recommended to set the traps under shade of the palm/tree canopy so as to maintain a uniform and sustained release of chemical in to the environment. A release rate of 3 mg ferruginol per day was recommended by Hallett *et al.*, 1999 while, trials conducted in Goa, showed that ferruginol sustained the efficiency of trapping the weevil even at a low dose of 0.48 mg per day

(Faleiro, 2005). In coconut plantations of Kerala with the PCI, Bangalore lure a release rate of 1.76 mg/day sustained the trapping efficiency over a period of 14 weeks (Jayanth, *et al.*, 2007).The release rate of commercial RPW pheromone lures was determined by Zada *et al.*, 2002 by using quantative GC analysis of secondary alchocol pheromones. Reports from Saudi Arabia suggest that both Ferrolure and Ferrolure+ (700 mg) exhausted in about 12 weeks during summer while the same lure was effective for 24 weeks during winter, when traps were set under shade (Faleiro *et al.*, 1999). In the coconut plantations of India Ferrolure+ and Ferrolure lasted for 150 and 84 days respectively, as compared to Tripheron which had field longevity of 100 days (Krishnakumar and Maheswari, 2003). However, Tripheron+ had a field longevity of 170 days and was superior to Ferrolure but at par with Ferrolure+ (Krishnakumar *et al.*, 2004). The CPCRI lure (78.5 mg) from India was reported to have a field longevity varying from 3 to 5 months (Faleiro *et al.*, 2004; Mayilvaganan *et al.*, 2003). Further, studies from India have reported a field longevity of 18 to 51 weeks for 400 mg and 1100 mg ISCA technologies lures, respectively and 15 and 35 weeks for 250 mg and 800 mg Chem Tica lures, respectively (Kalleshwaraswamy *et al.*, 2004). Under the agro-climatic conditions prevailing in Kerala, India the PCI, Bangalore lure (800mg) has field longevity of 450 days (Jayanth, *et al.*, 2007). A easy scoring technique to identify exhausted pheromone lures (Ferrolure) was devised by Faleiro *et al.*, 1999 and it was recommended that lures with less than five per cent of the chemical need to be replaced so as to maintain the trapping efficiency in area-wide RPW-IPM programmes.

Trapping Density

Oehlschlager, 1994 initially recommended a trapping density of one trap per hectare and one trap/100ha in mass trapping and monitoring programmes, respectively. However, mass trapping of RPW adults in pheromone based RPW-IPM programmes have been implemented at different densities ranging from 1.5 traps/ha in date plantations of Al-Hassa region in Saudi Arabia to high density trapping of 10 traps/ha of date plantations of Israel. Surveillance of this pest in these countries was done at one trap/ha and one trap/100ha in Israel and Saudi Arabia respectively (Soroker *et al.*, 2005; Abraham *et al.*, 2000). In date plantations of Egypt RPW was efficiently mass trapped at a trap density of 0.5 traps/ha. (Anonymous, 2004) while, in coconut plantations of Goa RPW was mass trapped a trap density of one trap per hectare (Faleiro,2005). However, weevil catches doubled by increasing the density to two traps per hectare in coconut (Faleiro and Satarkar, 2003b).Recently, Jayanth *et al.*, 2007 reported successful management of RPW in area-wide programmes by setting pheromone traps at 2.5 traps/ha. Rochat, 2006 recommended that RPW pheromone traps should be set apart or only at low density in areas with high infestation risks. In general, the agro-system involved, attractiveness of the lure, intensity of the pest, and the resources available to service the traps could influence the trapping density.

Impact of Pheromone Based RPW-IPM Programmes

Literature suggests that ferrugineol based pheromone traps attract and capture twice as many females as compared to adult male weevils (Oehlschlager, 1994; Hallett *et al.*, 1999; Abraham *et al.*, 1999b; Vidyasagar *et al.*, 2000a; Faleiro and Satarkar, 2003b; Soroker *et al.*, 2005, Jayanth *et al.*, 2007). Soroker *et al.*, 2005 reported that the preferential RPW female attraction to bait in the pheromone traps may be due to a number of factors *viz.* pressure on females to disperse in search of mates, food resources and oviposition sites and need further studies. As female weevils initiate damage to the palms by laying eggs that hatch into damage inflicting grubs, from the weevil management point of view trapping of more female weevils is beneficial. Studies on the reproductive status of female weevils captured in RPW pheromone traps have shown that most of these females are young, gravid and fertile (Abraham *et al.*, 2001; Faleiro *et al.*, 2003; Jayanth *et al.*, 2007). Pheromone trap captured female weevils on dissection were found to be gravid and had initiated egg development (Kalleshwaraswamy *et al.*, 2005).Further, Jayanth *et al.*, 2007 reported that 74 per cent of the pheromone trap captured female weevils had not started egg laying. These studies therefore highlight the potential benefits of using food baited pheromone traps in mass trapping programmes to suppress its population build up in the field.

Area-wide IPM programmes have been reported to be superior to farm-by-farm pest management and aim to reduce the pest population over a period of time, by taking into account the spatial and temporal distribution of the pest (Mumford, 1998; Yu and Leung, 2006). Due to the lethal nature of RPW and the high value of the palm species involved Faleiro, 2006b recommended that pheromone based area–wide RPW-IPM programmes need to be implemented at a low infestation level of just one per cent infested palms which needs sound decision making. Often area-wide management of RPW is implemented on the basis of trap captures or infestation reports. This can be inaccurate, as it may either under or over estimate the pest intensity in the field. Hence, based on the concept of sequential sampling Faleiro and Kumar, 2008 proposed decision making plans for rapid and accurate classification of RPW infestation in coconut plantations of India by inspecting palms in a sequence to detect infestation until a decision to either implement or not to initiate area-wide management of *R. ferrugineus* can be made. The sampling plans are based on the general hypothesis that 'area-wide management of RPW in not required'and is computed by taking into account its spatial distribution in the field, assumed action threshold values and a risk factor of making the wrong decision of either implementing the IPM programme when not required or not implementing when essential. RPW population in coconut plantations of Goa was found to be aggregated and followed the negative binomial distribution pattern (Faleiro *et al.*, 2002). Spatial distribution is one of the most important characteristic ecological properties of a species (Taylor, 1984) and serves as a basis for decision making to implement management strategies in the field (Bechinski and Pedigo, 1981). For any area-wide IPM stategy to be successful cooperation and participation of the farmers is essential (Yu and Leung, 2006). In the date growing oasis of Al- Hassa in Saudi Arabia, farmer participation in a 250 hectare date plantation in an RPW hot spot, resulted in significant reduction in infestation and hence farmer participatory RPW–IPM is highly recommended (Faleiro, 2008) in the over all strategy against this dreaded pest. For a pheromone based RPW-IPM strategy to succeed in date plantations of the middle-East, pheromone trapping has to be combined with an efficient programme to detect infestations in the field for which data on pheromone trap capture reports could be vital. Also, to overcome the disadvantages of pheromone trapping it would be advisable to take up periodic insecticidal cover sprays of young palms around 50-100m radius of the pheromone traps. Succeccful management of RPW in wide streches of date (El-Ezaby *et al.*, 1998; Abraham *et al.*, 2000; Vidyasagar *et al.*, 2000a, b; Soroker *et al.*, 2005; Oehlschlager, 2006; Al-Khatri, 2004) and coconut plantations (Rajapakse *et al.*, 1998; Faleiro, 2005; Sujata *et al.*, 2006; Jayanth *et al.*, 2007) have been reported from several countries. As suggested by Oehlschlager, 2005 identifying a suitable repellant for RPW could help strengthen push-pull strategies for this pest.

References

Abdallah, F. F. and Al-Khatri, S. A., 2005. The effect of pheromone, kairomone and food bait on attracting males and females of red palm weevil. *Egyptian J. Agri. Res.* 83: 169-177.

Abozuhairah, R. A., Vidyasagar, P. S. P. V. and Abraham, V. A., 1996. Integrated Management of Red Palm Weevil *Rhynchophorus ferrugineus* in Date Palm Plantations of the Kingdom of Saudi Arabia, *Proceedings*, XX International Congress of Entomology, 25-36, August, Firenze, Italy.

Abraham V. A., Mahmood Al Shuaibi, J. R. Faleiro, Reda A. Abozuhairah and PSPV Vidyasagar, 1998. An integrated approach for the management of red palm weevil *Rhynchophorus ferrugineus* Oliv.– A key pest of date palm in the Middle East. *Sultan Qaboos University Journal for Scientific Research, Agricultural Science* 3, 77-83.

Abraham, V.A. and Kurian, C., 1975. An integrated approach to the control *Rhynchophorus ferrugineus* F. the red weevil of coconut palm. *Proceedings*, 4th Session of the FAO Technical Working Party on Coconut Production, Protection and Processing, 14-25 September, Kingston, Jamaica.

Abraham, V.A., Abdulla Koya, K.M. and Kurian, C., 1989. Integrated management of red palm weevil (*Rhynchophorus ferrugineus* Oliv.) in coconut gardens. *J. Plantation Crops* 16 (Supplement), 159-162.

Abraham V. A., Nair, S.S. and Nair, C.P.R., 1999a. A comparative study on the efficacy of pheromone lures in trapping red palm weevil, *Rhynchophorus ferrugineus* Oliv. (Coleoptera: Curculionidae) in coconut gardens. *Indian Coco. J.* 30(7): 1-2.

Abraham V. A., Faleiro, J. R., Al Shuaibi, M. A., 1999b. Sex ratio of red palm weevil, *Rhynchophorus ferrugineus* captured from date palm plantations of Saudi Arabia using pheromone traps. *Indian J. Ent.* 61(2): 201-204.

Abraham, V. A., Faleiro, J. R., Al-Shuaibi, M. A. and Prem Kumar, T., 2000. A strategy to manage red palm weevil *Rhynchophorus ferrugineus* Oliv. on Date Palm *Phoenix dactylifera* L.–Its successful implementation in Al-Hassa, Kingdom of Saudi Arabia. *Pestology*, 24(12): 23-30.

Abraham V. A. and Nair, S.S., 2001. Evaluation of five insecticides for use in the red palm weevil pheromone traps. *Pestology* 25(1), 31-33.

Abraham, V. A., Faleiro, J. R., Al Shuaibi, M. A. and Saad Al Abdan., 2001. Status of pheromone trap captured female red palm weevils from date gardens in Saudi Arabia. *J. Trop. Agri.* 39: 197-199.

Ajlan, A.M. and Abdulsalam, K.S., 2000. Efficiency of pheromone traps for controlling the red palm weevil *Rhynchophorus ferrugineus* (Olivier) (Coleoptera: Curculionidae), under Saudi Arabia conditions. *Bulletin of the Entomological Society of Egypt* (*Economic Series*). 27: 109-120.

Al- Khatri, S. A., 2004. Date palm pests and their control. *Proceedings*, Date palm regional workshop on ecosystem-based IPM for date palm in Gulf countries, 28-30 March, Al-Ain, UAE, 84-88pp.

Anonymous, 1998. Final report of the Indian Technical Team (Part A),–Red palm weevil control project, Ministry of Agriculture and Water, Kingdom of Saudi Arabia, pp 1-65.

Anonymous, 2004. The Middle East red palm weevil programme, July 1998 to June, 2004 (Final report) published by The Peres Center for Peace, Israel, 62pp.

Borden, J.H., 1985. Aggregation pheromones, in G.A. Kerkut and L.I. Gilbert (eds). *Comprehensive Insect Physiology, Biochemistry and Pharmacology*, vol. 9. pp. 257-285, Pergamon Press, Oxford.

Boo, K.S.and Park, K.C., 195. Insect semiochemical research in Korea: overview and prospects.*Applied Entomology and Zoology*.40: 13-29

El-Ezaby, F., Khalifa, O. and El-Assal, A., 1998. Integrated Pest Management for the control of red palm weevil in the UAE Eastern region, Al-Ain. *Proceedings* First International conference on date palms, March, 1998, Al-Ain, UAE, 269-281pp.

El-Sayed,A.M., 195.The pherobase: Database of insect pheromones and semiochemicals. http://www.pherobase.com/index.htm

Esteban–Duran, J., Yela, J. L:, Beitia Crespo, F. and Jimenez Alvarez, A., 1998. Biology of red palm weevil, *Rhynchophorus ferrugineus* (Olivier) in the laboratory and field, life cycle, biological characteristics in its zone-of introduction in Spain, biological method of detection and possible control. (Coleoptera: Curculionidae: Rhynchophorinae). *Boletin de Sanidad Vegetal Plagas* 24: 737-748.

Faleiro J.R., 2008. Consultancy Report, submitted to the FAO of the United Nations on completion of the first 30 day IPM mission on red palm weevil at the National Date Palm Research Centre, Al Hassa, Saudi Arabia. 31pp

Faleiro, J. R and J. Ashok Kumar 2008. A rapid decision sampling plan for implementing area-wide management of red palm weevil, *Rhynchophorus ferrugineus*, in coconut plantations of India. 9pp, *J. Insect Sci.* 8:15.

Faleiro J. R., 2006a. Insight in to the management of red palm weevil *Rhynchophorus ferrugineus* Olivier: Based on experiences on coconut in India and date palm in Saudi Arabia. *Proceedings*, 1st International Workshop on red palm weevil, Valencia, Spain, 28-29, November, 2005. (Edt. *Fundacion Agroalimed*) ISBN :84-690-1742-X. 35-57.

Faleiro, J. R., 2006b. A review on the issues and management of red palm weevil *Rhynchophorus ferrugineus* (Coleoptera : Rhynchophoridae) in coconut and date palm during the last one hundred years. *Internat. J. Trop. Insect Sci.* 26(3): 135-154.

Faleiro J.R., 2005. Pheromone technology for the management of red palm weevil *Rhynchophorus ferrugineus* (Olivier) (Coleoptera : Rhynchophoridae)–A key pest of coconut. *Technical Bulletin* No.4, ICAR Research Complex for Goa.40pp.

Faleiro J. R. and Satarkar, V.R., 2005. Attraction of food baits for use in red palm weevil, *Rhynchophorus ferrugineus* Olivier pheromone traps. *Indian Journal of Plant Protection* 33(1): 23-25.

Faleiro J. R., Mayilvaganan, M., Nair, C.P.R. and Satarkar, V.R., 2004. Efficacy of indigenous pheromone lure for red palm weevil, *Rhynchophorus ferrugineus* (Olivier) (Coleoptera: Rhynchophoridae). *Insect Environment* 10(4): 164-166.

Faleiro J. R. and Satarkar, V.R., 2003a. Ferrugineol based pheromone lures for trapping red palm weevil *Rhynchophorus ferrugineus* Olivier (Coleoptera :Rhynchophoridae) in coconut plantations.). *Indian J. Plant Prot.* 31(1): 84-87.

Faleiro J. R. and Satarkar, V.R., 2003b. Standardizing pheromone trap density for mass trapping red palm weevil in coconut. *Proceedings, National Symposium on Frontier Areas of Entomological Research*, 5-7 November, New Delhi, 504pp.

Faleiro J. R., Rangnekar, P. A. and Satarkar, V. R., 2003. Age and fecundity of female red palm weevils *Rhynchophorus ferrugineus* (Olivier) (Coleoptera : Rhynchophoridae) captured by pheromone traps in coconut plantations of India. *Crop Protection* 22: 999- 1002.

Faleiro, J.R., Ashok Kumar J., Rangnekar, P.A., 2002. Spatial distribution of red palm weevil *Rhynchophorus ferrugineus* Oliv. (Coleoptera : Cuculionidae) in coconut plantations. *Crop Protection.* 21: 171-176.

Faleiro J. R. and Satarkar, V. R., 2002. Suitability of insecticides for use in red palm weevil pheromone traps. *Pestology* 26(5): 34-36.

Faleiro J. R., Abraham, V. A., Nabil Boudi, Al Shuaibi, M.A. and Premkumar, T., 2000. Field evaluation of different types of red palm weevil *Rhynchophorus ferrugineus* pheromone lures. *Indian Journal Entomology* 62(4): 427-433.

Faleiro J. R. and Mani Chellappan., 1999. Attraction of red palm weevil *Rhynchophorus ferrugineus* to different ferrugineol based pheromone lures in coconut gardens. *J. Trop. Agri.* 37: 60-63.

Faleiro, J. R., Mahmood Al Shuaibi, Abraham, V. A. and Premkumar, T., 1999. A technique to assess the longevity of the palm weevil pheromone (Ferrolure) under different conditions in Saudi Arabia. *Sultan Qaboos University Journal for Scientific Research, Agricultural Science* 4(1): 5-9.

Faleiro, J. R., Abraham, V. A. and Al Shuaibi, M. A., 1998. Role of pheromone trapping in the management of Red Palm Weevil. *Indian Coc. J.* 29(5): 1-3.

Gunawardena, N.E., 1994. Terpenes as potential semiochemical for coconut pest *Rhynchophorus ferrugineus* F. (Coleoptera :Curculionidae) an electroantennogram assay. *Journal of the Natural Sciences Council of Sri Lanka* 22(1): 35-42.

Gunawardena, N.E. and Bandarage, U.K., 1995. 4-methyl-5- nonanol (ferrugineol) as an aggregation pheromone of the coconut pest, *Rhynchophorus ferrugineus* F. (Coleoptera :Curculionidae): Synthesis and use in a preliminary field assay. *Journal of the Natural Sciences Council of Sri Lanka* 23(2), 71-79.

Hallett, R. H., Oehlschlager, A.C. and Borden, J.H., 1999. Pheromone trapping protocols for the Asian palm weevil, *Rhynchophorus ferrugineus* (Coleoptera: Curculionidae). *International J. Pest Mgt.* 45(3): 231-237.

Hallett, R.H., Crespi, B.J. and Borden, J.H., 2004. Synonymy of *Rhynchophorus ferrugineus* (Olivier), 1790 and *R. vulneratus* (Panzer), 1798 (Coleoptera, Curculionidae, Rhynchophoridae). *Journal of Natural History* 38(22): 2863-2882.

Hallett, R.H., Gries, G., Borden, J.H., Czyzewska, E., Oehlschlager, A.C., Pierce H.D., Jr.Angerilli, N.P.D and Rauf, A., 1993. Aggregation pheromones of two Asian palm weevils, *Rhynchophorus ferrugineus* and *R. vulneratus*. *Naturwissenschaften* 80, 328-331.

Hagley, E.A.C., 1965. Test of attractant for the palm weevil. *J. Eco. Ent.* 58: 1002-1003.

Jayanth, K.P., Thomas Mathew M., Narabenchi, G.B. and Bhanu, K.R.M., 2007. Impact of large scale mass trapping of red palm weevil, *Rhynchophorus ferrugineus* Olivier in coconut plantations in Kerala using indigenously synthesized aggregation pheromone lures. *Indian coconut Journal* 38(2): 2-9.

Kalleshwaraswamy, C.M., Jagadish, P. S. and Puttaswamy, 2004. Longevity and comparative efficacy of aggregation pheromone lures against red palm weevil, *Rhynchophorus ferrugineus* (Olivier) (Coleoptera, Curculionidae). *Pest Management in Horticultural Ecosystems* 10(2): 169-172.

Kalleshwaraswamy, C.M., Jagadish, P. S. and Puttaswamy, 2005. Age and reproductive status of pheromone trapped females of red palm weevil, *Rhynchophorus ferrugineus* (Olivier) (Coleoptera, Curculionidae). *Pest Management in Horticultural Ecosystems* 11(1): 7-13.

Krishnakumar, R. and Maheshwari, P., 2003. Efficacy of different pheromones in trapping the Red Palm Weevils *Rhynchophorus ferrugineus* (Oliv.). *Insect Environment* 9(1): 28.

Krishnakumar, R., Maheshwari, P. and Dongre, T.K., 2004. Study on comparative efficacy of different types of pheromones in trapping the red palm weevil, *Rhynchophorus ferrugineus* Oliv. of coconut. *Indian Coco. J.* 34(12): 3-4.

Kurian,C., Sathiamma, B., Sukumaran,A.S. and Ponnamma,K.N., 1979.Role of attractants and repellants in coconut pest control in India. Paper presented at the 5th session of the FAO technical working party, Manila.

Kurian, C., Abraham, V.A. and Ponnamma, K. N., 1984. Attractants, an aid in red palm weevil management. In: *Placrosym* V: 581-585.

Mayilvaganan, M., Nair, C.P.R., Shanavas, M. and. Nair, S. S., 2003. Field assay of locally synthesized ferrugineol for trapping *Rhynchophorus ferrugineus. Indian Coco. J.* 33(11): 8-9.

Mumford,J.D., 1998. Economics of area–wide pest control. Tan H.K editor. Area- wide control of fruit flies and other insect pests, pp39-48. Oxford University.

Muthiah, C. Natarajan, C. and Nair, C.P.R., 2005. Evaluation of pheromones in the management of red palm weevil in coconut. *Indian Coco. J.* 35(10): 15-17.

Nair, S.S., Abraham, V.A. and Nair, C.P.R 2000. Efficiency of different food baits in combination with pheromone lures in trapping adults of red weevil, *Rhynchophorus ferrugineus* Oliv. (Coleoptera : Curulionidae). *Pestology* 24(6): 3-5.

Nair,C.P.R. and Nair,S.S., 2003.Role of pheromones in the integrated management of coconut red weevil.*Indian Coco. J.* 33(12): 14&22-23

Nirula, K.K., 1956. Investigation on the pests of coconut palm, Part-IV. *Rhynchophorus ferrugineus. Indian Coco. J.* 9: 229-247.

Oehlschlager, A.C., 2006. Mass Trapping as a strategy for management of *Rhynchophorus* Palm Weevils. *Proceedings*, 1st International Workshop on red palm weevil, Valencia, Spain, 28-29, November, 2005. (Edt. *Fundacion Agroalimed*) ISBN :84-690-1742-X. 143-180

Oehlschlager, A. C., 2005. Current status of trapping palm weevils and beetles. Proceedings, Date palm regional workshop on ecosystem-based IPM for date palm in the Gulf countries.28-30 March 2004, Al- Ain, UAE. *Planter* 81(947): 123-143.

Oehlschlager, A. C, 1998. Trapping of date palm weevil. *Proceedings,* FAO Workshop on date palm weevil (*Rhynchophorus ferrugineus*) and its control, 15-17 December, 1998, Cairo, Egypt.

Oehlschlager, A. C, 1994. Use of pheromone baited traps in control of red palm weevil in the kingdom of Saudi Arabia. *Consultancy Report*- submitted to Ministry of Agriculture, Saudi Arabia, 17pp.

Oehlschlager, A.C., Chinchilla, C. M., Gonzalez, L. M., Jiron, L. F., Mexon, R. and Morgan, B., 1993. Development of a pheromone-based trapping system for *Rhynchophorus palmarum* (Coleoptera: Curculionidae). *J. Eco. Ent.* 86: 1381-1392.

Perez, A.L., Hallett, R.H., Gries, G., Oehlschlager, A.C. and Borden, J.H., 1996. Pheromone chirality of Asian palm weevil, *Rhynchophorus ferrugineus* (Oliv.) and *R. vulneratus* (Panz.) (Coleoptera :Curculionidae). *J. Chem. Ecol.* 22(2): 357-368.

Rajapakse, C.N.K., Gunawardena, N.E. and Perera, K.F.G., 1998. Pheromone baited trap for the management of red palm weevil *Rhynchophorus ferrugineus* F. (Coleoptera: Curculionidae) population in coconut plantations. *Cocos* 13: 54-65.

Rochat, D., 2006. Trapping: Drawbacks and Prospects- Need for more Research, *Proceedings*, 1st International Workshop on red palm weevil, Valencia, Spain, 28-29, November, 2005. (Edt. *Fundacion Agroalimed*) ISBN :84-690-1742-X. 99-104

Sujatha A, Chalapathirao Rao NBV and Rao DVR., 2006. Field evaluation of two pheromone lures against red weevil, (*Rhynchophorus ferrugineus* Oliv.) in coconut gardens in Andhra Pradesh. *J. Plant. Crops* 34(3): 414-416.

Soroker, V., Blumberg, D., Haberman, A., Hamburger-Rishad, M., Reneh, S., Talebaev, S., Anshelevich, L. and Harari, A.R., 2005. Current status of red palm weevil infestation in date palm plantations in Israel. *Phytoparasitica* 33(1): 97-106.

Vidyasagar, P.S.P.V., Al- Saihati, A.A., Al-Mohanna, O.E., Subbei, A.I. and Abdul Mohsin, A.M., 2000a. Management of red palm weevil *Rhynchophorus ferrugineus* Olivier. A serious pest of date palm in Al-Qatif, Kingdom of Saudi Arabia. *J. Plant. Crops* 28(1): 35-43.

Vidyasagar, P.S.P.V., Mohammed Hagi, Abozuhairah, R. A., Al-Mohanna, O.E., and Al- Saihati, A.A., 2000b. Impact of mass pheromone trapping on red palm weevil adult population and infestation level in date palm gardens of Saudi Arabia. *Planter* 76(891): 347-355.

Yadav,J.S. Prasad,A.R.Prasuna,A.L.and Jyothi, K.N., 193.Semiochemicals as potential tools in pest management.In Frointer Areas of Entomological Research (Eds. Subrahmanyan, B.B. Ramamurthy, V.V. and Singh, V.S. 506pp) 411-422p

Yu R.and Leung P., 2006.Optimal pest management : A reproductive pollutant perspective. *Internat. J. Pest Mgt.* 52: 155-166.

Zaid,A., De Wet,P. F., Djerbi, M.and Oihabi,A., 2002. Diseases and pests of date palm (Chapter XII). In: Date palm cultivation. Editor : A Zaid. FAO plant production and protection paper. 156, Review 1.

Chapter 7

Phytophagous Mites of Economic Plants of India

R.N. Singh and J. Singh
Department of Entomology and Agricultural Zoology, Institute of Agricultural Sciences, Banaras Hindu University, Varanasi – 221 005, U.P.

Introduction

Phytophagous mites have attracted the attention of acarologists globaly (Gupta, 1985 and 1991, Helle and Sabelis, 1985; Jeppson, *et al.*, 1975; Keifer *et al.*, 1982, Meyer, 1981; Singh and Mukherjee, 1989; Wilson *et al.*, 1998, and Walsh *et al.*, 1998). Predatory mites are more adversely affected than mite pests where broad-spectrum pesticides are used; consequently an outbreak of mite pest occurs (Van de vrie *et al.*, 1972). Many species of mites, which were rare and innocuous, have assumed the status of major pests. The outbreaks of many phytophagous mites on several crops have also attracted the attention of agricultural scientists (Dhooria and Sagar, 1975; Meyer, 1981; Penman and Chapman, 1988 and Sandhu *et al.*, 1987). The species of spider mites are capable of developing resistance to certain pesticides (Goodwin *et al.*, 1991; Thwaite, 1990 and Thwaite *et al.*, 1987).

Growers have started realising the destructive effect of mite infestation on their crops (Anonymous 1991 and 1993). The adopting of different agro climatic condition in agriculture and advance agricultural practices, including exhaustive usage of fertilizers, pesticides, irrigation and monocropping, and decreased genetic diversity has further aggravated this problem. The damage caused by many species is still innocuous and often mistaken. A total of 660 mite species are known to occur on various plants of India, out of which 319 are plant feeding and predatory in nature (Banerjee, 1988; Banerjee and Cranham, 1985; Bindra, and Singh, 1970; Gangwar and Lal, 1988; Ghai, 1964; Gupta, 1985 and 1991; Nassar, and Ghai, 1981; Singh *et al.*, 1987 and Singh and Mukherjee, 1989). The plant feeding mites are represented by 169 species whereas 148 species belong to predatory group. The major groups of plant mites belong to spider mite (> 69 species, Family: Tetranychidae), false spider mite (> 29 species, Family: Tenuipalpidae), gall or bud mite (> 64 species, Family: Eriophidae) and yellow or broad mite (> 4 species, Family: Tarsonemidae). Twenty-seven species of mites are major pests on various plants. Besides this, there are some predatory mites feeding on injurious mites. Among predatory mites majority (> 102 species) belong to Phytoseiidae and remaining (> 46 species) belong to other families of predatory group. The knowledge of important detrimental species is necessary when there is continuous fall in stability and sustainability of useful species due to rapid changes in the cultural and pest control practices, including indiscriminate use of pesticides. This has given a new atmosphere to flare up the mite population in the absence of predatory fauna. Very limited effort has been made to document the severity of the mite problem with changes in agriculture scenario and compounded nature of atmosphere. The present review is to visualize the magnitude of the mite menace and mite management in India.

Phytophagous Mites

In India 17 species of mites are considered as major pests and 30 as minor pests on different crops (Gupta, 1991 and Singh and Mukherjee, 1989). The major plant feeding mites (27 species) belong to families of spider mite (Tetranychidae), false spider mite (Tenuipalpidae), gall mite (Eriophyidae) and yellow or broad mite (Tarsonemidae). The important injurious mites of these families are listed in Table 7.1.

Tetranychidae

The tetranychid mites are widely spread and commonly known as spider mites. They are exclusively phytophagous in nature and many species are serious pests of agricultural crops, vegetables, ornamentals and fruits. Many mites are potential pests and their pest status has been described (Gupta, 1991; Jeppson *et al.*, 1975; Meyer, 1981 and Singh and Mukherjee, 1989. In Indian Tetranychidae consists of a total of 101 species under 2 subfamilies, 6 tribes and 10 genera (Gupta and Gupta, 1994). Confusing taxonomic status of many tetranychid mites makes it difficult to differentiate them among themselves. Dosse and Boudreaux (1963) mentioned that the economic importance of spider mites has made it necessary that good taxonomy be achieved. They further mentioned that the problems of mite taxonomy are still with us and workers must continue in making efforts at a better understanding of these mites. The harmful tetranychid mites of this family are mentioned in Table 7.1 (Singh *et al.*, 2000).

Tenuipalpidae

The false spider mites are known as tenuipalpid mites. Baker and Tuttle (1987) reported that false spider mites are widely distributed and are more numerous in warmer zones of the globe. Meyer (1979) reported 504 species in 21 genera for the world fauna. Many mites of this group are pests of economic plants, such as ornamentals and fruit trees (Jeppson *et al.*, 1975). These mites look like spider mites but do not spin silken web. Majority of spider mites is of great economic importance and infests fruit trees, vegetable crops etc. However, very little work has been done on these mites (Gupta, 1985). The false spider mites are flat, pear shaped, mostly reddish bright coloured, size 0.3 to 0.35 mm, slow moving and are normally found on the undersurface of the leaves near midrib or veins, twigs and fruits. Most specialised members of this family form plant galls within which they feed (Jeppson *et al.*, 1975).

In India genera *Brevipalpus* and *Tenuipalpus* are of great economic importance. At a time they cause serious injury to a wide variety of agricultural and horticultural crops. The first observable symptom on plant appears on the leaves as silvery areas that frequently become shrunken and brown. Seriously infested leaves become yellow and drop off from the plant. The prominent mites of this family are listed in Table 7.1.

Table 7.1: Important Mite Pests Causing Damage to Economic Plants

Sl.No.	*Family/Mite Species and Common Name*	*Major Host Plants*
	Tetranychidae	
1.	*Tetranychus urticae* Koch {Two-spotted mite (TSM)}	Okra, brinjal, beans, cucurbits, cotton and marigold
2.	*Tetranychus, neoealidonicus* Andre' (Red vegetable mite)	Vegetables and ornamental plants
3.	*Tetranychus ludeni* Zacher (Red-legged spider mite)	Vegetables and beans
4.	*Tetranychus macfarlanei* Baker and Pritchard (Spider mite)	Cucurbits
5.	*Eutetranychus orientalis* (Klein) (Oriental citrus mite)	Citrus
6.	*Eotetranychus hirsti* Pritchard and Baker (Fig spider mite)	Fig
7.	*Oligonychus indicus* (Hirst) (Sugarcane red spider mite)	Sugarcane and sorghum
8.	*Oligonychus coffeae* (Nietner) (Red tea mite)	Tea, coffee
9.	*Oligonychus mangiferus* (Rahman and Sapra) (Mango red spider mite)	Mango
10.	*Schizotetranychus andropogoni* (Hirst) (White patchy/ Sugarcane leaf spotted mite/Web mite)	Sugarcane and cereals
11.	*Panonychus ulmi* (Koch)	Pear apple
12.	*Petrobia latens* (Muller) (Brown wheat mite)	Wheat, barley and coriander
	Tenuipaldidae	
13.	*Brevipalpus phoenicis* (Geijskes) (Scarlet mite/ Reddish black flat mite/Leprosis mite)	Fruit trees and tea
14.	*Brevipalpus californicus* (Banks) (Citrus flat mite)	Citrus and tea
15.	*Larvacarus transitans* (Ewing) (Ber gall mite)	Ber (*Zyziphus mauritiana*)
16.	*Dolichotetranychus floridanus* (Banks) (Pineapple flat mite)	Pineapple
17.	*Raoiella indica* Hirst (Coconut red mite)	Coconut and areca nut
	Eriophyidae	
18.	*Aceria mangiferae* Sayed (Mango bud mite)	Mango
19.	*Aceria litchii* (Kiefer) (Litchi erineum mite)	Litchi
20.	*Aceria cajani* ChannaBasavanna (Pigeonpea sterility mosaic mite)	Pigeonpea
21.	*Aceria lycopersici* (Wolff) (Tomato/Brinjal erineum mite)	Brinjal and tomato
22.	*Aceria guerrerronis* (Keifer) (Coconut eriophyid mite)	Coconut
23.	*Acreia jasmine* ChannaBasavanna (Scarlet mite)	Jasmine
24.	*Acaphylla theae* (Watt) (Tea pink mite)	Tea
25.	*Eriophyes cernuus* Massee (Jujube (Ber) gall mite)	Jujube (*ber*)
	Tarsonemidae	
26.	*Polyphagotarsonemus latus* (Banks) (Silver mite/ Yellow mite/Broad mite)	Chilli, beans, potato and jute Brinjal
27.	*Steneotarsonemus spinki* Smiley (Rice sheath mite)	Rice

Eriophyidae

Eriophyidae are exclusively plant feeders and host specific is of economic importance. They are commonly known as gall, rust, bud, blister and erineum or simply eriophyid mites. Members of this family have two body regions, the gnathosoma and idiosoma. They have elongated worm like body shape with two pairs of legs. The body size varies from 0.1 to 0.5 mm. Eriophyid mite has simple life cycle but certain species has complicated cycle that includes alternation of generations. There are two forms, one is an over-wintering female called deutogynes and the other consists of both sexes known as protogynes. The symptoms caused by eriophyid mites vary from simple russeting to coupled gall formation. The galls may be simple leaf galls, bud galls and erineum. Witches broom, which is cluster of brush like eriophyid mites also cause growth of stunted twigs or branches on trees. Eriophyid mites also cause non-gall abnormalities such as leaf folding and twist blisters. Ghosh *et al.* (1989) described injury caused by gall mites. They recorded six types of plant injuries *i.e.*, big buds, curling/crinkling and shrinkage of leaves, discolouration, erineum, russeting and galls. Eriophyid mites also induce russeting and silvering of leaves. Keifer *et al.* (1982) described various symptoms of injuries caused by eriophyid mites on buds, shoots, stems, twigs, flowers and fruits. Large number of eriophyid mites is confined to one species of host plant. Some eriophyid mites act as vector of plant viral diseases. The important plant diseases transmitted by eriophyid mites are pigeonpea sterility mosaic disease, fig mosaic disease, wheat streak mosaic disease and sugarcane streak mosaic virus. About 1859 eriophyids belonging to 156 genera have been reported from different parts of the world (Boczek *et al.*, 1989; Davis *et al.*, 1982 and Keifer *et al.*, 1982). Extensive work has been done on the Indian eriophyids (ChannaBasavanna, 1966, 1971; Das and Chakrabarti, 1988; Gupta, 1985 and Mohanasundaram, 1981). The alarming eriophyid mites causing plant abnormaties in India are mentioned in Table 7.1.

Tarsonemidae

Tarsonemid mites are very small, ranging in length from 100 to 300m. Mature forms have a relatively hard and shiny integument just like glistering white or shiny silver piece. The body and posterior legs are rather sparsely be set with setae. The anterior pairs of legs especially their terminal segments are more densely clothed with setae and are often equipped with specialized sensory setae of various configurations and sizes. The tiny glossy mites are found on the undersurface of leaves as well as in association of fungus. Males often carry female deutonymphs. Some of these mites are important crop pests (Gupta, 1985; Meyer, 1981and Singh *et al.*, 1987).

Important Mite Pests of Agricultural and Horticultural Crops

Mite Pests of Vegetable Crops

Tetranychus urticae Koch (Two Spotted Mite, TSM)

T. cinnabarinus is synonym of *T. urticae.* Gupta (1985) described 59 synonyms of *T. urticae* (Dosse and Boudreaux, 1963; Dupont, 1979; Hill *et al.*, 1991; Meyer, 1981; and Van de vrie *et al.*, 1972). This species is very common and it can be found every part of the globe. Many species of *Tetranychus* are very serious mite pests in most parts of the world (Jeppson *et al.*, 1975). Indian acarologits are built a concept the *T. urticae* is found in temperate parts of the world and *T. cinnabarinus* in the tropical. It has a global distribution in the warmer parts and is an important pest of various crops like vegetables, ornamental plants, cereals, cotton, strawberry, deciduous fruits (Singh and Mukherjee, 1989). The population starts developing during late February to June and starts declining during July when monsoon comes (Gupta *et al.*, 1976; Pande and Sharma, 1979; Singh, 1994, 1995 and Singh and Singh, 1993).

Mali *et al.* (1983); Singh and Singh (1992) and Mukherjee *et al.* (1992) recorded this mite on under surface of the leaves and found feeding by stippling the leaf tissue. The severely infested leaves become completely yellow to brown and fall off from the plants. Silver to yellowish appearance of leaves with webbing is a common symptoms of this mite infestation. Van de Vrie *et al.* (1972) reported that neither *T. urticae* nor *T. cinnabarinus* produces silken threads or ballooning threads for transportation (Singh *et al.*, 2000).

Tetranychus neocaledonicus Andre (Red vegetable mite): *Tetranychus cucurbitae* is decribed as *Tetranychus neocalidonicus* (Jeppson *et al.*, 1975). Gupta and Nahar (1981) recorded a serious attack by this mite in the brinjal crop. They also reported this mite from sponge gourd, bitter gourd, beans, tomato and potato. Gupta (1985) reported this mite as one of the serious pests of vegetables in India. The mite of all stages remains confined on the undersurface of the leaves covered with webs. The symptoms of damages are like other tetranychid mites. Sharma and Pande (1981) recorded lowest population during January and peak during May. Lal (1982) recorded very low population during December, which increased thereafter. The peak population was recorded from January-April. Mukherjee and Singh (1993) recorded this mite throughout the year on fig and mango plant (Singh *et al.*, 2000).

Tetranychus ludeni Zacher (Red-legged Spider Mite)

Tetranychus ludeni is widespread in warmer parts of the globe. Singh and Mukherjee (1989) reported this species as most alarming mite pest of cowpea in Varanasi region. This species was earlier recorded as pest of vegetable crops from south India (Puttarudriah and ChannaBasavanna, 1959). ChannaBasavanna (1971) reported *Tetranychus ludeni* as an important pest of several vegetable crops in India. Meyer (1981) recorded the presence of this mite from about 70 plant species from South Africa. It was reported from vegetable crops. In the early infestation the yellowing of leaves can be observed and then necrotic patches. During heavy infestation of mites bronzing effects on leaves are very clear. This occurs at later part of fruiting. The crowding of mites on tips of fruits is common at 40°C or above. Authors during field surveys have recorded the hot spots/ballooning symptoms. This mite also occurs in mixed population with *T. urticae*. Identification of this mite is easy because it has dark red body with reddish legs. The females are oval in shape. The males are straw coloured with pointed opisthosoma. The life cycle of mite is of short duration. It may be completed in 10 days during summer months. Coates (1974) recorded about 111 eggs by a fertilized female and that progeny consists on an average of 84 females and 27 males. This is the only tetranychid mite in India known to be vector of plant viral disease, Dolichos Enation Mosaic Virus–DEMV (Rajagopalan, 1974).

Tetranychus macfarlanei Baker and Pritchard (Spider Mite)

The tetranychid mite, *Tetranychus macfarlanei* is a serious pest of vegetables and cotton in northern India. This mite builds up small colonies on the under surface of leaves. The sucking of plant cell sap by all the stages causes yellowing of leaves followed by drying. The infestation is more on mature crops than on the young ones. It disappears during the rainy season (Singh *et al.*, 2000).

Aceria lycopersici (Wolff) (Tomato/Brinjal Erineum Mite)

Aceria lycopersici (Wolff), tomato rust mite has a world-wide distribution and it attacks plants belonging to the family Solanaceae. It is an important pest of tomato and can easily kill the plants. Severe infestation may cause damage to chillies, potatoes and brinjals. This orange yellow mite is fusiform and microscopically small. It thrives in warm weather and has no over-wintering stage. In the winter it dies unless it has a suitable shelter. All stages of the rust mite are killed by temperature close to freezing. Favourable conditions for development are temperature of 26.5° C and a relative humidity of 30 per cent. Light and photoperiod have little effect on the development of this mite (Jeppson *et al.*, 1975 and Meyer, 1981). The initial symptoms of damage to tomatoes are the curling of the lowest leaves and appearance of a silver gloss on the lower surfaces. Later these leaves become bronze-coloured, withered and eventually die. The mites also feed on the stem and before damage is noticeable on the top most leaves, the lower most portion of the stem looses its surface hairs. The colour of the stem changes from green to brown, and cracks appear on the surface. As the population builds up and the mites continue to spread, the browning of the leaves and stem increases and defoliation takes place (Gupta, 1985 and Singh *et al.*, 1987). The symptoms of tomato rust mite can easily be confused with those of late blight, but the leaves don't wilt in the same way as with blight (Meyer, 1981).

Mite Pests of Field Crops

Oligonychus indicus (Hirst) (Sugarcane Red Spider Mite)

Oligonychus indicus is a harmful pest of sugarcane and sorghum. This species is widely distributed in India. It is grayish green mite with black blotches scattered over dorsum (Gupta, 1985 and Puttarudriah, 1947). In case of sugarcane, maize and sorghum, feeding of this mite causes the appearance of reddish spots, which increase with the severity of attack. These spots later coalesce to form large red patches and spread on the leaf surface to turn the colour of the leaf red. The attack of this mite on sugarcane is more in North India and less in South India and causes 20-30 per cent damage to the crop (Desai and Chavda, 1955). Sandhu *et al.* (1975) recorded increase in population after rains on some varieties of maize due to limited or least effect of rain on active stages and eggs. There are about 30 overlapping generations in a year; all stages occur together living in a delicate web of silk and breeding is continuous. The mite spreads by wind and also by crawling from plant to plant, when leaves touch each other. Females and immature stages are cause main damages accept the adult male. Soft leaf varieties of sugarcane are more susceptible to attack than those having hard leaf (Singh *et al.*, 1987 and 2000).

Schizotetranychus andropogoni (Hirst) (White Patchy/Sugarcane Leaf Spotted Web Mite)

Schizotetranychus andropogoni (Hirst) or white patchy or sugarcane leaf spotted mite or web mite is a very detrimental pest of sugarcane in North India and in Coimbatore (Agarwal, 1957 and Khan and Murthy, 1956). Recently in eastern India, this mite has been seen causing damage on paddy crop also (Gupta, 1985). Thousands of mites can be seen on the under surface of the leaves making small oval colonies covered with thin webs and are arranged irregularly on either sides of mid-rib. All the stages of mites in varying numbers (5-117) may be seen in the colonies numbering 1300-1500 per leaf. The feeding on leaves causes the appearance of white blotches. Later, the webs turn brown and finally are blown off. The affected leaves look sickly and gradually dry up. Infestation is more on ratoon crop than on newly planted crop (Banerjee, 1988; Gupta, 1975, 1985 and Singh *et al.*, 1987 and 2000).

Petrobia latens (Muller) (Brown Wheat Mite)

Recently identified brown wheat mite, *Petrobia latens* as a serious pest in dry land agriculture (Singh and Mukherjee, 1989). They reported high population of this mite resembles like aphids on coriander and wheat from Rajasthan State. Jain and Yadava (1988) recorded this mite on coriander during last week of February to mid June at Jobner (Rajasthan). Meyer (1981) described this mite as a pest of small grain from all over the world. She also described the diagnostic features, field identification, symptoms of damage and life cycle of the mite. It lays 50 to 90 eggs as non-diapause and 8 to 30 as diapause per female. Peairs (1994) reported in the extension bulletin the brown wheat mite spends the summer in the soil as a white egg resistant to hot, dry conditions. In the fall, as cooler, wetter conditions return, these eggs start to develop and hatch after 10-day incubation. Female mite mature after feeding on wheat for about two weeks and then lay round, red eggs that give rise to further fall (one or two) and spring (two or three) generations. Male have not bee observed in this species. As conditions return, white over-wintering eggs are produced. Both red and white eggs are placed on soil particles adjacent to wheat plants. Khan *et al.* (1969) reported about the infestation of this mite on wheat crop in Rajasthan Canal and Bhakra command areas. They reported that infested plants give a sickly yellowish or bronzing appearance. Gupta (1985) described *P. latens* as a serious pest of wheat in un-irrigated fields of M.P., Haryana, Delhi, Rajasthan and Punjab. He also described the different damage symptoms caused by this mite. The period of activity of this mite continues from early December to mid April with peak period during March (Banerjee, 1988 and Singh *et al.*, 1987). Singh and Bhatia (1983) reported that *P. latens* significantly reduced the yield of some barley varieties. They reported that aphid resistant variety of barley, DL 200 as highly susceptible to the mite with significant loss in yield (Singh *et al.*, 2000).

Aceria cajani Channa Basavanna (Pigeonpea Sterility Mosaic Mite)

This mite was first collected from Bihar in 1931 (Mitra, 1931). However it was named and described later (ChannaBasavanna, 1966). It is worm like eriophyid mite about 200-250mm long (Sheila *et al.*, 1988). This mite has a short life cycle of less than 2 weeks that include egg, 2 nymphal and adult stages. Eggs can be detected on growing tips of pigeonpea leaf, and are milky white, oval, translucent and slightly smaller than glands of trichomes. This mite is vector of sterility mosaic disease (SMD) of pigeonpea in India (Seth, 1962). A single mite vector can transmit SMD virus in as little period as 20 minutes (Reddy *et al.*, 1988). Nagia and Muniyappa (1995) found no correlation between density of mite and severity of the disease. The symptoms include general stunting and yellowing of infected plants, proliferation of branches, mosaic mottle of leaflets, reduction in leaflet size and partial or complete sterility of plants (Capoor, 1952; Nene, 1972 and Ramakrishnan and Kandaswamy, 1972). In northern India no population of this mite is seen during July but maximum increase is attained during April and May when crop is harvested or ready to harvest. The wild pigeonpea harbors very high population of this mite during May/June months of the year. The sterility mosaic disease (SMD) has practically reduced the cultivation of pigeonpea in some parts of Maharastra State. The Maharastra malady of SMD during 1990 season was 42 per cent compared to 1 per cent in 1980 (Zote *et al.*, 1991). Rathi (1983) reported colonies of eriophyid mite, *Aceria cajani* on lower surface of the leaves of *Oxalis circulata*, a perennial weed. An increased incidence of the disease has been observed in the state of Gujarat, Maharastra, Andhra Pradesh, Tamil Nadu and Karnataka (Reddy and Raju, 1993). The disease is also prevalent in Bangladesh and Sri Lanka, Myanmar and Nepal. Pigeonpea is the only known host of both the SMD pathogen and its mite vector, *A. cajani* (Reddy *et al.*, 1993 and Singh *et al.*, 2000).

Steneotarsonemus spinki Smiley (Rice Sheath Mite)

This tarsonemid mite is known as rice sheath mite and infests the leaf sheath causing brown discoloration. Infestation of this mite on panicles causes chaffy grains and also discolouration of filled grains (Srinivasa *et al.*, 2004). Rice sheath mite lives in colonies within the intercellular air spaces of the upper part of leaf sheath and occasionally in the basal part of the mid rib of leaf blades (Sogawa, 1977). Mite infestation in panicles resulted in poor exertion of ear heads and mite feeding on developing grains caused brown to black discolouration of the glumes externally and the ovary gets shrivel (Srinivasa *et al.*, 2004). This mite has been reported from Orissa (Rao and Das, 1977; Rao and Prakash, 1992) and Andhra Pradesh (Rao *et al.*, 2000). Mite required 17, 4 and 2.75 days to complete its development from egg to adult stage at 25°C, 28°C and 30°C, respectively. At 25–28 °C, the adult female and male, respectively lived for 15±1.0 days and 7.6±0.4 days, while both sexes lived for only 5 days at 30°C (Chen *et al.*, 1979).

Mite Pests of Fruit Plants

Eutetranychus orientalis (Klein) (Oriental Citrus Mite)

Eutetranychus orientalis (Klein) is known as oriental citrus mite and is a major pest of citrus. This mite has number of alternative hosts that include pear, peach, *Zizyphus*, cucurbits and cotton (Gupta, 1985). It is widely distributed in tropical and subtropical countries. The mite damages upper surface of fully-grown leaves, which results in the formation of grey spots. Infested leaves become chlorotic and may finally drop. Infested nursery plants and trees look bare due to dropping of the leaves. All active stages of the mite cause damage, and their cast off skins remain stuck to the leaf surface. High temperature, low humidity and bright sunny days are favourable for the population development of this species (Lal and Mukherjee, 1979). Dhooria and Bhutani (1983) recorded the seasonal incidence of *E. orientalis* and its predators. They recorded peak population on orange during May-June and September; during December to March negligible population was recorded. Dhooria (1984) studied the development of citrus mite, *E. orientalis* on different hosts. Development of this mite was much faster on castor. Pillai and Palaniswamy (1985) reported *E. orientalis* as a pest of cassava, feeding on the upper surface of the leaves. Ahuja (1994) reported this mite on castor with peak population during February and March in arid region of Rajasthan. Sandhu *et al.* (1982) reported this mite as a pest of cotton in Punjab. Dhooria (1994) recorded this mite as a serious pest of french bean in Punjab. This mite shows active growth on apple during March to June (Mukherjee and Singh, 1993 and Singh *et al.*, 2000).

Eotetranychus hirsti Pritchard and Baker (Fig Spider Mite)

Eotetranychus hirsti is an alarming pest of fig in Delhi, Punjab, West Bengal and Uttar Pradesh. This mite is host specific, hence known as fig spider mite. It is greenish yellow mite with black blotches all over dorsum. This is a serious pest of fig in India and Pakistan. Transparent green patches are produced on the under surface of leaves when viewed across light. The patches turn yellowish green, then brown with rough and dry texture. Such leaves along with fruits drop prematurely, often resulting in complete defoliation of plants. This mite normally colonises under surface of the leaves but in case of heavy infestation they may move to the upper surface of the leaves and fruits as well. The infested fruits get shriveled, fail to ripe and ultimately fall off. During November, the mites migrate to the terminal buds where they hibernate till February or early March (Bindra and Verma, 1966; Gupta, 1985; Mukherjee *et al.*, 1995 and Puttarudriah and ChannaBasavanna, 1959 and Singh *et al.*, 2000).

Oligonychus mangiferus (Rahman and Sapra) (Mango Red Spider Mite)

Oligonychus mangiferus is an important pest of mango and widely distributed in tropics and subtropics. This pest can also attack cotton and is the second most serious pest of pomegranate in Egypt (Jeppson *et al.*, 1975). This mite is dark red dorsolaterally and pales red mediodorsally and infests the upper surface of the leaves. It sucks sap from the leaves resulting in the appearance of characteristic yellowish spots. In case of mango the middle area of lamina becomes much yellowish with less browning whereas in case of black berry, the attack is more near the margin (Banerjee, 1988 and Gupta, 1985 and Singh *et al.*, 2000).

Panonychus ulmi (Koch), European Red Mite

The European red mite (ERM) is a major tree fruit pest, especially in the eastern U. S, and is considered by many growers to be their most important and sometimes most difficult pest to control. The mite was introduced to North America from Europe in the early 1900's and is now established in most fruit growing areas. The ERM will attack a number of different fruit crops and can cause extensive injury if left uncontrolled, but apple is usually the most important host. It will also attack cherry, nectarine, peach, pear, plum, and prune (Pfeiffer *et al.*, 2005).

The ERM overwinters as a fertilized egg. It is oval, bright red, and has a small stalk arising from the top, approximately the length of the diameter of the egg. Over-wintering eggs are deposited in-groups on the roughened bark area, especially around buds and fruit spurs. The summer eggs are globular, somewhat flattened, and pale green, almost white to greenish amber when first deposited, but slowly change to a reddish-orange just before hatch. Six-legged larvae hatch from the eggs which are initially a pale orange, darkening to a pale green as they feed. The larvae molt to eight-legged protonymphs, which are variable in color from a pale green to reddish brown with the dark green predominating. Pale spots at the base of the bristles are very faint, if present at all. The deutonymphs are variable in color from an amber color to a dark green, with a dark-green color predominating. Pale spots are distinct at the bases of the bristles. The eight-legged adult female mite has a globular body and is approximately 1/64th inch (0.4 mm) long, bright red to velvety brown in color, and has four rows of white hairs on its back. The adult male mite is smaller, has a pointed abdomen, and the color is straw yellow to reddish-yellow (Alston and Reding, 2005).

Over-wintering egg hatch in the spring is closely correlated with apple bud development and first occurs when 'Delicious' buds are in the pre-pink stage; hatch continues throughout the bloom period. The larvae immediately move to the young foliage and begin to feed (Hull and Beers, 1990). Before the mites transform to each succeeding stage, they go through a short quiescent period. Shortly before the adult female emerges, she releases a pheromone, which attracts a male. If mating ensues, female young are produced; if a female is unmated, she produces male offspring. Adult mites usually appear by petal fall. Female mites live up to an average of 18 days and each female is capable of laying approximately 35 eggs during her life span. The rate of development is temperature dependent. Under ideal hot conditions (average 80°F), a life cycle may be completed in 10 to 12 days, however, a generation normally takes 20 to 25 days to complete. Eight to 10 generations can occur during a normal growing season. ERM populations usually peak in July. Females begin laying over-wintering eggs in early to mid-August.

The mites injure the tree by feeding on leaves, destroying chlorophyll, and increasing respiration. This is accomplished by the insertion of the mite's mouth-parts into the leaf cells to withdraw the contents. All motile stages feed on the foliage. The lower leaf surface is preferred, but both leaf surfaces are attacked when populations are high. All apple varieties are susceptible to attack, but the mites appear to increase faster and to higher densities on 'Delicious' and 'Yorking'. Characteristic brown foliage, starting as a subtle cast to the green leaf, but in severe cases becomes bronze, results from heavy mite feeding. The most serious injury occurs in late spring and early summer when trees are producing fruit buds for the following season. Moderate to heavily injured trees (*i.e.* with 750 mite days or with 43 motile mites per leaf) produces fewer and less vigorous fruit buds (Marini *et al.*, 1994). Mites feeding on leaves also reduce their ability to manufacture enough photosynthates for desirable sizing of fruit. Injury effects from mite feeding are more severe during periods of drought stress. Late season high populations of mites can cause further indirect loss of fruit by depositing over-wintering eggs, especially in the calyx end (Hull and Beers, 1990).

Brevipalpus phoenicis (Geijskes) (Scarlet Mite/Reddish Black Flat Mite)

Brevipalpus phoenicis has been recorded as a polyphagous species of mite and found in many parts of the world (Ghai and Shenhmar, 1984 and Meyer and Smith, 1979). Sadana (1985) recorded this species on 79 species of plants. This mite is dark reddish and flat in shape with distinct yellow, black or green spots. Commonly this mite is found in the field with mixed population of *B. califomicus/ B. obavatus.* It is very difficult to determine the species responsible for symptoms. A large population of this mite is found on undersurface of leaves. Mostly mites are aggregated along the midrib. Infested leaves turn pale yellow and develop brownish patches. Such leaves drop off prematurely. In case of guava brownish patches appear on fruits. Jeppson *et al.* (1975) reported that high population in Florida occasionally caused diffuse chlorotic spotting of foliage of orange trees, symptoms that have been called phoenicis blotch. This mite completes its life cycle in about 3 weeks. Sadana and Kumari (1991) recorded that low temperature coupled with moderate humidity and less rainfall are very conducive for high population. Singh and Mukherjee (1989) reported this mite as a serious pest on guava. Sadana and Kumari (1991) mentioned that false spider mite, *B. phoenicis* is cosmopolitan in nature and infests many kinds of fruit trees. They recorded heavy infestation of this mite on guava in Punjab. The damage caused by the mite is most apparent in winter (Singh *et al.*, 2000).

Brevipalpus californicus (Banks) (Citrus Flat Mite)

Brevipalpus californicus (Banks) is the best known species of flat mite because it has long been known as a pest of citrus. This mite also infests variety of ornamental plants. This species has a worldwide distribution and sometimes causes extensive damage to crops and ornamental plants. It causes leprosis of citrus in various regions of the world and is a very important pest of tea in Sri Lanka, India and Java. This mite infests cotton, tea, tobacco and deciduous and subtropical fruits. This tenuipalpid mite has been observed to attack sweat lime plants in Ludhiana (Gupta *et al.*, 1971). Adult mites feed on the undersurface of leaves mostly along the mid-rib. In case of heavy infestation these mites congregate at the base of fruits and their feeding results in premature falling of fruits (Singh *et al.*, 1987). According to Meyer (1981), the mites are difficult to find because they lie flat against the leaf surfaces and fruit and because they move very slowly. These mites are flat and are difficult to see with naked eyes. The female has oval body, red, with a dark red pattern on the middle of the body. It is difficult to separate the three economically important *Brevipalpus* species that occur in South Africa mearly on the basis of colour, particularly *B. californicus* and *B. obovatus* look very similar. Males are more pointed posteriorly. Citrus flat mites always occur in citrus orchards but are difficult to locate because they are small and not always numerous. The life cycle takes 3 weeks time in normal condition. Mites prefer green to ripe fruit. A toxin or virus is transmitted to the leaves and fruit by the feeding of this mite and causes "pinhead" rust symptoms/nail head rust. In USA this symptom on twigs and branches is called "Florida scaly bark". Both types of symptoms are known by the term "Leprosis" in Argentina. Leprosis begin as small scabby lesions that are browning to bronze and corky. These may occur on leaves or fruits. As the damages increase, the lesions become proportionately larger, and glandular areas develop on the bark of twigs (Singh *et al.*, 2000).

Larvacarus transitans (Ewing) (Ber Gall Mite)

The unique character of this mite is the presence of 3 pairs of legs both in nymphal and adult males and females. Pritchard and Baker (1958) mentioned that unlike eriophyids, these mites approach the gall forming habit from eight-legged mite form and show only early stages of adaptation to life. The adaptations elongate the body of mites and reduce the rear pair of legs. The female can be differentiated from male having roundish opisthosoma, whereas the opisthosoma of male consists of a pair of terminal stylet like rods known as genital stylets. Average length of male is 244 microns whereas average length of female is 270 microns. The "Ber" trees (*Zizyphus mauritina*) are believed to be almost free from insects and diseases. Though presently in many orchards various pests are becoming alarming, one of such alarming problems is a false spider mite, *Larvacarus transitans*. Latif and Wali (1961) reported this species causing considerable damage to her orchards, especially to grafted plants. Ghai and Baker (1989) reported this on *Ziziphus mauritiana* as a major pest in Rajasthan area forming galls on the twigs. Latif and Wali (1956) and Pritchard and Baker (1958) reported it as widely distributed in India exclusively on *ber* and that mites remained within galls round the year except during rainy season when mites emerged for new infestation from the old galls. They further mentioned that during first phase, only males emerged from the unfertilized eggs resulting in their increased population compared to females whose numbers remained constant. The unfertilized females produce male only whereas fertilized females give normal progeny of males and females. Sharma and Kushwaha (1989) studied the incidence and varietals susceptibility of *L. transitans* on 5 varieties of *ber*. The result of their study indicates that the incidence of the mite commences from second week of October and continues up to second week of August. The maximum numbers of galls are found in May. Similarly, peak mite population occurs during May and gradually declines with the approach of high temperature and. rainy season in subsequent months. The galls rupture during mansoon and the mites after emerging from galls hide themselves in cracks and crevices of tree trunk, and start fresh infestation when new plant growth takes place after rainy season in September. Through this mite the trees are found infested an average 20 to 80 per cent in terms of galls. The twigs bearing blister galls bear less fruit in comparison to healthy twig (Singh *et al.*, 1996 and 2000).

Dolichotetranychus floridanus (Banks) (Pineapple Flat Mite)

Dolichotetranychus floridanus (Banks) was recorded first from South India (Karnataka), in 1971 (Gupta, 1985). The females have an elongated, oval body and are red to orange in colour. The body of the male is pointed posteriorly. The eggs are oval and orange. The six-legged larvae are oval and amber in color, with distinct dark red eyes. The pineapple flat mite attacks plants of all ages; it causes most of the damage to young plants. The mite feeds on the soft white tissues at the base of the plant. They produce rust-like spots that can be infected by bacteria and fungi, so that rotting of the buds may occur. Severely infested plants may remain small with no fruits (Gupta, 1985; Meyer, 1981 and Singh *et al.*, 2000).

Aceria mangiferae Sayed (Mango Bud Mite)

Aceria mangiferae was originally described from Egypt and is of economic importance on mangoes. This mite is popularly known as mango bud mite. The mango malformation-a serious malady of mango has been found associated with this mite. This malady has assumed an alarming magnitude in northern India and it is threatening the very existance of mango industry. It is widespread in Punjab, Delhi and Uttar Pradesh and also to some extent in Gujarat and Maharastra. It causes two types of malformations-the vegetative and floral malformation. The nursery seedling and young plants are often suffered by vegetative malformation. It is the floral malformation, which directly affects the productivity because of its failure to bear any fruit. Such families are characterized by a condensed mass of flower buds. Both types of malformation may also occur in the same branch. The varietal susceptibility has also been reported. Langara variety shows high incidence of floral malformation. This mite is more abundant during April and gradually reaches its peak in June (Agarwal and Singh, 1988). During rainy season the population of mite is least. Singh and Singh (1979) reported that in floral malformation the population of mite increases in the month of May whereas in case of vegetative malformation the mite population has a tendency to increase in June as compared to April and May. Verma *et al.* (1969 and 1974) clearly indicated that the eriophyid mites were not the cause. Many reasons have been suggested about the cause of malformation including feeding of mite, viral, fungal, imbalance of hormones and C/N ratio. According to present knowledge this disease is caused by fungus *Fusarium moniliforme,* the spores of which are transmitted by the vector mite *Aceria mangiferae* (Verma, 1983). The disease is globally widespread and has been reported from U.A.E., Pakistan, Israel, South Africa, Brazil, Mexico, Venezuela and Florida (Meyer, 1981).

Aceria litchii (Keifer) (Litchi Erineum Mite)

Aceria litchii (Keifer) is known as litchi erineum mite. Keifer (1940) described and named the litchi mite as *Eriophyes litchii* (Keifer). *Aceria* was further separated from *Eriophyes.* This mite is most important pest of litchi in India, Pakistan, Bangladesh and Hawaii. Thakur and Sharma (1990) reported that mite was active on litchi leaves throughout the year. The pest population increased from 30.1 to 93.1 per 2.5 cm^2. They recorded lowest pest population during December and highest in May. These mites develop erineum on under side of the leaves. In the beginning the leaf looks-more green than normal leaf but later turns light brown to deep brown or chocolate brown with velvety growth of hairs. The deep brown colour indicates that the erineum is drying and mites begin to abandon it for newer plant as their host. The affected leaves eventually become distorted or curled. All such leaves dry and fall off. Mite feeding also affects the growth of flower buds due to damage caused. The damage is more pronounced during flush growth periods. Sharma (1991) reported that previously *Aceria litchii* was known as the cause of litchi erineum. He later identified the hair like erineum due to parasitic alga, *Cephaleuros virescens.* It grooms in small yellow/green spots usually on ventral surface of the leaves, which turns reddish brown to chocolate brown at maturity.

Eriophyes cernuus Massee [Jujube (Ber) Gall Mite]

This mite causes woody gall usually on one side of jujube stem, ringing the base of the developing bud or shoots. Recently heavy infestation on jujube was recorded (Mukherjee *et al.*, 1994) at Varanasi through out the year except during July to August. The mite population commenced from September and reached its peak in June. Size of galls during different months showed similar trend as that of population. Gupta (1985) recorded maximum population of the mite on young galls.

Mite Pests of Plantation Crops

Oligonychus coffeae (Nietner) (Red Tea Mite)

This species is one of the serious pests of tea and coffee in India and Bangladesh (Banerjee and Cranham, 1985). This mite normally feeds on the upper surface of the leaves; in case of severe infestation during summer months, the lower surface and the young leaves are also infested. The mite lives under cover of silken webbing and is found throughout the year, but it attains economically significant density only from March to June. With the onset of monsoon rain in northeastern India, the mite disappears from May/June. The outbreak may also occur during September or October. The colour of mite is bright crimson, ventrally dark purplish or bronzy. The adult females are oval and dark red. The males are smaller and lighter in colour. All stages of mite suck the sap from leaves, which subsequently produces yellowish spots at a point and turn brown. The heavy infestation causes copper coloured appearance of leaves and ultimately followed by dropping. Attack of mite is more on old leaves than on the young leaves (Gupta, 1985). It has a short life cycle of two weeks during summer months while it takes 4 weeks during winter months (Singh *et al.*, 2000).

Raoiella indica **Hirst (Coconut Red Mite)**

This mite has been recorded from coconut and arecanut leaves from south India. The feeding of this mite causes reddish spots on the leaves. The mite is reddish in colour and colonizes the undersurface of the leaves in young and adult plants (Banerjee, 1988 and Gupta, 1985). Its population reaches at peak twice, in June and September (Sarkar and Somchoudhary, 1989 and Singh *et al.*, 2000).

Acaphylla theae **(Watt) (Tea Pink Mite)**

Pink mite, *Acaphylla theae* (Watt) is one of the important and injurious species in the family Eriophyidae. It is widely distributed in all tea growing areas. The extent of damage may not be as severe as caused by red spider mite (Banerjee, 1988). This mite is orange in color, sometimes with reddish tinge, spindle shaped, blunt anteriorly and progressively narrowed at posterior end (Gupta, 1985). It infests both the surfaces of leaves, petioles and tender stems causing discoloration of leaves. The affected leaves become pale coloured, often turn leathery in texture. In severe tea infestation the veins and midribs become pinkish; hence the mite is known as the pink mite. (Das, 1965; Das and Sengupta, 1958 and Watt, 1898). The average life cycle of this mite is around one week (Singh *et al.*, 2000).

Aceria guerreronis **(Keifer) (Coconut Eriophyid Mite)**

The Coconut eriophyid mite, *Aceria guerreronis* was first identified in Mexico during 1965. The mite is minute, 200-250 μ in size with two pairs of legs. It is pale and worm like. The life cycles of this *Aceria* group consist of egg, two larval instars and adult stage. The total life cycles lasts for 10-15 days. The mites suck sap from the soft tissues on the buttons. Initially triangular patches are seen close to the perianth. They live in the white tender portion covered by the inner bracts of the perianth. Continuously sucking of sap ultimately results in formation of brown patches. As the nutgrow they wait and longitudinal splits develop on the outer surface of the nuts. Severe infestation leads to poor nut growth, reduced copra content and cracking and hardening of fruit. Early stage of attack of this mite results in dropping of butons (Ganesan, 1999). This mite has recently attracted the attention of coconut growers of Kerala, Karnataka and Tamil Nadu states by causing large-scale damage to coconut. It causes premature dropping of young nuts, imparting discoloration or scars on the surface of nuts. At early stage this mite takes shelter beneath around the floral bracts. The injury begins around the floral bracts and extends down to the inner husks or fibers of coconut. The injured area becomes shriveled brown and scattered. The market value of shriveled coconut due to infestation of this mite is heavily reduced. Looking to the gravity, seriousness of the problem and aggressive demands of the growers, the Government has sanctioned Rs. 3.5 crores to the affected districts of Kerala for control of infestation of mites causing serious damage to coconut fruits. The expert committee has been set up by the Government to suggest the use of pesticides as well as biopesticides against this mite. This mite was never reported from India. It is presumed that this mite came to India through coconut germplasm from some Caribbean countries (Singh *et al.*, 2000).

Mite Pests of Spices

Polyphagotarsonemus latus **(Banks) (Silver Mite/Yellow Mite/Broad Mite)**

Earlier this species was known as *Hemitarsonemus latus* Dutta. Popularly it is known as the broad mite/yellow mite or citrus silver mite. Mite is small in size (0.25 mm), not visible without magnification, shiny, translucent white to very pale yellow in color. Mite feeds on underside of young, tender foliage. The damage shows up as bronzed rusted and distorted new growth with curled leaf margin. This mite has a worldwide distribution in the tropical and subtropical regions and attacks a wide range of agricultural crops, ornamental and indigenous plants and prefers warm humid conditions. It has been found damaging the cotton crop in South Africa since 1890 and damage is recognized as "acarinose" (Singh *et al.*, 2000).

Broad mite, *Polyphagotarsonemus latus* (Banks) has been found to be common in jute, potato, chilli and other plants in India. Injury to potato appears first as oily blackish spots on the under surface of young leaves. The under sides of the leaves turn reddish and plant becomes rusted. The leaf hairs become prominent and the leaves become wrinkled at their edges. Starting at the tip, the plants wither and auxiliary buds are produced which in turn are killed (Tambera disease). Similar effects are produced on red pepper where it, attacks in the seedling stage and prevents flower and fruit development, while later infestations cause flower drop-Murda disease of chilli (Kulkarni, 1922 and Mann *et al.*, 1920). Anh (1996) reported in ARC official report in 1992 that production of pepper and chilli was 9.683 million tons of which 4.263 million tons came from Asia (FAO, 1992). Estimated hot pepper production area is 1.6 million ha in Asia. India, Indonesia, China and Korea have the largest areas (AVRDC, 1993). Damages from diseases and insects influence pepper production. It has been reported from different parts of the world that nearly 300 different insect and mites species attack pepper plant during its growth period.

Mite Pest of Flowering Plant

Aceria jasmine Channa Basavanna (Scarlet Mite)

This jasmine affecting eriophyid mite commonly known as scarlet mite. In southern part of India the jasmine flowers have great economic importance and this flower completely devastated by the scarlet mite. Eriophyid mite is worm like with four legs and is 150-160 micrometers long and a piercing mouth. The abdomen is ringed with a number of micro tubercles and hairs. They lead to white, hairy growth on the surface of the leaves, stems and flowers buds. As a result, the plant's growth and flower yield are reduced. Reduce the damages by clip off and burn the leaves showing felt like growths. The resistant variety, *Jasminum ariculatum* minimizes the affect of this mite. Dicofol 25 ml and oxydemeton methyl 25 gram suggested in 10 liters of water are effective in controlling this mite (Manjunath, 1999).

Eco-friendly Approaches Towards Management

Jeppson *et al.* (1975) described that sulphur has been extensively used since 1920 to control mites. Eriophyidae and Tenuipalpidae are susceptible to sulphur. Most of the tetranychid mites are not readily controlled by sulphur. Still sulphur has been regarded as one of the very safe acaricides. The dinitrophenyl acaricides were used during 40s. DNOCHP that is popularly known by growers as DN-dry mix No. 1 had been used in Florida. Another commercial formulation DN 111 had been used for red spider mites.

The DN acaricides had many disadvantages like sulphur because they produced phytotoxic effects when applied during hot weather. During the late 40s, many chlorinated hydrocarbon acaricides appeared. DCPM (neotran) was found more effective than DN 111. The ovex and aramite were extensively used throughout the world for control of tetranychid mite infesting vegetables. The dicofol was introduced in 1952. This acaricide has been extensively used in tea gardens of India. Commercially this formulation is available as kelthane, hilfol etc. The acaricides like mores tan, cyhexatin (plictran) and chlorodimeform (galecron) which emerged after 60s are not being frequently used in India. Phosalone (zolone), dinocap (karathane), ethion (phosmite), tetradiform (tedion) are also not very popular acaricides or used moderately.

Insectoacaricides from the group of organophosphorus compounds are often used in India for mite control. These insecto-acaricides are dimeton (metasystox), disulphotonl thiodemeton (disyston), quinalphos (ekalux), thiometon (ekatin), formothion (anthio), phosphamidon (dimecron), endosulphon (thiodon), carbaryl (sevin) and dimethoate (roger). These are some primary insecticides and have been most useful for mite control. Synthetic pyrethroid (danitol) is also as effective as insecto-acaricides. Ray and Rai (1981) found phosphamidon as better insecticide for control of *Tetranychus neocalidonicus*. Raja *et al.* (1995) tested efficacy of certain chemicals to control sugarcane sheath mite, *Aceria sacchari*, a pest of ratoon sugarcane crop in Tamil Nadu. They found ethion and monocrotophos as effective chemicals to control sheath mite.

The mite control through acaricides is not very common in India. Still many farmers are using pesticides for wide coverage of pest control including mites. Singh and Singh (1992) evaluated some pesticides against *T. cinnabarinus*. Mukherjee *et al.* (1992) evaluated many insecticides/acaricides against *T. cinnabarinus*. Singh and Mukherjee (1989) reported that due to continuous use of dimethoate for pest coverage on cowpea resulted in outbreak of *Tetranychus ludeni* in Varanasi region. One of the components of integrated mite control is identification of eco-friendly acaricides, which can give safe coverage to predatory fauna also. This can be achieved by monitoring carefully the fauna of crop agroecosystem for several weeks after using these chemicals to make sure that sufficient predators survive. Therefore, use of safe miticide is always recommended at time of operation flux of outbreaks. Mansour and Ascher (1983) reported the effect of neem seed kernal extract on *T. cinnabarinus* particularly on fecundity and mortality. Mansour *et al.* (1987) reported the effect of neem seed kernal extract on *Phytoseiulus persimilis* and *T. cinnabarinus*. All extracts of neem were much more toxic to phytophagous mite in comparison to predatory mite. Patel *et al.* (1993) compared the efficacy of different pesticides including botanical pesticides and found that plant products gave better results against *T. cinnabarinus* and was comparable with the most effective pesticides/acaricides tested. Frequent application of acaricides from same chemical group hastens the development of resistance. Mansour *et al.* (2004) tested eco-friendly 16 plant extracts against *T. cinnabarinus* but found six best plant extracts which are repel the mites more than 75 per cent. Most of the chlorine-based acaricides are hazardous to the predators. Waiting period between last spray and predator release should not be less than 14 days and should not be used until predators have established in the plants. Kelthane and synthetic pyrethroids are very hazardous to predators and should not be used at all. The use of these chemicals should be restricted to spot spraying only.

Landa and Bohata (1999) estimated the compatibility of entomogenous fungus *Paecilomyces fumosoroseus* with natural insecticide based on azadirachtin and neem oil. The compatibility of two neem based insecticide with the of entomogenous fungus *Paecilomyces fumosoroseus* strain PFR 97-Apopka was studied. It was demonstrated that neither azadirachtin nor neem oil has a negative effect on the germination of conidia or growth and development of *Paecilomyces fumosoroseus*. Furthermore, a combination of the fungus neem oil or with azadirachtin usually enhanced efficiency against several pests e. g. *Trialeurodes vaporariorum*, *Aphis gossypii* or *Tetranychus urticae*. Both neem oil and azadirachtin might be applied as a tank mix with a biopreparation based on *P. fumosoroseus*. This combination could be successfully exploited as a part of an IPM for various greenhouse crops. Benuzzi and Santopolo (2001), Reported Naturalis: a bioinsecticide based on *Beauveria bassiana*. The use is discussed of this bioinsecticide on greenhouse crop to control whiteflies (*Trialeurodes vaporariorum* and *Bemisia tabaci*), some aphid and thrip species, and the spider mite *Tetranychus urticae*. Naturalis is thought to have a great potential in the framework of IPM programmes as it is completely compatible with beneficial organism and bumble bees which are often used in protected cultivation. The use of naturalis is also discussed on ornamental crops where pest populations developed tolerance of pesticides. Naturalis dose not contains genetically modified organism and is recommended for use in organic crops grown under plastic cover in south Italy. Zhang *et al.* (2001) A preliminary studies on pathogenicity to *Verticillium lecanii* (Zimn) Viegas to insect pests. *V. lecanii* V-816 strain was isolated from the body of *Anoplophora chinensis* adults. A formulation of the V-816 strain containing 1 per cent Tween-80 and $4x10^7$ spores/litre was spresd on *Brevicoryne brassicae, Plutella xylostella, Pieris brassicae* and *T. urticae.* Its pathogenicity to *T. urticae* and *B. brassicae* was significantly higher than that to *P. brassicae* and *P. maculipennis.* It had no pathogenicity to *Helicoverpa armigera* and *Ostrinia nubilalis.* The pathogenicity of the V-816 strain to insect pests increased with increase of spore content. The spore content of $4x10^5$/liter resulted in the mortality of 26.4 per cent of 2-3 instar *P. brassicae*, but the mortality reached 70.8 per cent when spore content increased to $4x10^8$/litre. The spore content of above $4x10^8$/litre did not result in an obvious increase of the mortality of the larvae. The pathogenicity of the V-816 strain to young larvae was greater than too older larvae. Temperature influenced the pathogenicity of the V-816 strain and the optimum temperature was 25-30°C. Fiedler, *et al.* (2002) reported the efficacy of entomopathogenic fungi *Verticillium lecanii, Beauveria bassiana* and *Paecilomyces burci;* the nematophagous fungus *Paecilomyces lilacinus;* and pesticides Fitoverm (abamectin) and Vertimec [abamectin] against the western flower thrips (*Frankliniella occidentalis*) and spider mite (*Tetranychus urticae*) was investigated in laboratory, pot and greenhouse experiments. Successful thrips and mite control was obtained using Fitoverm. Under laboratory condition, *V. lecanii* showed the greatest efficacy against the mite pest. The results showed that *P. lilacinus* could control mites and thrips efficiently. Omar and El-Khateeb (2002) Reported the effects of abamectin, a natural product produced by the soil microorganism *Streptomyces avermitilis,* azadirachtin a natural product from neem plants, *Beauveria bassiana* an entomopathogenic fungus, thiocyclam hydrogen oxalate a product from the marine annelid worm *Lumbrinereis* spp. and Fenazaguin conventional acaricide was evaluated against *Tetranychus urticae* infesting cowpea, *Vigna sinensis* [*V. unguiculata*], during 2000 and 2001 season in Egypt. The side effects of these materials on *Tetranychus urticae* associated predator; *Euseius scutalis* (predatory mite) and *Crysoperla carnea* (a predatory insect) were also evaluated. The obtained data revealed that azadirachtin was the most potent compound introducing the population size of *Tetranychus urticae* after two seasons. The general mean reduction percentage was 80.71 per cent followed by *B. bassiana* and abamectin (77.59 and 76.35 per cent respectively). However, thyocyclam and fenazaquin demonstrated low reduction percentages (55.96 and 52.26 per cent respectively). Fenazaquin and thyocyclam demonstrated unacceptable toxic effect on *E. scutalis.* The general mean reduction percentages were 72.11 and 63.14 per cent, respectively. However, azadirachtin exhibited the lowest toxic effect (12.35 per cent). The most harmful compound on the associated predator *C. carnea* was fenazaquin followed by thyocyclam, they reduce the predator population by 54.84 and 44.59 per cent respectively. The most secure compound was azadirachtin followed by *B. bassiana* and abamectin, reduction percentage were18.95, 23.44 and 27.31 per cent respectively. In conclusion, azadirachtin, *B. bassiana* and abamectin can be applied effectively against *T. urticae* with minimum impact on the non-target organism, *i.e.* the phytoseiid mite *E. scutalis* and thysanopterous insect *C. carnea.* They may be classified as integrated pest management programme. Compatible acaricides in the integrated pest management programme against *T. urticae* infesting snap bean. Jankevica and Cudare (2003) studied the investigation on entomophthoralean fungi in Latvia. The fermentative activity and biochemical composition (content of amino acid and fatty acids) of Latvian strains of entomophthoralean fungi were determined. Amount of amino acids depended on nutrient condition strains of *Conidiobolous obscurus, C. thromboides* and *Basidiobolus ranarum* exhibited high virulance on different aphid species, spider mites and flies. The *conodiobolous* strains caused 95 to 98 per cent mortality in the aphid *Aphis gossypii* within 48 hours *B. ranarum* caused 100 per cent mortality in the mite *Tetranychus urticae* from 12 to 24 hr following exposure to the fungus. Some strains were tested under greenhouse conditions against aphid and spider mites. Fungal strain that remained able to germinate after the process of dehydration was selected. Batta (2003), reported that the conidia of entomopathogenic fungus *Metarhizium anisopliae* were cformulated in an invert emulsion (water-in-oil formulation) with coconut/ soybean oil preparation providing the most stable emulsion layer (93 per cent) and lowest viscosity (27 cps). The fungal conidia remained viable in the formulation for 30.8 months with a 50 per cent reduction (half-life) in conidial viability after 4.6 months at 20±1°C. in comparision, dry non-formulated conidia held under the same conditions were non-viable after 2 months, with a half life of 0.5 months. The invert emulsion remained stable throughout the study. Under laboratory conditions, the fungus, at the concentration of $5x10^6$ conidia/ml, killed nymphs of the tobacco whitefly, *Bamisia tabaci*, with in 3 days of the treatment, with fungal outgrowth and sporulation observed on the cadavers with in 4-5 days. Whitefly mortality ranged from 66.7 to 100 per cent under laboratory conditions. When tested under field conditions, whitefly mortality ranged from 30.0 per cent to 92.2 per cent on eggplants. The fungus, at a concentration of $5x10^6$ conidia/ml also killed nymphs and adults of the red spider mites, *Tetranychus cinnabarinus*, with in 3-4 days of treatment under laboratory condition, with fungal outgrowth and sporulation observed on the cadavers with in 5 days. Spider mite mortality ranged from 58.3 per cent to 93.3 per cent under laboratory conditions. When tested in the field, spider mite mortality ranged from 25.9 per cent to 90.6 per cent on the eggplants. For both pest species, significantly higher levels of mortality were obtained when formulated conidia were compared to the non-formulated conidia or the control treatments, whether tested under laboratory or field conditions. When the formulation blank was applied as a control treatment, mortality was not significantly different from the distilled water-treated controls. The fungus shows great potential for management of both pest and activity appears to be enhanced when applied in a coconut/soybean oil formulation. Batta (2003), reported the, symptomology of tobacco whitefly and red spider mite infection with the entomopathogenic fungus *Metarhizium anisopliae* (Metsch.) Sorokin. The present research aimed at describing the microscopic and microscopic symptoms of infection with in the entomopathogenic fungus *Metarhizium anisopliae* on tobacco whitefly (*Bamisia tabaci*) and red spider mite (*Tetranychus cinnabarinus*). Conidial suspensions of the fungal strain Meta 1 were applied against larvae and pupae of *B. tabaci*, and adults and nymphs of *T. cinnabarinus* infesting aubergines (*cv.* Red Thin) under laboratory and simulated field conditions. Infected larvae and pupae of *B. tabaci* turned in black-greenish color due to the fungus attack 3 days after treatment and incubation at 20±1°C and 100 per cent RH using two types of incubation techniques. Infected adults and nymphs of *T. cinnabarinus* also turned a dark-brown color 3-4 days after treatment and incubation under similar conditions to *B. tabaci*. The same colour change in both pests due to the fungus infection under simulated field condition (19-30°C and > 90 per cent RH) was also obtained. Dark-brown fungal out growth and sporulation was observed on the cadavers of treated individuals with the fungus both pests 5 days after treatment and incubation under the above mentioned conditions. Microscopic examinations of infected individuals of *B. tabaci* or *T. cinnabarinus* have indicated that typical mycelia and conidia of *M. anisopliae* were observed on their surface. Moreover, dense mycelial growth of the fungus was also observed under higher microscopic magnifications in the hemocoel of attacked individuals of *B. tabaci* or *T. cinnabarinus*. The latter observation was realize after a careful superficial sterilization of the infected individuals of both pests, then mounting and clearing them in lacto phenol before examination. Overall result refers to the possibility of using the fungus in biological control of both pests. Simova and Draganova (2003) studied the virulence of isolates of entomopathogenic fungi to *Ttetranychus urticae* Koch. The virulence of four isolates of *B. bassiana* (Isolates 311,312,339 and 340) and one isolate each of *M. anisopliae* (isolate 17), *Paecilomyces farinosus* (isolate 112) and *Verticillium lecanii* (isolate 289) to the two-spotted spider mite, *Ttetranychus urticae,* was evaluated. Experiments were carryout under laboratory conditions with a population of the host reared on bean plant in pots. Spraying conidial suspentions of each entomopathogenic fungus (2x109 conidia/ml) treated the plants with mites. Virulence of the fungal isolates was estimated by the values of the median lethal time (LT_{50}) calculated by probit analysis. It was established that the two-spotted spider mite was susceptible to the examined fungal isolates of the entomopathogenic fungi, *B. bassiana, M. anisopliae, P. farinosus* and *V. lecanii.* All strains caused mycoses to the contaminated individuals of the host. Six from the seven examined isolates of the entomopathogenic fungi were highly virulent to the two-spotted spider mite. The isolate 339 of *B. bassiana* was the most virulent in experiments with the *T.urticae.* The calculated values of the median lethal time (LT_{50}) varied with in a narrow confidence interval with 95 per cent confidence limits from 1.293 to 1.420 days; the average value was 1.355 days. Isolates 312 and 340 of *B. bassiana* and 289 of *V. lecanii* were highly virulent to *T.urticae* but they could not be distinguished according to their virulence. The isolate 112 of *P. farinosus* was with a low virulence to the two-spotted spider mite. It was noticed in bioassays that the conidial suspensions of the examined fungal isolates had a repellent action to the host. Alston and Thomson (2004) reported the effect of fungicide residues on the survival, fecundity, and predation of the mites *Tetranychus urticae* (Acari: Tetranychidae) and *Galendromus occidentalis* (Acari: Phytosiidae). Representative fungicide from three or four families used for management of powdery mildew and other diseases in tree fruits were evaluated for their effects on a common spider mite and predator miter species. A modified Munger cell technique was effective in measuring the response of phytophagous and predaceous mites to fungicide residues on detached leaves in the laboratory. Demethylation-inhibiting (DMI) (imidazole, triflumizole and triazole (myclobutanil) and strobilurin (trifloxystrobin) fungicides were not toxic to female *T. urticae* and *G. occidentalis* [*Typhlodromus occidentalis*], and no sub lethal effects were found on fecundity and predation rate after 3-5 days exposure to residues. Benomyl, a benzimidazole fungicide, increased adult mortality and reduced fecundity for both mite species; however, it did not alter the predation rate of *G. occidentalis* that survived the lethal effects of benomyl and acaricide pyridaben were unimpaired in predation. Our results for benomyl substantiate those of earlier studies and provide evidence for non-toxic effects of DMI and strobilurin fungicide on mites. We propose that DMI and strobilurin fungicides are a good fit for integrated mite management programmes due to conservation of phytoseiid predatory mites. Men XingYuan *et al.* (2004) reported that the effects of pesticide application on pests (*Aphis gossypii, Tetranychus cinnabarinus* and *Tetranychus truncatus*) and predators (*Propyleaq japonica, Erigone* spp. And *Thomisus* spp.) were investigated in transgenic *Bt* cotton and nontransgenic cotton agroecosystem in 1999, 2000 and 2001, in Henan, China. Transgenic cotton did not caused changes in populations of acarids and did not reduce numbers of predators considerably; their effects on aphids were inconsistent. Although insecticides were not applied against the main pests, cotton bollworm (*Helicoverpa armigera*), on transgenic cotton, the total number of insecticide application in 3 years was no less than the total applied on nontransgenic cotton, because additional applications were required against sucking pests on transgenic *Bt* cotton. Pesticide applications decreased the numbers of aphids, acarids and predatory spiders significantly on both transgenic and nontransgenic cottons. The results suggest that the use of *Bt* cotton should be evaluated carefully in China. Shi and Feng (2004), studied the lethal effect of *Beauveria bassiana, Metarhizium anisopliae,* and *Paecilomyces fumosoroseus* on the eggs of *Tetranychus cinnabarinus* (Acari: Tetranychidae) with a description of a mite egg bioassay system. Fungal biocontrol agents, including 10 isolates of *Beauveria bassiana, Metarhizium anisopliae,* and *Paecilomyces fumosoroseus* were bio-assayed for their lethal effects on the eggs of the carmine spider mite, *Tetranychus cinnabarinus.* The eggs were naturally laid on detached fava bean leaves on agar in petri dishes at 25°C under a photophase of 12:12 L: D. Suspensions of aerial conidia produced on rice were sprayed onto the leaves, yielding three conidial concentration (No. conidia/mm^2) from low to high for each of the fungal isolates with 3 or 4 replicates (30-65 eggs/leaf or replicate) at each concentration. During a 12 day observation period after spray, the infected eggs shrunk in shape, than turned orange-brown for *B. bassiana* and *P. fumosoroseus* or dark-grey for *M. anisopliae*, and eventually had outgrowths of the sprayed fungus when maintained under moist condition. Egg moralities caused by fungal infection varied over time but had no change from day 8 or 9 onward. Differences of moralities on day 9 were significant among the fungal species or isolates tested. Based on the LC_{50} estimates determined by the concentration-mortality relationship of all isolates from probit analysis, *B. bassiana* Bb2860 with an LC_{50} of 548 conidia/mm^2 was highly infectious to *Tetranychus cinnabarinus* eggs, followed by *P. fumosoroseus* Pfr 116 (848 conidia/mm^2) and Pfr 153 (913 conidia/mm^2). Two- *M. anisopliae* isolates with LC_{50}s of 1717 and 1985 conidia/mm^2 were inferior to the three isolates. The other isolates were poor or very poor in ovicidal activity. Our results confirmed the ovicidal activity of the three fungal species and suggested the feasibility to search for more ovicidal isolates from fungal species that may serve as bio-control agents against spider mites such as *Tetranychus cinnabarinus*. The mite egg bioassay system developed in this study will be useful towards this goal. Shivaprakash *et al.* (2004). Reported the occurrence of an entomopathogenic fungus, *Beauveria bassiana* (Bals.) Vuillemin on red spider mites, *Tetranychus urticae* Koch (Acari: Tetranychidae) in India. During November December 2003, red spider mites (*T. urticae*) were surveyed for fungal infection under natural condition. A natural incidence of a new entomopathogenic fungus (deuteromycetes group) was recorded on the red spider mites in a nonchemically grown okra plot in Bangalore, Karnataka, India. The fungus was isolated and identified as the white muscardine fungus, *B. bassiana*, which had shown >80 per cent mortality in *in vivo* method. Chandler *et al.* (2005) Laboratory and green house evaluation of entomopathogenic fungi against the two-spotted spider mite, *Tetranychus urticae* (Acari: Tetranychidae), on tomato, *Lycopersicon esculantum*. Forty isolates of entomopathogenic fungi from six genera were assessed against the two-spotted spider mite, *Tetranychus urticae,* in a single dose, direct application laboratory bioassay on tomato leaflets. Only 3 isolates caused greater mortality than the control: these were *Metarhizium anisopliae* 442.99, *Hirsutella spp.* 457.99 and *Verticillium lecanii* 450.99. These isolates were assed in a multiple dose bioassay, together with three isolates cultured from commercial bio-pesticides as a follows: *Beauveria bassiana* 432.99 (cultured from 'Naturalis-L', Troy Biosciences, Phoenix, TX, USA): *Hirsutella thompsonii* 463.99 (cultured from Mycar, Abbott Laboratories USA); and *V. lecanii* 19.79 (used in Mycotal Koppert BV, The Netherland). *Beauveria bassiana* 432.99, *Hirsutella thompsonii* 463.99, *Metarhizium anisopliae* 442.99 and *V. lecanii* 450.99 were pathogenic to *T. urticae* in this bioassay. In addition, it was found that the mortality caused by *B. bassiana* 432.99 and Naturalis-L was increased when the mites were exposed to tomato leaflets sprayed previously with conidia suspensions, compare to spring the mites directly. In a glasshouse experiments sprays of *B. bassiana* 432.99, *H. thompsoni* 463.99, *M. anisopliae* 442.99, *V. lecanii* 450.99 and Naturalis-L reduce *T. urticae* populations in a tomato crop grown according to commercial practices. Naturalis-L reduced *T. urticae* numbers by up to 97 per cent. In a second glasshouse experiment, single sprays of Naturalis-L and the acaricide fenbutatin oxide (Torq) were compared as supplementary treatments to release of the predatory mites, *Phytoseiulus persimilis*. Supplementary sprays of fenbutatin oxide reduce the numbers the of *T. urticae* nymphs (80 per cent reduction), while Naturalis-L reduce numbers of *T. urticae* adults, nymphs and eggs (98 per cent reduction in all three cases). It is concluded that Naturalis-L has the potential to be used against *T. urticae* on glasshouse tomato crops. Tkaczuk *et al.* (2005). Studied about the fungi and mite pest of plants. Mite are plant pest mainly belonging to Tetranychidae, Eriophyidae, Tenuipalpidae, however species belonging to Tarsonemidae, Pyemotidae, Tydeidae, and Acaridae may also cause damaged to plants and spread infectious diseases of plants. Mites present economically important group of plant pests. The importance of use of biological control with in integrated pest management system for control of mites, especially in intensive crop production system, such as orchards, and protected and open field vegetable production, is discussed. Information is briefly included on natural enemies of mites, such as *Phytoseiulus persimilis* controlling pest in greenhouse, *Typhlodromus pyri*, a predator of *Panonychus ulmi* found in orchards, as well as *Anthocoris nemorum* and *Stethorus punctillum*. Information is presented on investigations conducted in Poland to study the possibility of using the entomogenus fungus *Neozygites floridana* for control of *Tetranychus urticae*, use of *Hirsutella* spp. for control of *Abacarus hystrix*., and description of new species *N. abacaridis* on *A. hystrix*. Information is included on development of fungal acaricides, *i. e.* Mycar based on *H. thompsonii* strain to control *Phylocoptruta oleivora* on citrus in Florida, the USA, and *Panonychus citri* in Australia. Use is also considered of Mycohit based on *H. thompsonii* for control of *Aceria guerreronis* on coconut. Sreenivas *et al.* (2005) reported the efficacy of entomopathogenic fungi against red spider mite, *Tetranychus neocalodonicus* Andre¢ (Acari: Tetranychidae). Laboratory studied conducted on the efficacy of entomopathogenic fungi against red spider mite, *Tetranychus neocalodonicus* showed their relatively pathogenicity. Among the three entomopathogenic fungi tested (*i.e. Beauveria bassiana, Metarhizium anisopliae* and *Verticillium lecanii*), treatment with *Metarhizium anisopliae* at $1.2x10^8$ CFU/ml registered higher percent mycosis but, it was statistically at par with the remaining two fungi. All the three evaluated fungi proved their superiority over water dipping/spraying and untreated control treatments both under laboratory and greenhouse conditions. Aghajanzadeh *et al.* (2006) studied the bioefficacy of six isolates of *Hirsutella thompsonii* Fisher against citrus rust mite, *Phyllocoptruta oleivora* Ashmead (Acari: Eriophyidae) and two spotted spider mite, *Tetranychus urticae* Koch (Acari: Tetranychidae). For studying the bio-efficacy of six isolates of *Hirsutella thompsonii* Fisher a fungal pathogen of eriophyid mites collected from different agro-climatic region of India, fungal suspension was sprayed on fruits and leaves infested with healthy citrus rust mite, *Phyllocoptruta oleivora* Ashmead and two spotted spider mite, *Tetranychus urticae* Koch. The results revealed that all isolates of *H. thompsonii* were effective against citrus rust mite HtEMC isolates of *H. thompsonii* brought about maximum mortality *of* 48.05 per cent ten days after treatment followed by HtCRMB (46.87 per cent), HtCRMK (44.27 per cent), HtCRMC (42.09 per cent), HtPMG (39.82 per cent) and HtPDBC (31.22 per cent), however, this differences was not significant statistically. These isolates were also found to be effective against adult two-spotted spider mite. HtEMC isolate of *H. thompsonii* brought about maximum mortality of 83.33 per cent after eight days of treatment, followed by HtCRMB (70.85 per cent), HtCRMK (39.48 per cent), HtPMG (33.05 per cent), HtCRMC (31.86 per cent) and HtPDBC (19.64 per cent). The HtCRMB isolates of *H. thompsonii* was found to be effective against nymphal stage of the two spotted spider mite with 88.85 per cent mortality six days after treatment. These results indicate that *H. thompsonii* isolates exhibit differences in their virulences, suggesting the requirement for screening the *H. thompsonii* isolates for their effectiveness against mite species before recommending as biological control agent.

Biological Control of Phtyophagous Mites

A list of important predatory mites is given in Table 7.2. Gupta (1987) listed 139 phytoseiid mites from India with their taxonomic categories. Hoy (1981) and Hoy *et al.* (1982) reviewed the latest Phytoseiidae and their biological control by mites. Biological control provides an environmentally safe, cost effective and energy-efficient means of pest control either alone or as a component of IPM (Hoy *et al.*, 1983). Phytoseiid mites are well accepted bioagents for phytophagous mite control. However, the families of Stigmaeidae, Erythraeidae, Tydeid, Anystidae, Cheyletidae and Hemisarcoptidae also contain species that are predacious (Laing and Knop, 1983). McMurtry (1983) reported that most of the efficient predators of phytophagous mites in orchards belong to family Phytoseiidae. Rijn and Houten (1991) reported about successful biocontrol of thrips by the predatory mite, *Amblyseius cucumeris*, whereas another mite *Amblyseius barkeri* is less successful in controlling thrips in sweet peper crops. Hluchy and Pespisil (1991) recorded tetranychid mites, *Tetranychus urticae* and *Panonychus ulmi* as a serious pest in Moravian/Czechoslovakia Vineyards and predatory mite *Typhlodromus pyri* was used successfully for biological protection of grone vines. Baier (1991) successfully used *Amblysieus barkeri* and *A. cucumeris* for the control of thrips and spider mite in Germany. McMurtry (1991) emphasised the use and augmentative release of *Phytoseiulus persimilis* for mite control on high value crops. This mite is most widely used species of predator for augmentative release and effectivity for a wide range of situations. Laing and Knop (1983) described the predatory mites from families Stegmaeidae, Erythraeidae, Tydeidae, Anystidae, Chetyletidae and Hemisarcoptidae. Lo *et al.* (1990) reported more than fifty species of spider mites in Taiwan and recorded thirty species of native predators as natural enemies of spider mite. Mori *et al.* (1990) described the use of native predator *Amblyseius longispinosus* and introduced predator *Phytoseiulus persimilis* against spider mite and concluded that combined action of both native and introduced predators gave better suppression of tetranychid population under experimental conditions. McMurtry (1989) reported the use of Coccinellid, *Stethorus picipes* as a main predator regulating abundance of *Oligonychus punicae*. Singh and Ray (1977) identified *Stethorus* beetle as a predator of *Tetranychus neocalidonicus*. Rasmy (1989) recommended biocontrol using predacious mites in mite pest management programme and suggested to follow IPM approach: (i) by utilizing naturally occurring predacious mites, (ii) by releasing predatory mites and (iii) by utilizing pesticide resistant strains of predacious mites (Singh *et al.*, 2000).

Spooner Hart (1991) reported about the alarming situation of the two spotted mite, *Tetranychus urticea*, a major pest of ornamental, fruit and vegetable crops. He mentioned that withdrawal of miticides in 1987 due to high level of resistance paved way for the use of integrated mite control. This utilises predatory mites as a major tool of mite control and reduced usage of pesticides. Use of the Chilean predatory mite, *Phytoseiulus persimilis*, a monophagous predator of tetranychid has revolutionized IPM in Australia and Europe. He further mentioned that within a period of twelve years *P. persimilis* has become one of the most widely used biological control agents in Australia and the backbone of many IMC programmes in horticulture. Osmelak and MacFarlane (1991) described the importance of *P. persimilis* for IPM of two-spotted mite in roses and carnations. Parker (1991) also emphasised on the wide spread use of IMC in flower crops.

Borthakur and Das (1987) reported the occurrence of three species of predacious mites *viz.*, *Agistemus* sp., *Exothorhis caudata* and *Cunaxa* sp. of tea mites. They further reported *Agistemus* sp. as an abundant predator on tea leaves in northeast India and expected it to be a very promising biocontrol agent for tea mite complex. Kumari and Sadana (1991) and Sadana (1985) identified *Amblyseius alstoniae* as a potential biological control agent for *Brevipalpus phoenicis* on grapevine, citrus and guava. Sadana and Sharma (1988) recorded *Amblyseius finlandicus* predating on *Oligonychus mangiferus* and *Eutetranychus orientalis, which* infest grapevine and citrus. Mallik *et al.* (1989) recorded interaction between *Tetranychus ludeni* and its two phytoseiid predators, *Amblyseius longispinosus* and *Amblyseius tetranychivorus*. They further recorded that *A. longispinosus* prefers egg followed by the younger immature stages, whereas *T. tetranychivorus* prefers adult and older nymphal stages. Jose *et al.* (1989) found predatory mite, *Amblyseius alstoniae*, predatory thrips, *Scolothrips indicus* and beetle, *Stethorus pauperculus* feeding on all stages of spider mites. Singh *et al.* (1989) listed natural enemy complex of *A ceria litchii* and found *Amblyseius* as most prevalent followed by species of *Typhlodromus* and *Agistemus*.

Table 7.2: List of Effective Predatory Mites

Sl.No.	Family/Predatory Mite	Prey Mite Host
	Phytoseiidae	
1.	*Amblyseius channabasavannai* Gupta	*Raoilla indica*
2.	*Amblyseius largoensis* (Muma)	*Oligonychus mangiferus*
3.	*Amblyseius alstoniae* Gupta	*Eutetranychus orientalis*
4.	*Amblyseius delhiensis* (Narayanan and Kaur)	*Brevipalpus phoenicis*
5.	*Amblyseius finlandicus* (Oudemans)	*Brevipalpus phoenicis* *Eutetranychus orientalis*
6.	*Amblyseius pruni* (Gupta)	*Eutetranychus orientalis*
7.	*Amblyseius longispnosis* (Evans)	*Oligonychus indicus*
8.	*Amblyseius multidentatus* (Swirski and Schechter)	*Tetranychus necocaledonicus*
9.	*Amblyseius tetranychivorus* (Gupta)	*Tetranychus ludeni* *Raoilla indica*
10.	*Typhlodromus homalii* Gupta	*Aceria mangiferae* *Eriophyes cernuus*
	Cunaxidae	
11.	*Cunaxa setirostris* (Hermann)	*Eutetranychus orientalis*
	Stigmaeidae	
12.	*Agistemus industani Gonzalex*-Rodriguez	*Eutetranychus orientalis*
13.	*Agistemus fleschneri* Summer	*Eutetranychus orientalis*
	Anystidae	
14.	*Anystis* sp.	*Oligonychus coffeae*
	Tydeidae	
15.	*Pronematus fleschneri* Baker	*Brevipalpus phoenicis*
	Ascidae	
16.	*Lasioseius terrestris* (Menon and Ghai)	*Oligonychus indicus*

Singh (1994) reported that *Phytoseiulus persimilis* Athias Henroit is very important and commercialised predatory mite and widely used for control of common injurious tetranychid mites. This mite was imported in India from the Glasshouse Crops Research Institute, Littlehampton, U.K. during 1984. Since then culture of this mite is being maintained at Project Directorate of Biological Control, Bangalore on *Tetranychus urticae,* which is most ideal. The bean plants are used for multiplication of *Tetranychus urticae.* The most preferred host of this predatory mite is *Tetranychus urticae* followed by *Tetranychus cinnabarinus, Tetranychus ludeni* and *Tetranychus neocaledonicus.* Krishnamoorthy (1989) found that introduced predator, *Phytoseiulus persimilis* feeds readily on *Tetranychus cinnabarinus, T. fifiensis, T. ludeni* and *T. neocalidoniclls.* Borah and Rai (1989) recorded *Amblyseius ovalis* for the first time as a predator of *Bemisia tabaci.* Bhagat and Patnaik (1989) recorded a neuropteran, *Micromus timidlls* as a predator of spider mite. Sufficient information is available on predatory fauna of mites but unfortunately none of the predatory mites/insects are commercialised so far in India for the use as bioagents at farmers field. Singh (1988) reported the methods for mass multiplication of predatory mites for large-scale production of *Amblyseius tetranychivorus.* Arbabi and

Singh (2002) reported *Agistemus industani* Gonzalez-Rodriguez (family- Stigmaeidae) is efficient predator of *T. ludeni* during its peak from March to June.

Acaricidal Management

Fortunately the acaricides resistance has not developed so far due to their limited use in India. However, many reports are available on pesticides resistance in spider mites (Brandenburg and Kenndy, 1987; Dittrich, 1975; Helle, 1965 Jeppson *et al.*, 1975 and Van de vrie, 1973). Herne *et al.* (1979) mentioned that spider mites are potential danger for many crops because of their ability to develop resistance to chemicals, which initially give effective control. Cranham and Helle (1985) discussed the pesticides resistance in Tetranychidae. Goodwin *et al.* (1991) reported about the miticide resistance in two spotted mite which had caused serious difficulties in maintaining product quality of flowers in Australia. They further pointed out that reliance of growers on miticides to provide economic control of two-spotted mite was a failure due to development of resistance. They suggested that chemical companies must undertake testing of their products against predatory mites and other biological control agents keeping in view the resistance problem and deleterious effects on non-target organisms. Hoy *et al.* (1980) and Hoy and Knop (1981) reported increased tolerance in phytoseiid mites against pyrethroids.

During recent years, the development of plastic culture particularly for growing vegetables and flowers under plastic cover or glass house conditions possibly have accelerated the problem of resistance in spider mites. Therefore, continuous watch is required to avoid the failure of a chemical product. Many times cross-resistance may also occur. Cardwell *et al.* (1987) suggested a rapid bioassay method that can detect buildup of resistance and help growers to avoid the use of acaricides. In India tea research units may follow up the bioassay method to detect acaricides resistance against *Oligonychus coffeae*, which is one of the serious pests in tea gardens (Banerjee and Cranham, 1985). Bhaskaran and Karthikeyan (1999) repoted the acaricidal management of *Aceria guerreronis* in young trees, sprying has to be done with dimethoate 1.0 ml in 1 litre of water with 0.5 ml to teepal. In older trees, above 10 years, rood feeding with monocrotophos 10 ml in 10 litres of water may be adopted after harvesting the matured nuts. If necessary, spraying or root feeding has to be repeated after one month. Kumar and Singh (2005) A field experiment were conducted in Varanasi, Uttar Pradesh, to evaluate efficacy of some miticides against mites of both the crops. Omite 57 per cent EC, Omite 57 per cent EC + Dhanuvit (1 mi/It.), Phosalone 35 per cent EC, Ethion 50 per cent EC, Gronim 0.015 per cent (Azadirachtin), Neem Seed Kernel Extract (NSKE) and Sulphur 80 per cent WP were evaluated at their recommended concentrations. The results revealed that Omite @ 2 ml/1 alone provided significantly best control of mites of both the crops, but addition of Dhanuvit @ 1 ml/1 (a surfactant) enhanced the efficacy of Omite, culminating in the mortality of mites 94.88, 98.77, 90.99 and 71.20 per cent on Okra and 86.77, 95.44, 88.66 and 67.13 per cent on Brinjal, after 1,3,7 and 14 days, respectively. Ethion and Phosalone were found only moderately effective and NSKE, Gronim and Sulphur have shown very poor control. Omite alone and Omite + Dhanuvit also enhanced the vigour and growth of plants. Kumar *et al.* (2005) A field experiment was conducted to test the efficacy of different acaricides against the Broad mite, *Polyphagotarsonemus latus* (Banks) during Rabi season of 2003 in Varanasi. The best results were given by Omite (0.228 per cent) combined with Dhanuvit (0.1 per cent), a surfactant, in which mortality in mites was 95.88, 98.99, 94.55 and 74.22 per cent after 1,3, 1 and 14 days respectively. This was followed by Omite alone which gave 90.66, 94.88, 68.77 and 50.55 per cent mite mortality after 1, 3, 7 and 14 days, respectively. Dhanuvit being a surfactant proved to be an excellent booster for increasing the efficacy of Omite. The other moderate treatments for their efficacy in a descending order were Ethion (0.05 per cent), Phosaione (0.07 per cent) and Sulphur (0.25 per cent).

References

Agarwal, L. and Singh, J., 1988. Present status of *Eriophyes mangiferae* in eastern Uttar Pradesh, India. In: *Progress in acarology*, Vol. 2. (eds. G.P. ChannaBasavanna and C.A. Viraktamath) Oxford and IBH, New Delhi, 109-114.

Agarwal, R.A., 1957. The surgarcane mite (*Schizotetranychus andropogoni* H) and its control. *Indian Sugar* 7: 394-399.

Aghajanzadeh, S.; Mallik, B. and Chandrashekar, S. C. 2006. Bioefficacy of six isolates of *Hirsutella thompsonii* Fisher against citrus rust mite, *Phyllocoptruta oleivora* Ashmead (Acari: Eriophyidae) and two-spotted spider mite, *Tetranychus urticae* Koch (Acari: Tetranychidae). *Pakistan Journal of Biological Sciences*, Faisalabad, Pakistan, 9 (5): 871-875.

Ahuja, D.B., 1994. Seasonal incidence and chemical control of oriental mite, *Eutetranychus orientalis* (Klein) on castor. *Indian J. Ent.* 56: 1-5.

Alston, D. G.and Thomson, S. V. 2004. Effects of some fungicide residue on the survival, fecundity and predation of the mites *Tetranychus urticae* (Acari: Tetranychidae) and *Galendromus occidentalis* (Acari: Phytoseiidae). *J. Econ. Ent.*, 97(3): 950-956.

Alston, Diane G. and Reding, Michael E. 2005. European Red Mites, *Panonychus ulmi*. Technical booklet, Utah State University Extention- Orchard IPM Series HG, USDA. 1-15.

Anh, N. H., 1996. Chilli pepper evaluation trial. *ARC report*. 1-4.

Anonymous, 1991. Mites of agricultural importance in India and their management. *Technology Bull. No.1, AICRP on Agricultural Acarology, Bangalore.*

Anonymous, 1993. A guide to the agriculturally important mites of India with illustrated keys and field keys for their easy identifications. *Technology Bull. No.2, AICRP on Agricultural Acarology, Bangalore.*

Arbabi, M. and Singh, J. 2002. Studies on *Agistemus industani* Gonzalez-Rodriguez (Acarina: Stigmaeidae), An efficient predator of *Tetranychus ludeni* Zacher on mulberry. *Acarina* 10(1): 85-89.

AVRDC, 1993. Breeding of Solanaceous and Cole crops, AVRDC, Shanhua, Tainan, Taiwan. 85-100.

Baier, B., 1991. Relative humidity-A decisive factor for the use of oligophagous predatory mites in pest control. In: *Modern acarology*, Vol. 2 (eds. F. Dusbabek and V. Bubva) Academia, Prague. 661-665.

Baker, E.W. and Tuttle, D.M., 1987. The false spider mites of Mexico (Tenuipalpidae: Acari). *U. S. Department of Agriculture, Technical Bulletin No. 1706.*

Banerjee, B., 1988. *An introduction to agricultural acarology*. Associated Publishing Co., New Delhi.

Banerjee, B. and Cranham, J.E., 1985. Control of Tetranychidae in crops-tea. In: *World crop pests, spider mites, their biology, natural enemies and control* (Eds. W. Helle and M.W. Sabelis) Elsevier, Amsterdam, 371-374.

Batta, Y. A. 2003. Production and testing of noval formulations of the entomopathogenic fungus *Beauveria bassiana* (Metschinkoff) Sorokin (Deuteromycotina: Hyphomycetes). *Crop Protection*, Oxford, UK; 22 (2): 415-422.

Benuzzi, M. and Santopolo, F. 2001. Naturalis: a bioinsecticide based on *Beauveria bassiana*. *Informatore Fitopatologico*, Italy. 51 (4): 61-64.

Bhagat, K.C. and Patnaik, M.C., 1989. Biological studies and host range of *Micromus finidus* (Neuroptera: Hemerobiidae). A predator of the common spider mite, *Tetranychus telarius* (Acari: Tetranychidae). In: *Progress in Acarology*, Vol. 2 (Eds. G.P. ChannaBasavanna and C.A. Viraktamath) Oxford and IBH, New Delhi, 381-384.

Bhaskaran, R. and Karthikeyan A., 1999. Mite infestation on coconut. *The Hindu*. February 11. 24.

Bindra, O.S. and Singh, G., 1970. Mite pests of ornamental plants in India. Pesticides 4: 17-24.

Bindra, O.S. and Verma, G.C., 1966. A study in the control of the fig mite, *Eotetranychus hirsti* Pritchard and Baker. *Jour. Res., Punjab Agric. Univ.* 3: 417-420.

Boczek, J.H., Shevtchenko, V.G. and Davis, R. 1989. *Genetic key to world fauna of eriophyid mites* (Acarina: Eriophyoidea). Warsaw Agricultural University Press, Warsaw, Poland.

Borah, D.C. and Rai, P.S., 1989. Potentiality of *Amblyseius ovalis* (Acari: Phytoseiidae) as a biological control agents on *Bemisia tabaci* (Homoptera: Aleyrodidae). In: *Progress in Acarology*, Vol. 2 (eds. G.P. ChannaBasavanna and C.A. Viraktamath) Oxford and IBH, New Delhi, 375-379.

Borthakur, M. and Das, S.C., 1987. Studies on acarine predators of phytophagous mites on tea in north-east India. *Two and A Bud* 34: 21-24.

Brandenburg, R.L. and Kennedy, G.G., 1987. Ecological and agricultural considerations in the management of two spotted spider mite (*Tetranychus urticae* Koch). In: *Agricultural zoology reviews*, Vol. 2 (ed. G.E. Russell) Intercept, Wimbome, 185-236.

Capoor, S.P., 1952. Observations on the sterility disease of pigeonpea pea. *Indian J. Agric. Sci.* 22: 271-274.

Cardwell, E.E. Grafton, Granett, J. and Dennehy, T. J., 1987. Quick tests for pesticide resistance in spider mites. *California Agriculture* July-August: 8-10.

ChannaBasavanna, G.P., 1966. *A contribution of the knowledge 'of Indian eriophyid mite (Eriophyoidea: Trombidiformes, Acarina)*. Univ. Agric. Sci. Bull. Bangalore, India.

ChannaBasavanna, G.P., 1971. The present status of our knowledge of Indian plant feeding mites. In: *Proceedings-3rd international congress of acarology*, Prague, 201-204.

Chandler, D.; Davidson, G. and Jacobson, R. J. 2005. Laboratory and greenhouse evaluation of entomopathogenic fungi against the two-spotted spider mite, *Tetranychus urticae* (Acari: Tetranychidae), on tomato, *Lycopersicon esculentum*. *Biocontrol Science and Technology*, Abingdon, UK; 15 (1): 37-54.

Chen, C. N.; Cheng, C.C. and Heiao, K. C., 1979. Bionomics of *Steneotarsonemus spinki* Smiley attacking rice plants in Taiwan. *Recent Adv. Acarol.*, 1: 111-117.

Coates, T.J.D., 1974. The influence of some natural enemies and pesticides on various populations of *Tetranychus cinnabarinus* (Boisduval.), *T. lombardinii* Baker and Pritchard and *T. ludeni* Zacher (Acari: Tetranychidae) with aspects of their biologies. *Entomology Mem. Dep. Agric. Tech. Servo Repub. S. Afr.* 42: 1-40.

Cranham, J.E. and Helle, W., 1985. Pesticide resistance in Tetranychidae. In: *World crop pests, spider mites, their biology, natural enemies and control* (eds. W. Helle and M.W. Sabelis) Elsevier, Amsterdam, 405-432.

Das, B. and Chakrabarti, S., 1988. Eriophyid mites (Acari: Eriophyoidea) of NorthEast India-some aspects of their evolution and host associations. In: *Progress in acarology,* Vol. 1 (Eds. G.P. ChannaBasavanna and C.A. Viraktamath) Oxford and IBH, New Delhi, 391-394.

Das, G.M., 1965. Pests of tea in northeast India and their control. *Tocklai Expt. Stn. Mem. No.* 27.

Das, G.M. and Sengupta, M., 1958. Observations on the pink mite *Acaphylla theae* (Watt) Keifer, of tea in north-east India. *J. Zool. Soc. India* 10: 39-48.

Davis, R., Flechtmann, C.H.W., Boczek, J.H. and Barke, H.E., 1982. *Catalogue of Eriophyid mites* (Acari: Eriophyoidea) Warsaw Agricultural University Press, Warsaw, Poland.

Desai, M.K. and Chavda, D.E., 1955. "Mite" (*Oligonychus*), as a causal agent for "*Ratada*" disease of sorghum. *Poona Agri. Coll. Mag.* 45: 138-141.

Dhooria, M.S., 1984. Development of citrus mite, *Euteranychus orientalis* (Acari: Tetranychidae) as influenced by age and surface of leaves of different hosts. *Indian J. Acarol.* 9: 82-88.

Dhooria, M.S., 1994. Effect of releasing different levels of citrus mite, *Eutetranychus orientalis* (Klein) (Acarina: Tetranychidae) on growth of french bean. *J. Insect Sci.* 7: 210.

Dhooria, M.S. and Bhutani, D.K., 1983. Seasonal incidence of citrus mite, *Eutetranychus orientalis* and its predators. *Indian J. Acar.* 7: 59-62.

Dhooria, M.S. and Sagar, P., 1975. An outbreak of *Tetranychus cinnabarinus* (Acarina: Tetranychidae) on squash-melon and its control. *J. Acat:* 1: 6-9.

Dittrich, V., 1975. Acaricide resistance in mites. Z. *Angew. Entomol.* 78: 28-45.

Dosse, G. and Boudreaux, H.B., 1963. Some problems of spider mite taxonomy involving genetics and morphology. In: *Advances in Acarology,* Vol. I, Cornell University Press, 343-349.

Dupont, L.H., 1979. On gene tlow between *Tetranychus urticae* Koch 1836 and *Tetranychus cinnabarinus* (Boisdmal) Boudreaux 1956 (Acari: Tetranychidae): synonym between the two species. *Entomol. Exp. Appl.* 25: 297-303.

Gangwar, S.K. and Lal, L., 1988. Phytophagous mites in the north-eastern hill region of India. *Trop. Pest Mang.* 34: 438-440.

Ganesan, S., 1999. Management of coconut mite. *The Hindu.* November 11. 7.

Ghai, S., 1964. Mites. In: *Entomology in India*: The Entomological Society of India, New Delhi, 385-396.

Ghai, S. and Baker, E.W., 1989. Notes on the gall forming mite, *Larvacarus transitans* (Tenuipalpidae) on *Ziziphus mauritiana* in India. In: *Progress in Acarology,* Vol. 2 (Eds. G.P. ChannaBasavanna and C.A. Viraktamath) Oxford and IBH, New Delhi, 81-89.

Ghai, S. and Shenhmar, M., 1984. A review of the world fauna of *Tenuipalpidae* (Acarina: Tetranychidae). *Oriental insects* 8: 99-172.

Ghosh, N.K., Das, B. and Chakrabarti, S., 1989. Injury by gall mites (Acari: Eriophoidea) to plants in north east India. In: *Progress in Acarology,* Vol. 2 (Eds. G.P. ChannaBasavanna and C.A. Viraktamath) Oxford and IBH, New Delhi, 135-140.

Goodwin, S., Gough, N. and Herron, G., 1991. Miticide resistance and field control in two-spotted mite, *Tetranychus urticae* Koch, infesting roses. In: Proceedings 1st National Conference, Australian Society of Horticultural Science, Macquarie University, Sydney, 325-328.

Gupta, S.K., 1985. *Hand book-plant mite of India.* Zoological Survey of India, Calcutta.

Gupta, S.K., 1987. *A taxonomic review of oriental Phytoseiidae with key to genera and species.* Records of the Z.S.I. Miscellaneous Pub. Occasional paper No. 95. Zoological Survey of India, Calcutta.

Gupta, S.K., 1991. The mites of agricultural importance in India with remark on their economic status. In: *Modern Acarology* Vol. 1 (Eds. F. Dusbabek and V. Bukva) Academia, Prague, 509-522.

Gupta, S.K. and Gupta, Y. N., 1994. A taxonomic review of Indian Tetranychidae (Acari: Prostigmata) with description of known species and keys to genera and species. *Memoirs of the Zoological Survey India* 18: 196.

Gupta, S.K., Dhooria, M.S. and Sidhu, A.S., 1976. Seasonal abundance of *Tetranychus telarius* (Linn.) on castor in the Punjab. *Oilseed Jour.* 6: 16-18.

Gupta, S.K. and Nahar, S.C., 1981. Plant mites (Acari) of agricultural importance in Bihar. In: *Contribution to acarology in India* (ed. G.P. ChannaBasavanna) Acarological Society of India, Bangalore, 6-11.

Gupta, S.K., Sidhu, AS. and Singh, G., 1971. Occurrence of a Tenuipalpidae mite on citrus in the Punjab and its control. *Indian J. Ent.* 33: 30-33.

Helle, W., 1965. Resistance in the acarina: mites. In: *Advances in acarology,* Vol. 2 (ed. J.A. Naegle) Comstock Publishers, Ithaca, New York, 71-93.

Helle, W. and Sabelis, M.W. (Eds.) 1985. *World crop pests, spider mites, their biology, natural enemies and control.* Elsevier, Amsterdam.

Herne, D.H.C., Cranham, J.E. and Easterbrook, M.A., 1979. New acaricides to control resistant mites. In: *Recent Advances in Acarology,* Vol. 1 (ed. J.G. Rodriguez) Academic Press, New York, USA.

Hill, R.L. and Donnell, D.J.O., 1991. Reproductive isolation between *Tetranychus lintearius* and two related mites, *T. urticae* and *T. turkestani* (Acarina: Tetranychidae). *Exp. Appl. Acar.* 11: 241-251.

Hluchy, M. and Pospisil, Z., 1991. Use of the predatory mite *Typhlodromus pyri* Schenter (Acari: Phytoseiidae) for biological protection of grapevine from phytophagous mites. In: *Modem Acarology,* Vol. 2 (Eds. F. Dusbabek and V. Bukva) Academia, Prague, 655-660.

Hoy, M.A. (Ed.) 1981. *Recent advances in knowledge of the Phytoseiidae.* Divi. of Agril. Sci., Uni. of Calif.

Hoy, M.A., Knop, N.F. and Joos, J.L., 1980. Pyrethroid resistance persists in spider mite predator. *Calif. Agri.* 34: 11-12.

Hoy, M.A. and Knop, N.F., 1981. Selection for and genetic analysis of permethrin resistance in *Metaseiulus occidentalis:* genetic improvement of a biological control agent. *Entomol. Exp. Appl.* 30: 10-18.

Hoy, M.A., Castro, D. and Cahn, D., 1982. Two methods for large scale production of pesticideresistance strains of the spider mite *Metaseiulus occidentalis* (Nesbitt.) (Acarine: Phytoseiidae). Z. *Angew. Entomol.* 94: 1-9.

Hoyt, S.C., Westigard, P.H. and Burts, E.C., 1978. Effects of two synthetic pyrethroid insecticides on the codling moth, pear psylla and various mite species in NorthWest apple and pear orchards. *J. Eco. Entomol.* 71: 431-434.

Hoy, M.A., Westigard, P.H. and Hoyt, S.C., 1983. Release and evaluation of a laboratoryselected, pyrethroid-resistant strain of the predacious) 1lite *Metaseiulus occidentalis* (Acari: Phytoseiidae) in southern Oregon pear orchards and a Washington apple orchard. *J. Econ. Entomol.* 76: 383-388.

Hull, L. A. and Beers, E. H. 1990. Validation of injury thresholds for European red mite (Acari: Tetranychidae) on 'Yorking' and 'Delicious' apples. *J. Econ. Entomol.* 83: 2026-2031.

Jose, V.T., Shah, A.H. and Patel, C.B., 1989. Feeding potentiality of some important predators of the spider mite, *Tetranychus macferlanei*, a pest of cotton. In: *Progress in Acarology*, Vol. 2 (Eds. G.P. ChannaBasavanna and C.A Viraktamath) Oxford and IBH, New Delhi, 357-360.

Jain, P.C. and Yadav, C.P.S., 1988. New record of brown wheat mite, *Petrobia latens* (Muller) on coriander. *Indian J. Ent.* 50: 396.

Jankevica, L. and Cudare, Z. 2003. Investigation on entomopathogenic fungi in Latvia. *Proceeding of the 8th European meeting of the IOBC/WPRS Working group Insect pathogens and insect parasitic nematodes, Athens, Greec. Bulletin. OILB/SROP*; 26 (1) 69-72.

Jeppson, L.R., Keifer, H.H. and Baker, E.W., 1975. *Mites injurious to economic plants*. Univ Calif. Press, California, Berkeley.

Keifer, H.H., 1940. Eriophyid studies. VIII. *Bull. Calif. Dept. Agric.* 29: 21-46.

Keifer, H.H., Baker, E.W, Kono, T., Delfinado, M. and Styer, W.E., 1982. *An illustrated guide to plant abnormalities caused by eriuphyid mites in North America.* U.S.D.A Hand book No. 573.

Khan, M.Q. and Murthy, D.V., 1956. *Dicanthium annulatum* Stapf. An important alternate host plant of jowar and sugarcane mites. *Indian J. Ent.* 18: 190-199.

Khan, R.M., Doval, S.L. and Joshi, H.C., 1969. Biology of brown wheat mite *Petrobia latens* (Muller). *Indian J. Ent.* 31: 258-264.

Krishnamoorthy, A., 1989. Development of *Phytoseiulus persimilis* (Acari: Phytoseiidae) on the carmine spider mite, *Tetranychus cinnabarinus* (Acari: Tetranychidae) at two temperature regimes. In: *Progress in acarology*, Vol. 2 (eds.G.P. ChannaBasavanna and C.A Viraktamath) Oxford and IBH, New Delhi, 369-374.

Kulkarni, G.S., 1922. The "murda" disease of chilli (*Capsicum*). *Trop. Agr. Peredenya* 58: 237.

Kumari, Meena and Sadana, G.L., 1991. Influence of temperature and relative humidity on the development of *Amblyseius alstoniae* (Acari: Phytoseiidae). *Experimental and Applied Acarology* 11: 199-203.

Kumar, S.; C. Mani and R. N. Singh. 2005. Field evaluation of omite and some other acaricides against broad mite, *Polyphagotasonemus latus* (Banks) on chilli under Varanasi conditions of eastern Uttar Pradesh. *Pestology*. XXIX (11): 61-63.

Kumar, S. and R. N. Singh. 2005. Suppression *Tetranychus macfarlanei* (Baker and Prichard) and *Amblyseius indicus* (Narayanan and Kaur) with certain acaricides. *Resistant Pest Management*, Michigan, USA. 15 (1): 12–15.

Landa, Z.and Bohata, A. 1999. Compatibility of entomogenous fungus *Paecilomyces fumosoroseus* with natural insecticides based on azadirachtin and neem oil. *Collection of Scientific Papers, Faculty of Agriculture in Ceske Budejovice. Series for Crop Sciences* 16 (2): 99-106.

Lang, J.E. and Knop. N.E 1983. Potential uses of predaceous mite other than Phytoseiidae for biological control of orchard pests. In: *Biological control of pests by mites* (Eds. M.A. Hoy, G.L. Cunningham and L. Knutson) University of California, Berkely, 28-35.

Lal, L. and Mukherjee, S.P. 1979. Observations of the injury symptoms caused by the phytophagous mites. *Zoologische Beitrage* 25: 13-17.

Lal, S.S., 1982. Influence of weather factors on the population of spider mites (Acari: Tetranychidae) on cassava in Kerala. *Indian J. Acar.* 7: 5-10.

Latif, A and Wali, M., 1961. Distribution, bionomics and description of *Larvacarus transitans* (Ewing) (Acarina: Phytoptipalpidae). *Pakistan J. Sci. Res:* 13: 77-87.

Latif. A. and Wali, M., 1956. Bionomics of "*Ber*" mite, *Larvacarus transitans* (Ewing). *In: Pro. 8th Pak Sci Conf., Lahore,* Pakistan.

Lo, Kang-Chen, Lee, Wen-Tai., Wu, Tze-Kann and Ho, Chyi-Chen., 1990. Use of predators to control spider mites (Acarina: Tetranychidae) in the Republic of China on Taiwan. In: *The use of natural enemies to control agricultural pests,* FFTC Book Sr. No. 40, 166-178.

Mali, AR., Gandhali, D.N. and Patil, AS., 1983. Heavy incidence of two spotted spider mite (*Tetranychus urticae* (Acarina: Tetranychidae) on grapevine and roses. *Acar. Newsl.* 12: 6-7.

Mallik, B., Krishnaswamy, H.S. and ChannaBasavanna, G.P., 1989. Mathematical models for the interaction between *Tetranychus ludeni* and its Phytoseiidae predators. In: *Progress in acarology,* Vol. 2 (eds. G.P. ChannaBasavanna and C.A Viraktamath) Oxford and IBH, New Delhi, 343-355.

Mann, H.H., Nagpurkar, B.D. and Kulkarni, G.S., 1920. The 'Tambera' disease of potato. *Agric. J. India* 15: 282.

Mansour, E A. and Ascher, K.R.S., 1983. Effects of neem (*Azadirachta indica*) seed kernel extracts from different solvents on the carmine spider mite, *Tetranychus cinnabarinus.* In: Proc. 2nd int. neem conf. Rauischholzhausen, 461-470.

Mansour, E, Ascher, K.P.S. and Omari, N., 1987. Effect of neem (*Azadirachta indica*) seed kernel extracts from different solvents on the predacious mite *Phytoseiulus persimilis* and the phytophagous mite *Tetranychus cinnabarinus. Phytoparasitica* 15: 125-130.

Mansour, F; Azalzeh, H; Saad, B; Tadmor, Y; Abo-Moch, F and Said, O., 2004. The potential of Middle Eastern flora as a source of new safe bio-acaricides to control Tetranychus cinnabarinus, the carmine spider mite. *Phytoparasitica.* 32 (1): 66-72.

Manjunatha, M., 1999. Mites reduces jasmine yield. *The Hindu* (March 18). 24.

Marini, R. P., Pfeiffer, D. G. and Sowers, D. S. 1994. Influence of European red mite (Acari: Tetranychidae) and crop density on fruit size and quality and on crop value of 'Delicious' apples. *J. Econ. Entomol.* 87: 1302-1311.

McMurtry, J.A., 1983. The use of Phytoseiid for biological control progress and future prospects. 1.n: *Recent advances in knowledge of the Phytoseiidae* (ed. M.A Hoy) Univ. of California Publication, USA, 23-48.

McMurtry, J.A., 1989. Utilizing natural enemies to control pest mites on citrus and avocado in California, U.S.A. In: *Progress in Acarology,* Vol. 2 (Eds. G.P. ChannaBasavanna and C.A. Viraktamath) Oxford and IBH, New Delhi, 325-336.

McMurtry, J.A., 1991. Augmentative releases to control mites in agriculture. In: *Modem acarology,* Vol. 1 (eds. F. Dusbabek and V. Bukva) Academia, Prague, 151-157.

Men XingYuan; Ge Feng; Edwards, C. A. and Yardim, E. N. 2004. Influence of pesticide applications on pest and predatory arthropods associated with transgenic *Bt* cotton and nontransgenic cotton plants. *Phytoparasitica*, 32 (3): 246-254.

Meyer, M.K.P. Smith., 1979. The Tenuipalpidae (Acari) of Africa, with keys to the world fauna. *Entomology Memp. Dep. Agric. tech. Servo Repub. Afr.* 50, 135.

Meyer, M.K.P., 1981. Mite pests of crops in southern Africa. *Sci. Bull. Dept. Agric. Fish Repub. S. Afr. No.* 397, 92.

Mitra, M., 1931. Report of the imperial mycologist. *Sci. Rep. Agr. Res. Inst., Pusa, 1929-30*, pp.58-71.

Mohanasundaram, M., 1981. New gall mites of the subfamily Nothopodinae (Acarina: Eiophyidae) from India. *Oriental Insects* 15: 145-166.

Mori, Hans, Saito, Yutaka and Nakao, Hiroshi. 1990. Use of predatory mites to control spider mites (Acarina: Tetranychidae) in Japan. In: *The use of natural enemies to control agricultural pests*, FFFC Book, Sr. No. 40, 142-156.

Mukherjee, I.N., Singh R.K. and Singh, J., 1992. Biology and chemical control of carmine spider mite, *Tetranychus cinnabarinus* (Boisd.) (Acarina: Tetranychidae) on greengram (*Vigna radiata*) in Varanasi. *Pestology* 16: 18-24.

Mukherjee, I.N. and Singh, J., 1993. Records of phytophagous and predatory mites associated with fruit plants in Uttar Pradesh. *J. Insect Sci.* 6: 134-136.

Mukherjee, I.N., Singh, R.K. and Singh, J., 1994. Incidence and control of jujube gall mite (*Eriophyis cernuus*) at Varanasi. *Indian J. Agril. Sci.* 64: 343-345.

Nangia, Neelu and Muniyappa, V., 1995. Screening for resistance to pigeonpea sterility mosaic and incidence of *Aceria cajani* (Acari: Eriophyidae). *J. Acarol.* 13: 75-79.

Nassar, O.A. and Ghai, S., 1981. Taxonomic studies on tetranychid mites infesting vegetable and fruit crops in Delhi and surrounding areas. *Oriental Ins.* 15: 333-396.

Nene, Y.L., 1972. A survey of the viral diseases of pulse crops in Uttar Pradesh. *G.B. Pant Univ. Agric and Tech., Pantnagar Exp. Stn. Bull. No.4.*

Osmelak, J.A. and MacFarlane, J., 1991. Integrated pest management of two spotted mite in roses and carnations. In: Proceedings 1st National conference, Australian Society of Horticultural Science, Macquarie University, Sydney, 337-344.

Omar, B. A. and El-Khateeb, H. M. 2002. Efficacy of some biopesticides against *Tetranychus urticae* infesting cowpea plants and their side-effects on certain predators. *Egypt Journal of Agricultural research*, 80 (3): 1157-1171.

Pande, Y.D. and Sharma, A.K., 1979. Evaluation of some pesticides for the control of *Tetranychus neocaledonicus*. In: *Contribution to Acarology in India* (ed. G.P. ChannaBasavanna) Acarological Society of India, Bangalore, 207-208.

Parker, R., 1991. The widespread use of integrated mite control in flower crops in S.E. Queensland. In: Proceedings 1st National Conference, Australian Society of Horticultural Science, Macquarie University, Sydney, 345-350.

Patel, C.B., Rai, A.B., Patel, M.B., Patel, A.J. and Shah, AA., 1993. Acaricidal tests of botanical pesticides in comparison to conventional acaricides/pesticides against red spider mites (Acari: Tetranychidae) on okra, brinjal and Indian bean. *Indian J. Ent.* 55 : 184-190.

Peairs, F. B., 1994. Mites in Wheat. Colorado State University Cooperative Extention 9/99. *Extension bullatine* No. 5.578. 1-4.

Penman, D.R. and Chapman, R.B., 1988. Pesticide-induced mite outbreaks: pyrethroids and spider mites. *Experimental mid Applied Acarology*. 4: 265-276.

Pfeiffer D. G.; Hull L.A.; Biddinger D. J. and Killian, J.C. 2005. European Red Mite, *Panonychus ulmi* (Koch) on apple indirect pests. Mid-Atlantic Orchard Monitoring Guide, NRAES, 152 Riley-Robb Hall, Ithaca, New York, 14853-5701.

Pillai, K.S. and Palaniswamy, M.S., 1985. Spider mites of cassava. *Tech. Bull.* Sr.1. Central Tuber Crops Research Institute, Trivandrum.

Pritchard, A.E. and Baker, E.W., 1958. The false spider mites (Acarina: Tenuipalpidae). *Univ. Calif. Publ. Ent.* 14: 175–274.

Puttarudriah, M., 1947. Field control of the leaf mite of 101a (*Andropogon sorghum*). *Mysore Agr. J. 26:* 17-19.

Puttarudriah, M. and ChannaBasavanna, G.P., 1959. A preliminary account of phytophagous mite of Mysore. *Proc. 1st All Indian Congr. Zool.*, Part 1. 2: 530-539.

Rajagopalan, K., 1974. First record of spider mite, *Tetranychus ludeni* Zacher transmitting *Dolichus* enation mosaic virus. *Curr. Sci.* 43: 488-489.

Raja, J., Rajendran, B., Rajsekaran, S. and Jhon Ambrose John, H., 1995. Efficacy of certain chemicals in the control of sheath mite of sugarcane. *Pestology* 19: 19-21.

Rathi, Y.P.S. 1983. *Studies on sterility mosaic disease on pigeon pea* (*Cajanus cajan* (L.) Millsp.). Technical report, G.B. Pant University of Agriculture and Tech., Pantnagar, India.

Ramakrishnan, K and Kandaswamy. 1972. *Investigations on viral diseases of pulse crops.* Final Technical Report, TNAU, Coimbatore, India.

Rao, Y. S. and Das, P. K. 1977., A new mite pest of rice in India. *Intl. Rice. Res. Newsl.*, 2:8.

Rao, J. and Prakash, A., 1992. Infestation of tarsonemid mite *Steneotarsonemus spinki* Smiley in Rice in Orissa. *J. Appl. Zool. Res.* 3: 103.

Rao, P.R.M.; Bhavani, B.; Rao, T. R. M. and Reddy, P. R., 2000. Spikelet sterlity/grain discolouration in rice in Andhra Pradesh, India. *Intl. Rice. Res. Newsl.* 25:40.

Rasmy, A.H., 1989. Prospects of biological control in integrated management of mite pests. In: *Progress in Acarology,* Vol. 2 (Eds. G.P. ChannaBasavanna and C.A. Viraktamath) Oxford and IBH, New Delhi, 337-341.

Ray, R. and Rai, L., 1981. Biology and control of *Tetranychus neocaledonicus* (Acari: Tetranychidae) on lady's finger at Varanasi. In: *Contribution to Acarology in India* (ed. G.P. ChannaBasavanna) Acarological Society of India, Bangalore, India, 41-56.

Reddy, M.V. and Raju, T.N., 1993. Some clues to increased incidence and seasonal variation of pigeonpea sterility mosaic in peninsular India. *International Pigeon pea Newsletter* 18: 22-24.

Reddy, M.V., Sheila, V.K and Nene, Y.L., 1993. *Cajanus scarabaeoids,* an alternative host of pigeonpea sterility mosaics pathogen and its vector *Aceria cajani. International Pigeon pea Newsletter* 18: 24-27.

Rijn, P.C.J. van and Houten, Y. M. Van., 1991. Life history of *Amblyseius cucumeris* and *A. barkaeri* (Acarina: Phytoseiidae) on a diet of pollen. In: *Modern Acarology,* Vol. 1 (Eds. F. Dusbabek and Y. Bukva) Academia, Prague, 645-670.

Sadana, G.L., 1985. *Plant feeding mites of India*. Kalyani Publishers, Daryaganj, New Delhi, India.

Sadana, G.L. and Sharma, N.K., 1988. Biology of predatory mite, *Amblyseius finlandicus* (Oud.) (Phytoseiidae: Acari). *J. Assam Sci. Soc.* 30: 47-60.

Sadana, G.L. and Sidhu, Rajinder. 1990. New species and new host record of Tenuipalpidae mites from Punjab (Acari: Tenuipalpidae). *Acarologia* 31: 357-360.

Sadana, G.L. and Kumari, Meena. 1991. Influence of temperature and relative humidity on the development of *Brevipalpus phoenicis* (Geijskes) (Acari: Tenuipalpidae). In: *Contribution to acarological researches in India* (Eds. A.B. Mukherjee, A.K Somchoudhary and P.K Sarkar) Kalyani, W. Bengal, 61-72.

Sandhu, G.S., Singh, B. and Dhooria, M.S., 1975. Effect of rain in population of *Oligonychus indicus* (Hirst) (Acarina: Tetranychidae) on different verities of maize (*Zea mays* L.) in Punjab, India. *Internat. J. Acarol.* 1: 10-13.

Sandhu, M.S., Gatoria, G.S., Sandhu, S.S. and Singh, S., 1982. Chemical control of tetranychid mite *Eutetranychus banksi* infesting cotton in Punjab, India. *J. Res. Punjab Agric. Univ.* 19: 127-129.

Sandhu, S., Chander, P., Singh, J. and Sidhu, AS., 1987. Effect of insecticidal sprays on the plant and secondary pest inductions in *Hirsutam* cotton in Punjab. *Agri. Ecosystems and Environment* 19: 169-176.

Sarkar, P.K and Somchoudhary, A. K., 1989. Influence of major abiotic factors on the seasonal incidence of *Raoiella indica* and *Tetranychus tijiensis* on coconut. In: *Progress in Acarology*, Vol. 2 (Eds. G.P. ChannaBasavanna and C.A Viraktamath) Oxford and IBH, New Delhi, 67-72.

Seth, M.L., 1962. Transmission of pigeon pea sterility by an eriophyid mite. *Indian Phytopath.* 15: 225–227.

Sharma, A and Kushwaha, K.S., 1989. Susceptibility of different varieties of *Ziziphus mauritinna* to *Larvacarus transitans* (Acari: Tenuipalpidae). In: *Progress in Acarology*, Vol. 2 (Eds. G.P. ChannaBasavanna and C.A Viraktamath) Oxford and IBH, New Delhi, 91-93.

Sharma, D.D., 1991. Occurrence of *Cephaleuros virescens*, a new record of leaf curls in litchi (*Litchi chinensis*). *Indian J. Agril. Sci.* 61: 446–448.

Sharma, H.S. and Pande, Y.D., 1981. Seasonal incidence of *Tetranychus* sp. on four improved varieties of brinjal. In: *Contribution to acarology in India* (ed. G.P. ChannaBasavanna) Acarological Society of India, Bangalore.

Sheila, Y.K, Manohar, S.K. and Nene, YL. 1988. Biology and morphology of *Aceria cajani. International Pigeonpea Newsletter* 7: 28.

Shivaprakash, T. H. M.; Ghosh, S. H. and Khan, H. K. 2004. Occurrence of an entomopathogenic fungus, *Beauveria bassiana* (Bals.) Vuillemin on red spider mites, *Tetranychus urticae* Koch (Acari: Tetranychidae) in India. *Insect Environment*, 10 (3): 103-104.

Shi WeiBing and Feng MingGuang 2004. Lethal effect of *Beauveria bassiana, Metarhizium anisopliae*, and *Paecilomyces fumosoroseus* on the eggs of *Tetranychus cinnabarinus* (Acari: Tetranychidae) with a description of a mite bioassay system. *Biological control* 30 (2): 165-173.

Singh, J., Gatoria, G.S. and Sidhu, A.S., 1987. Spider mite *Tetranychus cinnabarinus* (Boisd.) a secondary induced pest of cotton in Punjab. *Pesticides* 21: 43-44.

Singh, J. and Ray, R. 1977. *Stethorus* sp. (Coleoptera: Coccinellidae), a predator of *Tetranychus neocaledonicus* Andre' on okra at Varanasi. *Acarology Newsletter* 4: 5.

Singh, J. and Singh, R.K., 1979. Studies on mango bud mite (*Aceria mangiferae* Sayed.). *Pestology* 3: 20-26.

Singh, J., Singh, R.K., Mukherjee, I. N., Singh, R.N. and Agarwal, L., 1987. Mites of agricultural importance and their management in India. In: *Recent Advances in Entomology* (eds. Y.K. Mathur, A.K. Bhattacharya, N.D. Pandey, K.D. Upadhyaya and J.P. Srivastava) Gopal Prakashan, Kanpur, 169-185.

Singh, J. and Mukherjee, I.N., 1989. Pest status of phytophagous mites in some northern states of India. In: *The First Asia-Pacific conference of entomology* (APCE), the Entomology and Zoology Association of Thailand, 192-203.

Singh, J., Mukherjee, I.N. and Singh, R.K., 1996. Managing gall forming false spider mite in *ber*. *Indian Horticulture* 41: 30-31.

Singh, J.; Singh, R. N. and Rai, S. N., 2000. Expanding Pest Status of Phytophagous Mites and Integrated Pest Management. *IPM System in Agriculture* (eds-R.K. Upadhayay, K.G. Mukherjee and O. P. Dubey), New Delhi, India. 7: 1-29.

Singh, P., Somchoudhury, A. K. and Mukherjee, A.B., 1989. The influence of natural enemy complex on the population of *Aceria litchii* (Acari: Eriophyidae). In: *Progress in acarology*, Vol. 2 (Eds. G.P. ChannaBasavanna and C.A Viraktamath) Oxford and IBH, New Delhi, pp. 361-367.

Singh, R.N., 1994. Mite menace on vegetable crops in India. *Shashpa* 1: 59-74.

Singh, R.N., 1995. Mites of deciduous fruits and vegetables of eastern part of India and their economic status. *Adv. Agric. Res. India* 3: 179-193.

Singh, R.N. and Singh, J., 1992. Evaluation of some pesticides against carmine mite, *Tetranychus cinnabarinus* (Boisd.) on lady's finger. *Pestology* 16: 20-23.

Singh, R.N. and Singh, J., 1993. Incidence of *Tetranychus cinnabarinus* (Boisd.) (Acari: Tetranychidae) in relation to weather factors in Varanasi. *Pestology*.17 (8): 18-23.

Singh, S.P., 1988. Production and use of Phytoseiid mites. In: *First acarology workshop, G.A.U., Navasari*, 1-6.

Singh, V.S. and Bhatia, S.K., 1983. Reduction in yield of some barley varieties due to brown wheat mite infestation. *Indian J. Ent.*, 45: 190-193.

Simova, S. and Draganova, S. 2003. Virulence of isolates of entomopathogenic fungi to *Tetranychus urticae* Koch (Tetranychidae, Acarina). *Rasteniev, dni Nauki, Sofia*, Bulgaria. 40 (1): 87-90.

Spooner Hart, R., 1991. The use of predatory mite, *Phytoseiulus persimilis* for the control of two spotted mite *Tetranychus urticae* in horticultural crops. In: Proceedings 1st National Conference, Australian Society for Horticultural Science, Macquarie University, Sydney, 329-336.

Sogawa, K., 1977. Occurrence of the rice tarsonenmid mite at IRRI. *Intl.Rice.Res. Newsl.*, 2:17.

Sreenivas, A. G.; Ramanujam, B. R.; Prashanth Mohanraj; Nargund, V. B. and Shivaramu, K. 2005. Efficacy of entomopathogenic fungi against red spider mite, *Tetranychus neocalodonicus* Zacher (Acari: Tetranychidae). *Karnataka Journal of Agril. Sci.*, Dharwad, India 18 (4): 966-969

Srinivasa, N.; Prabhakara, H. and Malik, B., 2004. Rice sheath mite, *Steneotarsonemus spinki* Smiley (Acari:Taronemidae). Status paper. AINP-Agricultural Acarology (ICAR). 1-24.

Thakur, A.P. and Sharma, D.D., 1990. Influence of weather factors and predators on the populations of *Aceria litchii* Keifer. *Indian J. Pl. Prot.* 18: 104-112.

Thwaite, W.G., 1990. *Acaricide resistance and its management in Australian deciduous fruit orchards. Spider mites of tree fruit crops.* Meeting at Portland, Oregon, U.S.A.

Thwaite, W.G., Phimister, S.R. and Edge, V.E., 1987. Resistance management strategies for miticides. In: *Proceedings of the symposium on mite control in horticulture crops, orange,* Department of Agriculture, New South Wales, Australia, 99-101.

Tkcazuk, C.; Mietkiewski, R. and Balazy, S. 2005. Fungi and mite pest of plants. *Ochrona Roslin,* 50 (5): 34-38.

Van de vrie, M., 1973. Studies on prey-predator interactions between *Panonychus ulmi* and *Typhlodromus (A.) potentillae* (Acarina: Tetranychidae: Phytoseiidae) on apple in The Netherlands. In: *Proc. FAD conf. ecol. relations plant pest control, Rome,* 145-160.

Van de vrie, M., McMurthy, J.A and Huffaker, C.B., 1972. Ecology of tetranychid mites and their natural enemies: a review III. Biology, ecology and pest status and host-plant relation of tetranychids. *Hilgardia* 14: 342-432.

Verma, A., 1983. Mango malformation. In: *Exotic plant quarantine pests and procedures for introduction of plant materials,* pp. 173-188.

Verma, A, Lele, V.C., Majumder, P.K., Asha Ram, Sachchidananda, J., Shukla, U.S., Singh, G.C., Yadav, T.D. and Raychaudhuri, S.P., 1969. *Mango malformation.* ICAR Workshop Fruit Research, Ludhiana.

Verma, A, Lele, V.C., Raychoudhary, S.P., Ram, Asha and Singh, Asha. 1974. Mango malformation-a fungal disease. *Phytopath* Z. 79: 254.

Watt, G., 1898. The pests and blights of tea plant. Government Printing Press, Calcutta, India, 400-408.

Wilson, L.J. Bauer, L.R. and Lally, D.A., 1998. Effect of early season insecticide use on predators and outbreaks of spider mites (Acari: Tetranychidae) in cotton. *Bulletin of Entomological Research* 88: 477-788.

Zhang XianHong; he YunChun; Wang JianMing; Zhang ZuoGang and Li WenYing. 2001. A preliminary study on pathogenicity to *Verticillium lecanii* (Zimn) Viegas to insect pests. *Plant Protection* Beijing, China, 27 (1): 44-46

Zote, K.K., Mali, V.R., Mayee, C.D., Kulkarni, S.Y. and Mote, T.S., 1991. Outbreak of sterility mosaic of pigeon pea in Marathwada region, Maharahstra, India. *International Pigeon pea Newsletter* 1.4: 19-21.

Chapter 8

Simulation Modelling and Climate Change in Relation to Pest Management

Subhash Chander
Division of Entomology,
Indian Agricultural Research Institute, New Delhi – 110 012

Introduction

Pest management is a complex system involving many interrelated components such as crop, pests, natural enemies, beneficial organisms and non-target organisms subjected to man's production oriented interventions under a variable weather (Teng and Savary, 1992). On the advent of synthetic organic pesticides a lopsided approach of pest control was adopted where single pest was targeted with broad spectrum pesticides without due regard to other components of the agro ecosystem. This approach has been responsible for creation of several pest control and environment related problems, which led to the promotion of concept of integrated pest management in plant protection.

Pest management is generally considered as useful concept for sustainable crop protection (Rossing, 1993). It is also known as systems approach to tackle pest problems in agro ecosystems. Pest management involves intensive decision making requiring effective decision support tools. The development of an IPM scheme that fits into a sustainable farming system requires thorough analysis of the agro ecosystem system, which may be useful in identifying new control techniques by indicating intervention points (Rabbinge *et al.*, 1989; Boote *et al.*, 1993; Pinnschmidt *et al.*, 1994).

Models help to understand and analyze interrelationships among various components of a system. Pest management research is required to derive practical tools for developing tactics and strategies for pest management. Systems approach provides such tools in the form of simulation and decision models. In fact development, testing and evaluation of a simulation model are synonymous with application of systems approach in pest management (Teng and Savary, 1992).

Global warming concerns everyone these days as it poses potential threat to environment, and agricultural productivity and production throughout the world. Global climate change is the most important international issue being discussed extensively at present. It has implications for man's livelihood and his survival. Change in the global climate may affect the crop yields, incidence of pests and economic costs of agricultural production. Extreme weather events have caused severe crop losses and have taken heavy economic toll from farmers across the world. Roots *et al.* (2003) reported that the climate change is expected to have significant. impacts on the distribution, phenology and abundance of many species over the next few decades. Changes in community structure and extinction of some species is also expected (Thomas *et al.*, 2004). Plant-herbivore interactions are of particular significance because of their agricultural importance as well as their potential to affect ecosystem nutrient and carbon fluxes (Frost and Hunter, 2004). The probable effects of climate change on pests may include expansion of their ranges, change in population growth rates, increased period of activity, alteration in crop-pest synchrony and natural enemy-pest interaction, and changes in interaction among pest species (Sutherst, 1991). Present chapter details various aspects of modelling of crop-pest interactions as well as climate change impacts on insect populations.

Modelling Crop-Pest Interactions

Systems and Models

A system is a limited part of real world with inter-related components. It represents more than the mere addition of its components. The components of a system interact in such a way that a change in any component has repercussions for the whole system. The dimensions of a system depend upon the purpose of study. It may be a field, a farm, a village or a region. In crop protection, the following four main system types are distinguished. These are pathosystem, cropping system, farming system and agro ecosystem. Pathosystem includes the host and pest population. The effect of host plants on pest population dynamics and pest induced crop losses are analyzed here. Pathosystems are parts of cropping system that in turn consists of sequence of crops grown on a farm during the year. Cropping system thus include not only the crop protection aspects but also crop agronomic practices. Farming system consists of cropping system and other complementary enterprises such as cattle rearing, poultry, and fisheries etc. Finally, agro-ecosystem is an ecosystem, which is actively manipulated by man to serve his own needs in terms of getting maximum production.

A model is a simplified representation of a system. Basically models can be iconic or symbolic. Iconic models have physical resemblance to the system they represent *e.g.* model of a building while symbolic models do not resemble a system physically. Symbolic models can be further classified in to qualitative or quantitative models. Qualitative models represent a conceptual relationship among different variables of a system without its quantification *e.g.* experience about association of cloudy weather in winters with high aphid population on crops or reduction in aphid population on *rabi* crops by heavy rainfall. Quantitative models on the other hand represent a mathematical relationship among various variables of a system. These can be further classified as empirical or descriptive and mechanistic or explanatory models.

Empirical models are the regression relationships among various variables *e.g.* $y = a-bx$ represents a simple regression between yield of the crop (y) and pest incidence (x). The a and b are intercept and regression coefficient, respectively. The b signifies the change in crop yield with a unit change in pest incidence. These models do not explain the mechanism of yield loss due to pests. Such relationships are liable to vary spatially and temporally and are thus location specific and cannot be extrapolated.

On the other hand, mechanistic models explain the behavior of a system through underlying processes *e.g.* crop growth simulation model is based on various plant eco-physiological processes such as photosynthesis, respiration, growth, partitioning, water uptake, nutrient uptake etc. Similarly pest population simulation model takes in to consideration various bio-ecological processes like fecundity, sex ratio, migration, mortality factors and rate of conversion from one stage to subsequent stage etc.

The mechanistic models can be further classified in to static and dynamic models. Static models represent system behaviour that does not involve time *e.g.* a model used to represent relation between respiration and growth, using knowledge of underlying biochemical processes. A dynamic mechanistic model describes the changes in system behaviour with time. These can be further categorized into continuous and discrete models. Dynamic discrete mechanistic models describe the changes in the system in discontinuieties or in steps as in case of industrial production where products can be numbered. On the other hand, dynamic continuous mechanistic models account for gradual changes in the system with time. In agricultural systems all major elements such as leaf area, root length, weights of plant organs change only gradually with time in response to changes in weather, fertilizer or irrigation. Such systems are called dynamic continuous systems. Crop growth simulation models thus are dynamic continuous mechanistic models. When such models are run with weather data along with various inputs, these predict yield of crops at various intervals of time.

Crop Growth Simulation Models

Several tools of systems research related to computer and information technology are now available that can help in analyzing and tackling agricultural problems and crop growth simulation models are one among them. These models are based on quantitative understanding of the underlying crop physiological, meteorological and soil physical, chemical or biological processes and integrate the effect of soil, weather, crop, pests and management factors on growth and yield of crops.

Crop models are holistic, knowledge-based tools that are generic in nature and can be used for local and global applications. Together with other tools of systems approaches such as expert systems, decision support systems, databases, geographical information system (GIS) and remote sensing, these constitute a valuable tool of information technology that can facilitate integration of knowledge and its utilization by variety of stakeholders (Aggarwal *et al.*, 2004). Several crop growth simulation models *viz.* CERES-Rice (Ritchie *et al.*, 1986); CROPGRO models for grain legumes (Boote *et al.*, 1998); MACROS, a generic crop growth model (Pennings de Vries *et al.*, 1989); ORYZA for rice (Kropff *et al.*, 1994); ALFAPRO2 for alfalfa (Kanneganti, 1996); SUCROS, a generic model parameterized for wheat, potato, maize (van Keulen and Laar, 1997); WTGROS for wheat (Aggarwal *et al.*, 1994); INFOCROP, a generic crop growth model parameterized for rice, wheat and other crops (Aggarwal *et al.*, 2004) were developed with a purpose of looking in to opportunities of improving crop production.

The integrated pest management did not remain uninfluenced by developments in the area of crop growth simulation modeling. The adoption of simulation modeling in pest management commenced in the late 1960s. Before this most of the work in this area was carried out with empirical models. A number of single-species population dynamics simulation models came into being in the 1970s and early 1980s (Ruesink, 1976; Teng, 1985). Subsequently bitrophic population dynamics models involving natural enemy and pest species were also developed (Teng and Hofer, 1991). In late 1980s, crop and socio-economic factors were incorporated in pest models and as a result several decision support systems (DSS) have been developed since then. In its simplest form a decision support tool can be an economic threshold level, while a sophisticated form would be an interactive computer system that utilizes simulation models, databases and decision algorithms in an integrated manner (Teng *et al.*, 1998).

Great deal of work on modeling of crop-pest interactions was accomplished by coupling pests' effects to crop growth simulation models at physiological processes level and these coupled models were called crop-pest models (Alocilj, 1988; Pinnschmidt *et al.*, 1995; Sperow and Lybecker, 1998; Teng *et al.*, 1998). However, these models missed population dynamics component. Simultaneously work was going on coupling of pest models to crop models where both pest and crop could affect each other in dynamic mode (Rouse, 1988; Pinnschmidt *et al.*, 1990). Further attempts have been made to incorporate natural enemy population dynamics in the pest-crop models.

Crop growth simulation models take into consideration the physiological basis of pest damage for simulating crop-pest interactions (Pinnschmidt *et al.*, 1995). Being based on crop physiological and ecological processes, crop-pest models help to analyse multifactor interactions prevailing in agro-ecosystem and thus have potential for several applications in pest management.

Simulation

The process of developing simulation models and studying systems through them is called simulation. Basic steps of systems simulation are as follows.

Define Goals

Before any simulation work is attempted, the purpose of modeling should be clearly defined. All problems are not amenable to modeling. Similarly, many problems may not require a modeling solution.

Define System and its Boundaries

Since a system is a limited part of the reality it should have a boundary which would isolate it from its surrounding. In agricultural systems there are large number of climatic, soil, hydrologic, biotic, agronomic and socio-economic factors that influence crop growth and productivity. For understanding such complex systems, different levels of crop production in the order of descending productivity are considered.

Crop growth can be simulated at three levels of production *viz.* potential, attainable and actual production. Potential production is determined by plant genotype and climatological factors and there are no constrains of nutrients, water and pests. Plant genotype and climatological factors are called yield-determining factors.

Potential production gets reduced to attainable production level due to limitation of nutrients and water, which are called yield-limiting factors. There is not any pest constraint at this production level also. When besides nutrient and water constraints, there is pest attack, the yield gets further reduced from attainable to actual level. The pests are known as yield reducing factors. Therefore, it should be clearly defined as which level of production is to be simulated and what factors are to be considered.

Define Key Variables in the System

It is important to define the key variables within the system. In crop growth simulation models, state and rate variable approach is used. In this approach, it is assumed that the state of a system at any moment can be quantified and the changes in the system can be described by mathematical equations. The model variables can be distinguished as state, rate, driving variables and auxiliary. State variables describe the state or condition of the system such as biomass, leaf area, amount of nitrogen, water in the soil etc. Similarly in a pest population dynamics model, number of eggs, larvae, pupae and adult are the state variables. Each state variable is related to a rate variable, which describes the rate of change of the state variable *e.g.* rate of egg laying, rate of change of larval population etc. Driving variables are meteorological factors such as rainfall, solar radiation, temperature, relative humidity, wind speed etc. These are also called the forcing functions. These reveal the effect of the environment on the system. Finally, the stresses such as water stress, fertility stress or salinity stress are termed as auxiliary variables.

Quantification of Relationships

After defining the key variables of the system, the relationship between them is determined. Relational diagrams are used to indicate the relationships between variables. These diagrams show the relationship and feedback among the variables. Feedback is the change in rate variable based on state variable. Feedback can be negative or positive. If rate variable increases with an increase in state variable or vice-versa then feedback is termed as positive. On the other hand if rate variable decreases with an increase in state variable or vice-versa then feedback is called negative. The values of rate variables are calculated from field experiments and literature. These are then used to calculate state variables.

Calibration

After a crop simulation model has been developed, it needs to be calibrated. Calibration may be defined as the adaptation of crop growth simulation model to a particular experimental condition. The experimental data on crop phenology, agronomic management and weather conditions are used to calibrate a model. The model output (leaf area, biomass, yield) is then compared to experimental output. If these are not comparable then different parameters are adjusted such that model output becomes comparable to field output. The adjustment however be used with caution because it decreases the explanatory nature of the model. Crop-pest models can also be used to study the effect of pest incidence on crop growth and yield. In such cases, besides model calibration for crop growth and yield, it is also calibrated for pest damage

Validation

After calibration, the model is validated for various pest infestation treatments. The pest infestation levels observed in the field are input in the model. If model output is comparable to field output, then pest damage mechanisms are taken as validated. Validated model can be used for various applications in pest management.

Sensitivity Analysis

Sensitivity analysis is the procedure of increasing or decreasing the value of one parameter or a combination of parameters and simulating the results. If the change in output is large with change in the value of parameter then model is considered sensitive to the parameter. Sensitivity analysis can help in establishing the significance of different parameters.

Assessment of Crop Losses

Pest management can be seen as a decision making system (Walker, 1987). Decisions may be made daily by farmers, researcher considering research methods or by the administrator allocating resources at the national level. There are mainly two types of decision in pest management, tactical and strategic decisions (Zadoks, 1985). Tactical decisions are day to day decisions taken on farmers' level for a short time frame and smaller area while strategic decisions pertain to large areas and relatively longer period of time.

Assessment and extrapolation of yield losses are mandatory at both the strategic and tactical levels (Willocquet *et al.*, 1999). They are required at the strategic level for research prioritization and at tactical level for taking action or improving pest management. Crop loss models are often used to quantify the relationship between plant yield and pest incidence (Nutter, 1990).

Empirical Approach of Crop Loss Assessment

The crop-pest interactions have generally been quantified as damage functions (Pinnschmidt *et al.*, 1994). This methodology made use of an array of multivariate statistical methods (Savary *et al.*, 1996; Du *et al.*, 1997). An approach for handling several pests is the crop loss profile methodology using multiple regression (Pinstrup *et al.*, 1976; Chiarappa, 1980). Techniques such as principal components analysis, canonical correlation and multiple regression analysis have been applied to analyze complex, multivariate data generated by surveys (Stynes, 1980). Such empirical equations are simple models reflecting pest effects on crops but could not be used when weather, crop cultivar, soil type or management practices changed and were thus limited in their extrapolation value (Pinnschmidt *et al.*, 1995; Teng *et al.*, 1998). This is a weakness common to all empirical regression models (James and Teng, 1979; Chander and Singh, 2001). Regression analysis is a two-steps approach involving the pest incidence and crop yield without considering the effect of pests on plant physiological processes.

Mechanistic Approach to Crop Loss Assessment

To estimate the effect of pest infestations on different crop cultivars grown in different soil type and sites and under different management practices requires that physiological crop growth simulation models be used (Teng *et al.*, 1987). The mechanistic approach provided comprehensive simulation models that simulate potential, attainable and actual crop growth and yield. The CERES-Rice model (Ritchi *et al.*, 1986) developed under the International Benchmark Sites Network for Agrotechnology Transfer (IBSNAT) was the first model to be coupled with pest damage mechanisms (Pinnschmidt *et al.*, 1990). Further Pinnschmidt *et al.* (1995) proposed a generic approach to simulate the damage effects of single or multiple pests using the CERES-Rice model. Teng *et al.* (1998) used the CERES-Rice and CROPGRO models to exemplify coupling points for linking pest damage mechanisms to crop growth development.

The coupling points for damage mechanisms are located at the process level (photosynthesis, respiration, translocation) or at the state level (leaf area, leaf weight, stem weight, panicle weight) (Rabbinge *et al.*, 1994). Fabellar *et al.* (1994) simulated leaf folder feeding effects on yield of rice using MACROS model and leaf consumption for various combinations of larval density and crop age was predicted. Willocquet *et al.* (1999) tested the yield loss simulation (YLS) model for rice pests under different production systems of South Asia and Southeast Asia. Chander *et al.* (2002) also modelled the effect of stem borer on growth and yield of rice through YLS and MACROS model. InfoCrop (Aggarwal *et al.*, 2004), a generic crop growth simulation, has been coupled with important pest damage mechanisms and these have been validated for important insects, diseases and weeds of rice and wheat crops.

The assessment of crop losses with mechanistic approach is a three-step approach taking in to consideration, the pest incidence, pest damage mechanism and the crop yield (Figure 8.1). The pest incidence has to be provided as input in these models. This mechanism was termed as one-way

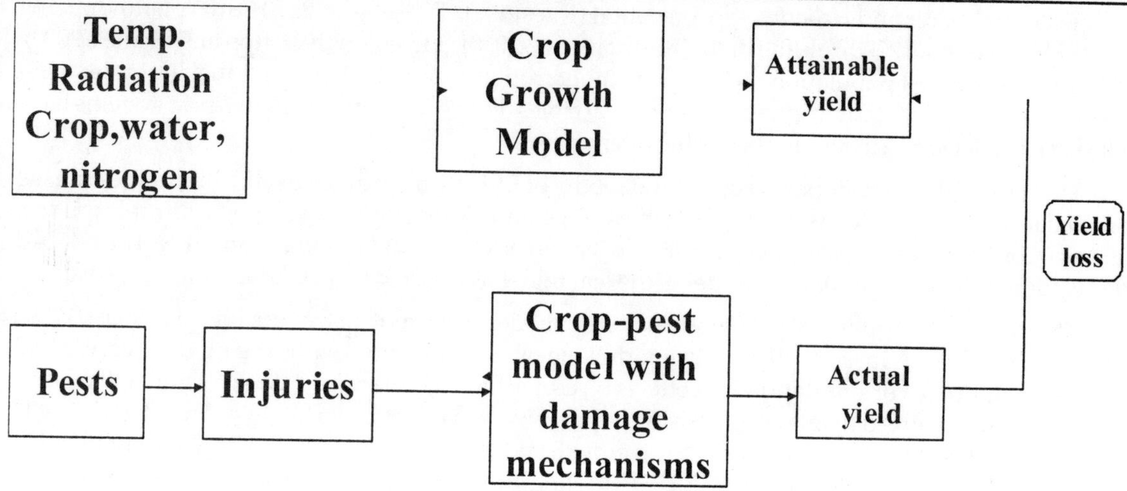

Figure 8.1: Assessment of Crop Losses Due to Pests with a Simulation Model

approach of simulating pest effects on crops as pest affected the crop without reverse being true. However, pest dynamics model that could simulate pest population under given environmental conditions can be linked to crop-pest simulation model (Benigno *et al.*, 1988; Pinnschmidt *et al.*, 1995; Teng *et al.*, 1998).

Damage Mechanisms

Damage mechanisms are the plant physiological processes affected by the pest injury. Major types of pest damage mechanisms are classified as germination reduction, stand reduction, light stealing, assimilation rate reduction, assimilate sapping, tissue consumption and tugor reduction (Rabbinge and Rijsdijk, 1981; Boote *et al.*, 1983; Aggarwal *et al.*, 2004). Accordingly the pests can be classified as germination reducers like mole crickets and seed maggots; stand reducers such as stem borer, cutworms, termites, damping off and wilts; light stealers such as rusts, blights, mildews and sooty mould; assimilation rate reducers like plant diseases; assimilate sappers such as aphids, plant hoppers, whiteflies, mites, thrips, and bugs; tissue consumers like defoliating beetles, grasshoppers, leaf folders, leaf miners, node borers, grain feeders and foliar diseases such as leaf blight, blast and sheath blight and turgor reducers such as nematodes and root pathogens.

Before modeling the effect of a specific pest on crop, it is necessary to quantify the relationship between the pest population and the process affected by it (Teng *et al.*, 1987). Some pests affect the crop in several ways *e.g.* brown plant hopper in rice acts as assimilate sapper as well as light stealer and foliar diseases act as light stealer, assimilation rate reducer and tissue consumer as well. All the important damage mechanisms of a pest are coupled to crop growth model at appropriate plant growth processes and then calibrated and validated through field experiments.

Simulation of Pest Population Dynamics

Work on developing population dynamics simulation models began in late 1960s. One of the first computer simulation programme, EPIDEM for early blight of potato was developed in 1969 (Waggoner and Horsfall, 1969). Subsequently, the population dynamics simulation models were developed for

European corn borer and *Heliothis* spp. on cotton (Ruesink, 1976; Teng, 1985); brown planthopper of rice (BPH) (Hao *et al.*, 2000); plume moth (Buttirani *et al.*, 2000); and onion fly (Otton and Homes, 2000). Majority of these pest population models were monotrophic type in which individual life cycle stages were driven by specific environmental or host variables. However, natural enemy-pest systems have also been modeled to account for bitrophic interactions.

Holt *et al.* (1989) developed a simulation model of BPH by including specific mortalities caused by predators and parasites as driving variables. Teng and Hofer (1991) developed a bitrophic model of Asian corn borer and *Trichogramma furnacalis*. Graf *et al.* (1992), and He and Wang (1999) developed population dynamics simulation model of the leaf folder as affected by predation and parasitism.

The process of coupling of pest models to crop models was undertaken during late 1980s to 1990 (Rouse, 1988; Pinschmidt *et al.*, 1990). The interlinking of pest simulators with crop models represented a two-way approach of simulating pest effects on crops. Pinnschmidt *et al.* (1995) formulated generic pest modules and linked them with 'Decision Support System for Agro-technology Transfer' (DSSAT) crop models to account for pest effects on crop growth and yield. Teng *et al.* (1998) developed population model for BPH and coupled it with CERES-Rice to estimate yield losses. There is a need to couple natural enemy effect in to pest-crop model for simulating tritrophic system.

Models Available for Coupling Pest Effects

Two types of model are available for linking pest damage mechanisms: (a) models simulating gross photosynthesis and (b) models simulating net photosynthesis

First type of models simulate gross photosynthesis. Net photosynthesis is then calculated by subtracting respiratory requirement of plants from gross photosynthesis. These are very detailed models and require a large number of inputs *e.g.* the Simple Universal crop Growth Simulator (SUCROS), generic models that were parameterized for many crops like wheat, potato and maize (Spitters, 1989); MACROS models (Penning de Vries *et al.*, 1989); the ORYZA models that were derived from SUCROS and MACROS (Kroff *et al.*, 1994); and the GRO models for crops like soybean, peanut and other leguminous crops.

Second type of models directly simulate net photosynthesis of the crop. These are relatively simple and require less number of inputs *e.g.* CERES (Crop estimation through resource and environment synthesis) models and InfoCrop (Aggarwal *et al.*, 2005).

Climate Change Impacts on Crop Pests

Slow and definite change in climate of a place is called climate change while year to year changes in weather conditions is known as climate variability. Climate change is a natural phenomenon and global climate has been changing since pre-industrial times. However, the atmospheric concentration of greenhouse gases (GHGs) *viz.* CO_2, CH_4, N_2O and CFCs has been increasing at an alarming rate in the recent past due to anthropogenic activities, which in turn are causing global warming. Concomitant to global warming are changes in rainfall pattern, increase in frequency of extreme climatic events such as floods and droughts and rise in sea level. According to the fourth assessment report of 'The Intergovernmental Panel on Climate Change (IPCC)', global average temperature has increased by 0.74°C over the last 100 years and temperature increase has been projected to be about 1.7-5.8°C by the year 2100.

The global climate change may impact food production in South Asia due to its large population, predominance in agriculture and its limited resource endowment. Elevated CO_2 concentration may stimulate biomass production in many crops, but increasing temperature in combination with drought could reduce yield by accelerating growth and maturation of arable crops. Temperate northern latitudes might experience an increase in productivity mainly due to the expansion of the growing season whereas at mid-latitude and southern latitude, effect may be more variable.

Temperature is the major factor in global climate change that directly affects insect development, reproduction and survival. Although insect responses to global climate change will vary, the effect of global warming in general has been predicted to increase intensity of herbivore pressure on plants. Changes in rainfall pattern also have implications for insect survival. It is being predicted that frequency of rainfall would decrease but its intensity would increase. This may lead to floods on one hand and long dry spells on the other. Small insects like aphids on wheat crop have been found to be adversely affected by rainfall (Bakhetia and Sindhu, 1983; Chander, 1998). More intense rainfall as project under climate change may thus reduce incidence of small pests on crops. On the other hand, heavy rainfall caused outbreak of armyworms (Lever, 1969; Chander and Sharma, 1997). Global climate change would have direct or indirect consequences on pests (Coakley *et al.*, 1999).

Direct Effects of Climate Change on Insects

Expansion of Habitat Range

Any increase in temperature is bound to influence the distribution of pests and diseases. It is predicted that 1°C increase in temperature would enable species to spread 200 km northwards or 140 km upwards in altitude (Parry *et al.*, 1989). Climatic warming will allow the majority of temperate insect species to extend their ranges to higher latitudes and altitudes. Many of these species will expand their geographical ranges to higher latitudes and altitudes, as has already been observed in a number of common butterfly species (Pollard *et al.*, 1995). Parmesan *et al.* (1999) documented pole-ward shifts in geographical ranges of up to several hundred kilometers in a variety of taxa. As the climate continues to warm, species with wide latitudinal distribution that already encounter considerable temperature variation, may be able to cope with changing climate better than species with restricted ranges particularly montane and cold-adapted species. However, known pest species if introduced in to new areas may encounter marginal conditions for their survival and suitable host plants may not be available. Besides natural enemies may also enter new areas thereby aiding in biological control of pests. There is a need to regularly observe activity of pests in different regions in terms of timing, population size and habitat ranges for drawing any meaningful conclusions.

Change in Migrating Behaviour

Minimum temperature plays important role in determining the global distribution of insect species rather than maximum temperature (Hill, 1987). Hence any increase in temperature will result in greater ability to over winter at higher latitudes. Cabbage butterfly migrates from hills to plains in winter and back to hills in summer (Atwal, 1986). The global warming may affect this migration and extend distribution of the pest further north. However, the activity of this pest may decline in plains.

Temperature thresholds for insect flight vary both among and within species with season and also with region (Wiktelius, 1981). Climate warming would advance the time of year at which the flight thresholds are first reached and increase the possibility of early immigration (Woiwood, 1997). In summer, upper thresholds might be reached more frequently, limiting flight opportunities. Likewise there are optimal and threshold temperatures for insect walking, an important factor in local redistribution. Below the optima, the climate warming would enhance movement but above it will be detrimental.

Aphids are most likely to be affected by temperature change as result of their low developmental threshold, low thermal constant and multivoltinism (Kiritani, 1997). In general, warmer conditions lead to rapid population build up in aphids, resulting in rapid alatae production. However, this may be modified by reduced wing production, which is a physiological effect of higher temperatures thus reducing the aphids' ability to disperse in response to crowding.

Changes in Over Wintering Success

With rise in temperature, onset of hibernation may be delayed while it may be suspended earlier than usual in spring thereby increasing period of activity of pests. Rice stem borer, *Scirpophaga incertulas*, becomes active with availability of rice crop in north India. With introduction of summer rice and increase in temperature, it may realize higher population levels. On the other hand, the increase in temperature in summer months may reduce the activity of some pests. Non-diapausing species of aphids such as *Myzus persicae*, which are able to over winter in their active stages showed increased survival in warm winters (Bale *et al.*, 1988). These can therefore colonize crops more quickly during spring and earlier flights are known to occur after milder winters (Harrington and Taylor, 1990).

Changes in Interspecific Interactions

The effect of climate change on species distribution and abundance could involve not only direct effect on each species individually in an ecosystem but also on species interactions (Davis *et al.*, 1998). Rapeseed-mustard is infested by two aphid species, *Lipaphis erysimi* and *Myzus persicae*, the former being dominant during severe winters while the latter during mild winters (Chander and Phadke, 1994). With rise in temperature, higher incidence of *Myzus persicae* may be witnessed. Such faunal shifts may also take place in other crops.

Effect on Pest-Natural Enemy Interactions

Climate change will have significant influence on biological control of pests. With increase in temperature, predators of aphids on *rabi* crops may become more active than at present and may help check aphid populations, which otherwise grow unhindered during peak winter months. Temperature increase from 17-22°C caused 10 per cent increase in aphid, *Sitobion avenae* reproduction on wheat while activity of its predator, *Coccinella septempunctata* was increased by 250 per cent simultaneously.

Changes in Population Growth Rates

In general insect species are active in wide range of temperature. The temperature range favourable for growth differs between species and is often related to climatic range and habitat. Within this favoured temperature range, temperature elevation increases the speed of development during the growth phase but the rate of increase differs between species. This again relates to ambient temperatures encountered in the field (Andrewartha and Birch, 1954). If after global warming, ambient temperature in a habitat remains within favourable range for insect species, then it would benefit the species by enhancing its growth rate otherwise it would have adverse effect on its population. Climatic warming would affect temperate annual and multivoltine species in different ways and to different degrees. In case of multivoltine species such as Aphididae and some Lepidoptera, higher temperatures would allow faster development rate probably allowing for additional generations within a year (Pollard and Yates, 1993).

Aphid population on barley was negatively related to January mean minimum temperature and February total rainfall (Chander *et al.*, 2003). The increase in minimum temperature and more intense rains as anticipated under climate change scenarios might thus reduce aphid incidence on barley.

Impact of global warming on the incidence of rice yellow stem borer and brown planthopper (BPH) under Delhi conditions was analyzed through comparison of pests' favourable temperature range with prevailing and projected ambient conditions. It was observed that 1-2°C rise in temperature would increase the incidence of the stem borer while it would adversely affect BPH.

Change in Pest-Crop Synchrony

Crop-pest synchrony may be altered due to climate change. There is a heavy incidence of stem borer and shoot fly in maize and sorghum during *kharif* season in north India. In south India, these pests are active during *rabi* season also but there is every likelihood that incidence of shootfly on sorghum would further increase.

After visualizing possible impact of climate change on pests, there is need to collect systematic information on pest behaviour under climate change scenarios so that theoretical considerations can be authenticated. This would then help in formulating strategies for mitigation and adaptation to climate change.

Indirect Effect through Host Plants

Effect of Increased CO_2

The elevated CO_2 will produce increased plant size and canopy density with high nutritional quality foliage and microclimate more conducive to pests. Under higher CO_2, there is an increase in C:N ratio that increases feeding of herbivores in order to derive more amino acids. At elevated CO_2, there is an increased partitioning of assimilates to roots in crops such as carrot, radish and sugarbeet. Due to more carbon storage in roots, losses from soil borne pests may be diminished under climate change (Coakley *et al.*, 1999). The climate mediated changes in individual plant biomass and size may give a good indication of resource availability for insect herbivores. Insect species richness has been shown to be strongly correlated with plant biomass and height as larger plants have increased structural complexity and greater range of resources that herbivores can utilize (Lawton, 1995).

Effect of Increased Temperature

Increase in temperature may alter host physiology and resistance. Elevated temperature may cause breakdown of temperature sensitive resistance. Moreover under stress conditions, the plant defensive system is lowered and they become more susceptible to pest attack. Conversely, in some forage species, there is increased lignification at higher temperatures (Wilson *et al.*, 1991) that can enhance the level of host resistance to pathogens.

Effect of Changes in Host Plant Distribution

Climate change will affect the geographical distribution of plant species and their growth patterns. Spatial changes of crops and cropping systems for sustained productivity are imminent under climate change scenario (Carter *et al.*, 1991), which would affect insect pest distribution. New pests may arise while some pests may become less important if global warming results in north ward shift of agro climatic zones and host plants migrate in to new regions (Coakley *et al.*, 1999).

Assessment of Impact of Climate Change on Pest Species

Impact of global climate change on crop productivity and pest population can be assessed through experiments as well as through crop growth models. For assessment of climate change impacts on crops, structures such as Open top chambers (OTCs), CO_2 rings, temperature gradient tunnels are used. Impact of CO_2 on insect population via host plants can be studied through OTCs and CO_2 rings. On the other hand, the impact of temperature rise on insect growth and development can be studied through temperature gradient tunnel or through growth chambers. Temperature gradient tunnels prove more suitable in case of sessile insects such as aphids, mealy bugs, scale insects etc. The OTCs and temperature gradient tunnels restrict natural to and fro movement of insects.

Mitigation of Climate Change

Mitigation is the adoption of measures against sources of greenhouse gases *i.e.*, to reduce emission of greenhouse gases. For mitigation of green house gases, consumption of fossil fuels in pest management activities should be reduced. Pesticides manufacturing, transportation and power spraying requires a lot of energy and by curtailing pesticide consumption, green house gas emission can be reduced.

Pest management can play important role in mitigation of global climate change as it stresses on judicious use of agricultural inputs such as fertilizers and pesticides. For example, transplanted rice is one of the major sources of methane from agriculture, another being dairy cattle. Continuous flooding of rice fields lead to emission of methane, an important green house gas and it also aggravates problem of plant hoppers on the crop. Likewise excess use of nitrogenous fertilizers stimulates pests' population and is also responsible for nitrous oxide emission. Therefore, alternate wetting and drying as emphasized in pest management, mitigates methane emission and also reduces plant hopper incidence on the crop.

Adaptation to Climate Change

Adaptation refers to adoption of measures to adapt to global climate change without suffering losses in biodiversity, natural resources and agricultural productivity. A few of the adaptation measures for combating pest menace under global climate change may be:

1. Global climate change would cause alteration in sowing dates of crops, which may alter host-pest synchrony. There is need to explore changes in pest-host interaction under early, normal and late sown conditions in order to recommend optimum sowing dates for reduced pest pressure and increased yield.
2. Elevated temperature may cause breakdown of temperature sensitive resistance and durability of resistance may be threatened due to increase in number of pest generations. This may lead to more rapid evolution of pest biotypes. This way many resistant cultivars may be lost. Breeding of cultivars for both pest and drought resistance will be better viable option.
3. Temperature tolerant strains of natural enemies of pests should be evolved to take their advantage under global climate change.
4. Climate change could affect efficacy of crop protection chemicals through, (*a*) changes in temperature and rainfall (*b*) morphological and physiological changes in crop plants (Coakley *et al.*, 1999). An increase in probability of intense rainfall could result in increased pesticide wash off and reduced pest control. Likewise, increased thickness of epicuticular wax layer under high CO_2 could result in slower or reduced uptake by host while increased canopy size may hinder proper spray coverage and lead to a dilution of the active ingredient in the host tissue. In contrast, increased metabolic rate at higher temperature could result in faster uptake by plants and higher toxicity to pests. The rates of pesticide application thus have to be modified according to new situations.

5. Reliable medium range weather forecast can help in proper timing of crop management practices such as sowing, irrigation, fertilizer application and pesticide application. This would increase efficiency of crop production and protection technology. Likewise pest forecasting can help in cautioning farmers about impending pest situation and adoption of preventive measures to avert pest problems.

Applications of Simulation Modelling in Pest Management

Crop-pest models are being used for various applications like rationalizing pesticide use, pest risk analysis, ecological pest zoning, assessing impact of climate change and pest forecasting.

Rationalizing Pesticide Use

Determination of economic injury levels (EILs)

Information on EILs can be used to obtain an optimal control of pests, avoid unwarranted pesticide application, saving producers unnecessary expenditure and conserving the environment (Archer and Bynum, 1992). The EIL is a highly dynamic entity may differ among geographic locations, plant growth stages, control expenditures and market prices (Girma *et al.*, 1993; Archer, 1994). Traditionally, the EILs for pests are based on empirical yield-infestation relationships that are site specific with a little scope for extrapolation. It would however be very expensive and time consuming to use field experiments to establish such yield-infestation relationships for different pest species, crops and locations.

Crop-pest simulation models can be used to establish location and weather-specific EILs thereby helping to increase the efficiency of field experiments substantially. Coupled models have been used for simulating EILs for bean leaf beetle infesting soybean (Nordh *et al.*, 1988), Russian wheat aphid damage in winter wheat (Chander *et al.*, 2006), and leaf folder (Satish *et al.*, 2007) and stem borer (Reji *et al.*, 2008) incidence in rice. These can also be used for establishing EILs for multiple pests associated with crop.

Formulation of ISO Loss Curves

Crop-pest models may be used to generate iso-loss curves, which may be used for pest monitoring and need-based intervention of control measures against pests (Figure 8.2). Daryaei *et al.* (2003) used InfoCrop model to devise iso loss curves for stem borer and bacterial leaf blight incidence in rice and stem borer and leaf rust incidence in wheat. Similarly simplified pest models or simplified decision rules from crop-pest models with economic values assigned to their outputs have been used for managing sweet corn common rust (Teng, 1987), and rice blast (Surin *et al.*, 1991).

Pest Risk Analysis

Simulation models can also be used for analyzing the risk associated with the introduction of exotic pests into new ecosystem. In general pest risk analysis (PRA) is split in to risk assessment and risk management. Risk assessment is a structured assessment of the risks of entry, establishment, spread and the potential socio-economic and environmental impact posed by exotic plant pests and pathogens. Risk management assesses the impact of the risk and decides whether it is sufficient to warrant phytosanitary measures and what these should be. PRA must be carried out for obnoxious pests and diseases of quarantine importance *e.g.* several countries including United Kingdom, the Netherlands, Canada, New Zealand and the United States have conducted PRA for *Phytophthora*

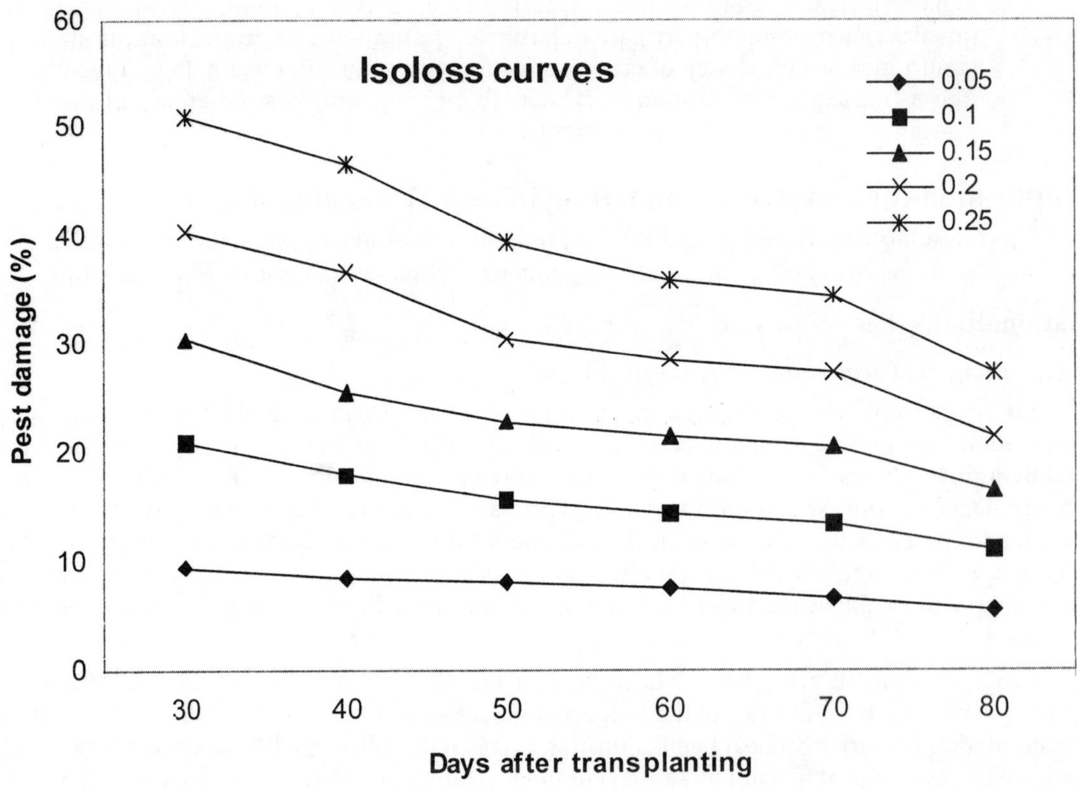

Figure 8.2: Iso Loss Curves Simulated with InfoCrop Model for Stem Borer Incidence on Rice

ramorus and PRAs formed the basis for implementing phytosanitary measures against this pathogen (RAPRA, 2005).

The foreign pests are under quarantine cannot be brought deliberately into new countries to test their damage potential. Therefore, models are very useful to evaluate situations where actual fieldwork on the pest is not possible (Teng, 1991). Yang *et al.* (1991) showed the possibility of the approach with a soybean rust model, SOYRUST, which when run with continental USA weather data predicted potential areas for epidemics. When model disease estimates were further linked to soybean crop model, the potential losses attributable to rust epidemics were determined.

Predictive Pest Zoning

Pest zonation is a concept that is particularly applicable for large area pest management, in which both tactics and strategies can be merged to achieve optimal management (Teng and Savary, 1992). Pest population dynamic model can be run with requisite weather data and probability of various levels of pest infestation for a site is determined (Yuen and Teng, 1990). Site predictions can then be extrapolated through GIS to carve out the zones of equal epidemic potential for a pest.

Historical climate data from sites have been used for analysing the favourableness of a site for specific wheat diseases (Coakley, 1990), while environmental risk assessment for tomato viruses have been carried out in Arizona using geostatistics and GIS (Teng and Savary, 1992). Song *et al.* (1994) used GIS to prepare spatial database to map pest distributions based on pest surveillance data and concluded that pattern of insect species changed over time. Chander *et al.* (2004) carried out ecological zonation of rice leaf folder, *Cnaphalocrosis medinalis* infestation in Haryana state of India and the state could be divided into zones of different epidemic potential for the pest. Likewise Reji (2007) developed pest-weather models between rice stem borer incidence and weather factors *viz.*, maximum temperature, minimum temperature and relative humidity at Pattambi, Coimbatore and Warngal, which were then extrapolated with GIS for the states of Kerala, Tamil Nadu and Andhra Pradesh. These states were divided in to zones of low, moderate, high and severe stem borer incidence (Figure 8.3). Knowledge of pest epidemic potential in different zones would allow strategic decisions with respect to selection of crop cultivars and appropriate management options.

Pest Forecasting

Models facilitate understanding of factors affecting outbreak and damage caused by pests. These can be used for predicting pest outbreaks and the performance of control measures and for making

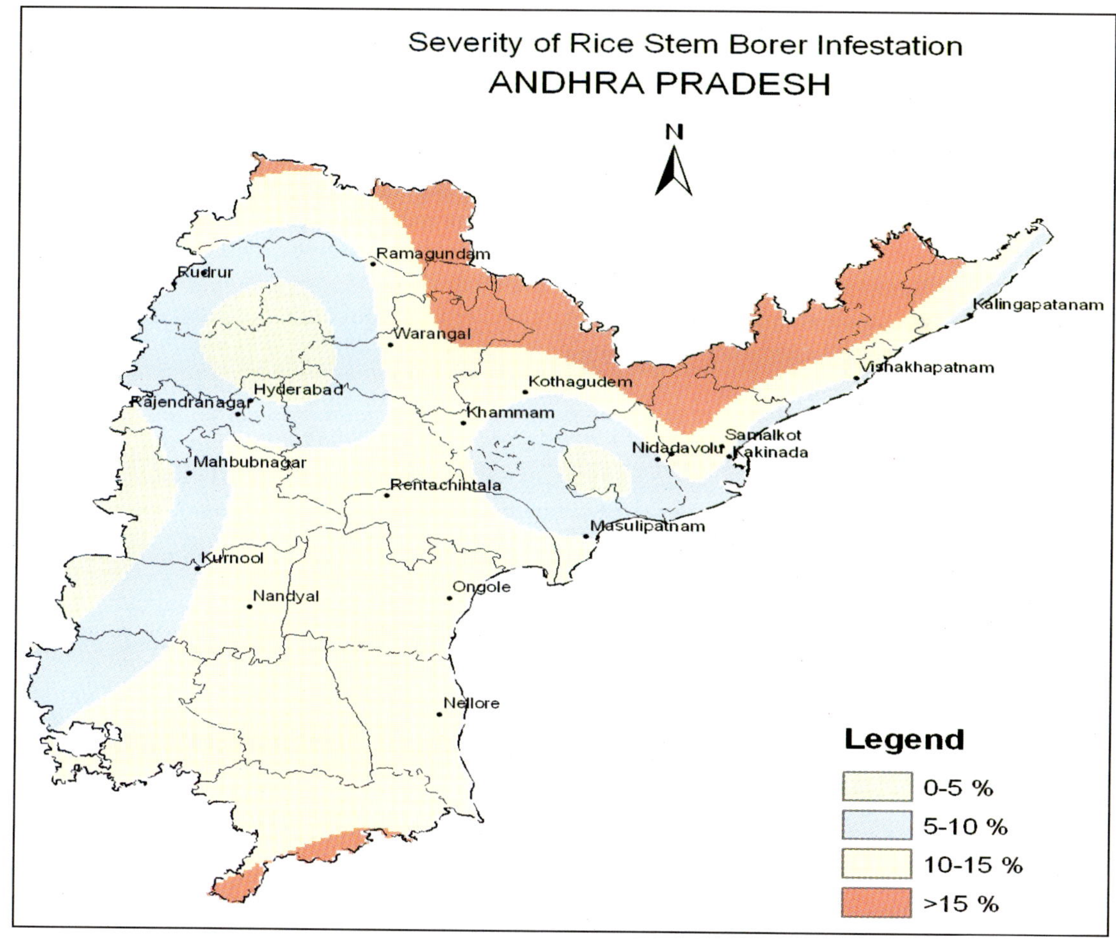

Figure 8.3: Ecological Zoning of Rice Stem Borer Incidence in Andhra Pradesh

recommendations. These models improved the understanding of the pest as an element of crop ecosystem and detect the crucial knowledge gaps in view of a holistic assessment of pest status (Graf *et al.*, 1992). The population dynamics models can be linked to crop growth model for predicting pest population peaks and thereby timing spray schedules based on economic thresholds (Benigno *et al.*, 1988).

EPIPRE (Epidemiology, Prediction and Prevention), a supervised control system of pests and diseases in wheat, has been successfully used for minimizing cost of plant protection and to reduce pesticide use (Zadoks, 1984). It is one of the earliest world wide attempts to develop computerized Integrated Pest and Disease management systems (IPDM). Reji *et al.* (2003) developed a population dynamics simulation model for rice stink bug *Leptocorisa acuta* using concept of thermal time and threshold of development and used it to analyze the impact of global warming on incidence of the pest incidence. It was revealedd that up to 2ºC rise in ambient temperature, the bug population would increase but further rise would adversely affect the pest.

Models have been used to evaluate control alternatives of pests such as sugarcane froghopper (Coakley, 1985), the gypsy moth and the coffee berry disease (Kranz and Hau, 1980).

Analysis of Impact of Climate Change on Pests and Crops

Impact of climate change on pests has been analyzed through empirical models (Coakley *et al.*, 1999) but these studies being location specific are limited in their application. Process based crop-pest simulation models will be thus more suited for this purpose.

The crop growth models have been used to analyze impact of climate change on crop growth and development. However, biotic stresses like pests have been ignored in these studies. Similarly population dynamics models have been used to assess impact of climate change on pest dynamics without associating it with crop productivity (Reji *et al.*, 2003). Pest dynamics-crop models will be useful tools to assess impact of climate change on pest dynamics and crop productivity both thereby accounting for its impact on crop-pest interactions as a whole.

Conclusions

Crop growth simulation models have been used for several applications in the area of pest management, which helped to increase the efficiency of field research greatly. These will be of even greater relevance in new emerging research areas such as climate change impacts on pests and crop yield, impacts of trangenics on environment, pest risk analysis for sanitary and phytosanitary requirements and pest forecasting. It is important to develop decision support systems based on simulation models to facilitate their application by users. Awareness regarding impacts of global climate change on crops and pests, and adoption of mitigation and adaptation measures can help in countering its adverse impacts on agricultural productivity and environment.

References

Aggarwal, P.K., Kalra, N., Chander, S. and Pathak, H., 2004. InfoCrop: A generic simulation model for annual crops in tropical environments, Indian Agricultural Research Institute, New Delhi, 132 pp.

Aggarwal, P.K., Kalra, N., Singh, A.K. and Sinha, S.K., 1994. Analyzing the limitation set by climatic factors, genotype, water and nitrogen availability on productivity of wheat 1. The model documentation, parameterization and validation. Field Crops Research 38: 73-91.

Andrewartha, H.G. and Birch L.C., 1954. *The distribution and abundance of Animals*. University of Chicago Press, Chicago.

Archer, T.L., 1994. Economic injury levels and chemical control of the Russian wheat aphid. In: Archer, T. L., Kroening, M. K., Simmons, C. L. (eds), Proceedings Sixth Russian Wheat Aphid Workshop, 23-25 January 1994, Fort Collins, Colorado State University, Colorado, USA, pp 97-106.

Archer, T. L. and Bynum, E. D. Jr., 1992). Economic injury level for the Russian wheat aphid on dryland winter wheat. *J. Econ. Entomol.*, 85: 987-992.

Atwal, A.S., 1986. *Agricultural pests of India and South-East Asia. Kalyani Publishers*, Ludhiana. 529 pp.

Bakhetia DRC and Sindhu, SS., 1983. Effect of rainfall and temperature on the mustard aphid. *Indian J. Ent.*, 45(2): 202-205.

Bale, J. S; Harrington, R. L and Clough, M.S., 1988. Low temperature mortality of the peach–potato aphid, *Myzus persicae. Ecological Entomology* 13: 121–129.

Benigno, E.A., Shepard, B.M., Rubia, E.G., Arida, G.S., Penning de vries, F.W.T., and Bandong, J.P., 1988. Simulation for rice leaf folder population dynamics in lowland rice, IRRI, Manila, Philippines, 8 pp.

Boote, K. J., Jones, J.W., Mishoe, J.W. and Berger, R.D., 1983. Coupling pests to crop growth simulators to predict yield reductions. *Phytopathology* 73: 1581-1587.

Boote, K.J., Jones, J.W. and Pickering, N.B., 1998. The CROPGRO Model for grain legumes. In Tsuji, G.Y. (ed.), *Understanging Options for Agricultural Production*, Kluwer Academic Publication, Dordrecht, the Netherlands.

Butturani, A., Tiso, R. and Molonari, F., 2000. Phenological forecasting model for *Cydia funebrana. Bulletin-OEPP* 30: 131-136.

Carter, T.R., Parry, M.L. and Porter, J.H., 1991. Climatic change and future agroclimatic potential in Europe. *International J. Climatology* 43(1): 251-269.

Chander, S., 1998. Infestation of root and foliage/earhead aphids on wheat in relation to predators. *Indian J. Agric. Sci.*, 68(11): 754–755.

Chander, S., Aggarwal, P. K., Kalra, N., Swarooparani. D. N. and Prasad, J. S., 2002). Assessment of yield loss due to stem borer in rice using simulation models. *J. Ent. Res.*, 26: 23-28.

Chander, S., Aggarwal, P.K. and Swarooparani, D.N.S., 2004). Agro-ecological zonation of leaf folder (Cnaphalocrosis medinalis) in Haryana. *Indian J Agric Sci* 74(8):455-457.

Chander, S., Ahuja, L.R., Peairs, F.B., Aggarwal, P.K. and Kalra, N., 2006). Modeling the effect of Russian wheat aphid, *Diuraphis noxia* (Mordvilko) and weeds in winter wheat as guide to management. *Agricultural Systems* 88(2-3): 494-513.

Chander, S. and Phadke, K.G., 1994. Incidence of mustard aphid, *Lipaphis erysimi* and potato aphid, *Myzus persicae* on rapeseed crop. *Ann. Agric. Res.* 15(3): 385-387.

Chander, S., and Singh, V.S., 2000. Distribution, economic injury level and sequential sampling of leaf folder on rice. *Indian J Agric Sci* 71: 768-771.

Chander, S., Singh, V. S. and Kalra, N., 2003. Aphid infestation on barley in relation to climatic variability. *Proc. National Symposium on Frontier Areas of Entomological Research*, 5-7 Nov., 2003 Entomology Division, IARI, New Delhi, pp. 37-38.

Chiarappa, L., 1980. *Crop loss assessment methods*, Supplement No. 3, FAO/CAB, Farnham, England 123 pp.

Coakley, S.M., 1985. Describing and quantifying the environment. *Plant Disease* 69: 461-466.

Coakley, S.M., 1990. Using historical weather and pest data for pest zoning. In: *Crop Loss Assessment in Rice* (ed. Teng, P.S.), IRRI, Manila, Philippines, pp 303-312.

Coakley, S.M., Scherm, H. and Chakraborty, S., 1999. Climate change and plant disease management. *Ann. Rev. Phytopathol.* 37: 399-426.

Coop, L.B., Croft, B., Murphy, C. and Miller, S., 1989. Economic analysis of African locust/grass hopper control. GHLISM–Analysis of the 1987 grasshopper campaign. International Plant Protection Centre, Oregon State University.

Daryaei, M.G., Aggarwal, P.K., Chander, S. and Singh, V.S., 2003. Use of simulation model to formulate iso-loss curves for stem borer and bacterial leaf blight incidence in rice. *Indian J. Ent.* 65(4): 532-543.

Davis, A.J., Jenkinson, L.S., Lawton, J.H., Shorrocks, B. and Wood, S., 1998. Making mistakes when predicting shifts in species range in response to global warming. *Nature* 39: 783-786.

Du, P.V., Savary, S. and Elazegui, F.A., 1997, 'A survey of rice constraints in the Mekong Delta', *International Rice Research Notes* 22.1: 43-44.

Fabellar, L.T., Fabellar, N.G. and Heong, K.L., 1994. Simulating rice leaf folder feeding effects on yield using MACROS. *International Rice Research Newsletter* 19.2: 15-16.

Girma, M., Wilde, G.E. and Harvey, T.L., 1993. Russian wheat aphid affects yield and quality of wheat. *J. Econ. Entomol.*, 86: 594-601.

Graf, B., Lamb, R., Heong, K.L. and Fabellar, K.L., 1992. A simulation model for the population dynamics of rice leaf folders and their interaction with rice. *J. Applied Ecology* 29: 558-570.

Harrington, R and Taylor, L. R., 1990. Migration for survival: fine scale population redistribution in an aphid *Myzus persicae*. *J. Animal Ecology*. 59: 1171–1193.

Hao, S., Cheng, X., Zhang, X., Hao, S.G., Cheng, X.N. and Zhang, X., 2000. Simulation on development and survival rate of brown plant hopper nymphs. *Chinese Journal Rice Science* 14: 157-160.

He, M. and Wang, S.S., 1999. Dynamic simulation of *Cnaphalocrosis medinalis* and its predators. *Natural Enemies of Insects* 12: 151-157.

Hill, D.S., 1987. *Agricultural insect pests of temperate regions and their control*. Cambridge Univ. Press, Cambridge.

Hill, D.S. and Dymock, J.J., 1989. *Impact of climate change on agricultural/horticultural systems*, DSIR, New Zealand.

Holt, J., Wareing, D.R., Norton, G.A. and Cook, A.G., 1989. A simulation of the impact of immigration of brown planthopper population dynamics in tropical rice. *J. Plant Prot. Tropics* 6: 173-187.

James, W.C. and Teng, P.S., 1979. The quantification of production constraints associated with plant diseases. In: *Applied biology* (ed. Coaker, T.H.). Academic Press, New York, pp 201-267.

Kanneganti, R., 1996. Crop models for DAFOSYM, cold hardiness and winter injury in alfalfa. In: *Proceedings of the Dairy Forage Research Centre, Conference with Dairy and Forage Industries* (ed. Satter, L.D.), USDA, Madison, WI, pp 111-118.

Kiritani, K., 1997. The low development threshold temperature and thermal constant in insects, mites and nematodes in Japan. Miscellaneous Publications of the National Institute of Agro-Environmental Sciences, 21, 1-72.

Kropff, M. J., van Laar, H.H. and Matthews, R.B., 1994. ORYZA1 an ecophysiological model for irrigated rice production. *SARP Research Proceedings*. Wageningen Agricultural University, Wageningen, Netherlands and IRRI, Manila, Philippines. pp 110.

Lawton, J. H., 1995. The response of insects to environmental change. In: *Insects in a changing Environment*. Academic Press, London.

Lever, R.J.W., 1969. Do armywoms follow the rain? *Wild Crops* 21: 351-352.

Nordh, M.B., Zavaleta, L.R. and Ruesink, W.G., 1988. Estimating multidimensional economic injury levels with simulation models. *Agricultual Systems* 26: 19-33.

Otton, M. and Hommes, M., 2000. Development of a simulation model for the population dynamics of onion fly in Germany. *Bulletin-OEPP* 30: 115-119.

Parmesan, C., Ryrholm, N., Stefanescu, C., Hill, J.K., Thomas, C.D., Descimon, H., Huntley, B., Kaila, L., Kullberg, J., Tammaru, T., Tennent, W.J., Thomas, J.A. and Warren, M., 1999. Poleward shifts in geographical ranges of butterfly species associated with regional warming. *Nature* 399: 579-583.

Parry, M.L., Carter, T.R. and Porter, J.H., 1989. The greenhouse effect and the future of UK agriculture. *J. Royal Agric. Soc.* 150: 120-121.

Penning de Vries, F.W.T., Jansen, D.M., ten Berge, H.F.M. and Bakema, A., 1989. Simulation of ecophysiological processes of growth in several annual crops. Pudoc, Wageningen, Netherlands pp 271.

Pinstrup, A.P., de Londono, N. and Infante, M., 1976. A suggested procedure for estimating yield and production losses in crops. *PANS* 22: 359-365.

Pinnschmidt, H.O., Teng, P.S., Yuen, J.E. and Djurle, A., 1990. Coupling pest effects to the IBSNAT CERES crop model for rice. *Phytopathology* 80: 997.

Pinnschmimidt, H.O., Batchelor,W.D. and Teng, P.S., 1995. Simulation of multiple species pest damage on rice. *Agric. System.*, 8: 193-222.

Pinnschmidt, H.O., Luo, Y. and Teng, P.S., 1994. Methodology for quatifying rice yield effects of blast. In: *Rice blast disease* (eds. Ziegler, R.S., Leong, S.A. and Teng, P.S.), CAB International, UK, pp 318-408.

Pollard, E and Yates, T.J., 1993. *Monitoring butterflies for Ecology and Conservation*. Chapman and Hall, London.

Pollard, E., Moss, D. and Yates, T.J., 1995. Population trends of common British butterflies at monitored sites. *J. Appl. Ecol.* 32: 9-16.

Rabbinge, R. and Rijsdijk, F.H., 1981. Disease and crop physiology: a modeler's point of view. In:*Effects of disease on the physiology of the growing plants* (ed. Ayres, P. G.), Cambridge University Press, Cambridge, U.K. pp 201-220.

Rabbinge, R., Rossing, W.A.H. and van der Werf, W., 1994. Systems approach in pest management: the role of production ecology. In (eds. Rjan, A. and Ibrahim, Y.) *Proceedings of the Fourth Intl. Conf. on Plant Protection in the Tropics*, 28-31 March 1994, Kuala Lumpur, Malaysia. pp 25-46.

Rabbinge, R., S. A. Ward and H. H. van Laar, 1989. *Simulation and systems management in crop protection.* CABO Simulation Monographs 32, Pudoc, Wageningen. 420 pp.

RAPRA, 2005. Pest risk analysis for *Phytophthora ramorum* for the EU (work package 8), http://rapra.csl.gov.uk/pra/index.cfm

Reji, G., Chander, S., Singh, V.S. and Satish, D.G., 2003. Simulation of population dynamics of rice gundhi bug. *Proc. National Symposium on Frontier Areas of Entomological Research*, 5-7 Nov., 2003, Entomology Division, IARI, New Delhi, pp 20-21.

Reji, G., Chander, S. and Aggarwal, P.K., 2008. Simulating rice stem borer, *Scirpophaga incertulas* Wlk., Damage for developing decision support tools. *Crop Protection* 27: 1194-1199.

Reji, G., 2007. Modelling yield loss due to rice stem borer and agro-ecological zonation. *Ph.D. thesis*, P.G. School, Indian Agricultural Research Institute, New Delhi.

Ritchie, J.T., Alocijia, E.C. and Uehara, G., 1986. IBSNAT/CERES Rice Model. *Agrotechnology Transfer* 3: 1-5.

Rodhe, A.L., 1990. A comparison of the contribution of various gases to the greenhouse effect. Science 248: 1217-1219.

Rossing, W.A.H., 1993. Modeling effects of insects and pathogens on growth and yield of field crops. In: *SARP Research Proceedings: Mechanisms of damage by stem bore, bacterial leaf blight and sheath blight, and their effects on rice yield* (eds. Rossing, W.A.H., Rubia, E.G., Heong, K.L., Heerati-Kasikorn, M., and Reddy, P.R.), DLO, Wageningen, IRRI, Los Banos. pp 22-32.

Rouse, D.I., 1988. Use of crop growth models to predict the effects of disease. *Annual Rev. Phytopathol.*, 26: 183-201.

Ruesink, W.G., 1976. Status of the systems approach to pest management. *Annual Rev. Entomol.*, 21: 27-44.

Satish, D., Chander, S. and Reji, G., 2007. Simulation of economic injury levels for leaf folder on rice. *J. Sci. Ind. Res.* 66: 905-911.

Savary, S., Elazegui, F. and Teng, P.S., 1996. A survey portfolio for the characterization of rice pests constraints, IRRI Discussion Paper Series No. 18, IRRI, Manila, Philippines.

Sperow, M. and Lybecker, D.W., 1998. Modeling the alfalfa stem nematode in irrigated alfalfa. *Agricultural Systems* 58: 555-570.

Spitters, C.J.T., 1989. Weeds: population dynamics, germination and competition. In: *Simulation and systems management in crop protection* (eds Rabbinge, R., Ward, S.A., and van Laar, H.H.), Pudoc, Wageningen, Netherlands, pp 182-217.

Stynes, B.A., 1980. Synoptic methodologies for crop loss assessment. In: *Crop loss assessment.* Misc Publ Univ Minnessota Vol 7, pp 166-175.

Surin, A., Aruniyaanrt, P., Rojanahusdin, W., Munkong, S., Dhitikiattipong, R. and Disthaporn, S., 1991. Using empirical blast models to establish disease management recommendations in Thailand. In: *Rice Blast Modelling and Forecasting*, IRRI, Los Banos, pp 69-74.

Teng, P.S., 1987. A comparison of simulation approaches to epidemic modeling. *Annual Rev. Phytopathol.*, 23: 251-279.

Teng, P.S., 1987. The systems approach to pest management. In: *Crop loss assessment and pest management* (ed Teng, P.S.), American Phytopathological Society Press, St. Paul, pp 160-167.

Teng, P. S., 1991. Modeling and non modeling approaches to assessing the impact of field releases of bioengineered organisms. *Phytopathology* 80: 1137-1142.

Teng, P.S., Batchelor, W.D., Pinnschmidt, H.O. and Wilkerson, G.G., 1998. Simulation of pest effects on crops using coupled pest-crop models: the potential for decision support. In: *Understanding Options for Agricultural production* (eds Suji *et al.*), Kluwer Academic Process, Great Britain, pp 221-226.

Teng, P.S., Calvero, S. and Pinnschmidt, H., 1991. Simulation of rice pathosystems and diseases losses. In: *Systems simulation at IRRI* (eds Penning de vries, F.W.T., Kropff, M.J., Teng, P.S., and Kirk, G.J.D.). IRRI Research Paper Series No. 151, pp 27-33.

Teng, P.S. and Hofer, J., 1991. Assessing the impact of biological control in plant protection. In: *Biological control in the tropics* (eds Ooi, P.A.C., Lim, G.S., and Teng, P.S.), Malaysian Plant Protection Society, Kuala Lumpur.

Teng, P.S. and Savary, S., 1992. Implementing the systems approach in pest management. *Agricultural Systems*, 40: 237-264.

Teng, P.S., Torres, C.Q., Nuque, F.L. and Calvero, S.B., 1987. Current knowledge on crop losses in tropical rice. In: *Proceedings of Intl Workshop on crop loss assessment* (ed. Teng, P.S.), pp 39-53.

Turner, J.R.G., Gatehouse, C.M. and Corey, C.A., 1987. Does solar energy control organic diversity-butterflies, moths and British climate? *Oikos*: 195-205.

Waggoner. P. and Horsfall, J.G., 1969. EPIDEm, a simulator of plant disease written for a computer', *Connecticut Agriculture Experiment Station Bulletin* No. 698, pp 1-80.

Walker, P.T., 198). Quantifying insect populations and crop damage. In: *Proceedings Intl Workshop on crop loss assessment in rice in south and southeast Asia* (ed Teng, P.S.), pp 55-65.

Wiktelius, S., 1981. Diurnal flight periodiciities and temperature thresholds for flight for different migrant forms of *Rhopalosiphum padi*. *Zeitschrift fur Ange Wande Entomologie* 92: 449–457.

Willocquet, L., Savary, S., Fernandez, L., Elazegui, F. and Teng, P., 1999. Simulation of yield losses caused by rice diseases, insects and weeds in tropical Asia. IRRI Discussion Paper Series No. 34, IRRI, 62 pp.

Wilson, J.R., Deinum, B., and Engels, F.M., 1991. Temperature effects on anatomy and digestibility of leaf and stem of tropical and temperate forage species. *Netherlands J. Agric. Sci.* 39: 31-48.

Woiwood, I. P., 1997. Detecting the effects of Climate Change on Lepidoptera. *J. Conservation* 1: 149–158.

Yang, X. B., Dowler, W.M. and Royer, M.H., 1991. Assessing the risk and potential impact of an exotic plant disease. *Plant Disease* 75: 976-982.

Yuen, J.E. and Teng, P.S., 1990. Assessing risk with stochastic dominance. *Phytopathology* 80: 997.

Zadoks, J.C., 1985. On the conceptual basis of crop loss assessment: the threshold theory. *Ann. Rev. Phytopathol.*, 23: 455-73.

Zadoks, J.C., 1984. 'EPIPRE: a disease and pest management system for winter wheat developed in the Netherlands'. *EPPO Bull.*, 11: 365-369.

Chapter 9

Information Technology in Integrated Pest Management

Niranjan Singh and H.R. Sardana
National Center for Integrated Pest Management (ICAR), Pusa Campus, New Delhi – 110 012

Introduction

Information Technology, or IT refers to systems for producing, storing, sending and retrieving digital files with suitable equipments. Digital files can be stored indefinitely without any loss of quality, they can be reproduced at almost no cost, and they can be transmitted almost instantaneously over great distances. These files can contain text, sounds and images, both still and moving. Most commonly IT is the physical form of a desktop computer connected to the internet with modem. Devices such as compact disk drives, digital cameras, scanners and colour printers can be connected to the computer to provide additional means for inputting and outputting of data. Milestone in the past development of IT include the launch of World Wide Web in 1991. The World Wide Web (WWW), is built on the idea that all files can be given an address, known as a uniform resource locator or URL. These addresses and software for browsing these addresses have created an enormous global information system. Within last decade, computers and the internet have become ubiquitous in government offices, universities, research organizations, development agencies and businesses around the world.

The application IT in Integrated Pest Management (IPM) has been one of the most exciting developments in the field over the last decade. IT has improved the efficiency of data collection, analysis, pest identification, control measure selection and pesticide application etc. IPM is a system for accomplishing a specific goal; minimizing the impact of pests by using a variety of control procedures and attempting to decrease the overall chemical input into the environment.

Relevance of IT for IPM

IPM has frequently been described as a knowledge intensive approach to farming, as opposed to chemical intensive. Information rather than pesticides is the key input for management practices at the farm level, and considerable expertise is required to collect and analyze relevant data. If farmers are to become experts, the field staff of government agencies and NGOs who advise and train farmers need more knowledge and skills than was required under earlier extension systems.

With the above in mind, we can distinguish between two different uses of IT in the context of IPM programmes. Firstly, IT support for decision making in pest management. Secondly, IT support for IPM training. The former has greater relevance in North America, Europe and Australia, where farmers are using computers, while the latter is more appropriate to the circumstances in developing countries.

The use of IT may support many steps of IPM, which depend upon accurate and timely availability of information. Implementation of IPM programmes can greatly benefit from the development of improved methods of accessing and disseminating information.

- *Step 1* Identifying the pest, its damage and life history
- *Step 2* Establishing economic threshold levels
- *Step 3* Monitoring, scouting and predicting populations
- *Step 4* Selecting and application of management practices
- '*Step 5* Evaluation of treatment efficacy

IT Tools Useful in IPM

The following are the major IT tools which may facilitate the implementation of IPM

- Databases
- Expert Systems
- Geographic Information System
- Internet

Databases

Computerized database is orderly collection of interrelated data in form of tables stored on a disk so that information required at any time may be easily searched and obtained. There are many advantages in storing the information in a computer database such as one-time data entry, larger and long term storage, ease and faster retrieval, Flexibility in analyzing data, meeting various information needs, simultaneous availability and sharing of data etc.

There may be different kinds of databases for different types of information. A pest database may contain information about the pests, their identification, symptoms, crops attacked, plant parts affected, time of appearance, control measure, natural enemies, ETL, conducive weather parameters etc. The pest database can be used for research, planning, extension and training activities. IPM being knowledge intensive, it is necessary to have huge amount of information prior to formulating IPM programme. Efforts are on for having a sound database on IPM by different agencies all over the world.

As IPM is knowledge intensive, it becomes necessary to have huge amount of information prior to formulating IPM programmes. Efforts are on for having a sound databases on IPM by different agencies all over the world.

Databases Developed for IPM in India

Pest Management Information System (PMIS) on Cotton, Basmati Rice, Chickpea, Mustard and Groundnut

National Centre for Integrated Pest Management (NCIPM) has developed database on cotton called "Pest Management Information System on Important Crops". The aim of developing PMIS is to bring all information on pests of cotton crop, their management, identification, and control measures available at one place so that complete information reaches to the end users. PMIS is the software incorporates information about pests of above-mentioned crops (insects, diseases, nematodes, weeds). High quality photographs of pests, symptoms, weeds are available for easy identification.

National Information System for Pest Management (NISPM)

This is an online pest monitoring system for Bt cotton. this is available at NCIPM website (www.ncipm.org.in).

NEMVAR

The NEMVAR database has been developed by NCIPM, which contains information about resistance of different varieties against various nematodes. Useful in identifying proper varieties in Nematode infected areas. The database was created with the help of AICRP on Nematodes and available in the form of floppy.

Nematode Management Information System

The database is also developed by NCIPM and contains information about identification, symptoms and management of various nematodes in India. The database was created with the help of AICRP on Nematodes.

Pesticide Advisor

Databases Developed for IPM in Other Countries

CABI Crop Protection Compendium

It gives outline information for 1000 insect pest, disease and weed species of the world and their natural enemies, with synonyms, host range, and geographical range (mapped). There are selected agricultural and pest management data for all countries. It has in-depth information on South-East Asia and the Pacific. The Internet address is http://pest.cabweb.org/cpc/cpchp.htm.

Database of Information Resources (DIR)

It is a continuously updated and expanded information retrieval and referral system for the worldwide electronic IPM resources. DIR has three main components including directories of Internet IPM resources, search engines, and informational databases/knowledge bases. Currently, DIR contains seven bibliographic databases, four knowledge bases and a database with more than 3,700 documented Internet resources and their hyperlinks searchable through its search engines. This database is available at http://ippc.orst.edu/DIR and via)

Codling Moth Information Support System (CMISS):

CMISS is a composite of knowledge base and informational and bibliographic databases on different aspects of coding moth taxonomy, biology, ecology, modeling and management.

Bibliographic Database on Cultural Control of Insect Pests

This database is a component of a website on Cultural Control of Pests (still under construction). The database is searchable by Author, Year, Publication (journal/book title) and Key Words.

Database of Microbial Biopesticides (DMB)

A knowledge base on using pathogens for the control of insect pest and plant diseases.

Bacillus thuringiensis - Based Biological Control of Insect Pests

A knowledge base on practical use of Bt in IPM.

Bibliographic Database

Systems Analysis and Modeling of Codling Moth Phenology and Population Dynamics: This bibliographic database currently contains more than 175 references (mostly journal/review articles, book chapters, conference papers, and technical reports). The database is searchable by Author, Year, Publication (journal/book title) and Key Words. You can add a new reference (not cited in our bibliography) to this database right from your own desktop.

Codling Moth Index (CMI)

A searchable database of worldwide codling moth literature from 1700–1997. CMI provides almost a complete coverage of codling moth literature (almost 6,000 references). This database is designed to allow convenient searches by authors, year of publication, titles, publication (journal/book title) Keywords. Citations from this database may include abstracts when available and can be freely used to create specialty databases or personal reprint indexes.

Database of IPM Textbooks (DIT)

This database provides listing of and useful information about IPM textbooks from all over the world. DIT Search Engine can be used to find IPM textbooks that contain information on a particular topic. This database also provides the whole contents of the books. For some books detailed contents and synopsis are also available.

IPMlit

This online searchable database offers a selection of published current research and technical papers focused on Integrated Pest Management (IPM) and related topics. These titles are selected from a wide array of technical and professional journals and other periodicals and are listed as a service to aid the worldwide IPM community sift through the thousands of paper and articles published each moth. IPMlit broadly groups listed papers by pest or tactic categories, *i.e.*, Biocontrol, Entomology, Nematology, Plant Pathology, Vertebrate management, and general.

Indigenous Crop Protection Practices in Sub-Saharan East Africa

A composite of knowledge bases and databases on the subject matter.

Contents

Knowledgebase

Indigenous Crop Protection Practices in Sub-Saharan East Africa, their Status and Significance Relative to Small Farmer IPM Programs in Developing Countries.

NTRAP

The Database of Natural and Traditional Pesticides - Natural and Traditional Products Used, or with Potential Use, for Crop Pest Control in Sub-Saharan East Africa.

Library of Indigenous Pest Control Knowledge

Bibliographic Database of Worldwide Literature on Indigenous Pest Control Practices and Natural Products.

Expert Systems

An Agricultural Expert System is a Decision Support System for Agricultural Extension Agents who has to decide what advice to be offered to farmers who have to decide what action to take based on it. It is one of the most efficient extension tool to take the technology from scientists to the farmers directly without any dilution of content which normally creeps in because of the number of agencies involved in normal technology transfer systems.

Decision support programs imitate an expert by involving a client in a problem-solving situation, often providing a recommendation in response to a client's request for help in making a decision. Expert systems have the potential to help farmers run their business more economically. The areas selected for use of Agriculture expert systems are applications to diagnosis the diseases and pests of various crops. In recent years, research and development of the expert system fields of agriculture domain have been paid much attention by many countries, not only by developed countries but also developing countries. The complexity of problems confronting farmers like yield loses, soil erosion, diminishing market prices from international competition, increasing chemical pesticides costs and pest resistance and economic barriers hindering adoption of farming strategies necessitates that they become expert managers of all aspects of their farming operations. On the other hand agricultural researchers need to address problems of farm management and discover new management strategies to promote farm success. Numerical methods have failed because understanding about crop systems are qualitative based on experience and cannot be mathematically represented. Expert System are Computer programs that are different from conventional computer programs as they solve problems by mimicking human reasoning processes, relying on logic, belief, rules of thumb opinion and experience.

Although there is no general standard for expert systems, most include:

- A knowledge base of domain facts and associated heuristics
- An inference procedure or control structure for utilizing the knowledge base
- A natural language user interface

The knowledge base component includes both Domain facts and Heuristics.

Domain–constitutes a body of information widely shared and generally publicly available within the domain.

Heuristic knowledge–is mostly privately and individually held. It includes rules-of-thumb, judgments, and sometimes, experience based guesses that typically characterize human expert level decision-making. In order for an expert system to solve a problem, a program must have both kinds of knowledge, facts and heuristics, in its knowledge base.

In addition to knowledge base, systems include an inference system or procedure, also commonly called the inference engine. This system contains the general problem solving approach deciding which heuristics are applied to the problem, accessing the appropriate rules in the knowledge base, executing the rule, and determining when an acceptable solution has been found. In effect, the inference system 'runs' an expert system. The remaining component part of the expert system permits bi-directional communication. It is known as the input/output system.

The most basic difference of expert systems with that of conventional computer programs is that expert systems manipulate knowledge while conventional programs manipulate data.

Expert Systems for IPM Developed in India

SOYAPEST

Web based expert system has been developed to identify active pest and to calculate the activity of active pest followed by their diagnosis. SOYAPEST uses the knowledge-based inferences for the activity calculation.

Rice-Crop

National Institute of Agricultural Extension Management (MANAGE) has developed an expert system to diagnose pests and diseases for rice crop and suggest preventive/curative measures. The rice crop doctor illustrates the use of expert systems broadly in the area of agriculture and more specifically in the area of rice production through development of a prototype, taking into consideration a few major pests and diseases and some deficiency problems limiting rice yield.

Expert System for Grapes

Indian Institute of Horticultural Research Institute, Bangalore has prepared the first software for use by the grape cultivators. This spontaneous response made them to undertake similar software for providing guidance to mushroom cultivators, which became extremely popular and a large number of growers using it regularly for getting solutions to their problems. The Institute has launched into an effort to give a comprehensive package of practices of about 148 horticulture crops for cultivation in the 4 Southern states of Kerala, Tamilnadu, Karnataka and Andhra Pradesh.

AGREX

Center for Informatics Research and Advancement, Kerala has prepared an Expert System called AGREX to help the Agricultural field personnel give timely and correct advice to the farmers. These Expert Systems find extensive use in the areas of fertilizer application, crop protection, irrigation scheduling, diagnosis of diseases in paddy and post harvest technology of fruits and vegetables.

Farm Advisory System

Punjab Agricultural University, Ludhiana, has developed the Farm Advisory System to support agri-business management. The conversation between the system and the user is arranged in such a way that the system asks all the questions from user one by one which it needs to give recommendations on the topic of farm Management. The inputs are encouraging and acceptance by farmers is very good.

Expert Systems Developed in Other Countries

COMAX

Provides information on integrated crop management in cotton. The system uses a combination of expert derived rules and results generated by the cotton crop simulation model.

GOSSYM

It requires external information such as weather data, soil physical parameters, soil fertility levels, and certain pest management information. Based on this input data, the system makes daily management decision recommendations.

POMME

Provides information about pest and orchard management of apples. External information such as weather data including forecasts and crop symptoms are utilized by the system to generate management decision recommendations.

PLANT/ds

Supports the diagnosis of soybean diseases and can be used by growers and country agents alike.

Expert Systems for Cotton Crop Management

This Expert System provides appropriate management recommendations to cotton growers.

CALEX

This is a blackboard based integrated expert decision support system for agricultural management, developed at University of California. Growers, pest control advisors, consultants and other managers, can use CALEX.

Weiping Jin Expert System

There are fertilizing, irrigating, spraying insecticide process and adopting other measures in crop management, which rely on crop state analysis. However, there is a lot of fuzzy phenomenon in crop state analyse. Then it is necessary to analysis or to measure these fuzzy phenomena by means of fuzzy set theory for crop management. An expert system that can deal with fuzziness has been developed for crop management in PR.China. This paper shows the principal construction of the expert system, CMES, for growers to manage their crop. CMES provides support for crop growth control system (CGCS), *i.e.*, advises growers about optimal population and structure of crop in planting stage and when and what to adopt measures on their crop to keep at optimal state, to avoid Infestation in various stage of growth and development of crop.

G I S Expert Systems by Naiqian Zhang

The expert system software WHEATWIZ was developed as an effective tool to assist Kansas farmers, extension workers, and agri-business personnel in variety selection for hard red winter wheat. The software offered three modes of operation. The concept of "prescription farming" or precision application of agronomic inputs may dramatically increase the attention a farmer gives to the spatial distribution of nutrients in the soil and methods of fertilizer application. By using a geographic information system (GIS) and a cotton model, GOSSYN, together as a spatial crop simulation, a methodology for investigating the implications of incorporating prescription-farming programs for cotton production is investigated.

CLIPS Expert System by L.L.M. Rhykerd

The expert system has been developed by Purdue University agricultural scientists. This form of artificial intelligence provides an extension tool, which will enable farmers to very economically use a computer program to reach conclusions concerning profitable alfalfa production that normally would require consultation with a forage expert. To date, the management considerations and the sequence in which they were built into the knowledge base are: soil drainage, soil pH, soil P test, soil K test, use of alfalfa crop, chemical weed control, expected longevity of stand, variety recommendation, method and rate of seeding and pure live seed.

Geographic Information System

A GIS is a set of computer programs that can store, use, and display information about places of interest. Information for any place of interest are soil types, rainfall and temperature patterns, land use, ownership patterns, roads, vegetation types, and topography. A GIS stores two types of data that are found on a map, the geographic definitions of Earth's surface features (spatial reference) and the attributes or qualities that those features possess. It is generally agreed that a true GIS is capable of several characteristic activities: the storage and retrieval of information with a spatial reference, as well as the input, analysis, and reporting of spatially referenced information in digital form.

GIS provides pest managers with the capabilities to store, retrieve, process, and display spatially referenced data. It seems only logical that GIS technology will be rapidly embraced because so many questions from insect ecology to pest management have a spatial component. Whether studying the dynamics of host or making prediction of pest problems. Now it is possible to identify two general areas where GIS technology has been used in entomology-applied insect ecology research and insect pest management. Within the general area of applied insect ecology, perhaps the major use of GIS is in the relation of insect outbreaks to environmental features of the landscape. The compilation and interpretation of spatially referenced insect and habitat data is a complex process, if for no other reason than the sheer volume of information. Although GIS software is designed to handle this complexity successfully, these systems often are not easy to use. In order to make a GIS more accessible to applied problems, GIS is increasingly being linked as a part of a larger decision support system (DSS). These systems typically use a GIS to manage habitat, geophysical, political, and census data. The DSS uses these data, along with other data, as input to mathematical models and other modeling methods to produce useful abstractions or recommendations. These outputs might be maps of high damage hazard or even maps of proposed control areas.

Dynamic sites that include interactive models, GIS (Geographical Information System) based decision systems, real-time weather, and market information are being rapidly developed and made available on the Internet. GIS is another form database with a difference, information are linked to their geographic locations. The use of geographic locations adds another dimension to the information and it is very convenient to analyze and interpret information and drafting policies using GIS.

Some specific applications of GIS in the field of IPM are given below.

Scouting, Monitoring and Mapping of Pest Populations

Scouting and monitoring pest populations and making control decisions based on those population estimates is one of the bases for IPM. The ability to accurately and precisely map the density and location of pest populations has obvious advantages, especially in regard to application of control tactics.

Monitoring programs can now digitally map the location of sample sites (taken in the field with a GPS) and the results can be used to interpolate the pest population over the sampled area. This is not a new technique; interpolative techniques, without GIS, have been used to estimate populations. However, GIS will help in not only estimate populations but can be linked to other similar areas throughout a region, resulting in regional maps, which estimate pest populations. This technique is used when constructing state or region wide estimates of pest populations or crop damage.

Preparation of Pest Distribution Maps

Another important application of GIS is to prepare county wise pest distribution maps. Several such maps are available in India as well as in other countries. Pest distribution maps are prepared using databases on pest, their intensity of attack on different locations over years. The maps for different years of same pest may be overlaid to obtain a composite map. This map will help in identifying hotspots of a particular pest in the country. Further, the map will also help in identifying spread of the pest, which will be very useful in quarantine purposes, and taking remedial actions well in advance. Pest Distribution maps of cotton and Rice crop in India were prepared at NCIPM for key pests of these crops, the hotspots are identified.

Precision Application of Agrochemicals

Using GIS to map the populations of weeds and plant pathogens offer some very tangible practical applications. Weeds especially are likely to re-occur in the same or neighboring area in a field. Consequently, mapping weed beds in a field during harvest and using this map to direct application of pre-emergent herbicide the following year can lower the amount of herbicide used in a production system.

Internet

The Internet is a worldwide network of networks and computers interconnected for information exchange. The advent of Internet has made the world into a real global village. The Internet has broken the geographic, cultural, religious, government barriers for interested people to come together and exchange information. Through Internet, one can retrieve information on any aspect. There is a large volume of useful IPM information available on the Internet, however, the information is scattered all across the globe. These resources range from topics such as pest identification, biology, control tactics, IPM definitions and basic concepts, to modeling and systems analysis. As awareness of the Internet increases worldwide, more people are participating not only as users of the information but also as creators of new information; as a consequence, the number of both IPM Internet servers and clients is increasing rapidly, perhaps slightly lagging but generally accompanying the exponential growth of the internet itself. Various newsgroups on IPM and agriculture are available on Internet, which facilitates exchange information on IPM. Timely information, such as state pest alerts and changing label information or spray recommendations can be available within minutes of the information being received by the Web site administrators.

The internet has a number of characteristics which make it highly suitable for this purpose:

It is Quick

The contents of a website can be revised far quicker than a printed booklet, and the delivery of the new content is immediate.

It is Persistent

The internet can be accessed at any time (compare this to fortnightly visits by extension agents).

It is Detailed

The vast scope of the internet makes it possible to find details which are relevant to the specific needs of individuals (compare this the simplified content of extension materials aimed at a large heterogeneous audience).

It is Open

There are very few controls on who can use the internet, both as senders or receivers. The internet doesn't stop people from accessing information because of their sex, race, class or age. And although most agricultural information is produced by government or commercial organizations, websites can also be created by NGOs, producer associations, community groups and individual farmers.

Internet-based decision support systems provide users with all the information they need to select the most appropriate pest control strategy, including means of pest identification, pest life cycles, sampling and decision-making criteria, sampling threshold calculations, and pest distribution models linked to weather monitoring systems. They also offer details of environmentally friendly biocontrol methods, as well as of available pesticides and their safety risks and environmental impacts. These applications could be further improved with the addition of 'intelligent' functions such as e-learning tools and dynamic simulations of crop ecosystems. IPM training materials for many crop systems are available on CD-ROM or online.

The transfer of research and extension information plays the key role in the promotion and adoption of IPM. Electronic communication provides an effective multidirectional exchange of information. Electronic extension systems provide 24-hours access to an inquirer of specific information to be used in planning and decision support. In fact, it is rapidly changing the way individuals exchange information and make decisions. Now it is possible for extension workers, researchers and educationists to deliver and receive information to and from via fax (both Internet- and telephone-based document delivery systems), multimedia programs, email, and the web. The emphasis is, however, beginning to shift from traditional one-way flow of information from research, then to extension, and finally to end-users of information, to the more egalitarian process where the pool of total experience and knowledge available in the community, from growers, industry, research and extension, is readily exchanged through electronic means, focused learning workshops, and increased on-farm applied research.

Types of Services Available on Internet

It was noted above that internet is a means to many different ends. Different groups of people require different types of information, and different mechanisms are used to get that information. A rough categorization is given below, with examples.

Mailing Lists

Email messages sent from a single source to multiple recipients are a cheap and quick way of distributing news. Currently there are no mailing lists specifically about IPM in Asia, but one list which is popular with IPM project staff around the world is the Pesticide Action Network Updates Service (PANUPS). Anybody can subscribe at this address: http://www.panna.org/panna/resources/panups.html. Another list for people who want to keep track of international meetings on the environment (including PIC, POPs and GMOs) is the Earth Negotiations Bulletin, produced by the International Institute for Sustainable Development (subscribe at http://www.iisd.ca/)

Bulletin Boards

Often called 'discussion groups', these are variants of the mailing list, which allow any member to send messages, which are received, by all other members. A number of academic and research-oriented groups exist. FPR-IPM is a discussion group run under the umbrella of the CGIAR System – Wide Initiative on IPM which links those with a special interest in using participatory research and learning methods to support IPM programmes (to join, write to a.braun@cgiar.org). The Rodent Pest Network is a bulletin board managed by CSIRO, which allows members to share information about rats and other rodents (http://www.dwe.csiro.au/research/progv/rodents/network.html).

On-line Databases and Libraries

Websites have various degrees of specialization. Some are useful because they provide only one type of information, but a lot of it. EXTOXNET is a database of information about pesticides maintained by the University of California at http://ace.orst.edu/info/extoxnet/. The Food and Fertilizer Technology Center, funded by the Governments of Japan, Korea, the Philippines and Taiwan, has created a large on-line library of scientific papers focusing on agriculture in Asia (http://www.agnet.org/). The ASEAN IPM Knowledge Network is even more specialized, providing a on-line catalogue of documents dealing with IPM in Asia; the documents are not available on-line, but can be requested by members of the network (http://asean-ipm.searca.org/)

Web-based Resource Centers

These websites integrate various types of information - news, training materials, directories of organizations and experts – but lack the depth of the on-line databases. The author of this article is Webmaster of www.CommunityIPM.org, which started as an activity of the FAO Programme for Community IPM in Asia. The site includes information about the FFS approach to IPM training in twelve countries. The Asia Office of the International Potato Center (CIP) also provides a range of information about activities in the Region, including scientific papers, training materials and contact addresses at http://www.eseap.cipotato.org. Perhaps the only example of a website at the national level which focuses on IPM is in China; the site, in both English and Chinese, is maintained by the Youth Committee of the China Society of Plant Protection at http://www.ipmchina.net/.

Organizational Home Pages

Government Departments, NGOs and development projects often create websites, which are on-line version of the organization brochure. These sites provide the organization with a presence and identity on the internet, but the information is rarely changed and has limited educational value. Examples include sites created by the Ministry of Agriculture in Malaysia (http://agrolink.moa.my/) the DANIDA project to Strengthen Plant Protection Services in Bangladesh (http://spps.hypermart.net/); and Thai Education, an NGO, which has pioneered IPM activities in Asian schools (http://thai-ed.org/).

Trends in IPM

Application of IT in IPM has already taken place all over the world and it will continue to make tremendous influence on future IPM strategies. As to increase agriculture production lots of inputs are being used and in this respect main emphasis is being given to reduce the yield losses due to different kinds of pests. Timely availability of right information for decision-making can either result in application of pesticides saving crop worth crores of rupees or in non-application of pesticides saving the cost of involved pesticides and the environment for being polluted besides saving the beneficial organisms. Thus in future it may not be possible to neglect the IT component in any IPM programme. However, in India, IT has made limited progress particularly in IPM infrastructure and other problems. But the situation is changing and it is changing fast towards good.

References

Conference on Information for Sustainable Agricultural Development Abstracts: NIC and Ministry of Agriculture, 1994.

Database of IPM Resources by Consortium for International Crop Protection (CICP), University of Illinois, Urbana, Illinois, USA and Integrated Plant Protection Center (IPPC), Oregon State University, Corvallis, Oregon, USA.

Expert Systems in Agriculture By J. Panduranga Rao.

Expert System in Developing Countries Practice and promise edited by Charles K Mann and Stephen R. Ruth, 1992.

Expert System Development Group PennState College of Agriculture Sciences.

Internet-based IPM Informatics and Decision Support Waheed I. Bajwa and Marcos Kogan from Integrated Plant Protection Center (IPPC) and Oregon State University, Corvallis.

I T challenges and opportunities proceedings of XXX Annual Convention of the Computer Society of India, Sep 1995.

Knowledge based systems in Agriculture Richard E. Plant and Nicholas D. Stave McGraw-Hill 1991.

Proceeding IFAC Workshop an Expert System and Agriculture. International Academic Publishers Beijing 1992.

Chapter 10

Fruit Flies and their Management

H.R. Sardana
National Centre for Integrated Pest Managemt, New Delhi – 110 012

Introduction

The diverse agro-climate zones in the country provide much scope for growing wide varieties of fresh fruits and vegetables in India. India is the second largest producer of fruits in the world with an annual production of 40 MT and holds second place in vegetable production (next to China) in the world. Fruits and vegetables play a vital role in maintaining subsistence for rural people, providing vitamins and other nutrients (as well as dietary diversity) at a time when nutritional quality is seen as of increasing importance as survival- level poverty continuous to decline in India (Anonymous, 2001). Of the total fruits and vegetable production in, Maharashtra, Andhra Pradesh, Karnataka, Tamil Nadu, Bihar, Gujarat, Uttar Pradesh, and West Bengal are leading states, producing 32.661 m tonnes fruits and 72.1 m tonnes vegetables annually.

Fruit fly in India is identified as one of the ten most serious problems of the entire agriculture and because of the polyphagous nature of their larvae, many species of it cause high economic losses in fruits and vegetables. Of 207 species of fruit flies in India, nine *i.e., Bactrocera cucurbitae, B. dorsalis, B. zonata, B.tau, B. diversa, B. correcta, Dacus ciliatus, Carpomyia vesuviana* (*ber fly*) *and Acanthiophilus helianthi* (seed fly) are identified to be the major and economically important.

Of the fruits, mango, sapota, guava and ber bear the heavy loss of Rs 2,916 crores (Rupees two thousand nine hundred sixteen only) approximately while among vegetables, cucurbits *viz.* bitter gourd, little gourd, ridge gourd, muskmelon, pumpkin and cucumber are also badly effected bearing the loss of about Rs 4,705 crores (Rupees four thousand seven hundred five only) annually due to havoc of fruit flies.

The flies lay their eggs under the skin ("sting") of young fruits of cucurbits or fruit trees. In suitable host plants, the egg will develop into a yellow maggot of a few millimeters in length. The maggot feeds on the pulp of the fruit. When the maggot is ready to pupate, it leaves the fruit and drops to the ground where it pupates. A week or two after pupation, the adult fly emerges and it will begin to lay eggs after another week.

The presence of fruit flies is best noticed on the symptoms of damage. Heavy attacks on young fruits lead to early fruit dropping or many rotten fruits on the plant. Cucurbit fruits often show a characteristic curving with a dark spot where the fruit was punctured during egg laying. In many cases, droplets of sap can be seen oozing from the puncture. When damaged fruits are cut open, the maggots can be seen in the fruit pulp.

Economic effects of pests species include not only direct loss of yield and increased control cost, but also the loss of export market and/or the cost of constructing and maintaining fruit treatments and eradication facilities. Thus, the destructive association with commercially grown fruits and vegetables crops makes them the subject of intensive agriculture research.

The seriousness of fruit fly damage in mango alone can be gauged from the fact that inspite of an increase in mango area by 37 per cent during the last decade *i.e.* between 1990-91 and 1990-2000, the production has decreased by 1.6 per cent. Besides, the post harvest losses the fruit fly alone causes are about 1.8 million tonnes equivalent to the monetary loss of Rs 1800 crores (Rupees one thousand eight hundred crores only) annually, which clearly necessitates the management of fruit fly in a country like India where fruits and vegetables are first step to commercial cultivation by subsistence farmers and the production is already much below the requirements.

Fruit flies controls either are absent or farmers rely on insecticides cover sprays which are neither cost effective nor sustainable. Health hazards and destruction of natural enemies are other draw backs. Apart from this there is need to understand how effective the responses are to what farmers apply to control fruit flies.

The scientists of UK and India have shown keen interest about the burning problem of fruit flies and a collaborative project was started, opting eight locations under five centres for field experiments, in three clusters in different zones of India and a knowledge review centre–NCIPM New Delhi (Sardana, Tyagi and Singh, 2005).

It is important for any effective pest control, to have knowledge about the species wise distribution and their significance, population dynamics, fluctuations, distribution patterns, environmental ecology, preference and survival among different hosts and host resistance particularly in a case such as Tephritids where species differ in their susceptibility to control. Area, production and productivity of major fruits and vegetables crops in India is given in Table 10.1 and 10.2.

Andhra Pradesh, Maharashtra, Karnataka and Uttar Pradesh are the leading states in the production of major fruits while Orrisa, Uttar Pradesh, Bihar, West Bengal and Maharashra are the leading states in the production of major vegetables. The fruit fly species *viz. B. dorsalis, B. zonatus, B. correcta, B. diversus* are most prevalent in above mentioned states restricting the production by causing heavy losses to the tune of about Rs 7000 crores (Rupees seven thousand crores only approximately) annually inspite of a large area coverage under fruits and vegetable crops. In these circumstances, there may be a positive approach towards increasing the production of fruits and vegetable of the country by managing fruit flies.

Table 10.1: Area, Production and Productivity of Major Fruit Crops in India

Fruits	*Area ('000 Ha)*	*Production ('000 tonnes)*	*Productivity (Tonnes/Ha)*	*Area (%) of Total*
Banana	482.8	16167.0	33.5	12
Mango	1522.6	10237.0	6.7	39
Citrus	496.6	4399.5	8.9	13
Papaya	70.1	1767.1	25.2	2
Guava	148.2	1631.5	11.0	4
Apple	239.8	1226.6	5.1	6
Pineapple	78.2	1221.1	15.6	2
Grapes	45.2	1056.9	23.4	1
Sapota	69.2	674.0	9.7	2
Litchi	53.6	412.0	•7.7	1
Others	680.1	6577.3	9.7	17
Total	**3886.4**	**45370.0**		

Source: Indian Horticulture Database, National Horticulture Board, New Delhi.

Table 10.2: Area, Production and Productivity of Major Vegetable Crops in India

Vegetables	*Area ('000 Ha)*	*Production ('000 tonnes)*	*Productivity (Tonnes/Ha)*	*Area (%) of Total*
Potato	1211.3	222242.7	183.5	31
Brinjal	472.1	7676.9	16.3	12
Tomato	458.7	7277.1	15.9	12
Cabbage	245.4	5617.1	22.9	6
Onion	448.9	4721.1	10.5	12
Cauliflower	256.3	4694.6	18.3	7
Okra	349.1	3344.6	9.6	9
Peas	319.3	3007.6	9.4	8
Others	2487.4	35339.8	14.2	64
Total	**6248.5**	**293921.5**		

Source: Indian Horticulture Database, National Horticulture Board, New Delhi.

Species of Fruit Flies and their Distribution

Geographical distribution of fruit flies may be the most effective information in controlling the harmful menace of pest species. It is imperative to know about such species which are trying to invade new areas and thus very important for quarantine purpose. Of about 4352 species or subspecies of fruit flies, classified in 481 genera in the world, 207 species of fruit flies under 71 genera, 13 tribes and four sub families are known from India. *Bactrocera dorsalis, B. zonata* and *B. diversa* are considered as major pest species as they infest a very large number of vegetable and fruit crops in India (Kapoor, 1970).

Information collected on previous work on fruit fly species distribution indicated that *B. dorsalis* was widely distributed in India on mango, guava, loquat, peach, banana, citrus and a variety of vegetables. In India, *B. zonata* has been reported to occur regularly in Uttar Pradesh, Punjab, Bihar, Madhaya Pradesh, Maharashtra, Tamil Nadu and Karnataka. *B. correcta* frequently attacks on jujube and tropical almond was found prevalent in Bihar, Tamil Nadu, Punjab and Madhya Pradesh in most of the studies. *B. cucurbitae* and *B. diversa* have been reported regularly on cucurbits and guava, respectively (Kumar and Lal, 1984).

Table 10.3: Species of Fruit Flies in India

Sl.No.	*Fruit Fly Species*	*Sl.No.*	*Fruit Fly Species*
1.	*Dacus brachycera*	32.	*Acanthonevra gravelyi*
2.	*D. eumenoides*	33.	*A. hemileina*
3.	*D. icariiformis*	34.	*A. imparata*
4.	*D. munroi*	35.	*A. inermis*
5.	*D. quadristriata*	36.	*A. vaga*
6.	*D. sphaeroidalis*	37.	*A. vidua*
7.	*D. crabroniformis*	38.	*Diarrhegma modestum*
8.	*Bactrocera (Polistomimetes) minax*	39.	*Hexacinia radiosa*
9.	*B. (Javadacus) aberrans pallescentis*	40.	*Phorelliosoma ambitiosum*
10.	*B. (Arodacus) cocciniae*	41.	*Rioxa parvipunctata*
11.	*B.* (A.) *trilineata*	42.	*R. sexmaculata*
12.	*B. (Bactrocera) affinis*	43.	*Themara maculipennis*
13.	*B. (B.) andamanensis*	44.	*Urophora stylata*
14.	*B. (B.) biguttata*	45.	*Acroceratitis ceratitina*
15.	*B. (B.) bangalorensis*	46.	*A. separata*
16.	*B. (B.) correcta*	47	*A. gladiella*
17.	*B. (B.) diaphora*	48.	*Acrotaeniostola apiventris*
18.	*B. (B.) parvula*	49.	*Anoplomus flexuosus*
19.	*B. (B.) occipitalis*	50.	*Callistomyia pavonina*
20.	*B. (B.) poonaensis*	51.	*Carpophthorella scutellomaculata*
21.	*B. (B.) scutellarius*	52.	*Chaetellipsis atrata*
22.	*B. (Polistomimetes) osciniae*	53.	*C. dispilota*
23.	*B. (Paradacus) pusaensis*	54.	*C. paradoxa*
24.	*B. (P.) watersi*	55.	*Galbifascia sexpunctata*
25.	*B. (Melanodaeus) citronellae*	56.	*Gastrozona balioptera*
26.	*B. (Parazeugodacus) bipustulata*	57.	*G. fasciventris*
27.	*B. (Zeugodacus) duplicata*	58.	*G. montana*
28.	*B. (Z.) gavisa*	59.	*Phaeospila varipes*
29.	*Acanthonevra dunlopi*	60.	*Phaeospilodes bambusae*
30.	*A. formosana*	61.	*Proanoplomus laqueatus*
31.	*A. fuscipennis*	62.	*P. vittatus*

Contd...

Table 10.3–Contd...

Sl.No.	Fruit Fly Species	Sl.No.	Fruit Fly Species
63.	*Sophira phlox*	100.	*Trypeta Buddha*
64.	*Taeniostola limbata*	101.	*T. indica*
65.	*T. vittigera*	102.	*Vidaliaceratophora*
66.	*Xanthorrhachis annandalei*	103.	*V. cervicornis*
67.	*X. assamensis*	104.	*V. fletcher*
68.	*X.assamensis*	105.	*V. melanonotum*
69.	*Oxyaciura monochaeta*	106.	*V. triceratops*
70.	*Sphaeniscus quadrincisus*	107.	*V. trigenata.*
71.	*Adrama apicalis*	108.	*Rhabdochaeta asteria*
72.	*A. austeni*	109.	*R. bakeri*
73.	*A. determinate*	110.	*R. gladifera*
74.	*Meracanthomyia intermedia*	111.	*Dictyotrypeta longiseta*
75.	*M. hotiensis*	112.	*Platensina amplipennis*
76.	*M. maculipennis*	113.	*P. fulvi.facies*
77.	*Ceratitis capitata*	114.	*P. tetrica*
78.	*Euphranta (Rhacochlaena) cassiae*	115.	*Indacilira basivitta*
79.	*E. (R.) crux*	116.	*Isoconia bifaria*
80.	*E. (R.) dissolute*	117.	*I. Reinhardi*
81.	*E. (R.) nigripeda*	118.	*Tephraciura basimacula*
82.	*Ichneumonosoma imitans*	119.	*T. pachmarica*
83.	*Indophranta humerata*	120.	*Tephrella decipiens*
84.	*A. discalis*	121.	*T. variegata*
85.	*A. e1ythraspis*	122.	*Acanthiophi.luslugubris*
86.	*A. riaxaeformis*	123.	*Actinoptera carignaniensis*
87.	*Acidiostigma apicalis*	124.	*A. formosana*
88.	*A. lucens*	125.	*A. montana*
89.	*Anomoia immsi*	126.	*Campiglossa cri.bellata*
90.	*A. mirabilis*	127.	*C. kumaonensi.s*
91.	*Chetostoma completa*	128.	*Paratephri.ti.s abstractus*
92.	*C. sarolensis*	129.	*Paroxyna gemma*
93.	*Chenacidiella bangaloriensis*	130.	*P. iracunda*
94.	*Hemilea cnidella*	131.	*P. lyncea*
95.	*H. praestans*	132.	*P. parvula*
96.	*Myiopardalis pardalina*	133.	*Pliomelaena udhampurensis*
97.	*Myoleja fossata*	134.	*Pliomelanena quadrimaculata*
98.	*Pseudacidia himalayensis*	135.	*Tephriti.s atocoptera*
99.	*Rhagoletis bezzianum*	136.	*T. darjeelingenss*

Contd...

Table 10.3–Contd...

Sl.No.	Fruit Fly Species	Sl.No.	Fruit Fly Species
137.	*T.ludhianaensis*	145.	*T. pteralis*
138.	*Trupanea asteria*	146.	*T. pentadactyla*
139.	*T. aucta*	147.	*T. proovi.ta*
140.	*T. augur*	148.	*T. stellata*
141.	*T. cosmia*	149.	*Craspedoxantha indica*
142.	*T.sirhindiensis*	150.	*C. octopunctata*
143.	*T. inaequabilis*	151.	*Orellia tribulicola*
144.	*T. keralaensis*	152.	*Ictericodes cashmerenisis.*

Source: Indian Fruit Flies (Kapoor).

The distribution of different species was analysed by various scientists to focus on some of the domestic and international quarantine which are necessary for the restriction of the entry of some fruit flies species in to the region where they have not been reported. Table 10.4 indicates that *B. latifrons, B. minax, B. oleae* are the species reported to cause damage in South India, Bhutan and J and K, respectively. Therefore, strict domestic quarantine mearures should be in place to check the spread of these in to other regions. *B. corambola* and *B. albighiga* have been reported in Andaman and Nicobar by past workers of fruit flies and it has been observed that by applying quarantine measures their entry in main land of India can be prevented (Ranganath and Veena Kumari, 1996; Ranganath, Veena Kumari and Prasad, 2000). The spread of another species, *B. nigrotibialis* should also be watched closely as many species of fruit flies are ready to enter within the states and from out side of India through imported fruits containing fruit flies larvae, infested fruits bags, transport etc. Data tabulated in Table 10.5 shows that occurrence of some species *viz. B. caudata, D. longistylus, Ceratitis capitata* etc in India is still controversial. Thus, there is need for serious survey work on the exact distribution of different species of fruit flies in the different regions of the country.

Table 10.4: Species of Fruit Flies Need to be Quarantined

Sl.No.	Species	Occurrence
1.	*Bactrocera latifrons*	South India
2.	*B. minax*	Bhutan
3.	*B. oleae*	Jammu and Kashmir
4.	*B. carambola*	Andaman and Nicobar
5.	*B. albighiga*	Andaman and Nicobar
6.	*Carpomyia pardalina*	–

Table 10.5: Species Waiting to Enter India/Doubtful Presence

Sl.No.	Species
1.	*B. carayae*
2.	*B. caudate*
3.	*Ceratitis capitata*
4.	*R. cingulata*

Loss Estimates in Different Hosts of Fruit Flies

For computing loss estimates due to fruit flies in different hosts, the country has been divided in to eight zones which were surveyed to observe the prevalence of different species of fruit flies. On the basis of survey estimates, the data on species wise prevalence of fruit fly in different eight zones has been tabulated. A critical look at the data (Table 10.8) indicated that among different species of fruit flies, *B. dorsalis* is widely distributed in all zones followed by *B. zonata* and *B. correcta* in orchard fruits (Gupta, Verma and Bhalla, 1990). In vegetables *i.e.* cucurbits, *B. cucurbitae* is widely distributed followed by *D. ciliatus*. *B. tau* was found prevailing only in Eastern Himalayan zones and the area of Bengal basin (Kapoor and Grewal, 1997).

Table 10.6: Economically important species of fruit flies on major fruits and vegetables

Host	*Fruit Fly Species*							
	B. dorsalis	*B. zonata*	*B. correcta*	*B. cucurbitae*	*B. diversa*	*B. tau*	*D. ciliatus*	*C. vesuviana*
Mango	+	+	+		+			
Guava	+	+	+	+	+			
Sapota	+	+				+		
Jujube			+					+
Cucumber				+		+	+	
Bitter gourd				+			+	
Little gourd				+			+	
Musk melon				+			+	
Water melon				+			+	

Table 10.7: Host Range and Distribution of Major Fruit Fly Species in India

Host range	*Fruit fly species*	*Distributuion*
Mango, guava, loquat, peach, apple, fig, quince, banana, plum, pomegreanate, citrus, apricot and a variety of vegetables.	*Bactrocera dorsalis* Hendel	Widely distributed
Peach, mango, guava, bael, sapota, tomato, toria, custard apple, brinjal, melon,ber	*B. zonata* Saunders	Uttar Pradesh, Punjab, Bihar, Madhya Pradesh, Maharashtra, Tamil Nadu, Karnataka
Citrus, ber, beal, peach, mango	*B. correcta* Bezzi	Bihar, Tamil Nadu, Punjab, West Bengal, Uttar Pradesh, Assam
Guava, citrus, banana, mango	*B. diversa* Coqu	Tamil Nadu, Karnataka, Maharashtra, Punjab, West Bengal, Uttar Pradesh, Assam
Mango, sapota, pumpkin, gourd, tomato	*B. hageni* de Meijeri	Assam, Sikkim, Bihar, Uttar Pradesh, Madhya Pradesh, Punjab, Tamil Nadu, Karnataka

The information collected from the results of various studies conducted during last 4-5 decades on the loss estimates caused by different species of fruit flies in different hosts indicated that some of the major fruits and vegetables *viz* mango, guava, ber, little gourd, muskmelon, pumpkin, ridge gourd, bitter gourd, cucumber etc bear the heavy loss annually (Kumar and Bhatt, 2002). The data on this aspect has been tabulated and analysed to compute the annual monetary loss in some of heavily damaged host crops of fruit flies.

Table 10.8: Relative Prevalence (Per cent) of Major Fruit Flies Species in Causing Loss in Different Zones

Zones	*Orchard Fruits*			*Cucurbits*		
	B. dorsalis	*B. zonata*	*B. correcta*	*B. cucurbitae*	*B. tau*	*D. ciliatus*
Himalayan Highlands West	28.0	71.2	–	95.2	–	4.76
Punjab/Ganga Plain	41.2	58.7	1.2	90.9	–	9.09
Arid/Semi-Arid West	58.8	41.1	–	89.2	–	10.7
West-Central/Lava Plateau	62.5	37.5	11.1	89.2	–	10.7
South	71.4	7.14	21.42	89.2	–	10.7
Semi-Arid/Semi-Humid East	38.5	57.4	4.02	96.1	–	3.84
Bengal Basin	44.4	55.5	–	95.2	4.76	–
Himalayan Highlands East	42.8	57.1	–	44.4	55.5	–

Average monetary loss of fruit flies in different hosts has been calculated on the basis of percentage average loss in yield per year, production per year and the recent price values (in rupees) of each fruit and vegetables (Table 10.9). Percentage loss in yield has been calculated by computing average of losses reported due to fruit flies in past studies by various workers in each host crop. Information on production per year and price (Rs/Kg) of each host crop are from the recent report of current IMFFI fruit fly key informant survey, March 2005 by John Stonehouse and Dr. Verghese. Multiplying the price (Rs/Kg) and total production/year (MT) of each host, the production in terms of rupees is calculated and finally the monetary loss has been computed. The losses of Rs 1800.00, 421.20, 31.50, 1736.00, 870.00, 504.00, 1368.00, 115.50, 112.00 crores have been calculated in mango, guava, jujube, cucumber, bitter gourd, little gourd, musk melon, pumpkin, ridge gourd, respectively. Over all loss due to fruit fly in above mentioned crops annually has been estimated to be Rs 6958.20 crore (Sardana, Tyagi and Singh, 2005).

Table 10.9: Nation Wide Monetary Loss Estimates in Different Hosts of Fruit Fly

Host	*Avgerage Approx. Loss (%)*	*Production/ Year (mt)*	*Price (Rs/kg)*	*Production (Crore Rs.)*	*Monetary Loss (Crore Rs.)*
Mango	18.00	10.00	10.00	10,000.00	1800.00
Guava	39.00	1.80	6.00	1,080.00	421.20
Jujube	35.00	0.15	6.00	90.00	31.50
Cucumber	31.00	2.80	20.00	5,600.00	1736.00
Bitter gourd	58.00	1.00	15.00	1,500.00	870.00
Little gourd	42.00	0.80	15.00	1,200.00	504.00
Musk melon	76.00	3.00	6.00	1,800.00	1368.00
Pumpkin	33.00	1.00	3.50	350.00	115.50
Ridge gourd	28.00	0.80	5.00	400.00	112.00
Total					6958.20

Population Dynamics and Environmental Ecology

Analysis of studies on population dynamics and environmental ecology of major species of fruit flies of economic importance revealed insignificant effect of relative humidity on population dynamics. However, significant positive correlation with temperature was found in most of the studies (Abraham *et al.*, 1998). Experiments conducted in mango to observe seasonal incidence revealed lowest infestation of *B. dorsalis* in April which gradually increased in May and June and slightly declined in July. Further experiments on hourly fluctuations indicated that catches of fruit flies were maximum in the afternoon with a peak between 16.00 and 17.00 hr (Bagle and Prasad, 1983).

Adults of *B. cucurbitae* overwinter in Nov.–Dec. and the pest was consistently found to be active on various cucurbits from Feb- Nov and most active from July to august (Bhatia and Mehta, 1968). The analysis of findings of various fruit fly workers revealed that the attack of *C. vesuviana* on ber starts around mid-oct and increased suddenly in mid November continuing until December. The females fruit flies preferred to oviposit on the distal or centre part of the fruit. *B. zonata* was observed maximum during the 3rd week of June, lowest during last week of Aug on peach. *B. correcta* on guava was found abundantly when temperature and RH were 26-30° C and 70-75 per cent, respectively. In some experiments *B. correcta* was also trapped in mango and sapota orchards. In mango, trapping peaked during the second fortnight of April and second fortnight of May while maximum trapping was seen in the first week of June in sapota.

In both the fruits, positive correlation was found with temperature and relative humidity. Ecological studies conducted in the current IMFFI project on population dynamics revealed the early peak appearance of orchard fly then those of melon fly. Further studies indicated that each fly type peaked first in their presumed preferred habitat until the preferred host fruits pass their production peak and after saturation, flies venture further a field in to non preferred habitats.

Host Preference, Survival and Host Plant Resistance

Analysis of literature on host preference by fruit flies and the resistance among the varieties of different hosts is of utmost importance so as to form an important component of integrated management of fruit flies. A critical look on the colected information (Table 10.7)on the above aspects in the last 3-4 decades in cucurbits indicated that bitter gourd, little gourd, round gourd and cucumber are most preferred host of *B. cucurbitae* as the maximum egg laying and development of maggots was found on these hosts. Ridge gourd, cucumber, bottle gourd, snake gourd and pumpkin were least preferred. Due to the combined influence of a tough rind and the bio chemical properties of the fruit, wild melon was preferred by *B. cucurbitae* as against of musk melon (Rajpoot *et al.*, 2002).

A critical insight into the studies conducted on the plant resistance further revealed that cultivars J 18-1, J 56-1 and Arka Jyoti of watermelon were found least preferred by *B. cucurbitae* (Pareek and Kavadia, 1995). Of bitter gourd cultivars, BG 14 was found most resistant/tolerant against *B. cucurbitae*. The different cultivars of pumpkin, sponge gourd, bottle gourd, and cucumber evaluated for their resistance. A critical analysis of last four decades shown that short green Kareli, small sugar, Pilibhit, Padmini, Sutton,s long white, IHR 79-2 and improved long green exhibited more tolerance/resistance against *B. cucurbitae* consistently over the years.

In case of fruits studies, *D. dorsalis* preferred mango, followed by guava and sapota. Of the various varieties of ber evaluated for their tolerance/resistance; Umran, Gola, ZG-2, Deshi alwar, Tas bataso, Iliachi and Tikadi were found to exhibit resistance continually against fruit flies. A deep insight on the data further indicated that infestation of fruit fly was positively correlated with fruit weight, pulp stone ratio, total soluble sugar and total sugar and it can be inferred that the varieties high in pulp contents, total soluble sugar, low acidity and total phenol were highly susceptible to fruit fly attack. Among guava cultivars, suskha¸ strawberry, Lucknow 46, chittidar were found to be most resistant/ tolerance consistantly indicated that high level of vit C, total soluble solid, and total phenol in the fruits were the factors that contributed to guava resistance to *B. correcta* (Arora *et al.*, 1998).

Of the mango hybrids evaluated for resistance against *B. dorsalis*, Nedogua, A.U. Rumani, Mehmood bahar, Neleshan gujrat, Arka punit sindhu, Majira, HY-16 and Neeludin shown multiple pest resistance including fruit fly (Kumar *et al.*, 2002).

Rearing and Artificial Diets for Fruit Fly

The effect of different nutrients and other elements used as artificial diet components, on reproductive potential of fruit flies has been observed in detail by various scientists. After critical analysis of different studies it was inferred that casein based diet incorporated with amino acids *viz.* phenylalanine and trytophan gave best growth of Tephritids larvae as against diet based on hydrolyzed yeast. The readymade salt composition of "Wesson;s salt mixture" consisting of 12 salts, has been regularly used in different experiments and it was further observed that omission of some salts *viz.* ferric sulphate, manganese sulphate from the artificial diet improved the oviposition, fertility and longevity of *B. cucurbitae* (Gupta and Anand, 1993). Thus these salts could be permanently removed from the diet and made more economical. The diet lacking in vitamin B, folic acid and biotin were responsible for reduced oviposition, oviposition period, longevity and egg viability (Kumar and Anand, 1992). Essential amino acids and minerals were found necessary for the sexual maturity of fruit flies. Alpha- trocopherol @ 5mg, ascorbic acid @1 mg/ml, sucrose @ 66.67 per (dry wt) and pH 5 were found to be optimal for longevity, total oviposition and egg viability.

Some cucurbits tried as a media for mass rearing of fruit flies in laboratories revealed, muskmelon as most suitable followed by ribbed gourd.

Further analysis on the studies conducted on the biology of fruit fly, reared in laboratory, indicated that adult emergence was highest and pre oviposition, oviposition, incubation, larval and pupal periods were minimum during the pre-monsoon (April-June) followed monsoon (July–Sep.) and post-monsoon (Oct.–Dec.) seasons.

Chemical Control of Fruit Flies in Fruit Crops and Vegetables

B. dorsalis and *C. vesuviana* are the species of fruit flies which cause an economic loss of about 2,252 crore rupees (Rupees two thousand two hundred and fifty two crores only) annually in mango, guava and jujube fruits nationwide. B. *zonata* has been reported to cause losses in apple and peach. Various insecticides *viz*, endosulphan, quinalphos, phenthioate, fenvalerate, decamethrin, monocrotophos, phosphamidon, dimethoate, methyl-parathion, malathion, oxedemeton methyl, carbaryl, econeem etc were consistently and frequently evaluated at different concentrations by different workers for the control of fruit flies in fruits (Kumar, 1995).

The analysis of results of various studies on insecticide use conducted during the last 50 years revealed that dimethoate (0.03-0.06 per cent), fenthion (0.05-0.1 per cent), deltamethrin, carbaryl (0.1-0.6 per cent) and methyl parathion (0.003-0.005 per cent) and malathion proved most effective as spray insecticides (Doharey, 1985). Waiting period for malathion, dimethoate and carbaryl were 2, 3 and 7 days and were found quite safe for use for fruit trees. Even 0.05 per cent malathion treated fruits were found safe for human consumption after washing for 30 sec on the very first day of treatment.

Soil application of fenitrothion @ 25kg/ha, under the canopy of trees resulted in reduction in adult emergence upto 80-95 per cent. In case of mango, fumigation of ethylene di bromide @ 20g/m^3 and 7 days, in two hours gave 100 per cent mortality. In various studies fenitrothion was found consistently effective followed by fenthion against *B. zonata* in peach fruits. Some studies were conducted particularly on apple and peach for the residue of chemicals applied for the control of *B. zonata*. The half life and effective life (period of protection) of fenitrothion were found higher on apple (5-6 and 23-27, respectively) than on peach (3 and 10-11days). The waiting period before the fruit could safely be consumed was 19-22 days for apple and 14-16 days for peach.

A close look on different experiments conducted at various locations for the control of fruit fly (*B cucurbitae*) in cucurbits indicated that insecticides *viz.* Malathion, fenthion, Endosulphan, dichlorovos, fenitrothion, quinalphos, leptophos, carbaryl, tetrachlorvinphos, methyl parathion, decamethrin, phosphomidon etc. has been frequently and consistently evaluated (Agarwal *et al.*, 1987). A further look on the dose of insecticides indicated that lower concentration of fenitrothion, decamethrin, tetrachlorvinphos, carbaryl, malathion and methyl parathion gave good mortality of fruit fly in cucurbits as spray insecticides (Gupta and Verma, 1979; Kavadia *et al.*, 1978). Decamethrin, fenitrothion and methyl parathion were found relatively most effective. Carbaryl was found superior than malathion in most of the trials. A close on the analysis of various studies conducted on residual effect of insecticides in cucurbits indicated that fenitrothion was highly persistent (12 days), followed by parathion methyl (7 days), malathion (7 days). However, fenitrothion residue decreased to below permissible level (0.3 ppm) with in 3 days of treatment and below detectable level with in 7-10 days. Safety interval for malathion, methyl parathion, fenitrothion and fenthion was found 2, 7, 9 and 5 days, respectively, in different studies. It can be inferred from the analysis of past studies that the above mentioned insecticides provided reasonable protection in field (Babu *et al.*, 2002).

Soil application of granular carbofuran @ 1.5 kg a.i/ha at the time of sowing gave effective control against fruit fly in cucurbits. Application of carbofuran at vining and flowering stage was not found desirable due to the high level of terminal residues in most of cucurbits (Thomas and Jacob, 1990).

Baits, Para-Pheromones and Colour Lures for the Management for Fruit Flies

In fruit crops, attractants *viz.* methyl eugenol, protein hydrolysate in combination with insecticides *viz.* malathion, fenvalerate, fenthion and dichlorovos at different concentrations have been consistently used by past workers and a close look on these experiments revealed that methyl eugenol @ different concentrations (0.33 to 2 ml) was regularly found effective attractant in combination of 0.05 per cent dichlorovos, malathion 50 EC and even 200 ml of water (Makhboor and Singh, 1998). Methyl eugenol was found more effective with DDVP as against malathion (Lakshmanan *et al.*, 1973). Addition of mango pulp in the above mentioned combinations increased trapping of fruit flies. Five sprays of the combination of fenvalerate (0.05 per cent)/fenthion (0.1 per cent) with (0.15 per.cent) protein hydrolysate at weekly interval was also found to be effective (Kumar *et al.*, 1976).

A critical look on the various experiments conducted in different zones of India under the Indo-UK IMMFI project in fruits indicated that methyl eugenol soaked wooden blocks were consistently effective than water traps of methyl eugenol. Further, among wooden blocks, plywood traps consistently proved effective material for trapping of fruit flies. Several experiments were conducted to enhance the trapping of the methyl eugenol soaked plywood blocks by adding some solvents. Increased trapping of fruit flies was observed by mixing ether and ethanol as they increased the soaking capacity of blocks.

A critical analysis of the studies conducted in the past on fruit fly management in cucurbits revealed that various kinds of baits *viz.* protein hydrolysate, methyl eugenol, gur, palm juice, dried mango juice, oil of citronella, molasses, jack fruit pulp have been used as attractants along with the insecticides *viz.* endosulphan, monocrotophos, quinalphos, malathion, fenitrothion, dichlorovos, carbaryl, carbofuran and were experimented at different concentrations for the best possible management of fruit flies (Gupta and Verma, 1982). The analysis of the results on the baits efficacy indicated that some of the combinations like fenitrothion 0.025 per cent + molasses 0.5 per cent, fenitrothion 0.025 per cent + protein hydrolysate 0.25 per cent, Jack fruit pulp + citronella oil + malathion were consistently found to be effective in trapping fruit flies. Although, malathion + 1 per cent gur was quite effective but carbaryl was found superior to malathion. Higher quantity of mixing of gur with any of the insecticides increased the infestation rather then managing fruit flies. Different combinations of insecticides and attractants *viz.* banana alone and with boiled and unboiled jaggery, protein hydrolysate, brewer's yeast, protein hydrolysate + autolysate experimented at 8 locations of different five zones experimented under current IMFFI project indicated that banana fruits impregnated with carbofuran at cut surface effectively managed fruit flies. Addition of boiled jaggery in banana increased the effectiveness of traps. Although, higher fruit flies trapped in banana traps but females trapped in protein traps were higher. Bait application proved consistently effective in case of all attractants (Protein hydrolysate, jaggery, banana etc) as compared to spraying (Dale and Jiji, 1997).

Of the pheromones evaluated in the Indo-UK fruit fly project, Cue lure was found effective than Tephritids lure. Home made wooden blocks soaked in Cue lure performed better than the ready made traps. Further, a close analysis of the studies conducted on the efficacy of the different substrates like wood, cotton wool and sponge soaked in pheromone indicated that non wood substrates like sponge and cotton wool caught flies for longer time.

References

Abraham Verghese, Devi, K.S. and Reddy, P.P., 1998. Relation between trap catch of *Bactrocera dorsalis Hendel* and abiotic factors. *Proc. of the First Nat. Sym. on Pest Management in Horti. Crop*15-18

Arora, P.K., Batra, R.C. Mehrotra, N.K. and Thind, S.K., 1998. Screening of some promising guava varieties against fruit fly. *Proc. Nat. Symp. Pest Management Hort. Crops Bangalore* : 43-44

Agarwal, M.L., Sharma, D.D. and Rahman, O., 1987. Melon fruit fly and its control. *Indian Hort.* 32(2): 38301

Babu, P.G, Reddy, D. J. and Jadhav, D. R., 2002. Comparative efficacy of selected insecticides against pests of watermelon. *Pesticide Research Journal*, 14 (2): 57-62.

Bagle, B.G. and Prasad, V.G., 1983. Effect of weather parameters on population dynamics of oriental fruit fly. *J. Ent. Res.* 7(2) : 95-98

Bhatia, S.K. and Mahta, Y., 1968. Influence of temperature on the speed of development of melon fly, *Dacus cucurbitae. Indian J. Ag. Sci.*, 40(9): 821-828

Dale, D. and Jiji, T., 1997. Pheromone trapping for the management of melon fly. *Proc. 9th Kerala Sci. Cong. Thiruvanathapuram.* 110.

Doharey, K.L., 1985. Efficacy of some insecticides against fruit flies. *Indian J. Ent.*, 45(4) : 465-469.

Gupta, J.N. and Verma, A. N., 1982. Effectiveness of fentrothion bait sprays against melon fly, *D cucurbitae* in bitter gourd. *Indian J. Agri. Res.* 16(1): 41-46

Gupta, J.N. and Verma, A.N., 1979. Relative efficacy of insecticides, as contact poisons, to the adults of melon fruitfly, *Dacus cucurbitae. Punjab Hort. J.* 41 (2): 117-120.

Gupta, D., Verma, A. K. and Bhalla, O.P., 1990. Population of fruit-flies (*Dacus zonatus* and *D. dorsalis*) infesting fruit crops in north-western Himalayan region. *Indian J. Agric. Sci:* 60 (7): 471-474

Gupta, K. and Anand, M., 1993. Effect of different constituent salts of HMW salt mixtures on growth and development of *Dacus cucurbitae* maggots in artificial diet. *Uttar Pradesh J. Zool.* 13(1): 71-75.

Kapoor, V.C., 1970. Indian Tephritidae with their recorded hosts. *Oriental Insects.* 49(2): 207-251.

Kapoor, V.C. and Grewal, J.S., 1997. A new species of *Actinoptera rodani* with distributional records of other fruit flies from India. *Entomological News* (*USA*)*:* 88(5-6): 148-150.

Kavadia, V.S., Gupta, H.C.L. and Pareek, B.L., 1978. Comparative efficacy and residues of insecticides applied singly or in mixture with attractants against fruit fly affecting long melon (*Cucumis utilissimus* Duthie and Fuller). *Pesticides,* 5(2): 183-187.

Kumar, A., Sigh, N.B., Yazdani, S.S., Sinha, R.P. and Mohammed, A., 1976. Efficacies of different baits against melon fly. *Proc. Bihar Acad. Agric. Sci.* 24(2): 71-74.

Kumar, K.K., 1995. Studies on the major pests of fruit crops and their management in mango. *IIHR Ann. Rep.*

Kumar, M. and Lall, B. S., 1984. Incidence of insect pests of litchi fruits in different localities of Bihar. *Bull. Ent.*, 25 (1): 98-99

Kumar, N.P. and Anand, M., 1992. Effect of vitamins on the growth and survival of *Dacus dorsalis* (Hendel) maggots. *Indian J. Ent.*, 54(2): 139-149

Kumar, S. and Bhatt, R.I., 2002. Crop losses due to fruit fly, *Bactrocera correcta* Bezzi in some commercial varieties of mango. *J. Applied Zool. Res.* 13(1): 64-65

Kumar, S., Nair, A.G. and Bhatt, R.I., 2002. Evaluation of promising and released mango hybrids for multiple pest resistance. *J. Applied Zool. Res.* 13 (1): 66-68.

Lakshmanan, P.L., Balasubramaniam, G. and Subramaniam, T.R., 1973. Effect of methyl eugenol in the control of the oriental fruit-fly *Dacus dorsalis* Hendel on mango. *Progressive Hort.*, 60 (7). 628-629

Makhboor, H.D. and Singh, T., 1998. Effective concentration of methyl eugenol for the control of guava fruit fly *Dacus dorsalis* Hendel in guava. *Ann. Plant Prot. Sci.*, 6(2): 165-169.

Pareek, B.L. and Kavadia, V.S., 1995. Screening of musk melon varieties against fruit fly *Dacus cucurbitae* under field conditions. *Indian J. Ent.*, 57(4): 417-420.

Rajpoot, S.K.S., Ali, S. and Rizmi, S.M.A., 2002. Relative population and host preference of fruit fly *Bactrocera dorsalis* on cucurbits. *Annals Pl. Prot. Sci.*, 10 (1): 62-64.

Ranganath, H.R., Veenakumari, K. and Prasad, G.S., 2000. Carambola fruit fly: Can we prevent its entry into mainland India? *Curr. Sci.*, 78(4): 373.

Ranganath, H.R. and Veenakumari, K., 1996. Some new records of fruit flies (Diptera–Tephritidae) from the Andaman and Nicobar islands. *Entomon*. 21(1): 95-97.

Sardana, H.R., Tyagi, A. and Singh, A., 2005. Knowledge review on fruit flies. NCIPM, New Delhi, 198 PP.

Thomas, C. and Jacob, S., 1990. Bioefficiency and residue dynamics of carbofuran against the melon fruit fly *Dacus cucurbitae* infesting bittergourd *Momordica charantia* in Kerala. *J. Ent. Res.*, 14(1): 30-34.

Chapter 11

Integrated Pests and Diseases of Temperate Fruit Crops and their Management

Farooq A. Zaki, Qazi Nissar Ahmad, M.A. Mantoo and Munazah Yaqoob

Sher-e-Kashmir University of Agricultural Sciences and Technology of Kashmir, Shalimar Campus – 191 121, Jammu & Kashmir

Introduction

India is the second largest producers of fruits and vegetables. The area under fruits alone is 32.05 lakh hectares with a production of 329.55 lakh tones. Jammu and Kashmir is the most important horticultural state with unique temperate fruit and European type of vegetable crop production. The total area during 2005-06 under the fruit crops in the state was 2.83 lakh hectares with a production of 1504.01 thousand metric tones (Table 11.1). Almost 45 per cent of economic returns under agriculture sector are contributed by horticulture, thereby adding 7 per cent to the Gross State Domestic Product (Anonymous, 2007). Among the fresh and dry fruits, apple and walnut, respectively, are the most important fruit crops in the Jammu and Kashmir. However, there is a wide gap between the present yield and the potential yield of fruits. One of the limiting factors for this gap in fruit production is severe attack of arthropod pests and diseases affecting yield and quality. Unlike agricultural crops, horticultural crops are grown as perennial monocultures, therefore, harbor the insect pest and disease problems that are entirely different and complex in nature. These two factors impose huge losses upon horticulture in terms of alleviated production, reduced quality and cost of management.

Table 11.1: Details of Major Horticultural Crops in Jammu and Kashmir during 2005-2007

	2005-06	2006-07
Area (lakh ha)	2.68	2.83
Production (thousand m t)	1419.16	1504.01
Yield (m t/ha)	5.26	5.31
Revenue (Crores)	20,00	21,12

Pests of Economic Importance in Temperate Fruits

A large number of pests have been observed on different horticultural crops with the effects ranging from mild damages to serious losses in production and quality. The arthropod pest spectrum of various crops in Kashmir (Tables 11.2–11.4) is very wide. San Jose scale, *Quadraspidiotus perniciosus* is one of the most serious and key pests on apple. Its first peak commences in the second or third week of May and second peak during second or third week of August and the 3rd generation overwinters. The pest is a sucking type, feeding on the young and old branches and twigs, reducing the vigour and vitality of trees, the surfaces of the branches appear ash grey, fruit is severe damaged rendering it unmarketable. Woolly apple aphid (*Eriosoma lanigerum*), a serious pest on apple commences during 2nd week of March continuing up to September. The pest migrates from roots to aerial portion and *vice versa* during spring and fall, with maximum dispersal during day between 4-6 pm. These form colonies on the bark of the tree trunk and the branches characterized by white cottony cushion and suck out the plant sap, weakening the tree and make it prone to the attack of other diseases. Apple green aphid (*Aphis pomi*) drains sap from the leaves, flowers, buds and terminals of apple trees reducing their vitalities. It remains active from 2nd week of April to last week of May. About 12 phytophagous mite species have been identified on apple in Kashmir (Rather and Masoodi, 1988) of which European red mite (*Panonychus ulmi*) and two spotted spider mite (*Tetranychus urticae*) are the most important. European Red Mite has emerged as a limiting factor in apple production in Kashmir. European red mite completes six generations with high biotic potential, starting its activity in mid spring and continuing till late autumn. The infestation of the mites causes yellowing and bronzing of the foliage, affecting the photosynthetic efficiency of the leaves and results in the premature leaf drop. The situation is worsened by the presence of *Alternaria* causing complete leaf blotch rendering the apple tree denuded (Zaki *et al.*, 2004). Stem borer (*Aeolesthes sarta*) and bark beetles (*Scolytus* sp.), the pests of both fruit as well as forest trees, have emerged as a serious problem for apple and walnut. Codling moth (*Cydia pomonella*) is a pest of international importance. However, in Jammu and Kashmir its incidence is confined to Leh and Kargil. This pest attacks many fruit trees *e.g.* apricot, pear, walnut, etc. besides all cultivars and varieties of apple. The pest inflicts direct damage to the apple and apricot rendering the fruit unfit for the human consumption and market. The extent of losses caused to apple crop alone is the tune of 50 per cent in Ladakh. Indian gypsy moth (*Lymantria obfuscata*), is a voracious polyphagous pest attacking many fruit and tree plantations including apple and walnut. Chaffer beetles (*Adoretus simplex, Brahmina coriacea, Holotrichia longipennis*) damage fruit crops especially High-density plantation of apple and walnut. Among 70 species of leaf feeding beetles recorded in temperate regions of India, scarabaeids, cetoniids, curculionids and chrysomelids in that order cause significant damage to pome and stone fruits. Walnut weevil (*Alcides porrectirostris*) prefers thin shelled varieties to the less valuable thick shelled walnut types and is restricted in its distribution to Tangdar area of Kupwara district. The kernel of the nut is damaged and the attacked fruit usually drops or is unmarketable. The dropping of the fruit starts from 2nd week of May till end of August. Another important pest associated with walnut is grey weevil (*Myllocerus* spp.). It feeds on foliage by biting holes in the lamina or by peripheral feeding on margins. The maximum damage is caused from mid May to mid June. Among leaf rollers, folders *Archips pomivora* is the most important. It causes maximum damage in June. It attacks foliage but injury on fruits and their rotting cause economic losses. An important pest of Red cherry and Black cherry is cherry fruit worm. The extent of damage if caused by this pest is predominantly visible when the cherry starts to ripen. The cherry is also infested by Cherry fruit fly if the fruit harvest is delayed in certain localities. Pomegranate fruit borer (*Deudorix epijarbas* Moor) a widely distributed and predominant pest of pomegranate limiting its cultivation and causing considerable loss to the crop. Almond mealy bug (*Drosicha dalbergia*) is a serious pest on almond which feeds on both aerial and underground parts of the plant, colonizing at the collar region of the stem, in the crevices and the wounded sites. Infested plants exhibit sickly appearance, thereby, reducing the quality and quantity of the fruit.

Table 11.2: Major Arthropod Pests under Kashmir Agro-Climatic Conditions

Pest	Scientific Name	Active Period	Remarks
San Jose scale	*Quadraspidiotus perniciosus*	March-Oct.	Serious key pest
Wooly apple aphid	*Eriosoma lanigerum*	Feb.-Nov.	Perennial pest
Gypsy moth	*Lymantria obfuscata*	April-June	Attaining a serious status
Apple green aphid	*Aphis pomi*	May-August	Perennial pest
European red mite	*Panonychus ulmi*	April-Nov.	Key and serious pest
Two spotted spider mite	*Tetranychus urticae*	April-Nov.	Key and serious pest
Short hole borer	*Scolytoplatipus* sp.	April-Oct.	Moderately serious pest on apple but most serious on walnut
Stem borer	*Aeolesthus sarta*	April-Nov.	Serious pest
Pear psylla	*Psylla pyricola*	May-Sep.	Minor pest
Leaf minor	*Lyonetia prunifoliella*	April-June	Moderately serious
Chaffer beetles	*Protacta neglecta*, *Adoretus* spp.	Grub attacks throughout the year. Adult from June-July	Serious, and moderately serious, respectively
Tent caterpillar	*Malacosoma kashmiriensis*	April-Nov.	Moderately serious
Walnut weevil	*Alcides porrectirostris*	April-Sep.	Most serious
Pin hole borer	*Scolytoplatipus* sp.	April-Oct.	Most serious
Cabbage butterfly	*Pieris brassicae*	March-Oct.	Serious
Mustard saw-fly	*Athalia proxima*	Feb.-May	Sporadic pest
Aphid	*Myzus persicae*, *Lipaphis erysimae*	Throughout the year	Most serious
Diamond backmoth	*Plutella xylostella*	June-Oct.	Moderately serious
Bark beetle	*Scolytus nitidus*	April-Oct.	Moderately serious
Cut worm	*Agrotis ipsilon*	May-June	Serious
Army worm	*Mythimna separata*	March-Sep.	Sporadic pest
Cabbage aphid	*Brevicoryne brassicae*	Throughout the year	Serious

Table 11.3: Arthropod Insect Pests Attacking Apple in Kashmir

Sl.No.	*Name of the Pest*	*Scientific Name*	*Order*	*Family*
1.	San Jose Scale	*Quadraspidiotus perniciosus*	Homoptera	Diaspididae
2.	Woolly apple aphid	*Eriosoma lanigerum*	Homoptera	Aphididae
3.	Leaf blister mite	*Phytoptus pyri*	Prostigmata	Eriophyidae
4.	Gypsy moth	*Lymantria obfuscata*	Lepidoptera	Lymantriidae
5.	European red mite	*Panonychus ulmi*	Prostigmata	Tetranychidae
6.	Two spotted spider mite	*Tetranychus urticae*	Prostigmata	Tetranychidae
7.	Shot hole borer	*Scolytoplatipus* sp.	Coleoptera	Scolytidae
8.	Bark beetle	*Scolytus niditus*	Coleoptera	Scolytidae
9.	Leaf miner	*Lyonetia prunifoliella*	Coleoptera	Phyllocnistidae
10.	Codling moth	*Cydia pomonella*	Lepidoptera	Tortricidae
11.	Stem borer	*Aeolesthes sarta*	Coleoptera	Cerambycidae
12.	Tent caterpillar	*Malacosoma kashmirensis*	Lepidoptera	Lasiocampidae

Table 11.4: Insect Pests Attacking Walnut in Kashmir

Sl.No.	*Name of the Pest*	*Scientific Name*	*Order*	*Family*
1.	Walnut weevil	*Alcides porrectirostris* (Marshall)	Coleoptera	Curculionidae
2.	Stem borer	*Aeolesthes sarta* (Solsky)	Coleoptera	Cerambycidae
3.	Pin hole borer or shot hole borer	*Scolytoplatipus* spp.	Coleoptera	Scolytidae
4.	Chaffer beetle	*Protecta neglecta* *Adoretus simplex*	Coleoptera	Scarabaidae
5.	Indian Gypsy moth or hairy caterpillar	*Lymantria obfuscata* (Walker)	Lepidoptera	Lymantriidae
6.	Walnut leaf defoliater	*Chaetoprocta odata* (Hewitson)	Lepidoptera	Lycaenidae

Diseases of Economic Importance in Temperate Fruits

The major diseases affecting the apple and other horticultural crops (pome and stone fruits) are mainly caused by fungi, and some are also caused by bacteria and viruses (Anderson, 1956; Johnson *et al.*, 1986; Singh, 2000; Sabiha *et al.*, 2006; Slippers *et al.*, 2007). Apple scab and pear scab caused, respectively, by *Venturia inaequalis* and *V. pirina*, affect leaves, fruits, petioles, and green twigs. Infections on the leaves and fruit are most common and obvious. Fruits are rendered unmarketable. Apple blotch caused by *Alternaria mali* occurs mainly on leaves but sometimes also on fruits and twigs. Severe infection causes pre-mature leaf fall. *A. mali* is a weak pathogen and its infection is facilitated by leaf injuries due to European red mite resulting in two-fold increase in leaf fall. Sooty blotch and fly speck caused by *Gloeodes pomigena* and *Schizothyrium pomi*, respectively, are also important diseases of apple and pear that occur exclusively on fruits, rendering them shabby and unmarketable. Either or both of the pathogens reportedly downgrades 5-30 per cent of the apple crop annually inspite of adequate fungitoxicant applications, causing losses upto 17 per cent or more (Williamson and Sutton, 2000; Batzer *et al.*, 2002; Sabiha *et al.*, 2006). Peak period occurs one month prior to harvest particularly in orchards with high humidity conditions. Powdery mildew is caused by *Podosphaera leucotricha* on apple and by *Sphearotheca pannosa* on stone fruits which include almond, apricot, peach, plum and cherry. Both the pathogens infect blossoms, petals, sepals, receptacles and peduncles but blossom infection is common and economically important as mildewed blossoms either fail to set fruit or produce small, stunted and / or rosette fruit. Root rot caused by *Dematophora necatrix* in apple trees has been increasingly imposing problem and resulting in the death of plants. This soil borne disease also spreads to the neighboring sites. Thus economic losses are likely to multiply in compound manner within and outside the infected orchards. A number of new canker diseases have appeared on pome and stone fruits due to increase in area under apple cultivation along with growing of certain susceptible varieties. However, stem and branch cankers cause huge losses through girdling of branches, limbs and dieback of twigs resulting in death of plants and are aggravated through the attack of bark beetles. Most frequently pathogens associated with these cankers include *Cytospora cincta, C. leucostoma, Botryospheara obtuse*, etc. Collar rot caused by *Phytophthora cactorum* exhibits the oozing of bark at the collar portion. Once infection is deep seated, xylem conduction is affected leading to die back is seen to occur. Shot hole caused by *Wilsonomyces carpophilus* is most common disease of stone fruits. Pathogen attacks blossom, fruit and twigs. In almond, pre-mature leaf fall occurs and after sometime, new flesh of leaves develop on infected plants. Thus, plant vitality is reduced which subsequently affects its ability to produce fruits in subsequent years. In apricot and almond, gummosis is a serious problem. Leaf curl caused by *Taphrina deformans* besides peach and apricot also occurs, though less commonly, in almond. Leaves become puffy, develop red tints and fall pre-maturely. *Cryptosporiopsis* blight caused by *Cryptosporiopsis* sp. attacks fruit buds, leaf buds, leaves and current year twigs of stone fruits especially on almonds. There is no fruit bud development and most commonly die back occurs. Brown rot occurs on almost all the stone as well as pome fruits and is caused by *Monilinia fructicola, M. fructigena and M. lexa*. Brown spots appear on blossoms and fruits. Infected fruits shrink and turn black to become mummified. Crown gall and hairy root are caused by *Agrobacterium tumifacience* and *A. rhizogenes*, respectively. They occur on both pome and stone fruit orchards and nurseries. Globular or irregular galls or tumor like enlargements appear on roots and crown (at or near the ground line). When they are accompanied by excessive root formation, the disease is termed as hairy root.

Pome and stone fruits, as reported elsewhere, are also prone to various bacterial and viral diseases which include bacterial fire blight of pome fruits (*Erwinia amylovora*), apple mosaic virus, apple russet ring virus, pear stony pit virus, apricot ring pox virus, cherry mottle leaf virus, cherry rasp leaf virus, cherry short stem virus, black line virus. They can cause leaf distortion and enations, fire blight, dieback, gummosis, hypertrophy, stock-scion incompatibilities, defoliation and even tree death. They can also affect size, shape and quality of fruit.

IPM for Major Insect Pests on Fruit Crops in Kashmir

Mineral Oil Sprays

Horticulture mineral oils that are used for the management of important fruit pests are physical toxicants and result in the death of insect pests due to asphyxiation. They are eco-friendly and bio-safe as they come from earth and are formed as a result of transformation of plankton due to the action of high pressure and high temperature over millions of years. Comparative bioefficacy of different summer spray oils against San Jose scale and European red mite (ERM) in Kashmir during 2004-2008 resulted in a maximum mortality per cent of about 87.71 and 94.15, respectively (Table 11.5). A number of such mineral oils, viz., ATSO spray oil, HP spray oil, Arbofine spray oil, Mak all season spray oil, etc. at 2 per cent and 0.75 per cent concentrations during winter and summer, respectively, have been found effective against the scale and mites (Sofi *et al.*, 2008, Zaki *et al.*, 2008a, Zaki *et al.*, 2008b).

Table 15.5: Comparative Bioefficacy of Different Summer Spray Oils Against European Red Mite (ERM) in Kashmir

Treatment	*Concentration (%)*	*Per cent Mortality*			
		Arabal (Srinagar)	*Hanjiwara (Baramulla)*		*Poolen Mean*
		2004	*2003*	*2004*	
PD Spray oil	0.75	83.53	92.42	82.00	86.83
	1.0	92.99	95.67	91.44	94.15
ATSO Spray oil	0.75	83.21	90.45	84.05	83.34
	1.0	89.20	96.74	88.18	92.45
Orchex-796 Spray oil	0.75	83.01	92.12	82.75	85.96
	1.0	93.35	94.69	87.09	91.71
Arbofine Spray oil	0.75	84.53	92.94	82.64	87.75
	1.0	92.35	95.24	91.08	94.01
HP Spray oil E	0.75	81.43		87.63	84.53
	1.0	90.18		88.40	89.29
HP Spray oil	0.75	77.98		81.21	79.59
	1.0	88.18		87.38	87.78
Sparrow 888 plus Spray oil	0.75	76.88		71.89	74.38
	1.0	88.69		80.20	84.44
Control		21.61		25.87	23.74

Each figure is mean of 15 replicates and each replicate is mean of 12 observations.

Chemical Control

The most critical issue on the use of chemical control is the precise timing of insecticidal application, which is important for achieving the desirable control of important insect pests. Every year SKUAST-K, Shalimar prepares a spray schedule (Annexure I) where the proper timing and application of particular pesticides is provided to farmers. By and large, two summer sprays of an acaricide/insecticide or insecticide-cum-acaricide is advocated during the third to fourth week of May and first to second week of July, respectively, to control and manage most of the arthropod pest of apple crop in Temperate Horticulture. Normally no acaricide/insecticide is recommended beyond that period, thereby ensuring that no acaricide/insecticide residues are in the fruit at the harvest. Need based sprays are recommended only if the mite and /or San Jose scale populations exceed economic threshold level during August or after harvesting.

Biological Control

A large number of natural enemies of the horticultural and vegetable pests have been recognized in Kashmir. The biological control agents associated with the important pests of horticultural crops are presented in Table 11.6. The important biocontrol agents include *Aphytis proclia*, *Encarsia pernisiosi*, *Chilocorus bijugus*, *Adalia tetraspilota*, *Coccinella septumpunctata* and *Chrysoperla carnae*, against San Jose scale (Anonymous, 1988); *C. septumpunctata*, *A. tetraspilota* and *Symnus grsoilis* (against Apple aphid); *Aphelinus mali*, *C. septumpunctata*, *C. bijugus*, *Amblyseius finlandicus*, *Phytoseius mixtus* and *Stethorus* sp. (against European red mite and Two spotted spider mite); *Scymenus gracilis* (against Peach leaf curl aphid). During 1989-90 studies on the population record of predators associated with woolly apple aphid in Kashmir revealed presence of *Aphelinus mali* an eulophid endoparasite and as many as 7 predators. During the year 2000, observations in respect of predators were same. However, during 2002 only three predators including *Chilocorus* sp., *Coccinella* sp. and *Phasoscymnus balteatus* were found. The prevalence of pests has been observed to directly influence the population status of its natural enemies. Appearance and population built-up as well as disappearance of coccinellid beetle and syrphid fly was found to follow the population built-up pattern of their host aphid.

Table 11.6: Biological Control Agents Reported from Jammu and Kashmir

Pest	*Pest status*	*Biological Control Agent*
San Jose scale (*Quadraspidiotus perniciosus*)	Serious key horticulture pest	*Aphytis proclia, A. paramculicornis, Azotia kashmirensis, Maristta camesi, Encarsia perniciosi* (Parasitoids); *Chilocorus bijugus, Pharoscymnus flexibilis, Cybocephalus sp., Sticholitis marqinalia, Adalia tetraspilota, Melachius Sp., Harmonia Sp., Coccinella septempunctata, Chrysoperla carnae* (Predators).
Apple aphid (*Aphis pomi*)	Perenial pest	*Coccinella septempunctata* Linnus, *Adalia tetraspilots* Hoe, *Hippodsima variegate* Goeze, *Harmonia breiti* Madar, *Pharoscymnus* sp., *Symnus grsoilis, Exochomus uropygialia, Plantynaspis sp.* Mulsant, *Propyles japonica* Thunberg, *Dropephylla almorensis* Cham, *Chrasoperis carnae* Stephen, *C. prestes* Banks, *Tomnoatethus* (*Ectemnus*) *paradoxus* Hetchinson, *Lygus sp., Episyrphus balteatus* DeGreer, *Eumerus elbirfrons, Metasyrphus confrator* Weid, *Eristalis tenax* Linnaeus, *E. arbustorum* Linnaeus, *E.correalis* Fabricius, *Scaeva pyrastri* Linnaeus, *Sphaerospheria ecripta* Linnaeus, *Aphidoletes aphidomyza* Rondani, *Pachyneuron* Sp.
Wooly apple aphid (*Eriosoma laniqurem*)	Perenial pest	*C. septempunctata, C. bijugus* (Predator); *Aphelinus mali* (Parasitoid)
European Red mite (*Panonychus ulmi*)	Key and Serious pest of apple	*Anystis baccurum* Linnaeus, *Eusius vignus* Rishi and Rather, *Phytoseius domesticus* Rather, *Euseius insanus* Khan and Chaudhri, *Amblyseius finlandicus* Oudemans, *Phytoseius mixtus* Chaudhri, *Euseius sp., Cheyletus sp., Agistemus sp.* (Acarine predators); *Hippodamia sp., Stethorus punctillum, Stethorus punctum, Chrysoperla carnea, Oligota sp., Anthocoris sp., Scolothrips sp.* (Arthropod predators)
Mealy plum aphid (*Hyalopterus arundinis*)	Moderately serious	*Coccinella undecimpunctata, Adalia tetraspilota, Hippodamia variegata, Scymenus gracilis, Episyrphus balteatus, Eumerus nepalensis, Sphaerophora indica, Syrphus* sp, *Anthocoris* sp, *Chrysoperla carnia, Trioxys rishii.*
Peach curl aphid (*Anuraphis helichrysii*)	Serious pest	*Scymenus gracilis*
Walnut aphid (*Callaphis juglandis*)	Serious pest	*Metasyrphus corollae*
Oystershell scale (*Lepidosaphes kirgisica*)	Minor	*Aphytis paramaculicornis*

Cultural Control

Burlapping

Burlapping is a technique of placement and wrapping of a gunny bag band or corrugated cardboard band around the tree trunk at 4-6 feet above ground level to provide artificial hibernating/ nesting sites. This practice is followed for the management of codling moth in Ladakh where the larvae are collected and destroyed at an appropriate time which reduces the primary infestation for the next season or the next generation. Gunny bag burlaps have been found to be more effective and economic because these can be re-used. Burlapping in integration with the insecticidal spray at Kargil under moderate infestation showed that the number of larvae trapped in corrugated cardboard in different insecticidal spray ranged between 2.3 larvae in fenvalerate treatment to 12.3 larvae in endosulfan as compared to 17.3 larvae in unsprayed control. The results indicated that even without insecticidal application the efficacy of this practice is quite satisfactory. Gunny bag burlaps have been found to be more effective and economic because these can be re-used. The practice of burlapping have also been found effective in the trapping, collection and destruction of caterpillars and eggs of Indian gypsy moth, which otherwise take shelter under the bark of the trees. Bark scrapping for the control of the gypsy moth and San Jose scale reduces the pest infestation.

Fruit Destruction

Destruction of infested fruits is one of the important cultural practices for the management of various insect pests. This method can bring up the effective control of fruit borers eg. Cherry fruit fly, Pomegranate borer, codling moth, etc. Department of Horticulture at Leh and Kargil had adopted a practice of destroying or burying the apple fruits infested with codling moth in farmers' fields and paying remuneration to farmers. The practice was carried out with an objective to destroy the larvae believed to be infesting the fruit and to reduce the larval population and subsequent incidence for the next year. However, during 1992-93 on the recommendation of the author the practice was stopped by the District Development Commissioner, Kargil inspite of the fact that 80 per cent of fallen fruits were found affected by the codling moth. This shall theoretically eliminate 35 per cent of primary infestation. However, our observation indicated that by the time the department initiated the process of fruit collection and destruction in the farmers' field the full fed larvae are already out of the infested fruit to take shelter and undergo hibernation beneath tree barks, brickbats, stone bunds, adjacent houses, etc. The destruction of affected but non infested fruits from which the larvae exit had remained of no practical value for larval destruction over a period of many years, therefore, rendered redundant (Personal communication).

Collection and Destruction of Wood

Wood is more susceptible to the infestation of stem borers, pin hole borers and short hole borers. Therefore, placing some logs of dry hard wood in the orchards may considerably prevent these orchards from infestation of such insect pests. The wood thus placed act as a trap strategy to these pests, which need to be collected and destroyed.

Training and pruning

Proper training and pruning of fruit trees is very important cultural practice as it helps in the management of economically important insect and non-insect pests *e.g.* Aphids, San Jose scale, European red mite, bark beetle, etc. Off season biology of the important pests is carried over on the tree branches and trunk/bark, therefore, pruning of infested parts reduces the initial pest load in the fruit orchards. The borer and bark beetle infested branches and twigs should be removed.

Fertilizer Management

It has been found that the use of fertilizers help in improving the fertility of soil, however, the balanced dosage of fertilizers is equally important. When the nitrogen is applied in higher amounts, it increases the quantity of sap but at the same time increases the infestation of sap sucking pests eg. Green aphids, black aphids, wooly aphid, European red mite, peach leaf curl aphid, San Jose scale, two spotted mite, almond mealy bug, etc. Balanced doses of fertilizers, micronutrients and higher doses of potassium were found to reduce the infestation of ERM and aphids.

Habitat Management

The ground cover management around fruit trees is another important cultural practice as it provides habitation to various predators *e.g. Amblyseius* sp. during application of pesticides and dormancy. It also harbours the prey for such predators during the off season. Therefore, in order to enhance the efficiency of biological control of European red mite, it is advised to allow some flora around the apple tree trunks.

Scrapping

Scrapping of San Jose scale infested twigs with khurpi is another technique, yet uncommon among the farmers but can help in reducing the infestation and further increment of scale population considerably.

Quarantine Control

The distribution of codling moth in India is restricted to Ladakh only. In order to prevent its accidental entry to other localities and particularly the temperate fruit growing states of Himachal Pradesh, Uttranchal, Arunchal Pradesh and Uttar Pradesh, the pest has been brought under domestic quarantine through an SRO promulgated by the Government of Jammu and Kashmir under " Destructive Insect Pest and Weeds Act, 1914 amended 1968" Clause-II of Government of India. The export of fresh fruit from Ladakh to other places is banned which is regulated by the Jammu and Kashmir State Department of Horticulture, at the Sonamarg post. The domestic quarantine has, therefore, checked its entry form Ladakh to other places. However, with the operating of Leh-Manali road and no quarantine check post on the highway there are all apprehensions of this pest getting entry into Himachal Pradesh through the fresh fruit transport by the tourists and exchange of plant materials. The issue needs to be reviewed seriously by the Government of Himachal Pradesh and the Government of India for Northern and N. E. states. Import of plant materials from other countries in the past without following proper regulations have introduced a number of pests and diseases not reported earlier. Root rot of apple in epidemic form in Kashmir and the report of dreaded blackline virus in walnut nursery at Zainpora, Kashmir which prompted the Department of Horticulture to destroy the plantations, are two examples of many unwanted troubles brought in due to poor quarantine. Fire blight caused by *Erwinia amylovora* reported infecting roses by Indian Institute of Horticulture, Bangalore (Personal communication), has probably been introduced in the country along with the rose plants imported for the new varietal introductions. The apple growing states of India should be vigilant on the introduction of rose plants infected with the bacteria through AICRP on Ornamental Crops, which may destroy apple crop on which the disease is not reported so far. The domestic quarantine measures need to be enforced on the movement of the rose plants in the apple growing states.

IDM for Major Diseases on Fruit Crops in Kashmir

IDM of pome fruit diseases in temperate horticulture is mainly based on following practices:

1. Orchard sanitation which includes:
   (*i*) proper removal and destruction of fallen leaves, fruits and twigs to reduce primary inoculum (Holb, 2006; Qazi *et al.*, 2008).
   (*ii*) removal of mummified/culled fruits to reduce the further spread of pre- and post harvest diseases/rots (Biegh and Qazi, 2005).

   Studies at SKUAST-K, Shalimar, indicated that sanitation provided an opportunity to delay start the spray programme around pink bud to petal fall stages without any reduction in disease control of leaf and fruit scab (Qazi, 2008). Furthermore, the delaying of first spray in sanitated orchard even upto petal fall stage provided more effective control as compared to initiating the spray proramme in unsanitated orchard as early as silver tip or green tip stages. Thus, a minimum of one initial fungicidal spray could be saved in bearing apple orchards by employing sanitation measures. Pre-leaf fall application of urea or fungicides in autumn is another strategy to minimize the primary inoculum of apple scab and leaf blotch pathogens (Vaseem and Qazi, 2001).
2. Use of spore traps to monitor maturity and release of spores of various disease causing pathogens.
3. Removal of alternate and collateral hosts.
4. Proper pruning and training of fruit trees (Sagar and Qazi, 2004). Remove mildewed twigs and shoots at least more than 10 cm below the last point of visible infection. Apply wound dresser like Bordeaux paste, Chaubatia paste or any copper paste on the cut surfaces.
5. Uses of disease forewarning devices like METOS to record infection periods which help in decision making for need based sprays and in turn help in avoiding the use of unnecessary fungicidal sprays. In one of our study (Qazi and Mir, 2002), METOS-based spray schedule comprising of 5 or 6 fungicidal sprays proved to be equally as effective as that of 10-14 days interval schedule (9 sprays) and recommended spray schedule (6 sprays) in controlling scab both on fruits and leaves. However, METOS-based spray schedule usually has been more of a curative type while as rest of the schedules are generally prophylactic or protective in nature. Although the continuous, frequent and prolonged use of systemic fungicides has to be minimized to avoid or restrict the risk of development of resistance in the pathogens (Koller *et al.*, 2004, Ishii, 2006; Nassreen and Qazi, 2008). The METOS-equipment takes into consideration, besides others, the amount of rainfall and the type of fungicide used in preceding spray.
6. Application of biocontrol agents like *Trichoderma* spp., *Bacillus subtilus, Chaetomium globosum,* etc. have been proved effective in controlling the root rot and collar rot diseases. They can also reduce primary inocula of foliar diseases during fall but these bio-agents do not prove to be effective as foliar application during growing season, due to various environmental conditions (MacHardy, 1996; Qazi and Sagar, 2004).
7. Post-harvest application of fungitoxicants and urea on apple enhances the decomposition of fallen leaves and minimizes the primary source of inoculum (Thakur and Sharma, 1999; Vaseem, and Qazi, 2001; Qazi *et al.*, 2005).

8) Promotion of disease resistant varieties would minimize the pesticidal usage and subsequent cost of production.
9) Application of effective fungitoxicants as per the spray schedule that is prepared and recommended by SKUAST- K, Shalimar every year (Annexure I; Anonymous. 2008). In another study at SKUAST-K, Shalimar, both systemic (hexaconazole 5EC @ 0.05 per cent, diniconazole 25WP @ 0.03) and non-systemic (captan 50WP @ 0.30, mancozeb 75WP @ 0.30, zineb 75 WP @ 0.30) alone and their combinations (hexaconazole + mancozeb 34.25 SL @ 0.15, hexaconazole 10WP + captan 70 WP @ 0.10 per cent) fungitoxicants effectively controlled the disease complex of sooty blotch and flyspeck (Sabiha *et al.*, 2006). However, hexaconazole and zineb treatments were most effective in restricting intensity of sooty blotch and flyspeck disease complex as low as 0.50 per cent compared to 17.50 per cent recorded in check.
10) Propagate nursery stock certified as bacteria/virus-free and ensure that both rootstock, scion and budwood are used from fruit plants free from any disease symptom as viruses / bacteria spread primarily through infected propagative material and insect vectors. Application of pesticides for vectors like aphids, leafhoppers, psyllids, mealybugs, mites, etc. would minimize the transmission of these viruses. Although bacterial fire blight disease has been reported only on roses in India. Thus, adequate quarantine measures should be strictly adopted to restrict the spread of this dreaded disease to pome fruits.

IDM of stone fruit diseases in temperate horticulture is mainly based on following practices:

1. Pruning of diseased twigs, shoots and wood during dormancy and ensure their destruction to minimize the source of primary inocula. Removal and destruction of mummified fruits.
2. Nursery sanitation to minimize the amount of primary inocula.
3. Rogue out the diseased plants especially in case of viral diseases.
4. Secure bud wood from trees free from diseases.
5. Wrap the well-fitted grafts with antiseptic tape.
6. Mummified fruits and prune out infected twigs or cankers from trees and orchards floor should be removed and their destruction ensured. This can be even done during the active growing season.
7. Diseased twigs should be pruned during the dormancy and burnt. In almond regular clipping of during growing season and dormant period is very essential.
8. Congested orchards should suitably be thinned out during dormancy.
9. Peach plants, which are more prune to leaf curl should not be planted in or near the other *Prunus* spp.
10. Better orchard management including clean cultivation, ploughing / hoeing, judicious manuring/fertilization should be ensured.
11. Use of resistant varieties.
12. Phenological studies like leaf fall, bud swell/bud burst, petal fall and fruit let and 15 – 21 days after fruit let are very important for chemical control.

    Following fungicides have been found giving better performance:

    (*a*) Copper oxychloride, (*b*) Captan, (*c*) Thiophenate methyl, (*d*) Dodine, (*e*) Penconazole

13. Give two sprays of penconazole 10 EC @ 0.5 per cent or copper oxychloride 50 WP @ 0.25 per cent at ending May and mid June.
14. Scratching of cankered areas and application of copper paste.
15. Paint the cut surfaces immediately.
16. Remove affected fruits as it appears
17. Remove all fallen ripe fruits during the season, do not wait until autumn.

Constraints in Management of Insect Pests of Fruit Crops in Kashmir

Despite spectacular progress made in the area expansion and increase in the fruit production, the pace of development has started stumbling down due to devastating emerging challenges in management of insect pests and diseases causing poor productivity and quality in apple/walnut orchards in Kashmir, thereby, putting enormous pressure on the sustainability of horticulture and export of fruits to international markets.

European red mite (*Panonychus ulmi*), stem borer and bark beetles on fruit plantation have emerged as most serious pest problems, due to drought like conditions in the valley during last few years. European red mite has become one of the limiting factors in apple orchards in recent past, posing a serious challenge in its management. San Jose scale and wooly aphid have also upsurged in recent past. The following factors have been attributed to the control failures against these pests:

Mineral Oil Sprays

Dormant spray oils which were earlier used regularly against San Jose scale and would also kill over wintering eggs of European red mite, had been partially abandoned due to ease of application of synthetic insecticides, high input and labour cost, hostile weather conditions during dormancy, shift in priorities from pests to apple scab disease control when it appeared in epiphytotic form, replacement of diesel oil by summer pesticide spray. Such an omission in some orchards has resulted in population explosion of the ERM to damaging proportions.

Disruptive Use of Pesticides

Out break of ERM, San Jose scale and wooly aphid has also been induced by intensive and indiscriminate use of broad-spectrum pesticides resulting in the destruction of natural enemies, which acted as strong population stabilizing factors. Besides dormant oil replaced by summer sprays for the control of San Jose scale, resulted in the decimation of predatory mites and lady bird beetles. Consequently in the absence of natural controlling factors the mite population and San Jose scale increased to alarming proportions. The predator activity in the pesticide managed and unmanaged orchards showed higher activity of predators with lower mite infestation in the latter in comparison to the former (Zaki and Chan, 2003).

Insect Resistance to Insecticides

Development of resistance of mites to pesticides is a major challenge influencing the changing pattern of chemical pest control strategies and is, therefore, the most cogent reason for an integrated approach to pest control in apple orchards. ERM has developed resistance to almost the entire-spectrum of acaricides and insecticides-cum-acaricides used in tree fruit production (Zaki *et al.*, 2005) in Kashmir. European red mite has developed 7-80 fold resistance to endosulfan, dimethoate, chlorpyriphos, methyl-o-demeton and fenazaquin.

Non-standard Pesticides

The sale of spurious pesticides has resulted in loading the orchards with disastrous chemicals, which besides their toxic hazards have changed the balance between pests and natural controlling factors. Such spurious pesticides besides being ineffective impose a sub lethal dose selection pressure on the key pests.

Nutritional Imbalance

There exists a direct relationship between levels of nitrogen and mite population. Most of the orchards do not receive balanced dose of fertilizers and urea seems to be predominant fertilizer applied to apple trees allowing succulence and predisposition to sucking pests.

Adverse Effect on Natural Enemies

The repeated applications of insecticides to temperate fruits with broad-spectrum pesticides have a destructive effect on beneficial insect populations. Fields and orchards under heavy treatment schedules have become veritable "biological deserts". The re-establishment of adequate complexes of natural enemies in such biological deserts may require few years after the use of insecticides/ acaricides is abandoned.

Pesticidal Poisoning of Pollinators

The poisoning of honey bees by the application of insecticides is a major problem of beekeepers especially where applications are made to crops during bloom period. Because of domestication honey bees have not developed the wide variety of detoxification enzymes as found in many other insects, therefore, are highly susceptible to pesticides. Wild bees such as bumble bees, *Megachile* and other bees are also important for pollination and are highly susceptible to insecticides.

Pre-mature Leaf Fall

The premature leaf fall in apple orchards has become an alarming syndrome in recent years. The problem is enhanced and further aggravated by *Alternaria* leaf blotch particularly under high temperature and higher humidity regimes. Higher mite population in the *Alternaria* affected orchards and concurrent presence of the two in the affected orchards lead to higher disease incidence, intensity and premature leaf fall.

Socioeconomic Constraints

Participation of people/farmers in taking up the management strategies seriously is very poor.

Kashmir as Agro-Export Zone

Declaration of the state as Agro-Export Zone for the apple and walnut fruits was one of the major challenge before growers, development officers and the agricultural scientists to figure in export map. The fruits for export have to be qualitative, competitive, standard grade and produced either organically or free of all pesticide residues. Apple concentrate being exported have been found free from objectionable pesticide residues. There is strong need to promote biological and genetic methods of pest control as vital components of IPM to further enhance the potential of apple and walnut export. The potential of walnut export from India is to the tune of 700 crores against Rs 100 crores currently being exported to other countries.

High-Density Plantation

High-density plantation with dwarf cultivars is an important strategy for enhancing productivity manifold but it has been found to be highly susceptible to pests. Therefore, new management strategies have to be developed for combating the pest in high-density plantation in order to enhance productivity. High density plantation have been found to be infested by a different pest spectrum in addition to the pest recorded in traditional orchards, therefore the pest and disease management in this system is more management, labour and cost intensive.

Constraints in Management of Diseases of Fruit Crops in Kashmir

Precision in time interval between two consecutive fungitoxicants in the back drop of bioedaphic factors.

1. Evaluation of spray schedule/ pesticides on non-target pathogens and pests is lacking.
2. Monitoring and management of development of resistance in pathogens to fungitoxicants is lacking.
3. Disease development models for appropriate forecasting / prediction to minimize the frequency of fungicidal application.
4. Inadequate availability and utilization of modern technology like computerized fore-warning devices comprising automatic electronic weather recorders (eg. METOS).
5. Limited germplasm or lack of disease resistant variety possessing commercial acceptance.
6. Limitations of biological control/ botanicals as foliar spray though they possess effectiveness for many soil borne pathogens.
7. Improper crop and quarantine regulation.
8. Lack of availability of quality inputs well in time.
9. Inadequate or improper extension programmes for implementation of effective management programme.
10. Use of spurious fungitoxicants.

References

Anderson, H. W. 1956. *Diseases of Fruit Crops*. First edition. McGraw-Hill Book Company, New York, Toronto and London. 501p.

Anonymous, 1988. Annual Report, Division of Entomology, SKUAST-K Shalimar, Technical Document No. 3, pp. 45-56.

Anonymous, 2007. *Economic Survey 2007-2008*. II Edition. Directorate of Economics and Statistics, Planning and Development Department, Jammu and Kashmir Government.

Anonymous. 2008. *Apple Spray Schedule for the Year 2008*. Department of Horticulture, Government of Jammu and Kashmir, Srinagar, J&K. 6p.

Batzer, J. C., Gleason, M. L., Weldon, B., Dixon, P. M. and Nutter, F. W. Jr. 2002. Evaluation of post harvest removal of sooty blotch and flyspeck on apples using sodium hypochlorite, hydrogen peroxide with peroxyacetic acid, and soap. *Plant Disease*, 86: 1325-1332.

Beigh, G. M. and Qazi, N. A. 2005. Post-harvest diseases of temperate fruits and their management. In: *Challenging Problems in Horticultural and Forest Pathology*. (R. C. Sharma and J. N. Sharma Eds.). Indus Publishing Company, New Delhi. 452 p.

Holb, I. J. 2006. Effect of six sanitation treatments on leaf litter density, ascospore production of *Venturia inaequalis* and scab incidence in integrated and organic apple orchards. *European Journal of Plant Pathology,* 115: 293-307

Ishii, H. 2006. Impact of fungicide resistance in plant pathogens in crop disease control and agriculture environment. *JARQ* 40: 205-211.

Johnson, J. F., Penrose, L. J. and Thwaite, W. G. 1986. *Deciduous Fruits Spray Calendar: Tablelands, Slopes and Coastal Highlands of N.S.W. 1985-86* (25th Edn.). Department of Agriculture, New South Wales.112 p.

Koller, W., Parker, D. M., Turechek, W. W., Avilla-Adame, C. and Cronshaw, K. 2004. A two-phase resistance response of *Venturia inaequalis* populations to the QoI fungicides Kresoxim-methyl and Trifloxystrobin. *Plant Disease* 88: 537-544.

MacHardy, W. E. 1996. *Apple Scab: Biology, Epidemiology and Management*. American Phytopathological Society, St. Paul, Minnesota, USA. 545 p.

Nassreen, F. and Qazi, N. A. 2008. Sensitivity of *Alternaria mali* to difenconazol in Kashmir valley. *Journal of Mycology and Plant Patholcgy*, 39: (in press).

Qazi, N. A. 2008. *Integrated pest and disease management validation technology on apple*. Progress Report, ICAR Funded Project, SKUAST-K, Shalimar. 15p.

Qazi, N. A. and Mir, N.A. 2002. Comparison of different spray schedules for control of *Venturia inaequalis* infection of apple cv. "Red Delicious" in Kashmir. *Applied Biological Research*, 4: 43-51.

Qazi, N. A. and Sagar, V. 2004. Fruit and foliar diseases of apple and their management. In: Compandium on "*Diseases of Temperate Fruits and Vegetables and their Management*". Division of Plant Pathology and Directorate of Extension Education SKUAST, Shalimar (J&K). pp.7-13.

Qazi, N.A., Ahmad. K., Beig, M. A., Munshi, N. and Khan, N. 2008. Impact of inoculum distance on the initiation and spread of apple scab infection [*Venturia inaequalis* (Cke.) Wint.] to isolated apple plantation. *Applied Biological Research,* 10: 40-43.

Qazi, N. A., Beig, M. A. and Ahmad, K. 2005. Impact of post-harvest urea application on primary inoculum and infection of *Venturia inaequalis* (Cke.) Wint. and plant behaviour of apple. *Applied Biological* Research, 7: 37-43.

Rather, A. Q. and Masoodi, M. A. 1988. Survey of mite pests and their natural enemies. In: *Annual Report*, Division of Entomology, SKUAST-K Shalimar, pp. 71-75.

Sabiha, A., Qazi, N.A., Ahmad, S., Ghani, M.Y., Vaseem, Y. and Badri, Z.A. 2006. *In vitro and in vivo* evaluation of fungitoxicants against sooty blotch and fly speck pathogens of apple. *SKUAST Journal of Research*, 8: 49-52.

Sagar, V. and Qazi, N. A. 2004. Root, collar and nursery diseases of apple and their management. In: Compendium on "*Diseases of Temperate Fruits and Vegetables and their Management*". Division of Plant Pathology and Directorate of Extension Education SKUAST, Shalimar (J&K). pp.14-18.

Singh, R. S. 2000. *Diseases of Fruit Crops*. Oxford and IBH Publishing Co. Pvt. Ltd., New Delhi. 310 p.

Slippers, B., Smit, W. A., Crous, P. W., Coutinho T. A., Wingfield, B. D. and Winafield, M. J. 2007. Taxonomy, phylogeny and identification of Botryosphaeriaceae associated with pome and stone fruit trees in South Africa and other regions of the world. *Plant Pathology*, 56: 128-139.

Sofi, M. A., Zaki, F. A., Dar, M. A. and Khan, Z. H. 2008. New approaches to San Jose scale control by newer oils alone and in combination with insecticides and their effect on grading of apple in Kashmir. *Environment and Ecology*, 26(3A): 1369-1372.

Thakur, V.S. and Sharma, R.D. 1999. Effect of urea on microbial degradation of apple leaf litter and its relationship to the inhibition of pseudothecial development of *Venturia inaequalis*. *Indian Journal of Agricultural Sciences*, 69: 147-151.

Vaseem, Y. and Qazi, N. A. 2001. Impact of single autumn application of chemicals on primary inoculum production of *Alternaria mali*, the causal agent of *Alternaria* leaf blotch of apple, in Kashmir. *Applied Biological Research*, 3: 23-27.

Williamson, S. M. and Sutton, T. B. 2000. Sooty blotch and flyspeck of apple: Etiology, biology and control. *Plant Disease*, 84: 714-724.

Zaki, F. A. and Chan, S. A. 2003. Management of European red mite (*Panonychus ulmi*) on apple in Kashmir. *SKUAST Journal of Research*, 5: 198-203.

Zaki, F. A. Dhobi, G. M. and Mantoo, M. A. 2004. Interaction of European red mite (*Panonychus ulmi* Koch) infestation with *Alternaria* sp. in premature leaf fall in apple (*Malus domestica* Borkh). In "*Bioresources: Concerns and Conservation*" (Kamili, A. N. and Yousuf, A. R. Eds.) pp. 119-126.

Zaki, F. A., Lone, A. H. and Mantoo, M. A. 2005. Resistance in European red mite (*Panonychus ulmi* Koch) to some pesticides on apple in Kashmir. In "*Proceedings of the 1st Congress on Insect Science: Contributed papers*" (Dilawari, V. K., Doel, G. S., Joia, B. S. and Chhuneja, P. K. Eds.), December 15-17, INSAIS, PAU, Ludhiana, pp. 161-163.

Zaki, F. A., Mantoo, M. A. and Parrey, M. A. 2008a. Bioefficacy of some newer acaricides against European red mite, *Panonychus ulmi* Koch (Acari:Tetranychidae) on apple crop, *Malus domestica* Borkh in Kashmir. *Science For Better Tomorrow*: 1-4.

Zaki, F. A., Mantoo, M. A., Parray, M..A., Khan, A. A., Rather, A. Q., Wani, N. A. and Wani, A. R. 2008b. Paraffinic mineral oils–A new eco-friendly agrochemical strategy for agricultural pest management. IUPAC sponsored *First International Conference on Agrochemicals Protecting Crop, Health and Natural Environment*, Jan.,08-11, 2008, New Delhi, India. Abstr. No. P-77.

Annexure I

Sher-e-Kashmir University of Agricultural Sciences & Technology of Kashmir
Directorate of Research Shalimar, Srinagar – 191 121, J&K (India)
Plant Protection Schedule for the Management of Pests and Diseases of Apple in Kashmir (Season-2008)

Spray	*Tree Stage*	*Fungicide/100 lit of Water*	*Insecticide/Acaricide/HMO per 100 lit of Water*
I	Green tip (Delayed Dormancy) (Mid March-Mid April)	–	Horticultural Mineral Oil (HMO) @ 2 litres (2 per cent)
II	Pink bud (Mid April-End April)	Myclobutanil 10 WP (30 g) or Dodine 65 WP (60g) or Bitertanol 25 WP (50g) or Fenarimol 12 EC (40 ml)	*** 3-4 days after II fungicidal spray in case spray I is missed. Chlorpyriphos 20 EC (100 ml) or Endosulfan 35 EC (140 ml) or Summer Spray Oil (750ml) or Fenazaquin 10 EC (40ml) or Herbal (200 ml)
III	Petal fall (80 to 100 per cent)	14-18 days after II spray Hexaconazole 5 EC (30ml) or Difenaconazole 25 EC (30ml) or Triadimefon 25 WP (50g) or Diniconazole 25 WP (40g)	3-4 days after III fungicidal spray Quinalphos 25 EC (100ml) or Methyl-o-Demeton 25 EC (80ml) or Phosalone 35 EC (140ml) or Summer Spray Oil (750ml) or Fenazaquin 10 EC (40ml)
IV	Fruit let (Pea size)	14-18 days after III spray Ziram 80 WP (200g) or Mancozeb 75 WP (300g) or Ziram 27 W/V (600ml) or Captan 50 WP (300g) or Propineb 75 (300g) or Mancozeb Flowable 35SL (300ml) or Zineb 75 WP (300g)	***3-4 days after IV fungicide spray Chlorpyriphos 20 EC (100ml) or Methyl-o-Demeton 25 EC (80ml) or Phosalone 35 EC (140 ml) or Dicofal 18.5 EC (108ml) or Summer Spray Oil (750 ml) or Fenazaquin 10 EC (40 ml)
V	Fruit development I	14-18 days after IV spray Bitertanol 25 WP (50g) or Dithionon 75 WP (75g) or Penconazole 10 EC (50ml) or Fenarimol 12 EC (40ml)	3-4 days after V fungicide spray Chlorpyriphos 20 EC (100ml) or Methyl-o-Demeton 25EC(80ml) or Dimethoate 30 EC(100ml) or Docofol 18.5 EC (108ml) or Summer Spray Oil (750ml) or Fenazaquin 10 EC (40ml) or Abamectin 1.8 EC (55.5ml)
VI	Fruit development II	14-18 days after V spray Hexaconazole 5 EC (30ml) or Myclobutanil 10 WP (30g) or Triadimefon 25 WP (50g) or Diniconazole 25 WP (40g)	–
VII	Fruit development III	14-18 days after VI spray Difenoconazole 25 EC (30ml) or Dodine 65 WP (60g) or Dithionon 75 WP (75g) or Penconazole 10 EC (50g)	3-4 days after VI fungicidal spray Chlorpyriphos 20EC (100ml) or Dimethoate 30EC (100ml) or Endosulfan 35EC (140ml) or Methyl-o-Demeton 75 EC (80ml) or Phosalone 35 EC (140ml) or Fenazaquin 10EC (40ml) or Abamectin 1.8 EC (55.5ml)

Contd...

Annexure I–Contd...

Spray	*Tree Stage*	*Fungicide/100 lit of Water*	*Insecticide/Acaricide/HMO per 100 lit of Water*
VIII	Fruit development IV	*14-18 days after VII spray Hexaconazole 5 EC(30ml) or Bitertanol 25 WP (50g) or Fenarimol 12 EC (40ml)	–
IX	Pre-harvest	** 25 days before harvest Ziram 80 WP (200g) or Mancozeb 75 WP (300g) or Ziram 27 W/V (600ml) or Captan 50 WP (300g) or Propineb 75 WP (300g) or Mancozeb Flowable 35 SL (300ml) or Zineb 75 WP (300g)	–
X	Post harvest	–	***** Ethion 50 EC(100ml) or Fenazaquin 10 EC (40ml) or HMO (75ml) or Chlorpyriphos 20 EC (100ml)

*: Need based spray if leaf spotting incidence is more than 20 per cent; **: Need based spray for long term storage of fruits; ***: Need based spray when 5-8 mites/leaf are observed; ****: Need based spray when 10 or more mites/leaf are observed; *****: Need based spray when 20 or more mites/leaf are observed during late August and spray at Fruit Development III (during August) has been missed.

Chapter 12

Sustainable Management of Insect Pests of Spice Crops

S. Devasahayam
Indian Institute of Spices Research,
Marikunnu P.O., Calicut – 673 012, Kerala

Introduction

Spices are high-value products obtained from diverse crops and are extensively used all over the world to add flavour and taste to human food besides being used in indigenous medicine. Spices are of considerable economic importance for the country, being major sources for foreign exchange among agriculture commodities and play a crucial role in the economy of various states were they are produced. Over 50 species of plants yield spices in India and among them black pepper (*Piper nigrum* L.) (Piperaceae), cardamom (*Elettaria cardamomum* Maton), ginger (*Zingiber officinale* Rosc.) and turmeric (*Curcuma longa* L.) (Zingiberaceae) account for considerable share of export earnings. Though India is a leading producer of spices in the world, the productivity of these crops is low due to various factors among which infestation by insect pests is a major factor.

A number of insect pests have been recorded on spice crops in India causing severe economic losses (Table 12.1). Many of these insect pests were controlled by application of chemical pesticides alone earlier. Excessive and indiscriminate use of pesticides could result in pesticide residues in the produce, mortality of pollinators, natural enemies and other non-target organisms, development of resistance and environmental pollution. Since spices are export-oriented products, it is also important that the levels of pesticide residues are kept below tolerance limits in view of the stringent standards set by the importing countries. There has been a renewed interest in developing environment–friendly pest management schedules in sustainable agriculture in recent years, reflecting the increasing concern over pesticide misuse. The nature of damage caused by major insect pests of black pepper, cardamom, ginger and turmeric and the advances in sustainable management of these insect pests involving cultural and biological means, use of resistant varieties and plant products and need-based application of pesticides, are highlighted here.

Table 12.1: Major Insect Pests of Spice Crops

Crop/Insect Pest	*Order/Family*
Black Pepper (*Piper nigrum* L.)	
Pollu beetle (*Lanka ramakrishnai* Prathapan and Viraktamath = *Longitarsus nigripennis* Mots.)	Coleoptera : Chrysomelidae
Mussel scale (*Lepidosaphes piperis* Gr.)	Homoptera : Diaspididae
Coconut scale (*Aspidiotus destructor* Sign.)	Homoptera : Diaspididae
Top shoot borer (*Cydia hemidoxa* Meyr.)	Lepidoptera : Tortricidae
Leaf gall thrips (*Liothrips karnyi* Bagn.)	Thysanoptera : Phlaeothripidae
Cardamom (*Elettaria cardamomum*)	
Cardamom thrips (*Sciothrips cardamomi* Ramk.)	Thysanoptera : Thripidiae
Shoot and capsule borer (*Conogethes punctiferalis* Guen.)	Lepidoptera : Pyralidae
Root grub (*Basilepta fulvicorne* Jacoby)	Coleoptera : Chrysomelidae
Ginger (*Zingiber officinale* Rosc.) and Turmeric (*Curcuma longa* L.)	
Shoot borer (*Conogethes punctiferalis* Guen.)	Lepidoptera : Pyralidae
Rhizome scale (*Aspidiella hartti* Ckll.)	Homoptera : Diaspididae

Black Pepper

Black pepper, the dried mature berries of *P. nigrum*, a perennial climbing vine, is one of the oldest spice known to mankind and has been termed as the 'king of spices'. The crop is grown in about 2,57,020 ha in India mainly in Kerala and Karnataka with a production of about 79,640 tonnes annually. Around 60 species of insects have been recorded on the crop in India, and among them *pollu* beetle, scale insects, top shoot borer and leaf gall thrips are important (Devasahayam 2000a).

Pollu Beetle

The *pollu* beetle (*Lanka ramakrishnai* Prathapan and Viraktamath=*Longitarsus nigripennis* Mots.) is the most destructive insect pest of black pepper in India. The pest infestation is higher in the plains and midlands in Kerala and in endemic areas up to 40 per cent of the crop is reported to be lost due to the pest infestation (Devasahayam 2000a).

Bioecology

The adult *pollu* beetle feeds on tender shoots, leaves and spikes resulting in black patches on tender shoots and spikes and small irregular circular holes on tender leaves. Severely infested leaves and spikes often rot and drop due to invasion of secondary microorganisms. The damage caused by the larva (grub) is more serious, and it bores into tender spikes and berries and feeds on the internal contents. The infested spikes develop necrotic patches and the berries turn black and crumble when pressed (the hollow nature of the infested berries is termed as *pollu* in Malayalam) (Figure 12.1) (Devasahayam 2000a).

Figure 12.1: Tender Berries of Black Pepper Damaged by *Pollu* Beetle Larvae

The adult *pollu* beetle measures about 2.5 mm x 1.5 mm in size, the head and thorax being yellowish brown and the abdomen brown; the elytra (fore wings) is black. Fully-grown grubs are creamy yellow and measure about 5.5 mm in length. There are many overlapping generations in a year. The adults are seen on black pepper vines throughout the year in the field though their population is generally higher during September-October (Premkumar 1980; Babu 1994).

Management

An integrated strategy involving regulation of shade and spraying of insecticide and neem-based product is to be adopted for the management of *pollu* beetle. Regulation of shade in the plantation by pruning branches of support and shade trees before the onset of the monsoon helps in reducing the build up of pest population. The spraying schedule to be adopted includes spraying the vines during July and October with quinalphos (0.05 per cent) or spraying quinalphos (0.05 per cent) during July followed by Neemgold (0.6 per cent) (neem-based insecticide) during August, September and October (Premkumar *et al.*, 1986; Devasahayam and Koya 1999).

Resistance

Black pepper cultivars show wide variation in their reaction towards *pollu* beetle in the field. Forty four cultivars of black pepper maintained at Pepper Research Station, Panniyur (Kerala) were screened against *pollu* beetle among which, Kalluvally Type II was less susceptible (Pillai and Abraham 1979). Four cultivars, six wild species of *Piper* and two inter-specific hybrids were identified to be resistant to the pest from the germplasm collections of IISR, Calicut. These resistant lines are being utilized in breeding programmes for developing pest resistant varieties against the pest (Sasikumar *et al.*, 1998; Devasahayam 2000a).

Natural Enemies

Very few natural enemies have been recorded on *pollu* beetle in the field and they include an entomophagous nematode (Mermithidae), predatory spider (Araneae) and *Oecophylla smaragdina* Fabr. (Fomicidae). However, the level of parasitization/predation by these natural enemies is quite low and they may not have a major role in the regulation of the pest population in the field (Devasahayam and Koya 1994a).

Plant Products

Leaf extracts of *Chromolaena odorata* L., *Azadirachta indica* A. Juss and *Strychnos nux-vomica* L., root bark extracts of *Uvaria narum* Blume and *U. hookeri* King, neem oil, neem seed kernel extract and custard apple (*Annona squamosa* L.) seed extract posses appreciable antifeedant activity against *pollu* beetle, indicating their potential in utilizing them in IPM schedules (Babu *et al.*, 1996; Devasahayam and Anandaraj 1997; Devasahayam and Leela 1998).

Scale Insects

Scale insects are major insect pests of black pepper at higher altitudes in Kerala, Karnataka and Tamil Nadu. Among the various species of scale insects recorded on the crop, the mussel scale (*Lepidosaphes piperis* Gr.) and coconut scale (*Aspidiotus destructor* Sign.) are the most common (Koya *et al.*, 1996).

Bioecology

The mussel scale encrusts main stems, lateral branches, mature leaves and berries of black pepper vines resulting in chlorotic patches, yellowing and drying of leaves (Figure 12.2). Younger vines often succumb to the pest during the course of 1-2 years when the infestation occurs on the main stem. On older vines the infested lateral branches wilt and dry resulting in vacant spaces in the canopy. The coconut scale infests mature leaves resulting in chlorotic patches on them and sometimes also infest berries (Figure 12.3) (Koya *et al.*, 1996).

Adult females of *L. piperis* are elongated and dark brown measuring 3-4 mm in length. Adult females of *A. destructor* are circular and light yellow measuring 1.5-2.0 mm in diameter. Males are weak-bodied with a pair of wings and do not live long and die after mating. Scale insects are observed in the field throughout the year but their population is higher during the post monsoon and summer months (Koya 1994). *A. destructor* is highly polyphagous and has been known to infest more than 20 economically important plants in India. *L. piperis* occasionally infest cassava (*Manihot esculenta* Crantz) stems in the field and storage (Devasahayam 2000a).

Management

Spraying of dimethoate 0.1 per cent may be undertaken on affected vines after clipping off severely infested branches after harvest of berries; the spraying has to be repeated after 21 days if the infestation persists (Devasahayam and Koya 1994b; Rajkumar *et al.*, 2001).

Natural Enemies

Various predators and parasites have been recorded on scale insects in the field among which *Chilocorus* spp., *Pseudoscymnus* spp. (Coccinellidae), *Aphytis* sp. and *Encarsia* spp. (Aphelinidae) have been identified as major natural enemies (Devasahayam and Koya 1994a; Selvakumaran *et al.* 1996). Methods were standardized for mass rearing of *C. nigrita* (Fab.) and *C. circumdatus* (Gyllen.) in the

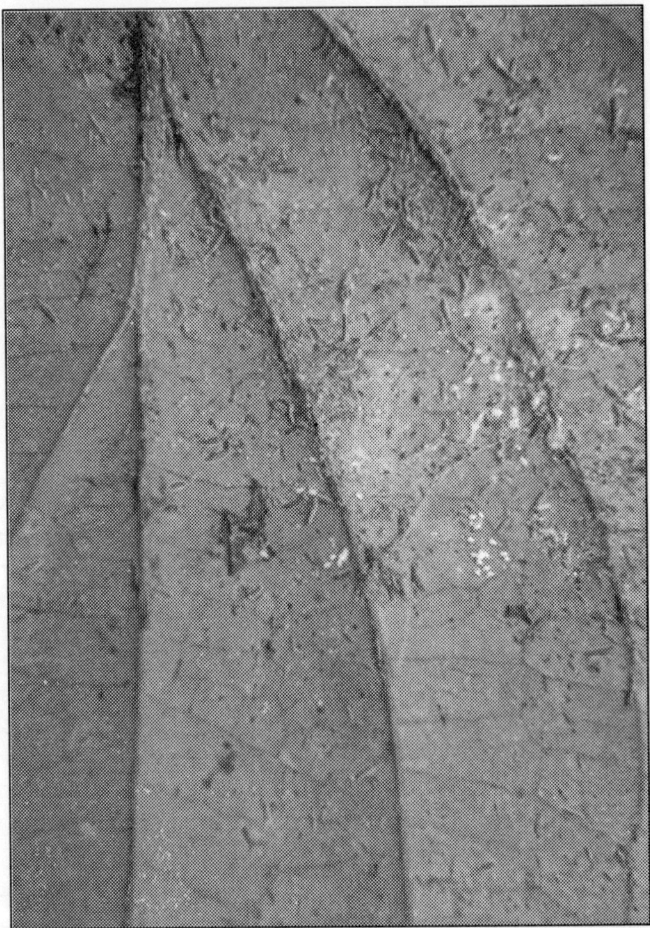

Figure 12.2: Leaf of Black Pepper Infested with Mussel Scale

laboratory. Evaluation of these predators in the field indicated that their release brought down the population of *L. piperis* and *A. destructor* (Devasahayam 1998).

Plant Products

Neem products such as neem oil 0.3 per cent, Neemgold 0.3 per cent and natural products such as fish oil rosin 3 per cent were promising for the management of scale insects and safer to the coccinellid predators (Devasahayam 1998).

Top Shoot Borer

The top shoot borer (*Cydia hemidoxa* Meyr.) is a serious pest in young black pepper plantations in all black pepper areas. The larvae bore into tender terminal shoots and feed on internal tissues resulting in decay and blackening of infested shoots affecting the growth of the vine (Figure 12.4). In some young plantations even up to 100 per cent infestation of new shoots are observed (Devasahayam and Koya 1993a).

Figure 12.3: Leaf of Black Pepper Infested with Coconut Scale

Bioecology

The adult is a tiny moth with a wingspan of 10-15 mm with crimson and yellow fore wings and grey hind wings. Fully-grown larvae are greyish green and measure 12-15 mm in length. The pest infestation is higher during September-October when numerous succulent shoots are available on the vines (Visalakshi and Joseph 1965).

Management

Spraying quinalphos (0.05 per cent each) on tender terminal shoots of young vines is effective in controlling top shoot borer infestation. The spraying has to be repeated to protect emerging new shoots especially during July to October (NRCS 1990).

Natural Enemies

Hexamermis sp. (Mermithide) and *Apanteles cypris* Nixon (Braconidae) were identified as major natural enemies parasitizing larvae of top shoot borer and play an important role in lowering the population of the pest in the field (Devasahayam and Koya 1993b; 1994c).

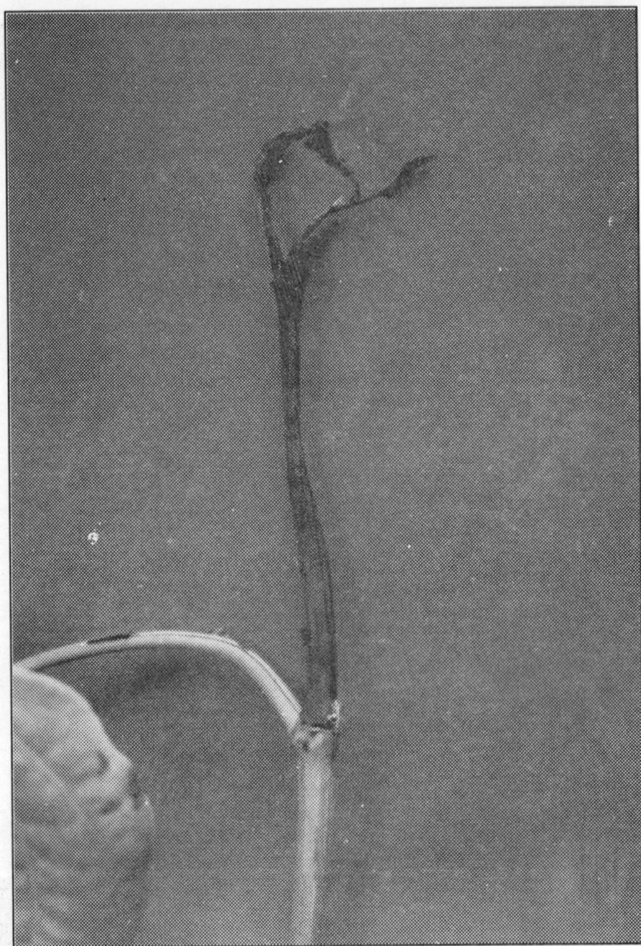

Figure 12.4: Tender Shoot of Black Pepper Damaged by Top Shoot Borer Larvae

Leaf Gall Thrips

Leaf gall thrips (*Liothrips karnyi* Bagn.) are serious insect pests of black pepper on younger vines at higher altitudes and also on rooted cuttings in nurseries in the plains, in Kerala, Karnataka and Tamil Nadu (Devasahayam 1991).

Bioecology

Leaf gall thrips infest tender leaves causing the leaf margins to curl downwards and inwards resulting in the formation of marginal, tubular galls. As the infested leaves grow they become crinkled, malformed and reduced in size and may also turn chlorotic later at the margins (Figure 12.5). The growth of younger vines and rooted cuttings in the nursery are affected in severe cases of infestation. Serious infestations on older vines affect the formation and development of spikes (Devasahayam 1991).

The adults are black with the distal segments of the antenna and legs light lemon yellow, and measure 2.5-3.0 mm in length; the larva the pupae are creamy white. Leaf gall thrips initially infest

Figure 12.5: Tender Leaves of Black Pepper Infested with Leaf Gall Thrips

tender leaves and remain in them till the leaves become mature. The pest population is higher in the field during June-September when numerous tender leaves are produced on the vines (Visalakshi and Joseph 1967; Devasahayam 1991).

Management

Spraying dimethoate 0.05 per cent during emergence of new flushes may be undertaken on younger vines and in nurseries for managing the pest infestation (Devasahayam 1990).

Natural Enemies

A few predators have been recorded on leaf gall thrips in the field among which *Montandoniola moraguesi* (Anthocordiae) and *Andothrips flavipes* (Phlaeothripidae) are more common and widely distributed. Both these species are voracious feeders on all stages of the pest and play an important role in lowering the population of the pest in the field (Devasahayam 1991).

Cardamom

Cardamom, known as the 'queen of spices', is the dried fruit of a perennial rhizomatous herb *E. cardamomum* and is mainly cultivated in Kerala, Karnataka and Tamil Nadu at higher altitudes. The total production of cardamom in the country is about 10,190 tonnes from an area of about 65,780. More than 60 insect species have been recorded on the crop in India, and among them cardamom thrips, shoot and capsule borer and root grub are important (Gopakumar and Chandrasekhar 2002).

Cardamom Thrips

Cardamom thrip (*Sciothrips cardamomi* (Ramk.) is the most widespread and destructive pest of cardamom occurring in all cardamom areas damaging shoots, panicles and capsules. The pest is known to cause up to 48 per cent crop loss (Gopakumar and Chandrasekhar 2002).

Bioecology

The adults and larvae of cardamom thrips lacerate the tissues of leaves, shoots, panicles, flowers and immature capsules and feed on the exuding sap resulting in shedding of flowers and immature capsules and scab formation on mature capsules (Figure 12.6). The infested capsules loose their aroma and the formation of seeds are also affected (Gopakumar and Chandrasekhar 2002).

Figure 12.6: Cardamom Panicle and Capsule Damaged by Cardamom Thrips

The adults are minute and greyish brown measuring 1.5–2.0 mm in length. Reproduction also takes place parthenogenetically. The pest population builds up rapidly during the post monsoon and summer months and declines with the onset of rains. The pest population is also higher in thickly shaded areas in the plantation. A number of zingiberaceous plants generally growing in and around cardamom plantations serve as alternate host plants (Gopakumar and Chandrasekhar 2002).

Management

Regulation of shade in the plantation by pruning branches of shade trees and removal of alternate host plants in the vicinity of plantations helps in reducing the build up of cardamom thrips population in the field. The pest generally occurs and breeds within left sheaths in the lower portion of the pseudostems especially during summer. Thus pruning of leaf sheaths during February-March before spraying operations also reduces the pest population and increase the efficacy of insecticides. Five to seven rounds of spraying of insecticides like chlorpyriphos (0.04 per cent), fenthion (0.05 per cent), phenthoate (0.05 per cent), phosalone (0.07 per cent) or quinalphos (0.025 per cent), during March, April, May, August and September are required for controlling the pest infestation in Kerala (NRCS 1989; Varadarasan 1992).

Resistance

The Mysore type of cardamom is highly susceptible to cardamom thrips compared to the Malabar type. At Mudigere (Karnataka), among the 73 elite clones screened for tolerance to cardamom thrips, three clones, namely, D-514, D-769 and CI-754 were relatively tolerant to the pest in the field with less than 10 per cent infestation on the capsules (Parvathi and Chandrappa 1993). Screening of 44 cardamom accessions at Saklesphur (Karnataka) indicated that two accessions, namely, Malabar Local (28.7 per cent infestation) and SKP-97 (31.4 per cent infestation) were moderately tolerant to the pest (Singh *et al.*, 1996).

Natural Enemies

Very few natural enemies have been recorded on cardamom thrips; *Chrysopa* sp. and anthocorid bugs which are general predators, have been observed to predate on adults and larvae in the field (Varadarasan 1995).

Plant Products

Various neem based products such as neem cake, Nimbicidine 3 per cent, Multineem 0.3 per cent, neem seed kernel extract 5 per cent and Nimbex 0.3 per cent were evaluated at Mudigere (Karnataka) against cardamom thrips; however, the results were not promising (AICRPS 2001).

Shoot and Capsule Borer

The shoot and capsule borer (*Conogethes punctiferalis* Guen.) infests cardamom plants in nurseries and plantations. Yield losses of 70-80 per cent have been recorded in severely infested plantations (Varadarasan 1995).

Bioecology

The earlier stages of larvae of shoot and capsule borer bore into panicles and immature capsules and the later stages into pseudostems and feed on the internal tissues (Figure 12.7). The presence of bore holes with extruding frass on the pseudostems and capsules and the withered central shoot are characteristic symptoms of pest infestation (Gopakumar and Chandrasekhar 2002).

Figure 12.7: Cardamom Shoot Infested by Shoot and Capsule Borer

The adult is a medium-sized moth with a wing span of about 20 mm; the wings and body are orange-yellow with minute black spots. The pest infestation is noticed throughout the year in the field but its population is higher during September-October and January-February coinciding with panicle, capsule and shoot formation stages (Varadarasan 1991). The shoot borer is highly polyphagous and has been recorded on more than 35 host plants including several economically important plants in India (Devasahayam and Koya 2005).

Management

Removal and destruction of alternate host plants in and around cardamom plantation, removal of infested suckers as indicated by extrusion of frass, during September-October when the infestation is less than 10 per cent, and collection and destruction of adults which are generally observed on the undersurface of leaves have been suggested for reducing the population of shoot and capsule borer in the field. Later, spraying of fenthion (0.075 per cent) or quinalphos (0.075 per cent) twice, during January-February and September-October is to be undertaken for controlling the pest infestation (NRCS 1989).

Resistance

PV-1, a Malabar type of cardamom, was identified to be resistant to the shoot and capsule borer at Pampadumpara (Kerala) (Rajkumar *et al.*, 2002).

Natural Enemies

Twenty genera/species of hymenopterous parasitoids have been recorded on the shoot and capsule borer among which, *Eriborus trocheanteratus* (Morely), *Xanthopimpla australis* Kr., *Friona* sp. and *Agrypone* sp. (Icheumonidae) are important (Varadarasan 1995).

Plant Products

Various neem based products such as neem cake, Nimbicidine 0.3 per cent, Multineem 0.3 per cent, neem seed kernel extract 5 per cent and Nimbex 0.3 per cent were evaluated at Mudigere (Karnataka) against shoot and capsule borer; however, the results were not promisng (AICRPS 2001).

Sex Pheromones

The presence of sex pheromones in *C. punctiferalis* has been demonstrated by many workers. Trials on the efficacy of sex pheromones in the field on cardamom indicated that (E)-10-hexadecenal and (Z)-10-hexadecenal and a blend of these two compounds at 9:1 concentrations was promising (Chakravarthy and Thygaraj 1980)

Root Grub

The root grub is a serious pest of cardamom plants in nurseries and plantations. Crop losses of 29 per cent–66 per cent have been reported in various regions (Varadarasan 1995).

Bioecology

The larvae of root grub feed on roots and rhizomes and in severe cases of infestation, the entire root system is eaten away (Figure 12.8). The infested plants turn yellow and become stunted; severally infested plants, especially seedlings, succumb to the pest attack. Seedlings damaged by the root grub

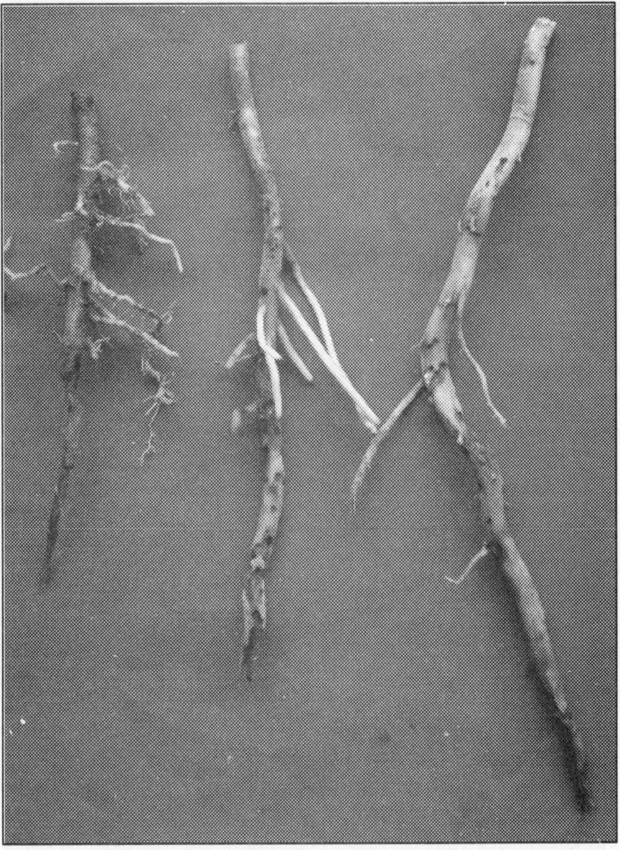

Figure 12.8: Cardamom Roots Damaged by Root Grub

are subsequently infected by secondary pathogens resulting in rotting (Gopakumar and Chandrasekhar 2002).

The adult beetles are metallic blue, bluish green or greenish brown measuring about 5.0 x 2.5 mm in size. The fully grown grubs are stout and 'C' shaped, pale white and measure about 1 cm in length. The adults emerge in large numbers from the pupae after the receipt of pre-monsoon showers (April-May and September-October). The adults are polyphagous and feed on a number of trees in and around cardamom plantations such as jack, Indian almond, mango, guava, cocoa, *Ficus* spp. and *Erythrina* sp. (Thyagaraj *et al.*, 1991).

Management

Collection and destruction of adult beetles during peak periods of emergence is effective in reducing the pest population in the field. The beetles emerge in large numbers especially during April-May after the summer showers and are to be collected using insect collection nets and destroyed. Along with collection of adults, application of phorate 10 G @ 20–40 g per clump or chlorpyriphos (0.075 per cent) twice a year during May-June and September-October synchronizing with emergence of adults and egg laying periods is effective for the management of the pest (Gopakumar and Chandrasekhar 2002).

Natural Enemies

The natural enemies recorded on the root grub include the green muscardine fungus *Metarrhizium anisopliae* (Metsch) Sorokin (Hyphomycetes) and the entomogenous nematode (*Heterorhabditis* sp.) (Heterorhabditidae) infesting grubs, and the white muscardine fungus *Beauvaria bassiana* (Bals.) Vuill. (Hyphomycetes) infesting adults in the field. These pathogens are highly pathogenic to the host under laboratory conditions and also in the field (Varadarasan 1995).

Ginger and Turmeric

Ginger (*Z. officinale*) and turmeric (*C. longa*) are herbaeceous plants and their dried rhizomes yield the spice of commerce. India produces about 3,97,990 and 7,16,840 tonnes of ginger and turmeric respectively, from an area of about 1,00,270 and 1,61,230, respectively. Kerala, Meghalaya and Arunachal Pradesh are the major ginger producing states whereas, Andhra Pradesh, Tamil Nadu and Orissa are the major turmeric producing states in India. More than 45 species of insects have been recorded on ginger and turmeric in India and many of them are common on both the crops. Among the various insect pests, shoot borer and rhizome scale are important (Devasahayam and Koya 2005; 2007).

Shoot Borer

The shoot borer (*C. punctiferalis*) is the most serious insect pest of ginger and turmeric. The larvae bore into pseudostems and feed on the internal shoot resulting in yellowing and drying of infested pseudostems (Figures 12.9 and 12.10). The presence of bore hole on the pseudostem through which frays is extruded and the withered central shoot is a characteristic symptom of pest infestation (Devasahayam and Koya 2005; 2007).

Bioecology

The adults are medium sized moths with a wingspan of 18–24 mm; the wings and body are pale straw yellow with minute black spots. Fully-grown larvae are light brown with sparse hairs and measure 16–26 mm in length (Jacob 1981). The pest is observed in the field throughout the crop season

Figure 12.9: Ginger Shoot Infested by Shoot Borer

and its population is higher during September-October in Kerala. The shoot borer is highly polyphagous and has been recorded on more than 35 host plants including several economically important plants in India (Devasahayam and Koya 2005; 2007).

Management

Spraying malathion 0.1 per cent at monthly intervals during July to October is effective for the management of shoot borer on ginger (Koya *et al.*, 1988). A sequential sampling strategy for monitoring the pest infestation in a field as a guidance for undertaking control measures has also been formulated (Koya *et al.*, 1986). An integrated strategy involving pruning and destroying freshly infested shoots during June to August and spraying of insecticide such as malathion 0.1 per cent during September to October has also been suggested for controlling the pest infestation on ginger (IISR 2001). Spraying of commercial products of *Bacillus thuringiensis* such as Dipel 0.3 per cent at 21 day intervals during July–October is also effective for the management of the pest (Devasahayam 2000b).

Figure 12.10: Turmeric Shoot Infested by Shoot Borer

On turmeric, spraying of malathion 0.1 per cent during July to October at 21 day intervals is effective in controlling the pest infestation (IISR 2005a). Spraying of commercial products of *B. thuringiensis* such as Dipel 0.3 per cent at 21 day intervals during July–October is also effective for the management of the pest (Devasahayam 2002).

Resistance

A few ginger and turmeric lines less susceptible to the pest in the field have been identified. In ginger, among the 25 cultivars screened, the pest infestation was minimum in Rio-de-Janeiro (Nybe and Nair 1979). In turmeric, Dindigam Ca-69 (Sheila *et al.*, 1980) and Mannuthy local (Philip and Nair 1981) were relatively less susceptible to the pest in the field at Vellanikara (Kerala). Velayudhan and Liji (2003) recorded the incidence of the shoot borer on 489 accessions of turmeric belonging to 21 morphotypes at Vellanikkara and the lowest incidence was observed in morphotype 2. Kotikal and Kulkarni (2001) screened eight genotypes of turmeric in the field at Belgaum (Karnataka) and all of them were susceptible with more than 10 per cent damage.

Natural Enemies

Various natural enemies including mermithid nematode (*Hexamermis* sp.), hymenopterous parasitoids (*Xanthopimpla australis* Kr. (Ichneumonidae), *Apanteles taragamae* and *Myosoma* sp. (Braconidae) and general predators such as spiders, earwigs and asilid flies have been recorded on shoot borer infesting ginger and turmeric and play an important role in lowering the population of the pest in the field (Devasahayam 1996).

Rhizome Scale

The rhizome scale infests rhizomes of ginger and turmeric both in the field and in storage. In the field, the pest infestation is generally seen during the dry post monsoon season and severely infested plants wither and dry. In storage, the pest infestation results in shriveling of buds and rhizomes; when the infestation is severe, it adversely affects the sprouting of rhizomes (Figures 12.11 and 12.12) (Devasahayam and Koya 2005; 2007).

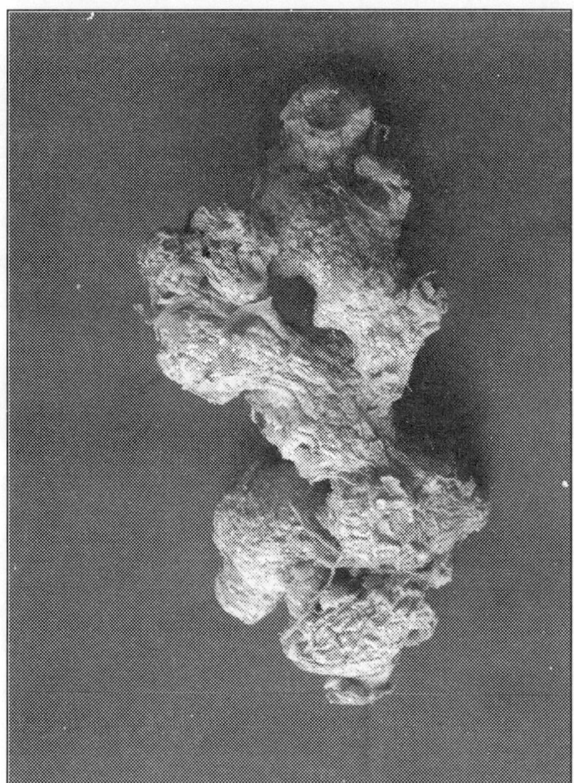

Figure 12.11: Ginger Rhizome Infested by Rhizome Scale

Figure 12.12: Turmeric Rhizome Infested by Rhizome Scale

Bioecology

The adult females are minute, circular and light brown to grey measuring about 1.5 mm in diameter. Females are ovo-viviparous and also reproduce parthenogenetically. The life cycle from egg to adult is completed in about 30 days (Jacob 1982). Rhizome scale also infest elephant foot yam, yams, tannia and taro (Devasahayam and Koya 2005; 2007).

Management

Timely harvest and discarding of severely infested rhizomes during storage reduces further spread of the pest infestation in storage. Dipping of seed rhizome in quinalphos 0.075 per cent after harvest and storage in dry leaves of *Strychnos nux-vomica* + saw dust in 1 : 1 proportion is effective in controlling rhizome scale infestation on ginger and turmeric (IISR 2005b).

Resistance

Eighty seven turmeric types were observed to be free of pest infestation in the field at harvest among the 191 types screened against rhizome scale at Coimbatore (Tamil Nadu) (Regupathy *et al.*, 1976). Velayudhan and Liji (2003) recorded the incidence of the rhizome scale on 489 accessions belonging to 21 morphotypes at Vellanikkara (Kerala) and 80 accessions were free of infestation.

Natural Enemies

Two species of hymenopterous parasitoids namely, *Physcus* (*Cocobius*) *comperei* Hayat (Aphelinidae) and *Adelencyrtus moderatus* Howard (Encyrtidae) are important natural enemies of the pest (Devasahayam 1996).

References

(AICRPS) All India Coordinated Research Project on Spices 2001. Annual Report 2000–01. National Research Centre for Spices, Calicut, 178 pp.

Babu, S., 1994. Some aspects of biology of *Longitarsus nigripennis* Mots. (Coleoptera: Chrysomelidae), a serious pest on black pepper *Piper nigrum* L. *Entomon* 19: 159–161.

Babu, P. B. S., Padmaja, V. and Hishan, A., 1996. Isodesacetyluvaricin-insect antifeedant against *Longitarsus nigripennis* Mots. *Indian J. Exptl. Biol.* 34: 377–379.

Chakravarthy, A. K. and Thyagaraj, N. E., 1980. Evaluation of selected synthetic sex pheromones of the cardamom shoot and fruit borer, *Conogethes punctiferalis* Guenee (Lepidoptera : Pyralidae) in Karnataka. *Pest Manage. Trop. Ecosyst.* 4: 78–82.

Devasahayam, S., 1990. Field evaluation of insecticides for the control of leaf gall thrips (*Liothrips karnyi* Bagnall) on black pepper. *Entomon* 15: 137-138.

Devasahayam, S., 1991. Bioecology and Control of Marginal Gall Forming Thrips (*Liothrips karnyi* Bagnall) Infesting Black Pepper. Final Report of Research Project, National Research Centre for Spices, Calicut, 25 pp.

Devasahayam, S., 1996. Biological control of insect pests of spices. In: Anandaraj, M. and Peter, K. V. (Eds.). *Biological Control in Spices*. Indian Institute of Spices Research, Calicut. pp. 33-45.

Devasahayam, S., 1998. Biological Control of Scale Insects Infesting Black Pepper. Final Report of ICAR Ad-hoc Research Scheme, Indian Institute of Spices Research, Calicut, 25 pp.

Devasahayam, S., 2000a. Insect pests of black pepper. In: Ravindran, P. N. (Ed.). *Black pepper Piper nigrum*. Harwood Academic Publishers, Amsterdam. pp. 309-334.

Devasahayam, S., 2000b. Evaluation of biopesticides for the management of shoot borer (*Conogethes punctiferalis* Guen.) on ginger (*Zingiber officinale* Rosc.). In: Ramana, K. V., Eapen, S. J., Babu, K. N., Krishnamurthy, K. S. and Kumar, A. (Eds.). *Spices and Aromatic Plants. Challenges and Opportunities in the New Century*. Indian Society for Spices, Calicut, pp. 276–277.

Devasahayam, S., 2002. Evaluation of biopesticides for the management of shoot borer (*Congethes punctiferalis* Guen.) on turmeric. In: Rethinam, P., Khan, H. H., Reddy, V. M., Mandal, P. K. and Suresh, K. (Eds.) *Plantation Crops Research and Development in the New Millenium,* Coconut Development Board, Kochi, pp. 489–490.

Devasahayam, S. and Anandaraj, M., 1997. Black pepper (*Piper nigrum* L.) research. In: Narwal, S. S., Pauro, P. and Bisla, S. S. (Eds.) *Neem in Sustainable Agriculture.* Scientific Publishers, Jodhpur, pp. 117-122.

Devasahayam, S. and Koya, K. M. A., 1993a. Effect of top shoot borer (*Cydia hemidoxa*) infestation on young vines of black pepper (*Piper nigrum*). *Indian J. Agric. Sci.* 63: 762–763.

Devasahayam, S. and Koya, K. M. A., 1993b. Seasonal incidence of hymenopterous parasites of top shoot borer (*Cydia hemidoxa*) infesting black pepper. *J. Entomol. Res.* 17: 205-208.

Devasahayam, S. and Koya, K. M. A., 1994a. Natural enemies of major insect pests of black pepper (*Piper nigrum* L) in India. *J. Spices Arom. Crops* 3: 50–55.

Devasahayam, S. and Koya, K. M. A., 1994b. Field evaluation of insecticides for the control of mussel scale (*Lepidosaphes piperis* Gr.) on black pepper. *J. Entomol. Res.* 18: 213–215.

Devasahayam, S. and Koya, K. M. A., 1994c. Seasonal incidence of *Hexamermis* sp. (Dor., Mermithidae) parasitising larvae of top shoot borer *Cydia hemidoxa* Meyr. (Lep., Tortricidae) on black pepper. *J. Appl. Ent.* 117: 31-34.

Devasahayam, S. and Koya, K. M. A., 1999. Integrated management of insect pests of spices. *Indian J. Arecanut Spices Med. Plants* 1: 19–23.

Devasahayam, S. and Koya, K. M. A., 2005. Insect pests of ginger. In: Ravindran, P. N. and Babu, K. N. (Eds.). *Ginger-The Genus Zingiber*. CRC Press, Washington. pp. 367–389.

Devasahayam, S. and Koya, K. M. A., 2007. Insect pests of turmeric. In: Ravindran, P. N., Babu, K. N. and Sivaraman, K. (Eds). *Ginger-The Genus Zingiber.* CRC Press, Washington, pp. 169-192.

Devasahayam, S. and Leela, N. K., 1998. Evaluation of plant products for antifeedant activity against pollu beetle (*Longitarsus nigripennis* Motschulsky), a major of black pepper. In: Reddy, P. P., Kumar, N. K. K. and Verghese, A. (Eds.). *Advances in I P M for Horticultural Crops.* Association for Advancement of Pest Management in Horticultural Ecosystems, Bangalore, pp. 172-174.

Gopakumar, B. and Chandrasekhar, S., 2002. Insect pests of cardamom. In : Ravindran, P. N. and Madhusoodanan, K. J. (Eds.). *Cardamom-The Genus Elettaria.* Taylor and Francis, London, pp. 179–206.

(IISR) Indian Institute of Spices Research, 2001. *Annual Report 2000.* Indian Institute of Spices Research, Calicut, 155 pp.

(IISR) Indian Institute of Spices Research, 2005a. *Turmeric (Extension Pamphlet)*, Indian Institute of Spices Research, Calicut, 8 pp.

(IISR) Indian Institute of Spices Research, 2005b. *Annual Report 2004-05* Indian Institute of Spices Research, Calicut, 83 pp.

Jacob, S. A., 1981. Biology of *Dichocrocis punctiferalis* Guen. on turmeric. *J. Plantn. Crops* 9: 119–123.

Jacob, S. A., 1982. Biology and bionomics of ginger and turmeric scale *Aspidiotus hartii* Green. In: Nair, M. K., Premkumar, T., Ravindran, P. N. and Sarma, Y. R. (Eds.). *Proceedings, National Seminar on Ginger and Turmeric.* Central Plantation Crops Research Institute, Kasaragod, pp. 131-132.

Kotikal, Y. K. and Kulkarni, K. A., 2001. Reaction of selected turmeric genotypes to rhizome fly and shoot borer. *Karnataka J. Agric. Sci.*, 14: 373-377.

Koya, K. M. A., 1994. *Studies on Coccids Infesting Black Pepper*. Final Report of Research Project. National Research Centre for Spices, Calicut, 19 pp.

Koya, K. M. A., Balakrishnan, R., Devasahayam, S. and Banerjee, S. K., 1986. A sequential sampling strategy for the control of shoot borer (*Dichocrocis punctiferalis* Guen.) on ginger (*Zingiber officinale* Rosc.) in India. *Trop. Pest Manage.* 32: 343–346.

Koya, K. M. A., Devasahayam, S., Selvakumaran, S. and Kallil, M., 1996. Distribution and damage caused by scale insects and mealybugs associated with black pepper (*Piper nigrum* Linn.) in India. *J. Entomol. Res.* 20: 129-136.

Koya, K. M. A., Premkumar, T. and Gautam, S. S. S., 1988. Chemical control of shoot borer *Dichocrocis punctiferalis* Guen. on ginger *Zingiber officinale* Rosc. *J. Plantn. Crops* 16: 58–59.

(NRCS) National Research Centre for Spices 1989. *Cardamom Package of Practices*. National Research Centre for Spices, Calicut, 22 pp.

(NRCS) National Research Centre for Spices 1990. *Annual Report 1989–90*. National Research Centre for Spices, Calicut, 61 pp.

Nybe, E. V. and Nair, P. C. S., 1979. Field reaction of turmeric types to important pests and diseases. *Indian Cocoa Arecanut Spices J.* 4: 107–109.

Parvathi, C. and Chandrappa, H. M., 1993. Preliminary screening of cardamom (*Elettaria cardamomum* Maton) elite clones for thrips (*Sciothrips cardamomi* Ramk.) damage in the field. *J. Spices Arom. Crops* 2: 81–82.

Philip, J. and Nair, P. C. S., 1981. Field tolerance of ginger types to important pests and diseases. *Indian Cocoa Arecanut Spices J.* 2: 109–111.

Pillai, G. B. and Abraham, V. A., 1979. Intensity of pollu beetle *Longitarsus nigripennis* Mots. (Coleoptera: Chrysomelidae) infestation in different cultivars of black pepper *Piper nigrum* L. In : Venkataram, C S (Ed.). *Proceedings, PLACROSYM-II 1979*. Indian Society for Plantation Crops, Kasaragod, pp. 289–293.

Premkumar, T., 1980. Ecology and Control of Pepper Pollu Beetle *Longitarsus nigripennis* Motschulsky (Chrysomelidae: Coleoptera). *PhD Thesis*, Kerala Agricultural University, Vellayani, Trivandrum, 160 pp.

Premkumar, T., Banerjee, S. K., Devasahayam, S. and Koya, K. M. A., 1986. Effect of different insecticides on the control of pollu beetle *Longitarsus nigripennis* Mots., a major pest of black pepper (*Piper nigrum* L.). *Entomon.*, 11: 219-221.

Rajkumar, A. J, Backiyarani, S. and Murugan, M., 2002. PV-1, a pseudostem borer tolerant cardamom variety. *Insect Environ.*, 7: 155–156.

Rajkumar, A. J., Kurien, P. S., Backiyarani, S. and Murugan, M., 2001. Efficacy of conventional insecticides for suppression of black pepper mussel scale, *Lepidosaphes piperis* Gr. (Diaspididae: Hemiptera). *Entomon.*, 26 (Special Issue): 113–114.

Regupathy, A., Santharam, G., Balasubramanian, M. and Arumugam, R., 1976. Occurrence of the scale *Aspidiotus hartii* C. (Diaspididae : Hemiptera) on different types of turmeric *Curcuma longa* Lin. *J. Plantn. Crops*, 4: 80.

Sasikumar, B., Chempakam, B., Remashree, A. B., Devasahayam, S., Dhamayanthi, K., Ravindran, P. N. and Peter, K. V., 1998. Characterization of two interspecific hybrids of *Piper*. *J. Hort. Sci. Biotechnol.* 74: 125–131.

Selvakumaran, S., Mini, K. and Devasahayam, S., 1996. Natural enemies of two major species of scale insects infesting black pepper (*Piper nigrum* L.) in India. *Pest Manage. Hort. Ecosyst*. 2: 79-83.

Sheila, M. K., Abraham, C. C. and Nair, P. C. S., 1980. Incidence of the shoot borer (*Dichocrosis punctiferalis*) Guen.) (Lepidoptera, Pyraustidae) on different types of turmeric. *Indian Cocoa Arecanut Spices J.* 3: 59–60.

Singh, J., Sudarshan M R and Kumaresan D., 1996. Evaluation of promising selections of cardamom (cv. Malabar) for thrips *Sciothrips cardamomi* Ramk. (Thysanoptera : Thripidae) tolerance. *J. Plantn. Crops* 24 (Suppl.): 283–285.

Thyagaraj, N. E., Chakravarthy, A. K., Rajagopal, D. and Sudarshan, M. R., 1991. Bioecology of cardamom root grub *Basilepta fulvicorne* Jacoby (Eumolpinae: Chrysomelidae: Coleoptera). *J. Plantn. Crops* 18 (Suppl.): 302–304.

Varadasan, S., 1991. Dynamics of life cycle of cardamom shoot borer, *Conogethes punctiferalis Guen. J. Plantn. Crops* 18 (Suppl.): 302–304

Varadarasan, S., 1992. Insect pest management in cardamom. In : Sarma, Y. R., Devasahayam, S. and Anandaraj, M. (Eds.). *Black Pepper and Cardamom-Problems and Prospects*. Indian Society for Spices, Calicut, pp. 51–54.

Varadarasan, S., 1995. Biological control of insect pests of cardamom. In : Ananthakrishnan, T. N. (Ed.). *Biological Control of Social Forest and Plantation Crops Insects*. Oxford and IBH Publishing Company Private Limited, New Delhi, pp. 109–119.

Velayudhan, K. C. and Liji, R. S., 2003. Preliminary screening of indigenous collections of turmeric against shoot borer (*Conognethes punctiferalis* Guen.) and scale insect (*Aspidiella hartii* Sign.). *J. Spices Arom. Crops* 12: 72-76.

Visalakshi, A. and Joseph, K. V., 1965. The biology of the pepper shoot borer *Laspeyresia hemidoxa* Meyr. (Eucosmidae: Lepidoptera). *Agric. Res. J. Kerala* 3: 48–50.

Visalakshi, A. and Joseph, K. V., 1967. Biology of *Gynaikothrips karnyi* Bagnall the marginal gall forming thrips of pepper. *Agric. Res. J. Kerala* 5: 79–81.

Chapter 13

Weed Management in Field Crops: Traditional to Modern Approaches

T.K. Das
Division of Agronomy, Indian Agricultural Research Institute, New Delhi–110 012

Introduction

Weeds have been existing ever since the man started domesticating/cultivating plants and animals around 10,000 B.C.–the most significant single occurrence in human history (Macneish, 1964). This was the beginning of farming and the farmers began to replace hunters about that time, a process continues till today. Weeds have been recognized as a problem since then and the battle against weeds is a never ending one and often the costliest agronomic input for the success in crop production. They consume much of the farmers' time, which, otherwise, would have been used for other crop production activities. Still weeds remain as the most underestimated crop pests in tropical agriculture although they cause higher reduction/loss in the yields of crops than other pests and diseases. Between humans and continuing food supply, there stand four natural hazards, *e.g.* weather, weeds, insect pests and plant diseases. Sometimes these hazards work independently and many times they work hand in hand (Muzik, 1970).

Of the total annual loss of agricultural produce from various pests in India, Yaduraju (2006) reported that weeds roughly account for 37 per cent, insects for 29 per cent, diseases for 22 per cent and other pests for 12 per cent. Weeds also decrease quantity and quality of produce/food, fibre, oil, forage/fodder, animal products (meat and milk) and cause health hazards for humans and animals. Sahoo and Saraswat (1988) from analysis of 10 years data (1978-1988) reported annual crop yield losses to the tune of 9.3 million tonnes in cereals, 0.8 million tonne in pulses, 0.6 million tonne in oilseeds and 7.2 million tonnes in fibres valued at Rs. 2799.00 crores occurring annually in India. Estimates also say that weeds generally reduce yield by 15-30 per cent in wheat, 30-35 per cent in rice and 18-85 per cent in maize, sorghum, pulses and oilseeds. They also on an average remove 30-40 kg N, 10-15 kg P_2O_5 and 20-40 kg K_2O per hectare (Alagh, 1988).

Controlling composite culture of weeds in the crop field is a complex phenomenon, which is governed by a host of factors, namely the type of soil, temperature, rainfall, relative humidity, age of weed flora and their composition/distribution, polarity, formulation and chemical nature of the herbicide. Similarly affected is the herbicide selectivity to crop, which is based on plant, soil, environment and herbicide interaction. Spectrum and efficiency of weed control are likely to be lower in the summer season under dry conditions. The same fate applies to pre-emergence herbicide under such conditions. However, crop selectivity may be greater/higher.

History of Weed Control Methods

Since adoption, weed control has undergone several modifications (Table 13.1) in terms of addition, superimposition/integration or withdrawal of weed control methods as per the changes in perspective of world agriculture in a phased manner (Radosevich *et al.*, 1997). There has been a phenomenal progress in its methodologies and approaches in recent years. The world has entered into an era of "integrated pest/weed management," which emphasizes not to exercise a single method, but a suitable combination of methods based on their compatibility, feasibility and sustainability in certain agro-ecosystem.

Table 13.1: Stages of Weed Control Practices Adopted/Developed in the World (Hay, 1974; Radosevich *et al.*, 1997; modified by the Author)

Sl.No.	Weed Control Practices Developed	Approx. Years
1.	People started cultivation and used to remove weeds by hand	Since10,000 B.C.
2.	People started to use primitive hand tools to till the land and control weeds	Since 6,000 B.C.
3.	Animal-drawn implements like harrows	Since 1,000 B.C.
4.	Improved animal-drawn implements like ploughs, cultivators, blade hoes, rotary hoes, finger weeders, row-weeders etc and fuel-powered mechanical implements like tractor-drawn ploughs, cultivators, blades, discs etc. have been put into use for tillage and weed control.	Since 1920 A.D.
5.	Chemical weed control using salts of inorganic acids or inorganic compounds exercised mainly in non-cropped areas	From 1880s–mid 1940s A.D.
6.	Biological control of weeds conceptualized	Since 1930 A.D.
7.	Chemical control using organic herbicides such as 2, 4-D, MCPA came into effect	Since 1947 A.D.
8.	Phenomenon of allelopathy was defined by Molisch in 1937, but researches initiated towards evaluation/detection of allelopathy across weeds and crops	Since 1960s
9.	Concept of integrated weed management (IWM)	Since 1970s

Tull (1731, 1733) for the first time proposed row planting for crops to maintain optimum plant population and facilitate hoeing by horses towards weed control. Later two great changes in the history of weed control, one being the replacement of manual weeding by animal power for physical weed control in the 18th century and the other being replacement of physical method of weed control by herbicides in the 20th century took place in the developed countries(Swarbrick and Mercado, 1987). These changes resulted in a considerable increase in food production and marked reduction in rural labour forces in those countries.

Approaches in Weed Management

Several approaches such as physical (manual and mechanical), cultural/ecological/cropping and competition, biological, chemical and allelopathy (Figure 13.1) are in use towards management of weeds. They may be uniformly applicable to all crops, but there is selective adoption of these methods

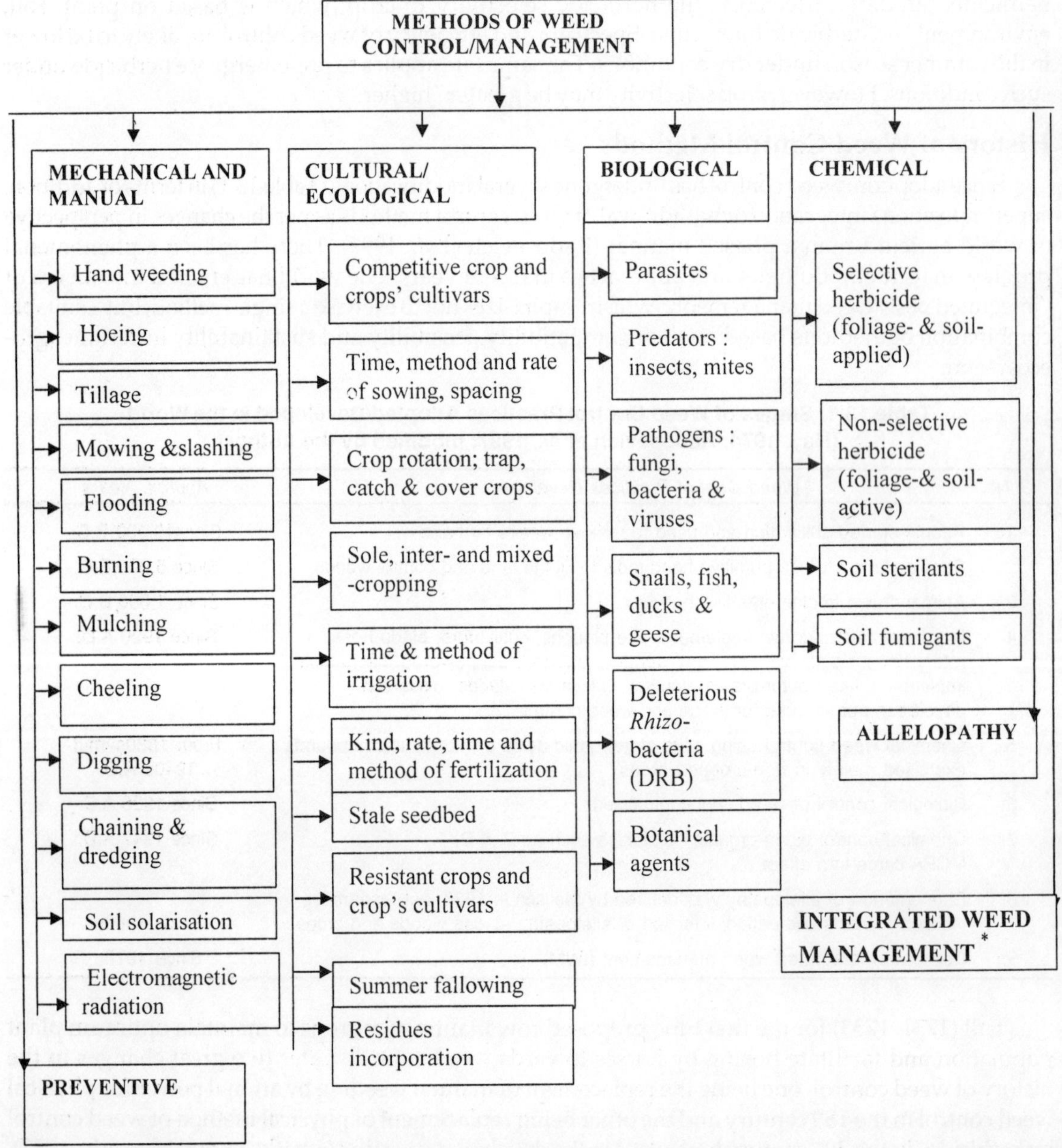

Figure 13.1: Methods of Weed Control

as per the requirements and feasibility in a crop and/or cropping systems. The determining factors for choice of these methods are critical period of weed competition, duration of crops, season of growing, economic importance of the crops and the most important one is socio-economic condition of the farmers.

Knowledge of weed seed characteristics, morphology, ontogeny, nature of competition and degree of association with crops are pre-requisite for suggesting some efficient weed control measures. It makes the users or weed researchers quite acquainted with the nature and spectrum of weeds existing in the crop fields and accordingly guides them to adopt or to recommend certain measures. Perennial weeds are generally more difficult to control than annuals. Many times deep tillage advocated cannot reach to deep roots of perennials and dig out the perennating structures. Systemic herbicide even cannot reach to the roots and vegetative structures lying deeper in soil.

Prevention Approaches

The preventive measures usually do not offer remedy over the already existing population and diversity of weeds in the crop fields, but they focus on the prevention of further introduction of weeds from different external sources/agents as well as perpetuation of weeds in the forth-coming years from the existing stands of weeds in crop fields. The preventive methods/measures are depicted in Figure 13.2.

Farmers should use pure and clean seeds of crops as possible. This does no not add seeds of the existing or new weed species to soil seed bank and reduces seed rain and further dissemination of weeds (Table 13.2). *Phalaris minor* problem cropped up gradually in wheat because of admixture since mid-sixties when Mexican dwarf varieties carrying *Phalaris* seeds were introduced in India (Agrawal, 1995; Jaya Kumar and Jagannathan, 2003). Similar reason is put forward for weed problem in maize (Rezene, 1984, 1985).

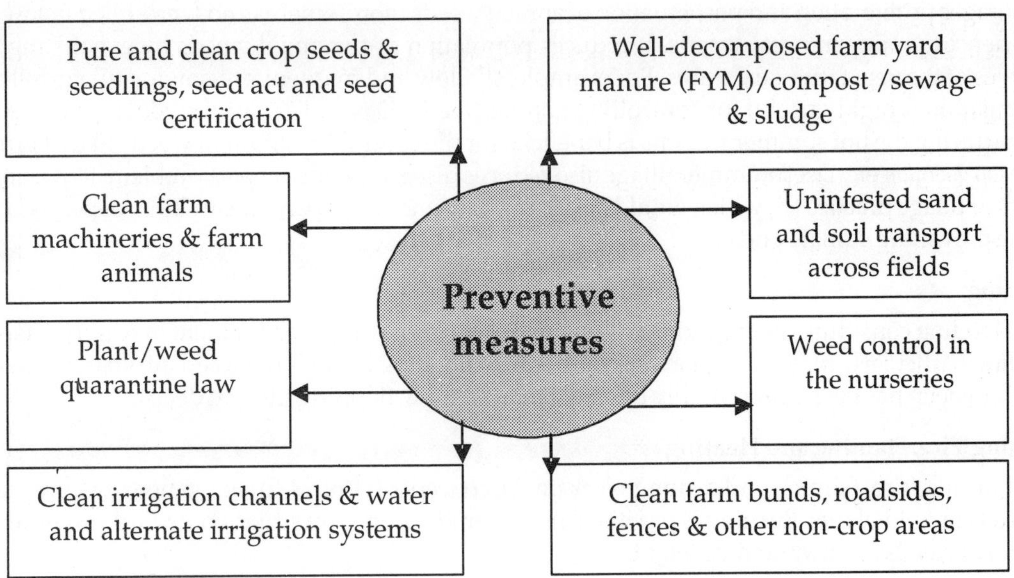

Figure 13.2: Different Weed Prevention Approaches/Measures

Table 13.2: Impurity of Farmer's Crop Seed from Different Sources (Bengtsson, 1983)

Crop	Other Crop Seed (%)	Weed Seed (%)	Inert Matter (%)	Weed Seed (No./kg crop seed)	
				Avena sp.	Lolium sp.
Wheat	0.6	1.4	0.8	84	439
Barley	0.6	1.3	1.9	800	692
Teff	0.0	2.1	0.3	103	118
Broad bean	0.2	0.7	0.0	0	0
Field pea	0.8	1.2	0.6	247	139
Linseed	0.2	5.6	4.1	276	691

Weed Quarantine Law enforces isolation of an area where a serious weed has established and prevents further movement of the weed into a non-infested area. An enactment is always required to check movement of noxious and pernicious weeds such as *Striga* sp., *Orobanche* sp., *Parthenium hysterophorus, Eichhornia crassipes, Chromolaena (=Eupatorium) odoratum, Salvinia molesta, Lantana camara* etc. It could be both inter-state and inter-country movement.

Physical (Manual and Mechanical) Methods

Physical (manual and mechanical) methods include hand weeding; hoeing and harrowing; tillage; mowing and slashing; flooding; burning and flaming; cheeling and digging; chaining and dredging; mulching; soil solarization; electromagnetic radiations (Figure 13.1).

Tillage

Tillage breaks, cuts or tears off weeds and exposes them to desiccation by sun. Frequent tillage encourages proliferation and perpetuation of annual weeds more, whereas no/zero-tilled fallow land experiences more perennial weeds. It reduces population of perennial weeds by exhausting food reserves of the vegetative structures. For example, shallow and frequent pre-sowing tillage followed by irrigation is highly useful for controlling annual weeds (Das and Yaduraju, 2001), whereas deep tillage during the hot summer months is beneficial for the control of perennial weeds like *Cyperus sp, Cynodon dactylon* etc. Deep summer tillage also reduces disease and insect pest problem. However, the choice of tillage practice (s), which would be appropriate under a particular situation, depends largely on the prevailing climate and soil.

Flooding

Flooding constitutes an important mechanism/aspect of weed management in rice. The depth of flooding particularly plays a role. Greater the depth of flooding lower is the weed infestation. Flooding 10-20 cm deep has been found highly effective than shallow flooding (2-5 cm deep).

Burning/Fire, Flaming and Heating

Burning/fire, flaming and heating work on the concept of control/management and prevention of weeds or wild plants. Burning is practised mainly under non-crop situation towards non-selective control of weeds or unwanted vegetation.

Mulching

Mulching has enough bearing towards weed suppression in cropped and non-cropped situations. The farmers can easily adopt mulching. Organic wastes or crop residues may serve dual purposes–

weed control as well as addition of organic matter to soil. However, the availability of enough mulch materials is a big bottleneck. For *in situ* availability of adequate mulch materials, appropriate crop rotations including periodic fallowing with grass or legume covers, are essential. Mulching materials having wider C:N ratio, may cause chlorosis of leaves showing nutrient deficiency during the first 1-2 seasons of adopting no-tillage practice without N or with low rate of N application. Kang *et al.* (1980) opined that no-tilled maize cultivation should not be recommended without N application in low fertile soil.

Soil Solarization

Solarization for a minimum period of 2 weeks during hot summer months (May and June in India) is sufficient to control weeds, but it may be continued to several weeks together for prolonged effect (Katan *et al.*, 1976; Kumar *et al.*, 1993; Yaduraju, 1997). Surface soil temperature may increase up to 55-60°C due to solarization during hot summer months, which proves lethal to many weed seeds and vegetative propagules, insects, nematodes and disease pathogens and causes them to die. It may control weeds in crops in the wet-season (*kharif*) as well as subsequently in the winter (*rabi*). It also controls soil-borne plant pathogens like *Fusarium, Verticillium, Pythium, Rhizoctonia, Sclerotium, Phytophthora* (Abdel Rahim *et al.*, 1988; AlMasoum *et al.*, 1998). Combination of solarization and herbicide, however, would be the best choice towards weed control (Yaduraju *et al.*, 1999).

Cultural/Ecological Approaches/Method

Cultural/ecological method exploits crop's competitiveness, environment and management practices towards smothering of weeds. Fryer (1983) stated that "a good crop is the best weed killer." Therefore, the farmers should adopt a good crop husbandry to raise a good crop. It includes crop species; crops' variety/cultivar; sowing of crop-time, method, rate and spacing; crop rotation; cropping practice, scheduling of irrigation; fertilization–kind, time, method and rate; stale seed bed technique; stimulation of germination; host plant's resistance (resistant crops and crop's cultivar); summer fallowing; residue incorporation into soil; trap and catch crops (Figure 13.1).

Competitive Crop Species or Cultivars

Crop species differ in their germination pattern, tillering, branching, nature and orientation/inclination of leaf, root growth, relative growth rate, stature/height, total growing duration etc. (Delorit and Ahlgren, 1967), and, therefore, variation in their competitive and weeds smothering ability is quite apparent. Alfalfa because of its perennial growth character consistently suppresses weeds more effectively than annual medic (Wells, 1969). Barley is more competitive to weeds than wheat mainly because of greater root proliferation at the initial 20-25 DAS and it also contains an allelochemical, *e.g.* gramine, which influences/intensifies the competition with weeds (Pavlychenko and Harrington, 1934; Pavlychenko, 1937 &1940; Bell and Nalewaja, 1968a, 1968b). Similarly, sorghum possesses allelochemical HCN (hydrocyanic acid) in shoot and foliages and allelopathic to *Abutilon theophrasti, Amaranthus hybridus, Setaria viridis, Bromus pectinatus* (Putnam, 1985), while in maize major source of allelochemicals is root exudates and maize is allelopathic to *Chenopodium album, Amaranthus retroflexus.* Several cases of competitive differences among weeds and crops have been reported (Kolbe, 1977; Kondap *et al.*, 1980; Rao and Shetty, 1981; Tiwari and Bisen, 1985; Singh *et al.*, 1992). Cousens *et al.* (1991) observed that barley, against wild oats, was more competitive than wheat at Long ashton (South-West England), but the reverse was true at Brooms Barn (Eastern England).

Crop cultivars too vary in their ability to compete with weeds, basically on the same principles as crop plants do (Rao and Shetty, 1976; Sandhu *et al.*, 1981; Gill and Mehra. 1981; Balyan and Malik,

1990; Malik and Balyan, 1990). Pawar (1994) observed that a semi-dwarf variety of wheat, Kundan (HD 2285) performed better than C 306, a tall variety, under sub-optimal levels of irrigation and N and had an equal competitive ability to suppress weeds as that of C 306. In rice, tall and dwarf cultivars showed a wide range of variation in their competitive abilities against weeds (Ghosh and Sarkar, 1975; Sarkar and Ghosh, 1977; Moody and De Datta, 1977). Under conditions of varying K availability (Siddiqi *et al.*, 1985) and P supply (Konesky *et al.*, 1989), barley cultivars exhibited considerable difference in their ability to compete with wild oats.

Time, Method and Rate of Sowing and Spacing

Time of sowing of a crop also influences weed competition. Under crop-weed ecosystem in every field, there exists a peak period of germination of seasonal weeds, which most often coincides with the germination or very early growth (seedling stage) of crop. Weeds pose more competition when wheat is sown on October than on November or December (Kolar and Mehra, 1992). Similarly late planting of wheat in December than its normal sowing in mid November in North and North-western India reduces *Phalaris minor* problem (Yaduraju and Ahuja, 1992). Similarly maize, cotton sown 15 days before the onset of usual monsoon rains with pre-sowing irrigation proved beneficial towards reduction of weed competition. Duary (1998) observed *Phalaris minor* more competitive in November-sown than in December-sown wheat.

In wheat, FIRBS (furrow-irrigated raised bed system) is of recent introduction from CIMMYT, Mexico and has been found useful in reducing overall weed including *Phalaris minor* competition in wheat mainly on the raised bed, but the furrows remain highly populated with weeds. *Cris-cross* sowing also has definite bearing in the reduction of weed competion in wheat. Transplanted rice generally encounters less weed problem than the upland direct-seeded one.

Seeding rate and per cent viability and germination of seeds usually determine crop density. Normally higher the density of a crop, lower is the weed competition and *vice-versa*. Angiras and Sharma (1993) reported that increasing wheat plant density by way of reducing row spacing from 20 cm to 15 cm could reduce the dry weights of *Lolium* and *Phalaris* by 11.9 per cent and 18.3 per cent, respectively. Duary (1998) reported that increase in wheat seed rate from 100 to 150 kg/ha was beneficial in higher *Phalaris minor* density plots since they recorded yield increase greater (10.5-16.9 per cent) than in weed free situation (4.8-4.9 per cent). Rogers *et al.* (1976) observed that for obtaining maximum yield of cotton planted in narrow row spaing of 53 cm, the period of weed control was 6 weeks, whereas the period of weed control was 14 weeks when cotton was planted in wider row spacing of 106 cm.

Crop Rotation

Crop rotation is an age-old, time-honoured and multi-tested useful practice. It is highly effective against parasitic weeds such as *Striga hermonthica/asiatica* (mainly in sorghum and maize), *Orobanche ramosa* (in *Brassicas* and solanaceous crops), *Orobanche cernua* (in tobacco), *Orobanche crenata* (in faba bean), *Cuscuta chinensis* (in alfalfa), *Cuscuta epilinum* (in linseed), *Cuscuta campestris* (in niger) (Parker, 1979; Parker and Riches, 1993) and crop-associated weeds like *Avena ludoviciana/fatua* and *Phalaris minor* (in wheat), *Cichorium intybus* and *Coronopus didymus* (in Egyptian clover/berseem), *Echinochloa colona/crusgalli* (in rice) (Gupta, 1998). Mustard/gobhi sarson, vegetable pea, potato, Egyptian clover/berseem if adopted in sequence after rice during winter season, *Phalaris* problem could be reduced to a great extent in North-western wheat belt of India (Singh and Singh, 2006).

Inter-cropping/Mixed Cropping

Inter-cropping/mixed cropping with compatible crop mixture reduces weed competition (Table 13.3) *vis-a-vis* increases total yield. Inter-cropping short-season crops (*e.g.* maize, melon) with longer-season crops (*e.g.* yam, cassava) prevents weeds from adapting to the growth cycle of either of the crops (Akobundu, 1978). Low-growing crops (melon, sweet potato) with maize, cassava and yam as inter-crops suppressed weed growth in Nigeria. Maize + green gram is also a useful combination (Bantilan and Harwood, 1973; Moody, 1973 &1979). Several other inter-croppings like sugarcane + green gram/black gram, maize + soybean/black gram, sorghum + cowpea/desmodium, wheat + gram are reported beneficial towards weed control.

Table 13.3: Mean Weed Biomass (t/ha) Growing in Assoiation with Sole-Cropped and Inter-cropped Maize and Mung Bean (Bantilan and Harwood, 1973)

Sl.No.	Treatments	Mean Weed Biomass (t/ha) Under Sole Cropping		Mean Weed Biomass (t/ha) Under Inter-cropping (maize + mung)
		Maize	Mung	
1.	Unweeded	2.06	0.85	0.28
2.	Inter-row cultivation	1.96	0.78	0.42
3.	Butachlor (0.6 kg/ha)	1.26	0.15	0.15
4.	Butachlor (1.2 kg/ha)	0.71	0.31	0.08
5.	Butachlor (2.4 kg/ha)	0.35	0.03	0.04
6.	Hand weeding	0.06	0.02	0.03

Time, Method and Rate of Fertilization

The time, method and rate of fertilization may alter crop-weed balance in favour of crop. Unger (1984) reported that the weed *Centella asiatica* occupying 35 per cent of the vegetation cover in the non-fertilized plots, was completely suppressed once fertilizer was applied.

The P- and K-fertilizers affect weed diversity, but they alone, unless some specific weeds do not influence weed growth much if N is absent. They, however, invariably affect crop growth whether N present or not. However, nitrogen influences equally crop and weeds particularly grassy weeds. Some broad-leaved weeds, namely *Melilotus alba/indica/parviflora, Chenopodium album, Medicago denticulata* are influenced much by P and K application. However, their growth may be entirely smothered by tall-growing wheat plants with enough supply of N (Das and Yaduraju, 1999a and 1999b). Sometimes application of acidic fertilizers like ammonium sulphate, ammonium chloride, urea may acidify soil and discourage leguminous weeds (Williams, 1976). All fertilizers (N, P and K) should be banded, side-dressed or placed below the crop rows (Yaduraju and Ahuja, 1997) before sowing. There, however, should not be any contact between fertilizer and crop seeds.

Generally higher the rate of fertilizer particularly N, lower is the weed growth. In wheat, *Phalaris minor* growth was reduced when N was applied at 150 kg/ha than at 120 kg/ha (Gautam, 1992; Das and Yaduraju, 2008). Similar reduction was also observed in the germination and growth of *Striga, Hygrophilla difformis* and perennial *Chara zeylanica* (Saraswat, 1989) with increased levels of N. Wimschneider *et al.* (1990) reported that increase in fertilizer levels led to disproportionately higher competition for nutrients from wild oats and the loss in wheat yield rose from 13 per cent to 23 per cent.

Stale Seedbed Technique

Stale seedbed technique is a cultural-cum-preventive measure. It is the easiest way of exhausting weed seed bank from soil. Unknowingly the farmers practise it partially when they plough the field repeatedly for 4-5 times before sowing of a crop and kill emerging or emergent weeds in every successive ploughing. Engstroem (1974) reported that weed population was reduced by 40-50 per cent and wheat grain yield increased by 50.7 per cent in stale seedbed compared to conventional seedbed. The land was prepared 3-4 weeks before sowing with minimum soil disturbance followed by paraquat application. Subsequent application of hand weeding or herbicide (MCPA) was again useful towards drastic reduction of weed growth (Table 13.4).

Table 13.4: Effect of Conventional and Stale Seedbed Techniques on Yield and Weed Flora in Wheat (Engstroem, 1974)

Sl.No.	Seedbed Techniques	Supplementary Weed Control Treatments		
		No Weeding	Hand Weeding	MCPA
		Wheat grain yield (kg/ha)		
1.	Conventional	1280	1900	1680
2.	Stale seedbed	1930	2230	2530
		Broad-leaved weeds weight (per cent)		
1.	Conventional	100	13	30
2.	Stale seedbed	61	9	18
		Grassy weeds weight (per cent)		
1.	Conventional	100	35	149
2.	Stale seedbed	50	23	42

Crop Plants' resistance to Weeds and Herbicides

The resistance research in weed science generally focuses/encompasses two major aspects, *e.g.* a) host plants' resistance (development of crop's cultivars, which is resistant to some parasitic weeds); and b) herbicide resistant crops/transgenics (development of crops resistant to some effective non-selective herbicides through biotechnological tools). In recent years, researches for resistant cultivars of crops against certain parasitic weeds have geared up in African and Asian countries (Table 13.5) and many varieties of crops resistant to *Striga*, *Orobanche* and *Cuscuta* have been developed or on the anvil. Simultaneously researches for development of transgenic crops have also become very fast in the world particularly by the Multinational Companies and as a result, a number of transgenic crops have evolved within a short period of 7-8 years through biotechnological tools (Table 13.6).

Weed Management in Field Crops

Generally under weedy/unweeded conditions of a crop, the population of composite weeds exceeds the population of a crop per unit area (Table 13.7). This is, however, not a rule. Weed competition is usually greater in pulses and oilseeds than in cereals if we ignore or overlook grass weed competition in cereals. Grass weed competition is severe in cereals mainly because of their similar growing habits. The population size, relative time of emergence and plant stature/height play a major role in the over-all competition. There is inverse relationship between the height of crop plants and growth of weeds or weed competition as a whole. Therefore, cereals having normally a tall stature are in an advantageous position than the pulses and oilseeds with regard to smothering of overall weeds. However, grass weeds having close association with cereals in terms of similar growth habit, phenology, nutrient demand etc. may pose more problems in cereals than in pulses and oilseeds.

Table 13.5: Crop's Varieties Resistant to Some Parasitic Weeds

Sl.No.	*Parasitic Weed Species*	*Crops and Resistant Varieties*	*Country*
1.	*Striga hermonthica*	Sorghum (S 1561, S 1477, S 1511–from ICRISAT, Hyderabad, India)	Africa
2.	*Striga hermonthica*	Sorghum [Yeju (ICSV 111), Gobiye (P 9401), Abshir (P 9403), Jigurty, Key # 8574 (from Sirinka Agric. Research Centre, Sirinka, Woldiya, Ethiopia)]	Africa (Ethiopia)
3.	*Striga hermonthica*	Maize (HR 2001, Gibe 1)	Africa (Ethiopia)
4.	Orobanche sp	Mustard (Durgamani* shows mild resistance in India); Fababean (F-402); Pepper (Maor, Odem)	–
5.	Cuscuta sp	Alfalfa (LLC 6* and LLC 7*–moderately tolerant); greengram (M 2*); blackgram (T 9*)	India

* Varieties reported herein are not transgenics or genetically modified rather they show as such or in-built tolerance.

Table 13.6: Herbicide-Resistant Crops in the World

Sl.No.	*Herbicide-Resistant Crops*	*Herbicides Resisted*	*Trademark Designation*	*Agrochemical or Seed Company*	*Commercial Status (Year)*
1.	Canola	Bromoxynil	BXN CANOLA	Rhone Poulenc	Europe 1995
	(*Brassica*	Glufosinate-AM	LIBERTY LINK CANOLA	AgrEvo	Canada and Europe 1995
	carinata)	Glyphosate	ROUNDUP READY RAPE	Monsanto	Canada 1997, Europe 1998
2.	Corn	Glufosinate-AM	LIBERTY LINK CORN	AgrEvo	USA 1997
	(*Zea mays*)	Glyphosate	ROUND UPREADY CORN	Monsanto, DeKalb	USA and Canada 1998
		Imidazolinones	IMI CORN	Pioneer, Ciba, Asgrow	USA 1997, Australia 1998-99
		Sethoxydim	SR CORN	BASF/DeKalb	USA and Brazil 1997
3.	Cotton (*Gossypium*	Bromoxylil	BXN COTTON	Rhone Poulenc, Calgene	USA 1997
	hirsutum)	Glufosinate-AM	LIBERTY LINK COTTON	AgrEvo	USA 2000
		Glyphosate	ROUNDUP READY COTTON	Monsanto	USA 1997
		Sulfonylureas	19-51A COTTON	DuPont	USA 1997
4.	Rice (*Oryza sativa*)	Glufosinate-AM	LIBERTY LINK RICE	AgrEvo	USA and Asia 2000-01
5.	Soybean (*Glycine max*)	Glufosinate-AM	LIBERTY LINK SOYBEAN	AgrEvo	USA 1998, Brazil 1998-99
		Glyphosate	ROUNDUP READY SOYBEAN	Monsanto, Asgrow	USA, Brazil and Argentina 1997
		Sulfonylureas	STS SOYBEAN	DuPont	USA 1993

Contd...

Table 13.6–Contd..

Sl.No.	Herbicide-Resistant Crops	Herbicides Resisted	Trademark Designation	Agrochemical or Seed Company	Commercial Status (Year)
6.	Tobacco	Bromoxynil	BXN TOBACCO		Europe 1997-98
	(*Nicotiana tabacum*)	Chlorsulfuron	CHLORSULFURON-RESISTANT TOBACCO	–	–
		Asulam	ASULAM-RESISTANT TOBACCO	–	–
7.	Sugarbeet (*Beta vulgaris*)	Glufosinate-AM	LIBERTY LINK SUGARBEET	–	Europe 1999-2000
		Glyphosate	ROUNDUP READY SUGARBEET	–	Europe 1997-98
8.	Rapeseed	Glyphosate	GLYPHOSATE-RESISTANT RAPESEED	–	–
9.	Tomato	Glyphosate	GLYPHOSATE-RESISTANT TOMATO	–	–
10.	Potato	Asulam	ASULAM-RESISTANT POTATO	–	–

Table 13.7: Density of Crop and Weeds in Unweeded Farmer's Field (Bengtsson, 1983)

Crops	Crop plants/m²	Weeds/m²	Crops	Crop plants/m²	Weeds/m²
Wheat	305	1115	Field pea	80	1335
Barley	215	1205	Linseed	280	2290
Maize	5	1340	Oats	100	1195
Sorghum	8	1235	Potato	6	735
Broad bean	40	1200			

Rice

Rice is grown in different cultures/conditions. A variety of weeds owing to different rice cultures usually prevail in rice fields across the countries of the world. Transplanting has certain bearing in weed control efficacy over direct-seeding. On yield front too, transplanted rice proves superior to direct-seeded one. The tall *indica* varieties are more competitive against weeds than dwarf *japonica* varieties, but dwarf varieties are higher yielder than tall varieties. Therefore, there is more reliance to dwarf varieties by the farmers although they encounter heavy weed problem.

Yield Reduction

15-20 per cent (transplanted rice), 30-35 per cent (direct-seeded puddled rice) and more than 50 per cent (direct-seeded upland rice).

Critical Period

15-30 DAS for direct-seeded upland rice and 30-60 days after transplanting (DAT) for lowland transplanted rice.

Weed Management

Nursery

Weed management in the rice nursery is a pre-requisite to combat weed problem in the main field. In the close dense population of rice seedlings particularly under broadcast sown nursery, mechanical weeding is very difficult to execute in the nursery. Even manual weeding proves less efficient if grass weeds and sedges similar to rice seedlings grow with rice in the nursery. Then chemical weeding becomes the ultimate resort/option for weeding in the nursery.

Chemical Control

1. Butachlor, pendimethalin or benthiocarb @ 1.0-1.5 kg/ha at 5-7 days after sowing (DAS). There should be sufficient moisture at the time of application of herbicide in the dry nursery. Otherwise, irrigation should be provided immediately after application of herbicide. The question of moisture, however, does not arise under wet nursery situation.
2. Pretilachlor @ 0.3-0.4 kg/ha or 0.75 kg/ha with surfactant as pre-emergence at 3-5 DAS.
3. Pyrazosulfuron-ethyl @ 0.015-0.020 kg/ha pre-emergence at 1-3 DAS for broad-spectrum weed control.
4. Cyhalofop-butyl @ 0.19-0.28 kg/ha as post-emergence at 1-2 leaf stage of rice for excellent grass weed control.

Direct-Seeded Upland (Normally Rainfed) and Direct-Seeded Puddled Rice

1. Mechanical/manual weeding is difficult to be carried out in direct-seeded broad-cast upland rice. If it is a direct-seeded drilled/line-sown rice, mechanical weeding may be executed. Beushening, a sort of blind tillage is practiced by the farmers in direct-seeded broad-cast upland rice, which controls weeds *vis-à-vis* put the rice plants in rows.
2. Butachlor, benthiocarb or pendimethalin @ 1.0-1.5 kg/ha as pre-emergence (after sowing but before emergence of rice). Moisture in soil is again a pre-requisite for greater herbicidal effects. In addition one hand weeding may be provided at 30-35 DAS.
3. Pretilachlor @ 0.3-0.4 kg/ha or 0.75 kg/ha with surfactant as pre-emergence at 3-5 DAS.
4. Pyrazosulfuron-ethyl @ 0.015-0.020 kg/ha pre-emergence at 1-3 DAS for broad-spectrum weed control or early post-emergence at 10-15 DAS.
5. Cyhalofop-butyl @ 0.19-0.28 kg/ha as post-emergence right from 1-2 leaf stage of rice until 60 DAS before hervest for excellent grass weed control.
6. Bensulfuron-methyl (Londax 60 DF®) @ 50-60 g/ha at 20 DAS.
7. Almix [metsulfuron-methyl (10 per cent) + chlorimuron-ethyl (10 per cent) + 0.2 per cent surfactant] @ 4 g/ha at 3-5 DAS as pre-emergence or at 20 DAS as post-emergence.
8. Oxadizon @ 0.5–0.75 kg/ha as pre-emergence or early post-emergence to control a broad-spectrum of weeds.
9. 2,4-D @ 0.75-1.0 kg/ha as post-emergence at 30 DAS to control broad-leaved weeds and few sedges. If hand weeding is provided at 30-35 DAS, there is no need of this treatment. The susceptibility of certain rice varieties may not be ruled out. They, of course, recover soon from the injury. The tall *indica* rice varieties are more tolerant to 2,4-D than dwarf *japonica* varieties and the growth of rice plants at the time of application of 2,4-D is a determining

factor towards selectivity of 2,4-D to rice plants. Selectivity lowers at very low and very high temperatures.

10. Metsulfuron-methyl @ 4-8 g/ha as post-emergence at 30-35 DAS may be a substitute of 2,4-D in controlling broad-leaved weeds. It, however, hardly controls sedges.

Transplanted Rice (Low Land Flooded or Irrigated Conditions)

1. *Puddling*: There is truly no substitute of puddling as far as the early establishment of rice seedlings, water retention/management and weed control are concerned. This is quite effective to eliminate the existing stand of weeds and to increase water holding capacity of land particularly under a situation where a standing water of few centimetres (2-5 cm) almost all through the growing period, is a pre-requisite for transplanted rice cultivation.
2. Manual/hand weeding is troublesome and labour-intensive. The labourers sometimes leave/skip *Echinochloa colona/crusgalli/glabrescens* and wild rice (*Oryza sativa* var *fatua*) unweeded since these weeds form mimicry with rice plants at the early stages of growth. They are very competitive with rice at the early stages and chemical weeding proves superior.
3. Butachlor, benthiocarb or pendimethalin @ 1.0-1.5 kg/ha at 3-5 days after transplanting (DAT) on saturated soil. Water usually standing in rice field should be taken out before applying herbicides and there should not be flooding with water for the next 3-4 days after application.
4. Pretilachlor @ 0.3-0.4 kg/ha or 0.75 kg/ha with surfactant as pre-emergence at 3-5 DAS.
5. Propanil is a good herbicide as far as the control of *Echinochloa sp* and *Panicum sp* is concerned. Propanil @ 2.0-3.0 kg/ha as post-emergence spray on the young weeds at 2 weeks after planting (WAP) is quite effective to control a wide array of annual grass weeds particularly under flooded condition. However, it does not have residual weed control ability.
6. Pyrazosulfuron-ethyl @ 0.015-0.020 kg/ha pre-emergence at 1-3 DAS for broad-spectrum weed control or early post-emergence at 10-15 DAS.
7. Cyhalofop-butyl @ 0.19-0.28 kg/ha as post-emergence right from 1-2 leaf stage of rice until 60 DAS before hervest for excellent grass weed control.
8. Bensulfuron-methyl (Londax 60 DF®) @ 50-60 g/ha at 30 DAT.
9. Almix [metsulfuron-methyl (10 per cent) + Chlorimuron-ethyl (10 per cent) + 0.2 per cent surfactant] @ 4 g/ha at 3-5 DAS as pre-emergence or at 25-30 DAS as post-emergence.f) Anilofos @ 0.40-0.43 kg/ha at 3-5 DAT. It controls mainly grass weeds.
10. 2,4-D @ 0.75-1.0 kg/ha at 30-35 DAT may be applied on saturated soil to control annual broad-leaved weeds and sedges. Usually 2,4-D is recommended as post-emergence at 30-35 DAT where broad-leaved weeds and sedges constitute a sizable population in the overall composition of weed flora.
11. Oxadiazon @ 0.5–0.75 kg/ha as pre-emergence or early post-emergence on puddled soil where direct seeding has been done and is quite effective against annual grasses particularly *Echinochloa crusgalli* and *Commelina benghalensis*.
12. Bentazon (1.0-2.0 kg/ha) as post-emergence at almost all stages of rice plants in transplanted rice.
13. Metsulfuron-methyl @ 4-8 g/ha as post-emergence at 30-35 DAS/DAT for controlling broad-leaved weeds.

14. *Alga control*: Sometimes undesired alga may grow luxuriantly in low-land rice field, which cause suffocation due to more anaerobic condition and prevent root growth and tillering of rice. Therefore, it needs to be removed manually or destroyed chemically. Copper sulphate ($CuSO_4.5H_2O$) or copper oxychloride @ 10 kg/ha may be recommended to control algal growth in rice. Nitrofen is also quite effective for algal control. However, if there are beneficial blue green alga living together with undesired ones, it is better that no herbicide is applied.

Integrated Approach

Direct-Seeded Upland Rice

1. Good crop husbandry + pre-emergence herbicide (pendimethalin, butachlor, benthiocarb, anilofos, pretilachlor) + beushening/blind tillage.
2. Good crop husbandry + pre-emergence herbicide (pendimethalin, butachlor, benthiocarb, anilofos, pretilachlor) + hand weeding at 30-35 DAS.
3. Good crop husbandry + residue incorporation/green manuring + post-emergence herbicide (propanil, bentazon, 2,4-D) or hand weeding or cultivation at 30-35 DAS.

Direct-Seeded Puddled Lowland Rice

In this culture, pre-germinated seeds are broadcast into puddled soil with standing water. Hand weeding and cultivation are impractical. Therefore, herbicides and other cultural practices need to be adopted. High seed rate reduces weed infestation. Residue incorporation (Water hyacinth, Parthenium) and green manuring are also promising. Their combinations with herbicide may play a workable role.

1. Good crop husbandry + residue incorporation/green manuring + high seed rate + pre-or post-emergence herbicide.

Transplanted Lowland Rice

1. Good crop husbandry + residue incorporation + clean seedling + pre-emergence herbicide + hand weeding at 30-35 DAT (days after transplanting).
2. Good crop husbandry + green manuring + clean seedling + pre-emergence herbicide + hand weeding at 30-35 DAT. If *in situ* green manuring is practised concurrently with rice transplanted, pre-emergence herbicide should be selective to both rice and green manure crop. Otherwise, post-emergence herbicide may be adopted after incorporation of green manure crop into soil. Row transplanting necessarily becomes the pre-requisite for *in situ* green manuring in transplanted rice culture.
3. Good crop husbandry + pre-emergence herbicide + hand weeding or rotavation at 30-35 DAT.
4. Good crop husbandry + residue incorporation/green manuring + clean seedling + high-density planting or skip row planting + pre-emergence herbicide + fish cultivation (under lowland condition).
5. Good crop husbandry + hand weeding at 15-20 DAT + post-emergence herbicide at 30-35 DAT.
6. Good crop husbandry + pre-emergence herbicide + shallow depth (5-10 cm) of standing water + hand weeding at 30-35 DAT.

7. Good crop husbandry + shallow depth (5-10 cm) of standing water + post-emergence herbicide at 30-35 DAT. A few centimeters of standing water during the growing period of rice helps the weed control process in a large way.

Wheat and Barley

Yield Reduction

10-25 per cent

Critical Period

In dwarf wheat, first 30-45 days after sowing (DAS) and in tall wheat 35-50 DAS.

Weed Management

1. *Proper seedbed preparation*: Proper seedbed preparation enhances smooth germination of wheat and eliminates early weed competition.
2. *Pure and clean seeds*: The importance of pure and clean crop seeds cannot be ignored. In India, wheat seeds are invariably mixed with *Phalaris* seeds. *Avena sp, Lolium sp* also remains mixed with wheat seeds. This leads to the spread of weeds in other areas and adds to the problems of weed control. Table 13.8 deals with or describes such problem of weed seed admixture with crop seeds.

Table 13.8: Number of *Avena* and *Lolium* Seeds Sown Together with Crop at a Given Sowing Rate (Anonymous, 1979a)

Crops	Sowing Rate (kg/ha)	Avena Seeds (no./ha)	Lolium Seeds (no./ha)
Wheat	125	10,500	54,875
Barley	85-100	68,000-80,000	58,820-69,200
Teff	25-30	2,575-3,090	2,950-3,540
Field pea	120	29,640	16,680
Linseed	25-35	6,900-9,660	17,275-24,185

3. *Hand weeding*: Two hand weedings at 20-25 and 40-45 DAS are adequate for wheat. Even sometimes the first weeding is sufficient for annual weed control usually present in wheat.
4. *Stale seedbed technique*: Stale seedbed technique has immense use towards weed control unless sowing of wheat is delayed. This is least possible when wheat is grown in sequence of other crops particularly rice, maize, pigeon pea, even soybean for less time available for the adoption of such practice. However, this could readily be practised in fallow lands or lands where no crop was taken in the previous season.
5. *Method, time and rate of sowing*: Furrow-irrigated raised-bed system (FIRBS), skip-row planting, criss-cross or bi-directional method of sowing have some bearing on yield of wheat over conventional method of sowing. The same proposition holds true to a small extent with regard to weed control. This, however, needs to be combined with other methods particularly herbicides for better weed control. Higher seed rate up to a certain level (125 kg/ha under drill-sowing and 150 kg/ha under broad-casting) also proves beneficial towards weed suppression in wheat.

6. *Rate, time and method of fertilization particularly nitrogen*: Among the major plant nutrients, nitrogen has got weed, particularly broad-leaved weed smothering ability by selectively boosting up of growth of wheat plants. In low to moderate infestation of grassy weeds particularly *Avena sp* and *Phalaris sp*, higher doses of N are useful to some extent to suppress their growth, but not under heavy infestation. Under heavy infestation, the initial higher population of these grass weeds may suppress wheat.
7. *Trichoderma viridae* (a fungus) is reported from National Research Centre for Weed Science (NRCWS), Jabalpur, India for selective bio-control of *Phalaris minor* through colonization in roots. Mechanism is similar to that how deleterious *Rhizo*-bacteria (DRB) and phytopathogenic bacteria work.

Chemical Control

Annual Grass Weeds Control

1. If it is ensured that there is no resistant population of *Phalaris minor* in certain areas, isoproturon @ 0.75-1.0 kg/ha as post-emergence at 25-30 DAS may be applied for broad-spectrum control of weeds since isoproturon is highly economical, cost-effective and widely available herbicide.
2. Diclofop-methyl (Illoxan® 360 g/litre EC and 280g/litre EC) @ 0.9-1.0 kg/ha as post-emergence at 25-30 DAS (4-5 leaf stage of wheat) to control annual grass weeds namely wild oat, *Lolium, Rottboellia, Phalaris* (at 1-4 leaf stage) could be useful. The grasses should be at the actively growing stage. This should be applied from emergence to 4 fully expanded leaves in barley.
3. One herbicide often recommended for pre-emergence application in wheat is pendimethalin (Stomp® 300 g/litre EC and 500g/litre EC) @ 0.75-1.0 kg/ha to control grass weeds mainly *Phalaris* and some broad-leaved weeds. Wheat should be sown at least 3 cm deep. It hardly controls *Avena sp*. Pendimethalin can, however, be applied to control weeds under inter-cropping situations of wheat with mustard, chickpea, lentil, pea.
4. Tralkoxydim @ 300-400 g/ha or chlorsufuron @ 20-30 g/ha also control grass weeds in wheat. Tralkoxydim has little broad-leaved weed killing activity and chlorsulfuron brings about stunting or retarded growth in grasses (*Avena sp* and *Phalaris sp*).
5. Flamprop-isopropyl L (Suffix BW® 200g/litre EC) @ 0.6 kg/ha as post-emergence from end tillering to second node of wheat to control actively growing wild oats. Do not apply within 7 days of applying hormone weed killers.

Resistant Phalaris minor Control

1. Clodinafop-propargyl (Topik®) @ 60-80 g/ha, sulfosulfuron (Leader®) @ 20-40 g/ha or fenoxaprop-*p*-ethyl (Puma super®) @ 100-120 g/ha as post-emergence at 28-35 DAS effectively controls *Avena ludoviciana* and *Phalaris minor*. They are also very effective to control isoproturon-resistant *Phalaris minor*. They, except sulfosulfuron, do not control any broad-leaved weeds.

However, continuous use of these herbicides for last 7-8 years to control isoproturon-resistant *Phalaris minor* has led to the development of cross-resistance in some of these *Phalaris* biotypes to fenoxaprop-p-ethyl, clodinafop-propargyl and sulosulfuron in some parts of Punjab and Haryana. In those areas, thee herbicides should not be recommended to control isoproturon-resistant population of *Phalaris minor*. On the contrary, a new herbicide, pinoxaden, which at 40-50 g/ha controls cross-resistant populations of *Phalaris minor* effectively, should be recommended.

Annual Broad-Leaved Weeds Control

1. 2,4-DEE (ethylester formulation) @ 0.5 kg/ha applied at 30-35 DAS of wheat is highly effective against most annual broad-leaved weeds. A few perennial broad-leaved weeds, *e.g. Cirsium arvense, Convolvulus arvensis* and sedges particularly *Cyperus esculentus* could also be suppressed to some extent by this treatment. Some varieties (*e.g.* Arjun/HD 2009, Kundan) may produce negligible ear malformation.
2. If there is no resistant population of *Phalaris minor* in certain areas, isoproturon @ 0.75-1.0 kg/ha as post-emergence at 25-30 DAS may be recommended for broad-spectrum control of weeds.
3. Post-emergence application of metsulfuron-methyl (Algrip®) @ 4-8 g/ha at 30-35 DAS when wheat is at 5–6 leaf stage.
4. Bentazon (Basagran®480 g/litre EC) @ 1.0-1.5 kg/ha as post-emergence at any growth stage of wheat but usually at 30-35 DAS may control annual broad-leaved weeds and to some extent *Cyperus esculentus*. It should not be applied if rain is expected within 8 hours.

Broad-Spectrum Weed Control

1. Tank-mixes, ready-mixes or followed-by/sequential application
   (*a*) Tank-mix of isoproturon + 2,4-D @ 0.75 + 0.25 kg/ha, respectively or ready-mix Isoguard *plus*® (which contains isoproturon and 2,4-D together) as post-emergence at 28-30 DAS may be advocated for broad-spectrum control of weeds. However, their sequential application (first grass killer and then broad-leaved killer after one week) proves more efficient than tank-mixes.

   (*b*) Tank-mix of clodinafop-propargyl + 2,4-D/isoproturon/metsulfuron @ (60 g/ha + 0.25/0.5/0.004 kg/ha, respectively) as post-emergence at 28-30 DAS,

   or

   Tank-mix of sulfosulfuron + 2,4-D/isoproturon/metsulfuron @ (30 g/ha + 0.25/0.5/0.004 kg/ha, respectively) as post-emergence at 28-30 DAS,

   or

   Tank-mix of fenoxaprop-*p*-ethyl + 2,4-D/isoproturon/metsulfuron @ (120 g/ha + 0.25/0.5/0.004 kg/ha, respectively) as post-emergence at 28-30 DAS.

   Clodinafop however should not be used in barley. Tank-mixes sometimes prove little phytotoxic to wheat. Therefore, it is better if one grass killer (clodinafop-propargyl, fenoxaprop-p-ethyl or sulfosulfuron) is applied first and then broad-leaved killer as follow–by application after one week. While clodinafop-propargyl and fenoxaprop-p-ethyl are exclusive grass killers, sulfosulfuron has little balanced/broad-spectrum activity on both annual grass and broad-leaved weeds.

   (*c*) A ready-mix Atlantis 3.6 WDG (mesosulfuron 30 g/litre and iodosulfuron 6g/litre with safener) is also recommended for broad-spectrum weed control in wheat.

Integrated Approach

1. Good crop husbandry + clean seeds + zero tillage/closer row spacing/criss-cross sowing/ skip row planting/FIRBS (Furrow-irrigated raised-bed system) + pre- or post-emergence herbicide (broad-spectrum or herbicide mixture).
2. Good crop husbandry + post-emergence herbicide (broad-spectrum or herbicide mixture). If *Phalaris minor* has developed resistance, clodinafop-propargyl, fenoxaprop-*p*-ethyl or sulfosufuron may be recommended in mixture or followed by application with 2,4-D, isoproturon or metsulfuron-methyl for broad-spectrum weed control. Crop rotation, *e.g. rice–mustard/pea* instead of *rice–wheat* sequence may prevent the resistant as well as susceptible *Phalaris minor* to appear in the field.
3. Good crop husbandry + pre-emergence herbicides, namely pendimethalin or post-emergence herbicide (broad-spectrum or herbicide mixture) + hand weeding [for controlling problematic weeds namely *Convolvulus arvensis* (Field bind weed), *Asphodelus tenuifolius* (Wild onion), *Cirsium arvense* (Canada thistle), *Carthamus oxyacantha* (Wild safflower) and *Cannabis sativa* (Bhang)] later.

Maize

Yield Reduction

A yield loss to the tune of 20-30 per cent has been reported in India. However, yield reduction varies across places, cropping conditions and systems. Rainy season maize generally encounters more weed problem and thereby higher yield loss than winter maize.

Critical Period

Early crop establishment to early reproductive stage *i.e.* tasseling, usually first 60 DAS.

Weed Management

1. Proper seed bed preparation and pure and clean seed
2. *Time of sowing*: Sowing maize in the last week of June with application of pre-sowing irrigation may bypass to a great extent the initial weed competition under North Indian condition. Sowing maize in the first week or first fortnight of July may experience severe weed competition than when sown on the last week of June.
3. *Hand weeding*: Frequency of rainfall determines the frequency of hand weeding to be adopted in the maize field during rainy season. Two hand weeding at 25-30 and 55-60 days after sowing (DAS) in moderate rainfall areas, whereas three weeding at 20-25, 40-45 and 60-65 DAS in high rainfall areas may be quite useful.
4. *Inter-cropping practices*: Maize can be inter-cropped with a number of crops such as soybean, green gram, black gram, cowpea and a successful inter-cropping has considerable bearing towards reduction of weed competition. Cowpea inter-cropping has more smothering capacity to weeds, but its growing beyond a certain period may prove competitive to maize.
5. *Soil solarization*: Soil solarization practised during hot summer months (mainly on June and July) for a period of 4-6 weeks with transparent polyethene film of 50-100 m thickness is a highly useful towards control of almost all annual weeds in maize.

Chemical Control

1. Atrazine as pre-emergence @ 1.0–1.5 kg/ha is quite effective to reduce weed competition. Use low dose on sandy soil. It may be mixed with other herbicides to control broad-spectrum weeds or should be followed by one hand weeding at 30-35 DAS. It could also be used as post-emergence, but lower dose should be applied.
2. Alachlor @ 1.5-2.0 kg/ha as pre-emergence controls many annual grasses and few broad-leaved weeds. *Rottboellia* is resistant to it. It may be incorporated to control *Cyperus esculentus* to some extent. Use low dose on light soils. Mix it with other broad-leaved killers to control broad-spectrum weeds or it should be followed by one hand weeding at 30-35 DAS. Apply to finely tilled moist soil or when rain is expected within 10 days. It is a short lived molecule in soil.
3. Metolachlor @ 1.5-2.0 kg/ha as pre-emergence controls many annual grasses and *Cyperus esculentus* to some extent in certain conditions, but not *Rottboellia*. It may be mixed with other herbicides to control broad-spectrum weeds or should be followed by one hand weeding at 30-35 DAS.
4. Pendimethalin @ 1.0-1.5 kg/ha as pre-emergence controls many annual grasses and few broad-leaved weeds. *Rottboellia* may be controlled. Use low dose on light or low organic matter soils. It may be mixed with atrazine or other broad-leaved killer herbicides to control broad-spectrum weeds or should be followed by one hand weeding at 30-35 DAS. Long dry spell after its application may reduce degree of weed control.
5. 2,4-D @ 0.5-0.75 kg/ha as post-emergence is useful for controlling some sedges and broad-leaved weeds including parasitic *Striga*. It should be applied when maize is 20-30 cm tall avoiding drift and its contact with upper leaves. Maize cultivars may vary in their tolerance.
6. In the inter-cropping situations of maize with other crops, there must not be made any blind/blanket recommendation of herbicides. Rather the herbicides or the mixtures of herbicides selective and safe to both the crops are to be recommended. For example, if there are inter-croppings like maize + pulses (green gram, black gram, cowpea, pigeon pea, soybean, haricot bean) and maize + oilseeds (groundnut, sunflower), pendimethalin, which is selective to all the crops mentioned above may be recommended. In maize + legume inter-cropping, butachlor @ 1.5-2.0 kg/ha may be applied (Bantilan and Harwood, 1973). Alachlor in certain legume inter-croppings may also be recommended. Atrazine, however, in no inter-cropping situation involving legume and pulses or oilseeds may be advocated.
7. Ready-mixes
   (*i*) Alachlor + atrazine (Lasso/atrazine® 350 + 200 g/litre FW) @ 4.0-5.0 kg/ha (product basis) as pre-emergence controls almost all annual grasses and broad-leaved weeds but not *Rottboelia*. Lower dose advocated if soil contains less than 40 per cent clay. Pre-requisite is fine tilth and moist soil at application or rain within 10 days.

   (*ii*) Atrazine + metolachlor (Primagram® 250 + 250 g/litre FW) and atrazine + metolachlor (Primextra® 200 + 300 g/litre FW) @ 3.0–4.0 kg/ha (product basis) applied as pre-emergence and up to 3 leaf stage of weeds and pre-emergence or early post-emergence to crop before weeds are 3 cm tall. *Rottboellia* is resistant and moderate control of *Cyperus esculentus*.

8. *Tank-mixes*: Pre-emergence application of the tank-mixes of atrazine + pendimethalin, atrazine + alachlor or atrazine + metolachlor @ 1.0+1.0 kg/ha each is quite effective and shows broad-spectrum activity against grasses and broad-leaved weeds. Sufficient moisture is a pre-requisite for greater herbicidal efficacy. Use low dose in light soils.

Sorghum

Yield Reduction

20-30 per cent, but variable with management practice condition, varieties and across locations.

Critical Period

Early vegetative to early reproductive stage. Generally 30-60 DAS is taken as most critical.

Weed Management

1. Pure and clean seed and fine seed bed free from stubble and residues are prerequisite for early crop stand establishment. Sorghum seeds are small and seedlings initially look thin and small. Annual weeds may surpass their growth and suppress the crop easily.
2. Manual weeding at 25-30 and 45-60 days after sowing (DAS).
3. *Inter-cropping practice*: Sorghum + cowpea inter-cropping has been proved world over as an important practice towards smothering of weeds including the deadly weed *Parthenium hysterophorus* (Besufekad, 2001; Besufekad *et al.*, 2000) and parasitic *Striga*. Cowpea inter-cropping also controls stem borer and shoot fly of sorghum. Therefore, inter-cropping has enough bearing not only towards weed control but also for the control of other insect pests and diseases of sorghum. The choice of an appropriate variety performing well under inter-cropping system pays a lot towards success of the system.

Chemical Control

1. Pre-emergence application of atrazine @ 1.0-1.5 kg/ha offers a good control of annual broad-leaved weeds and some grass weeds. It, however, may be followed by one hand weeding at 30-35 DAS to control late-emerging annual weeds and perennial weeds. Atrazine @ 0.25-0.50 kg/ha may also be applied as post-emergence at 25-30 DAS at 3-4 leaf stage of sorghum and pre-emergence and early post-emergence to weeds before they are 4 cm tall. Apply lower dose on sandy soils.
2. *Tank-mixes*: Pre-emergence application of the mixture of atrazine + metolachlor, atrazine + terbutryne or atrazine + pendimethalin @ 1.0+1.0 kg/ha each controls annual grasses and broad-leaved weeds very effectively. Sole pendimethalin as against its tank-mix with atrazine may prove little phytotoxic to sorghum seedlings. Similarly sole alachlor or metolachlor poses phytotoxicity, which could be amended using safeners like 1,8-naphthalic anhydride (NA).
3. Ready-mixes
   (*i*) Atrazine + metolachlor (Primagram® 250 + 250 g/litre FW) @ 2.0 kg/ha (product basis) as pre-emergence and up to 3 leaf stage of sorghum can control annual grasses and broad-leaved weeds and *Cyperus esculentus* to some extent. *Rottboellia* is not controlled. It needs incorporation into soil under dry conditions, but moisture should be restored in soil within a short period.

(*ii*) Atrazine + terbutryne (Gesaprim Combi® 250 + 250 g/litre FW) @ 2.0-3.0 kg/ha (product basis) exclusively as pre-emergence to both crop and weeds controls broad-spectrum of weeds.

4. In the inter-cropping situations like sorghum + pulses (green gram, black gram, cowpea, pigeon pea, soybean, haricot bean) and sorghum + oilseeds (groundnut, sunflower), pendimethalin, which is selective to sorghum as well as inter-crops mentioned above is to be recommended at 0.75–1.0 kg/ha as pre-emergence.
5. Post-emergence application of 2,4-D @ 0.75-1.0 kg/ha at 30-35 DAS where broad-leaved weeds including *Striga* cast a problem. 2,4-D also controls sedges like *Cyperus esculentus* to some extent.

Integrated Approach (Maize and Sorghum)

1. Good crop husbandry + pre-emergence herbicide (atrazine, pendimethalin or atrazine + pendimethalin) + hand weeding at 35-40 DAS for broad-spectrum control of weeds, although atrazine + pendimethalin mixture is capable of controlling a broad-spectrum of weeds.
2. Good crop husbandry [time of sowing of crops (sowing in the last week of June with irrigation to suppress over-all weed growth under North-Indian condition or sowing in early July to reduce *Striga* infestation under Ethiopian condition)] + fertilizer (mainly N) management (placement along the crop rows, higher N dose for *Striga* control, split application) + pre-emergence herbicide (atrazine, pendimethalin or their mixture) + hand weeding at 35-40 DAS for maize.
3. Good crop husbandry + inter-cropping [maize inter-cropped with soybean, green gram, black gram or cowpea (with sorghum)] + pre-emergence pendimethalin for both maize and sorghum + hand weeding at 40-45 DAS (if at all required). Once this sort of combination of methods has been adopted, hand weeding may not be so useful towards weed control in the concurrent season as for reduction of seed bank and weed growth in the following years/ seasons.
4. Soil solarization after final preparation of land during hot summer months + good crop husbandry (mainly sowing with little soil disturbance, recommended time, method and rate of fertilization, suitable control measure for shoot borer/shoot fly, proper irrigation etc.) + pre-emergence herbicide (if at all required) + hand weeding (with little soil disturbance) at 35-40 DAS.

Pearl Millet

Yield Reduction

On an average 20-30 per cent yield reduction is recorded, but under humid rainy season and extremely weedy situations, it may be quite high. Das and Yaduraju (1995) have reported 72 per cent or more yield loss in pearl millet due to its initial slow growth. It picks up growth, starts tillering and increases in height after 25-30 DAS and becomes more competitive against weeds. The field should be kept free from weeds at least for the initial 25-30 DAS.

Critical Period

Initial 30-45 DAS.

Weed Management

1. *Proper seedbed preparation*: Pearl millet has small seeds. Therefore, fine and properly levelled seedbed is required for proper placement of seeds at required shallow depth for smooth germination of pearl millet, which necessarily helps in the early establishment of crop. It is a tillering crop and early weed competition adversely affects tillering in pearl millet, which may be reduced even to nil under extremely weedy situations.
2. *Proper depth of sowing*: Proper depth of sowing is to be maintained since pearl millet seeds are very small. Seeds must be placed at 1.0-2.0 cm depth of soil and not below that. Otherwise, germination must not be achieved to the level of 100 per cent and crop stand would be non-synchronous, non-uniform and patchy, which cannot compete well with species-wise diverse group of plants so-called weeds. In case of dry seeding practised, irrigation should follow immediately.
3. Generally two hand weedings at 20-25 and 35-40 DAS are adequate to eliminate weed competition and boost up crop growth, which can smother weeds coming up later.
4. Pearl millet may be inter-cropped with pulses *viz.*, green gram, black gram, cowpea, French bean and oilseeds such as groundnut, sunflower under paired-row system of cropping and weed competition could be reduced considerably.

Chemical Control

1. Atrazine, propazine or simazine @ 0.5 kg/ha as pre-emergence will control annual broad-leaved weeds along with some grasses. Moisture in the surface soil is a pre-requisite for their optimal bio-efficacy. Atrazine because of its greater water solubility, however, may prove more efficient than simazine in controlling weeds under rainfed dryland situation.
2. Pendimethalin @ 0.75-1.0 kg/ha as pre-emergence controls annual grasses and some broad-leaved weeds. It should be applied in moist soil for higher efficacy. In the inter-cropping situations between pearl millet and pulses or oilseeds, it may be recommended.
3. 2,4-D @ 0.75-1.0 kg/ha as post-emergence may be recommended at 25-30 DAS if broad-leaved weeds are predominant in the field. It may also control *Cyperus esculentus* to some extent. Avoid herbicide drift.

Wet Season (*Kharif*) Pulses

Green Gram, Black Gram and Cowpea

Yield Reduction

Yield reduction due to weeds is to the tune of 50 per cent as usual, but in extreme conditions/ cases up to 90 per cent or complete failure of crop.

Critical Period

First 30-35 days after sowing (DAS).

Weed Management

1. Two hand weedings at 20-25 DAS and 35-40 DAS in the first 35-40 DAS usually give an effective control of weeds if no pre-planting or pre-emergence herbicide applied. The last/ second hand weeding may be withheld in cowpea depending on the growth of crop, *i.e.* if growth found satisfactory at the first 25-30 DAS.

Chemical Control

Pre-Plant Incorporated Treatment

1. Fluchloralin @ 0.75-1.0 kg/ha controls annual grasses and few broad-leaved weeds.
2. Trifluralin @ 0.75-1.0 kg/ha controls annual grasses and few broad-leaved weeds.

The above pre-plant incorporation treatments may be followed by one hand weeding at 25-30 DAS to control late-emerging annual as well as perennial weeds, namely *Cynodon dactylon* and *Cyperus rotundus*.

Pre-emergence Treatment

1. Pendimethalin @ 0.75-1.0 kg/ha to control mainly annual grass weeds and few broad-leaved weeds.
2. Alachlor @ 1.0-2.0 kg/ha to control mainly annual grass weeds and few broad-leaved weeds.
3. Metolachlor @ 0.5-0.75 kg/ha to control mainly annual grass weeds and few broad-leaved weeds.
4. Terbutryne @ 1.0-1.5 kg/ha to control mainly annual broad-leaved weeds and few grass weeds

The above pre-emergence treatments may be followed by one hand weeding at 25-30 DAS to control late-emerging annuals as well as perennial weeds, namely *Cynodon dactylon*, *Cyperus rotundus*.

Post-emergence Treatment

1. Bentazon @ 0.75-1.0 kg/ha at 2-trifoliate leaf stage of green gram is highly efficient for the control of standing population of broad-leaved weeds and few sedges such as *Cyperus esculentus*.
2. Fluazifop-*p*-butyl at 0.125-0.250 kg/ha is also recommended for controlling both annual and perennial grass weeds, namely *Cynodon dactylon* (Bermuda grass) in pulses.

Pigeon Pea

Critical Period

First 60 days after sowing (DAS).

Weed Management

1. Two hand weedings or hoeings at 20 days interval at 25-30 DAS and 45-50 DAS are quite useful towards weed control in pigeon pea if economics permit.

Chemical Control

Pre-plant Incorporated Treatment

1. Trifluralin or fluchloralin @ 0.75-1.0 kg/ha to control mainly annual grasses and some broad-leaved weeds.

Pre-emergence Treatment

1. Pendimethalin @ 0.75-1.0 kg/ha for mainly grass weed control.

2. Alachlor @ 1.0-2.0 kg/ha.
3. Terbutryne @ 0.75-1.0 kg/ha.
4. Linuron @ 1.0-1.5 kg/ha for controlling annual broad-leaved weeds and few grasses. It has longer residual action and, therefore, in this case, the supplementary hand weeding may be delayed for one week or so if situation so occurs.

The above pre-plant incorporation or pre-emergence treatments may be supplemented with one hand weeding at 35-40 DAS to control late-emerging annuals as well as perennial weeds, namely *Cynodon dactylon*, *Cyperus rotundus/esculentus*.

Post-emergence Treatment

1. Fluazifop-*p*-butyl (Fusilade® 125 g/litre EC) at 125 g/ha could be used for the control of annual and perennial grass weeds. It should be applied on actively growing weeds and in perennial weeds at least 3 leaves per shoot should be present.

Soybean

Yield Reduction

Yield reducation due to weeds is upto 96 per cent.

Critical Period

Early seedling to early flowering of soybean (till 60 DAS in India).Weeds present during first 4, 8 and 12 weeks after sowing may cause yield reduction to 49 per cent, 56 per cent and 75 per cent, respectively (Ethiopia).

Weed Management

1. Fine seed bed preparation will lead to uniform germination and later stand establishment of crop. This imparts greater competing ability to crop against weeds. Some soybean varieties are highly competitive against weeds if plant population maintained at required level right from germination. However, loss of soybean seed viability often leads to poor germination, some seed- and soil-borne diseases at seedling stage cause mortality of seedlings and then heavy infestation of weeds result in poor stand of soybean.
2. Narrow row spacing lower than 40 cm is useful to stimulate crop's competitiveness against weeds.
3. Two hand weedings at 20-25 and 40-45 DAS prove sufficient to control weeds in soybean.

Chemical Control

Pre-plant Incorporated Treatment

1. Fluchloralin, trifluralin or nitralin @ 0.75-1.0 kg/ha for controlling annual grasses and few broad-leaved weeds. Use lower dose in sandy or sandy loam soil.

Pre-emergence Treatment

1. Metribuzin @ 0.4-0.5 kg/ha for controlling grasses and broad-leaved weeds. The application should be made as immediately as possible after sowing of soybean. If application is delayed and by that time some soybean seeds started germination, metribuzin may prove phytotoxic to even soybean.

2. Pendimethalin (Stomp®) @ 0.75-1.0 kg/ha followed by one hand weeding at 30-35 DAS to control late-emerging annual weeds such as *Guizotia scabra, Nicandra physalodes* and *Commelina benghalensis*. Soil moisture is a prerequisite for its greater efficacy.
3. Alachlor (Lasso®) @ 1.0-1.5 kg/ha, metolachlor (Dual®) @ 0.75-1.0 kg/ha or propachlor @ 2.0 kg/ha followed by one hand weeding at 30-35 DAS to control late-emerging annual weeds. These herbicides should be applied to moist soil for greater efficacy.
4. Clomazone (Command®) at 0.75–1.0 kg/ha for broad-spectrum weed control followed by one hand weeding at 30-35 DAS.
5. Chlorimuron-*p*-ethyl (Kloben®) @ 8-12 g/ha + one hand weeding at 30-35 DAS for broad-spectrum weed control.
6. Imazethapyr @ 150 g/ha offers a good control of broad-spectrum of weeds.

Post-emergence Treatment

1. Imazethapyr @ 100 g/ha is a good post-emergence treatment for broad-spectrum control of weeds including Trianthema portulacastrum, Digera arvensis (not controlled by pendimethalin) and Cyperus rotundus.
2. Fenoxaprop-*p*-ethyl ((Whip Super®) @ 100 g/ha at 14-21 DAS for controlling annual grass weeds only.
3. Another "-fop/-prop" herbicide, fluazifop-*p*-butyl (Fusiladeâ 125 g/litre EC) at 200-250 g/ha could be used for the control of annual and perennial grass weeds. It should be applied well ahead of soybean canopy closes and to actively growing weeds and in perennial weeds at least 3 leaves per shoot should be present. Do not use higher rate in soybean. Cultivation prior to spraying will facilitate re-sprouting and greater/better weed control.
4. Lactofen (Cobrac®) @ 90-120 g/ha as post-emergence at 14–21 DAS for controlling annual broad-leaved weeds only.
5. Bentazon (Basagranc®) @ 0.8-1.0 kg/ha for broad-leaved weed control at 1-4 trifoliate leaf stage of soybean and 3-5 leaf of weeds. It controls many annual broad-leaved weeds and it should not be applied under hot temperatures greater than 21°C and if rain expected within 8 hrs.
6. Chlorimuron-*p*-ethyl (Kloben®) @ 8-10 g/ha as early post-emergence at 15-20 DAS may be applied if no pre-emergence application has been made. Otherwise, the time of application may be extended.

Integrated Approach to Wea Seasons (*Kharif*) Pulses

1. Soil solarization + good crop husbandry + pre-emergence herbicide (if at all required) + hand weeding with minimum soil disturbance at 30-35 DAS for controlling late-coming annuals and perennial weeds such as *Cynodon dactylon* and *Cyperus esculentus/rotundus*.
2. Repeated summer cultivation followed by irrigation or with residues (wheat, maize) incorporation + good crop husbandry + pre-emergence herbicide + hand weeding at 30-35 DAS.
3. Good crop husbandry + pre-emergence herbicide + hand weeding at 30-35 DAS.
4. Stale seed bed + good crop husbandry + early post-emergence herbicide (at around 15-20 DAS in the selected pulses, *e.g.* soybean) or hand weeding at 20-25 DAS.

Winter (*Rabi*) Pulses and Legumes

Chick Pea/Gram/Bengal Gram

Yield Reduction

Nearly 40-50 per cent

Critical Period

First 40-45 days from sowing

Weed Management

1. *Proper seed bed and depth of sowing*: Seeding at 8-10 cm depth may prevent *Fusarium* wilt in chick pea. Cloddy bed with proper moisture is useful for chick pea.
2. Two hand weedings at 20-25 and 40-50 DAS under irrigated condition, while one hand weeding at 30-35 DAS under un-irrigated condition may be sufficient to control weeds in chick pea.

Chemical Control

Pre-Plant Incorporated Treatment

1. Fluchloralin or trifluralin @ 0.75-1.0 kg/ha may be recommended for controlling annual grasses and some broad-leaved weeds under both irrigated and un-irrigated dry situations provided sufficient moisture present in soil at the time of application.

Pre-emergence Treatment

1. Pendimethalin @ 0.75 kg/ha followed by a shallow weeding at 30-35 DAS. It should be pre-emergence to crop and weeds both.
2. Alachlor @ 1.0-1.5 kg/ha followed by a shallow weeding at 30-35 DAS.
3. Prometryne @ 1.0 kg/ha for broad-leaved weeds followed by a shallow hand/hoe weeding at 30-35 DAS.
4. Terbutryne @ 0.5 kg/ha for broad-leaved weed control followed by a shallow hand/hoe weeding at 35-40 DAS. In irrigated chick pea, however, crop may be injured due to simazine application.
5. Linuron @ 0.5 kg/ha for grass weed control and usually preferred under heavy soils.

Field Pea

Yield Reduction

Yield reducation due to weeds is upto 15-67 per cent.

Critical Period

Early establishment to beginning of flowering. Yield reduction to the tune of 20, 43 per cent and 67 per cent due to weed competition during first 4, 7 and 10 WAS, respectively.

Weed Management

1. Proper seed bed and moisture for even and smooth germination. Also equally important is its line sowing since farmers still now follow broad-casting.

2. Frequency of manual weeding varies depending on the level of weed infestation. Two hand weedings during the first 40 DAS prove sufficient to eliminate early weed competition.

Chemical Control

Pre-Plant Incorporated Treatment

1. Fluchloralin @ 0.75-1.0 kg/ha as pre-plant incorporated treatment for broad-spectrum weed control.

Pre-emergence Treatment

1. Pendimethalin (Stomp® 500 or 300g/litre FW) @ 0.75-1.0 kg/ha as pre-emergence to both crop and weeds can control many annual grasses and few broad-leaved weeds. However, *Avena spp* (wild oat) control is not achieved by its application. Therefore, if *Avena spp* (wild oat) is a pre-dominant weed in pea, diallate may be recommended as pre-plant incorporation.
2. Prometryne (Gesagard® 500 g/litre FW) @ 0.75-1.0 kg/ha controls annual broad-leaved weeds and few annual grasses. It should be pre-emergence to crop and weeds both. In light sandy soils, it may pose phytotoxicity to crop and accordingly the dose may be reduced.

Post–emergence Treatment

1. MCPB @ 0.75-1.0 kg/ha as post-emergence at the 5-6 leaf stage of field pea for controlling broad-leaved weeds including *Cirsium arvense* (Canada thistle). Certain precaution needs to be exercised in its use. For example, the volume rate should be reduced as low as possible and should not exceed 250 litres/ha for ground application. This may likely prevent herbicide from falling on the soil and its possible conversion/bio-activation to MCPA, which proves phytotoxic to pea.
2. Dinoseb (Na or amine salt) @ 2.0 kg/ha at 10-15 cm tall pea plants for controlling annual broad-leaved weeds. It should not be applied to peas which are treated earlier with TCA for controlling grass weeds. Otherwise, pea receives phytotoxicity from dinoseb. Dinoseb can also be applied as pre-emergence.

Lentil

Yield Reduction

Yield reducation due to weeds is nearly 50 per cent.

Critical Period

Weed infestation is more or less similar to that in field pea. Therefore, critical period of weed competition remains similar, *i.e.* early establishment to early flowering.

Weed Management

1. Fine seed bed with sufficient moisture is ideal for smooth germination of lentil. This will help in maintaining the required crop plant population.
2. Two hand weedings at 20-25 and 35-40 DAS prove sufficient to control weeds under irrigated condition.

Chemical Control

Pre-plant incorporated Treatment

1. Fluchloralin @ 0.75-1.0 kg/ha for controlling annual grasses and few broad-leaved weeds.

Pre-emergence Treatment

1. Pendimethalin (Stomp®) 500 or 300g/litre FW) @ 1.0 kg/ha to control annual grasses particularly *Phalaris sp* and some broad-leaved weeds. It should be pre-emergence to crop and weeds both.
2. Prometryne (Gesagard®) 500 g/litre FW) @ 1.0 kg/ha to control many broad-leaved weeds and few annual grasses. It should be followed by a shallow hand or hoe weeding at 35-40 DAS to control late-emerging weeds, *e.g. Guizotia scabra*. It should be pre-emergence to crop and weeds both. Application on light sandy soils may prove risky but dose may be reduced under unavoidable situations.
3. Linuron @ 0.5 kg/ha for grass weed control and usually preferred under heavy soils.

Post–emergence Treatment

1. Diclofop-methyl (Illoxan® 360 g/litre EC) @ 0.5-0.75 kg/ha as post-emergence at three trifoliate leaf stage when lentil plants are 5-6 cm tall controls annual grasses like *Phalaris sp*. Do not apply if grasses are past the four-leaf stage.
2. Fluazifop-p-butyl (Fusilade® 125 g/litre EC) @ 250 g/ha as post-emergence at 3 trifoliate leaf stage of lentil controls annual and perennial grasses. Herbicidal drift should be avoided and it should not be mixed with other herbicides.

Integrated Appraoch of Winter (*Rabi*) Pulses

Integrated weed management is more of a concept rather a method of weed control. The following options have been formulated assuming certain workable combinations of methods of weed control.

1. Stale seed bed + good crop husbandry (*e.g.* higher seed rate, line sowing, time of first irrigation delayed, adequate insect pests and disease control measures, etc.) + pre-plant incorporated or pre-emergence herbicide + hand weeding at 30-35 DAS for late-emerging annuals and perennial weeds.
2. Residue (maize/sorghum) incorporation + good crop husbandry (*e.g.* higher seed rate) + pre-emergence herbicide + hand weeding at 30-35 DAS for late-emerging annuals and perennial weeds.

Weed Management in Oilseeds

Groundnut

Yield Reduction

Even up to 80 per cent under extreme situation. Maintaining proper plant population is a prophylactic measure, which equips crop with sizable density to occupy the field earlier than weeds. As a result, weed competition reduces to a great extent.

Critical Period

First 40-60 DAS depending on the situations.

Weed Management

1. Fine seed bed free from weeds and stubbles.
2. One hand weeding at 20-25 DAS before pegging initiates is to be executed invariably in every crop of groundnut if pre-planting or pre-emergence herbicide not applied. The adoption of second hand weeding once pegging started appears inhibitory to groundnut yield, but under unavoidable situation of severe weed infestation, hand pulling of weed may be advocated at 40-45 DAS with minimum soil disturbance.
3. Earthing-up and inter-row cultivation are useful practices in groundnut and may be advocated at the early stage around 25-30 DAS before peg formation and not any way after this period. Earthing-up loosens the soil and helps in better development of groundnut pegs/pods.

Chemical Control

Pre-plant Incorporation Treatment

1. Fluchloralin @ 0.75-1.0 kg/ha for controlling annual grass weeds and some broad-leaved weeds.
2. Trifluralin @ 0.75-1.0 kg/ha for controlling annual grass weeds and some broad-leaved weeds.
3. Nitralin @ 1.32 kg/ha can also be applied to control wide-spectrum of weeds.

Pre-emergence Treatment

1. Pendimethalin @ 0.75-1.0 kg/ha for controlling many grass weeds and few broad-leaved weeds. It should be pre-emergence to crop and weeds both. It controls *Trianthema portulacastrum* effectively, which is a dominant broad-leaved weed during pre-rainy and rainy seasons in India.
2. Alachlor (Lasso® 480 g/litre EC) @1.0-1.5 kg/ha controls annual grasses and some broad-leaved weeds. It should be applied to moist soil and pre-emergence to crop and weeds both. Seeding depth of crop should be at least 5 cm.
3. Metolachlor (Dual® 500 and 720 g/litre EC) @ 0.75-1.0 kg/ha as pre-emergence to crop and weeds both for controlling annual grasses and a few broad-leaved weeds.
4. Terbutryne (Igran® 500 g/litre EC) @ 1.0 kg/ha for controlling many annual broad-leaved weeds and some grasses. Do not use on sandy or low clay soils. Soil moisture is pre-requisite for its application.
5. Oxadiazon @ 0.75-1.0 kg/ha is also recommended for weed control in groundnut.

Post-emergence Treatment

1. Fluazifop-*p*-butyl (Fusilade® 125 g/litre EC) @ 0.125 kg/ha at 13-15 DAS and 0.250 kg/ha at 25–30 DAS can be applied for controlling annual and perennial grasses, namely *Cynodon dactylon* (Bermuda grass).
2. Bentazon (Basagran® 480 g/litre EC) @ 1.0 kg/ha at 1-4 trifoliate leaf stage of crop controls many annual broad-leaved weeds. Do not apply under hot temperatures greater than 21°C and when rain is expected within 8 hrs.
3. Diclofop-methyl (Illoxan® 360 g/litre EC) @ 0.75-1.0 kg/ha controls many annual grasses.

Sunflower

Critical Period

First 6-8 WAS or 40-60 DAS. After this period, sunflower is very competitive by means of its vigorous growth.

Weed Management

1. Good/fine seed bed for smooth and uniform germination.
2. Two hoeings or hand weedings at 20 days interval at 25-30 DAS and 45-60 DAS. If variety is of long duration, three hand weedings may also be required.
3. Sunflower is useful for rotational propose. It can be used in rotation with cereals, particularly wheat and barley, where there is severe infestation of grassy weeds. It is also useful as a trap crop for *Striga* control. Sunflower stimulates *Striga* to germinate but is not attacked by it. This technique could be adopted under low *Striga*-infestation in maize and sorghum.

Chemical Control

Pre-Plant Incorporated Treatment

1. Trifluralin @ 0.5-1.0 kg/ha + one hand weeding at 30-35 DAS.
2. EPTC @ 2.0-3.0 kg/ha + one hand weeding at 30-35 DAS.

Pre-emergence Treatment

1. Pendimethalin (Stompa® 300 and 500 g/litre EC) @ 0.75-1.0 kg/ha for controlling many grass weeds and few broad-leaved weeds. It should be followed by one hand weeding at 30-35 DAS to control late-emerging weeds as well as perennial weeds.
2. Alachlor (Lasso® 480 g/litre EC) @ 1.0-1.5 kg/ha + one hand weeding at 30-35 DAS. Apply to moist soil or moisture should be provided to soil within 10 days of application under dry conditions. Dose should be lower in sandy soils.
3. Metolachlor (Dual® 720 g/litre EC) @ 1.0-1.5 kg/ha controls many annual grasses and some broad-leaved weeds. It should be followed by one hand weeding at 30-35 DAS.
4. Prometryne (Gesagard® 500g/litre FW) @1.5-2.0 kg/ha for controlling annual broad-leaved weeds and some grasses. It could also be used as early post-emergence to seedlings. It is not to be applied to sandy soils and moisture is a pre-requisite for its application. It should be followed by one hand weeding at 30-35 DAS.
5. Terbutryne @ 1.5-2.0 kg/ha + one hand weeding at 30-35 DAS.

Post-emergence Treatment

1. Fluazifop-p-butyl (Fusilade® 125 g/litre EC) @ 0.125-0.250 kg/ha as post-emergence before crop canopy closes and weeds are actively growing. Mix with other broad-leaved killers to have broad-spectrum control.

Rapeseed and Mustard

Critical Period

First 30-40 DAS.

Weed Management

1. Fine seed bed with enough moisture for smooth and even germination and early crop establishment.
2. One hand weeding at 25-30 DAS as well as thinning at this time will facilitate rapeseed and mustard plants to pick up faster growth to smother weeds coming up later.
3. *Inter-cropping*: Inter-cropping with wheat, barley, chick pea in India and with maize and sorghum and less frequently with teff in Ethiopia may be a good proposition for dual benefit–weed control and some extra yield from inter-crop. However, the sown proportion of main and inter-crop has to be adjusted judiciously. Rapeseed and mustard should necessarily be the inter-crop when grown with cereals (wheat, barley) and main crop when grown with chick pea.
4. *Mulching*: Waste residues of previous *kharif* crops may be spread in between the rows of rapeseed and mustard, which may control weeds, conserve moisture and produce organic matter. However, the residues need to be checked periodically whether there is any termite infestation/attack.

Chemical Control

Pre-plant Incorporated Treatment

1. Fluchloralin or Trifluralin @ 0.75-1.0 kg/ha for controlling annual grasses and some broad-leaved weeds.

Pre-emergence Treatment

1. Pendimethalin @ 0.75-1.0 kg/ha followed by one hand weeding at 30-35 DAS if at all required. Sufficient soil moisture is a pre-requisite for its efficacy.
2. Alachlor @ 1.0-1.5 kg/ha can control many annual grasses along with some broad-leaved weeds. Soil should be moist at the time of application or irrigation moisture must be provided to soil within 10 days of application.
3. Napropamide (Devrinol® 500 g/kg WP) @ 1.0-1.5 kg/ha against annual broad-leaved and grass weeds.
4. Oxyfluorfen (Goal®) is selective to mustard and rapeseed. It can be applied @ 0.20-0.25 kg/ha in 500 litres of water at 30-35 DAS of mustard, where *Orobanche* sp. (Broom rape) poses problem.

Post-emergence Treatment

1. Isoproturon as early post-emergence @ 0.5 kg/ha at 4-5 leaf stage of crop for broad-spectrum control of weeds (both grass and broad-leaved weeds) particularly wild oats.
2. Dicamba may be applied when crop is at least 5 cm tall, but has not come to flowering if difficult-to-control weeds like *Cirsium arvense* (Canada thistle) invade rapeseed and mustard.

Integrated Approach (Oilseeds)

1. Good crop husbandry + pre-emergence herbicide + hand weeding at 30-35 DAS (for problematic weeds)
2. Good crop husbandry + pre-emergence herbicide + earthing-up (for groundnut)

3. Good crop husbandry + pre-emergence herbicide/early post-emergence herbicide namely isoproturon (mustard and rapeseed) + hand weeding at 30-35 DAS (if at all required).

Sugarcane

1. *Blind hoeing*: A mechanical control of weeds can be executed by hoeing blindly along the rows of sugarcane before the cane sprouts are visible on the soil surface. Weeds emerged before sugarcane sprouted or on the process of germination may be exposed/uprooted and killed.
2. *Hand weeding/earthing-up*: Three to four manual weedings may be sufficient to keep early sugarcane crop free from weed competition during the first 4-5 months after planting. Earthing-up usually practised after canes sprouted and have become sufficiently tall, has dual advantage to sugarcane. It controls weeds as well as holds sugarcane plants erect. It provides anchorage to sugarcane plants.
3. *Inter-cropping*: Inter-cropping with green gram/soybean/chick pea/black gram etc is very useful. It has dual advantage. It controls weeds at the early stages, when the growth of sugarcane is slow and provides extra yield from the inter-crops. This is a sort of parallel multiple cropping where juvenile phase of one crop (sugarcane) starts when another crop (inter-crop) has reached to almost senescence stage. Herbicide may be recommended based on the selectivity to both crops. Under inter-cropping with black gram/soybean/groundnut, pre-emergence pendimethalin (0.75- 1.0 kg/ha) or thiobencarb (1.0–1.5 kg/ha) may be recommended.
4. Now-a-days filter pressmud application/spreading has been reported a useful practice towards weed control in sugarcane although filter pressmud basically acts as a manure and soil amendment.
5. Flaming renders a good control of weeds, but it requires smooth surface and it does not have any residual effect. It is, however, expensive and not economical on the ground of costly equipment, fuel requirement and application cost.

Chemical Control

Pre-emergence Treatment

1. Atrazine @ 1.0-2.0 kg/ha should be applied after and not before any due post-planting irrigation. However, sufficient moisture in soil is a prerequisite for its efficacy. However, mixture of atrazine + pendimethalin each at 0.75 + 0.75 kg/ha as pre-emergence would be more useful towards broad-spectrum control of weeds.
2. Another substitute of blind hoeing in sugarcane is the tank-mix application of atrazine @ 0.25 kg/ha and 2,4-D @ 1.0 kg/ha a couple of days before crop emergence. *Cyperus spp* and broad-leaved weeds fall susceptible to this mixture.
3. Metribuzin @ 0.75-2.0 kg/ha is also useful.
4. Alachlor @ 1.0-3.0 kg/ha is also useful.

The effect of pre-emergence herbicides continues to exist there for 4-6 weeks depending on the nature of herbicides. Therefore, for achieving a good control of weeds in sugarcane, a sequential measure with hand weeding or post-emergence herbicide is highly essential. However, if inter-cropping has been undertaken, earthing-up after harvest of inter-crop may be more useful, which serves dual purpose–weed control and anchorage to crop.

Post-emergence Treatment

1. Directed spray of 2,4-D @ 0.5 kg/ha after the emergence of sugarcane will control many broad-leaved weeds. Also parasitic *Striga* could be effectively controlled by 2,4-D.
2. Glyphosate (0.25-0.5 kg/ha) and paraquat (0.25-0.5 kg/ha) are non-selective and highly effective chemicals. Therefore, they should be applied at very low rate by careful directed spray with less volume of water and when the sugarcane plants are enough grown-up to fairly tolerate these herbicides. It is likely when they are nearly 3 months old.

Integrated Approach

1. Good crop husbandry + pre-emergence herbicide + hand weeding at 35-40 DAS.
2. Good crop husbandry + inter-cropping with black gram + pre-emergence pendimethalin + earthing-up.
3. Good crop husbandry + inter-cropping with black gram/green gram + pre-emergence pendimethalin + earthing-up/post-emergence herbicide.

Cotton

Critical Period

30-60 DAS (Ethiopia) and 50-60 DAS (India)

Yield Reduction

70-75 per cent (Ethiopia) and 45-85 per cent (India)

Weed Management

1. Deep summer ploughing (usually 40 cm deep), digging and removal of underground parts, *e.g.* rhizomes, stolons of perennial weeds during land preparation and the field may be left for sun-drying without planking/levelling. It would be very effective if deep tillage/ploughing is practised after every irrigation (Das and Yaduraju, 2001).
2. Timely sowing and early weeding right from 20 DAS onward (Table 13.9) are useful towards achieving lower weed competition and higher crop yield. Hand weeding twice or thrice beginning from 20-25 DAS under irrigated cotton is highly useful if herbicide is not applied. In hill-planted cotton, cross cultivation is also practised.

Table 13.9: Effect of Different Frequencies of Hand Weeding on Cotton Yield (Anonymous, 1972)

Frequency of Hand Weeding	*Seed Cotton Yield*	
	(kg/ha)	*(Per cent)*
Once at 20 days after sowing	1550	100
Twice at 20 and 30 days after sowing	1880	116
Thrice at 20, 30 and 45 days after sowing	1920	117
Four times at 20, 30, 45 and 60 days after sowing	2290	147

3. Closer row spacing without any compromise on cotton yield may be useful. Therefore, judicious adjustment in the intra-row spacing without changing the plant population per unit area may be tried for quick canopy coverage.
4. Inter-cropping with onion, green gram/mung bean, lady's finger/okra/bhindi, radish is quite remunerative and takes care of weeds during the growing period of cotton. Paired row system may be adopted.
5. Flame cultivator may be used when cotton stems are 6 mm in diameter. The flat type burner using volatile gases such as propane or butane is suggested. The cost of fuel for making flame is the main hindrance to its use/adoption in the field particularly under Indian condition where majority farmers are small and marginal/poor.
6. Now-a-days filter pressmud application/spreading has been reported a useful practice towards weed control in cotton apart from its role as a manure and soil amendment.

Chemical Control

Pre-plant Incorporated Treatment

1. Trifluralin (Treflan® 480 g/litre EC) @ 1.5-2.0 kg/ha for controlling annual grasses and some broad-leaved weeds. Trifluralin reduces lateral root development in cotton in the treated zone, but does not affect fibre yield. Immediate incorporation into upper 5-10 cm soil helps to achieve greater weed control efficacy. Use lower rate on sandy soils. It can be used simultaneously in tank-mixes with prometryne, fluometuron, diuron and norflurazon. It is resistant to leaching.
2. Fluchloralin @ 0.75-1.5 kg/ha for controlling annual grasses and some broad-leaved weeds.

Pre-emergence Treatment

1. Diuron (Karmex® or Diumex [a]800 g/kg WP) @ 1.0-1.5 kg/ha for the control of annual grasses and broad-leaved weeds. Do not use on sandy soils. Do not use with insecticide seed treatment.
2. Pendimethalin @ 0.75-1.0 kg/ha for control of grasses and broad-leaved weeds.
3. Metolachlor (Dual® 720 g/litre EC) @ 1.0-1.5 kg/ha for control of annual grasses and some broad-leaved weeds. It may also control *Cyperus esculentus*.
4. Alachlor @ 1.0-2.0 kg/ha for the control of mainly annual grass weeds and some broad-leaved weeds. It may control to some extent *Cyperus esculentus* (yellow nutsedge).
5. Fluometuron (Cotoran® 800 g/kg WP) @ 1.5 kg/ha controls many annual broad-leaved weeds and some grasses. The soil should be moist at the time of its application. It may be mixed with metolachlor or trifluralin. Avoid applying in sandy soils. Systemic insecticides should not be used with it.
6. Norflurazon (Zorial® 800 g/kg WP) @ 1.0-2.0 kg/ha controls most annual weeds and can suppress *Cyperus spp*. It has long residual action in soil and, therefore, crop other than cotton should not be planted in sequence.
7. Prometryne (Gesagard® 500g/litre FW) @ 1.0-1.5 kg/ha can control many annual broad-leaved weeds and some grasses. Sufficient moisture should be present in soil on its application. Avoid applying in sandy soils. Its mixture with metolachlor also proves efficient.

8. In cotton, a mixture of trifluralin + diuron (1.0+1.0 kg/ha, respectively) and trifluralin + prometryne (1.0+1.0 kg/ha respectively) may be advocated for effective weed control.
9. Any pre-planting or pre-emergence herbicide + hand weeding at 30-35 days after herbicide application is also quite useful.

Post-emergence Treatment

The soil remains dry during summer. Therefore, the pre-planting or pre-emergence herbicide application goes many times an ineffective treatment. Besides there may occur flushes of germination of weeds after every irrigation in irrigated cotton and one pre-emergence or pre-planting application may not be sufficient to tackle the late-emerging weeds. Therefore, hand weeding or application of post-emergence herbicide becomes a necessity in cotton.

1. Diuron (pre-emergence) + diuron (post-emergence) : Sometimes diuron @ 0.75 kg/ha as pre-emergence followed by diuron @ 0.75 kg/ha as post-emergence offer a good control of a variety of weeds.
2. Fluazifop-*p*-butyl (Fusilade® 125 g/litre EC) @ 0.25 kg/ha as post-emergence applied before cotton canopy closes controls many annual and perennial grasses. If perennial grasses pose problem, higher dose @ 0.75 kg/ha may be advocated for their control. It has slow action and should not be mixed with other herbicides.

Integrated Approach

1. Combination of deep ploughing/tillage, pre-emergence or pre-plant incorporated herbicides (mentioned above) and hand weeding at 35-40 DAS would be an efficient option.
2. Good crop husbandry + pre-emergence or pre-plant incorporated herbicides (mentioned above) + hand weeding at 35-40 DAS.

References

Abdel-Rahim, M. F., Satour, M. M., Michail, K. Y., El-Eraki, S. A., Grinstein, A., Chen, Y., and Katan, J., 1988. Effectiveness of soil solarization in furrow-irrigated Egyptian soils. *Plant Disease*, 72:143-146.

Agrawal, R. L., 1995. *Seed Technology*. 2nd edn. Oxford and IBH Publishing Co. Pvt. Ltd. New Delhi, 829 pp.

Akobundu, I. O., 1978. Weed control strategies for multiple cropping systems of the humid and sub-humid tropics. In : I. O. Akobundu (ed.). *Proceedings, Weed and Their Control in the Humid and Sub-humid Tropics*. pp 80-95. IITA, Ibadan. Nigeria.

Al Masoum, A. A., Hashim, A. A., Jaafer, K. and Al Asaal, A., 1998. Effect of two mulch types for solarization on soil temperature. AMA, *Agricultural Mechanization in Asia, Africa and Latin America*, 29 (4). 73-75.

Alagh, Yoginder K., 1988. Inaugural Address on "Pesdcides and Indian Agriculture." In: *National Seminar on "Changing Pest Situation in the Current Agricultural Scenario* of *India*," June 14-16, New Delhi.

Angiras, N. M. and Sharma, V., 1993. Effect of cultural manipulations and weed control methods on crop-weed competition in wheat. *Indian Journal of Weed Science*, 23 (3&4) : 6-10.

Anonymous, 1972. Cotton Irrigation and Weeding Trials. *Progress Report (April, 1971–March, 1972)*, IAR Melka Werer Research Station, Addis Ababa, Ethiopia, pp 48-51.

Anonymous, 1979. *Handbook on Crop Production in Ethiopia*. IAR, Addis Ababa, Ethiopia.

Balyan, R. S. and Malik, R. K., 1990. Competing ability of wheat varieties with wild oats (*Avena ludoviciana*) under late-sown conditions. In: *Abstracts of Papers. Biennial Conference*, Indian Society of Weed Science, March 4-5, *p* 151. JNKVV, Jabalpur, M.P., India, 218 pp.

Bantilan, R.T. and Harwood, R. R., 1973. Weed management in intensive cropping systems. IRRI Saturday Seminar, 28th July, 1973. (Cited in : Moody, K. (ed.)., 1979. *Weed Control in Tropical Crops*. Weed Science Society Philippines, Manila. 136 pp.

Bell, A. R. and Nalewaja, J. D., 1968a. Competition effects of wild oats in flax. *Weed Science*, 16: 501-504.

Bell, A. R. and Nalewaja, J. D., 1968b. Competition of wild oats in wheat and barley. *Weed Science*, 16: 505-508.

Bengtsson, M. I., 1983. Rural development research and agricultural innovations. In: *Report No. 115*. Swedish University of Agricultural Sciences, Deptt. of Plant Husbandry, Uppsala, Sweden.

Besufekad Tadesse, Das, T.K. and Yaduraju, N.T., 2000. Integrated weed management in sorghum with particular reference to *Parthenium hysterophorus* L. In: *Abstract, National Seminar* on " Sustainability of Weed Control Options for the New Millennium." December 20-21, Annamalai University, Annamalai Nagar, T.N., p 30.

Besufekad Tadesse., 2001. *Studies on Integrated Management of Weeds in Sorghum with Special Emphasis on Parthenium hysterophorus L. M.Sc.Thesis*, New Delhi, India: Division of Agronomy, Indian Agricultural Research Institute, New Delhi.

Cousens, R. D., Weaver, S.E., Martin, T. D., Blair, A. M. and Wilson, J., 1991. Dynamics of competition between wild oats (*Avena fatua* L.) and winter cereals. *Weed Research*, 31: 203-210.

Das, T. K. and Yaduraju, N. T., 1995. Crop-weed competition studies in some *kharif* crops: ll, nutrient uptake and yield reduction. *Annals of Plant Protection Sciences*, 3 (2): 95-99.

Das, T. K. and Yaduraju, N. T., 1999a. Role of fertilizers on crop-weed competition in wheat as affected by irrigation. *Fertilizer News*, 44 (3): 25-30.

Das, T. K. and Yaduraju, N. T., 1999b. Effect of weed competition on the growth, nutrient uptake and yield of wheat as affected by irrigation and fertilizers. *Journal of Agricultural Science* (Cambridge), 133 : 45-51.

Das, T. K. and Yaduraju, N. T., 2001. Comparing several crop husbandry practices with soil solarization for weed control and crop yield in soybean (*Glycine max*)–broccoli (*Brassica oleracea convar botrytis var italica*) cropping system. *Indian Journal of Agricultural Sciences*, 71 (4): 284–286.

Das, T. K. and Yaduraju, N. T., 2008. Periodicity in germination of *Phalaris minor* Retz. and *Avena ludoviciana* Dur. (personal communication; unpublished).

Delorit, R. J. and Ahlgren, H. L., 1967. *Crop Production*, New Jersey, pp 557-596.

Duary, B., 1998. Ecophysiological analysis of competition between littleseed canary grass (*Phalaris minor* Retz.) and wheat (*Triticum aestivum* L.) *Ph. D. Thesis*, Division of Agronomy, IARI, New Delhi, India, 118 pp.

Engstroem, E., 1974. Crop Production Trials 1967-1972. Crop and Pasture Section. *CADU Publication No. 97*. Addis Ababa, Ethiopia.

Fryer, J. D., 1983. In : *Recent Advances in Weed Research*. W.W. Fletcher (ed.) CAB, London.

Gautam, K. C., 1992. Influence of nitrogen levels and crop densities on growth of weeds and wheat. In: *Abstracts of Papers, Annual Weed Science Conference*, Indian Society of Weed Science, March 3-4, Hisar, p 55.

Ghosh, A. K. and Sarkar, P. A., 1975. Nitrogen, variety and spacing effects on light-transmission ratio and weed intensity in transplanted rice. In : *Abstracts 5th Asian Pacific Weed Science Society Conference*, pp 120.

Gill, H. S. and Mehra, S. P., 1981. Growth and development of *Phalaris minor*, *Chenopodium album* and *Melilotus indica* in wheat crop ecosystem. In: *Procedings 8th Asian Pacific Weed Science Society Conference*, pp 175-179.

Hay, J. R., 1974. Gains to the growers from weed science. *Weed Science*, 22 : 439-442.

Jaya Kumar, R. and Jagannathan, R., 2003, *Weed Science Principles*. Kalyani Publishers, New Delhi, 338 pp.

Kang, B. T., Moody, K. and Adesina, J. O., 1980. Effects of fertilizer and weeding in no-tillage and tilled maize. *Fertilizer Research*, 1 s: 87-93.

Katan, J., Greenberger, A., Alon, H. and Grinstein, A., 1976. Solar heating of polyethene mulching for the control of diseases caused by soil borne pathogens. *Phytopathology*, 66: 683-686.

Kolar, J. S. and Mehra, S. P., 1992. *Changing scenario of weed flora in agro-ecosystems of Punjab*. In : Changing Scenario of Our Environment. G. S. Dhaliwal *et al.* (eds). PAU, Ludhiana, India : 252-262.

Kolbe, W., 1977. Long term studies on relations between weed cover and yield increase with chemical weed control (1967-76). *Pflanzenschutz-Nachrichten Bayer*, 30 : 121-137.

Kondap, S. M., Reddi, M.R. and Ramakrishna, K., 1980. Studies on yield reduction due to *Cyperus rotundus* in different crops. In: *Abstracts of papers, Annual Conference, Indian Society of Weed Science*, p 25.

Konesky, D.W., Siddiqi, M.Y., Glass, A. D. M. and Hsiao, A.I., 1989. Wild oat and barley interactions : varietal difference in competitiveness in relation to phosphorus supply. *Canadian Journal of Botany*, 67 : 3366-3371.

Kumar, B., Yaduraju, N. T., Ahuja, K. N. and Prasad, D., 1993. Effect of soil solarization on weeds and nematodes under tropical Indian conditions. *Weed Research*, 33: 423-429.

Macneish, R. S., 1964. The origin of new world cultivation. *Sci. Amer.*, 211 : 29-37.

Malik, R. K. and Balyan, R. S., 1990. Competing ability of wheat varieties with wild oats (*Avena ludoviciana*) under normal-sown conditions. In : *Abstracts of Papers. Biennial Conference*, Indian Society of Weed Science, March 4-5, *p*151-152. JNKVV, Jabalpur, M.P., India, 218 pp.

Moody, K., 1973. Weed control in tropical grain legumes. In: *Procedings First IITA Grain Legume Improvement Workshop*, Ibadan, Nigeria, pp 162-183.

Moody, K., 1979. Weed control in tropical legumes. In : Moody, K. (ed.). *Weed Control in Tropical Crops*. Weed Science Society, Philippines, Manila, pp 112-146.

Moody, K. and DeDatta, S. K., 1977. Integration of weed control practices for rice in tropical Asia. Paper presented at a *Workshop* on *Weed Control in Small Scale Farms* during the 6th *Asian Pacific Weed Science Society Conference*, Jakarta, Indonesia.

Muzik, T. J., 1970. *Weed Biology and Control.* McGraw-Hill Inc. New York, 273 pp.

Parker, C., 1979. Integrated weed control in sorghum. *FAO Plant Production and Protection Paper No., 19,* Rome, Italy, pp 110-119.

Parker, C. and Riches, C. R., 1993. *Parasitic Weeds of the World: Biology and Control.* CAB International. Wallingford, UK.

Pavlychenko, T. K., 1937. Quantitative study of the entire root systems of weed and crop plants under field condition. *Ecology,* 18: 62-79.

Pavlychenko, T. K., 1940. Investigations relating to weed control in western Canada. Impt. Bur. Pastures and forage crops. Wales, Great Britain. *Herbage Pub. Serv. Bull. 27.*

Pavlychenko, T. K. and Harrington, J. B., 1934. Competition efficiency of weeds and cereal crops. *Canadian Journal of Research,* 10 : 77-94.

Pawar, L. G., 1994. Effect of irrigation and nitrogen on crop-weed competition in tall and dwarf wheat. *Ph. D. Thesis,* Division of Agronomy, IARI, New Delhi, India, 109 pp.

Putnam, A. R., 1985. Weed allelopathy. Chapter 5, pp 131-155. In : *Weed Physiology, Vol. 1, Reproduction and Ecophysiology,* S. O. Duke (ed.). CRC Press, Boca Raton, Florida.

Radosevich, S. R., Holt, J. S. and Ghersa, C. M., 1997. *Weed Ecology: Implications for Management.* 2nd edition. John Wiley and Sons, Inc. New York, 589 pp.

Rao, A. N. and Shetty, S. V. R., 1976. Some biological aspects of intercropping systems on crop-weed balance. *Indian Journal of Weed Science,* 8 : 32-43.

Rao, A. N. and Shetty, S. V. R., 1981. Investigation of weed suppressing ability of some cropping system in relation to canopy development and light interception. In: *Proceedings* of the 8th *Asian Pacific Weed Science Society Conference.* pp 357- 364.

Rezene, F., 1984. The problem of parasitic weeds in Ethiopia. Paper presented at the Ethiopian Agricultural Research Conference (EARC), April 19-21, Addis Ababa.

Rezene, F., 1985. A review of weed science research activities on wheat and barley in Ethiopia. In : Tsedeke Abate (ed.). *A Review of Crop Protection Research in Ethiopia. Proceeding of the 1st Ethiopian Crop Protection Symposium.* Feb., 4-7, IAR, Addis Ababa, Ethiopia, pp 121-149.

Rogers, N. K., Buchanan, G. A. and Johnson, W. C., 1976. Influence of row-spacing on weed competition with cotton. *Weed Science,* 24 : 410-413.

Sahoo, K. M. and Saraswat, V. N., 1988. Magnitude of losses in the yields of major crops due to weed competition in India. *Pesticide Information,* 14: 2-9.

Sandhu, K.S., Kolar, J.S. and Singh, G., 1981. Studies on cultural control of wild oats (*Avena ludoviciana* Dur.), a grassy weed in wheat. *Indian Journal of Weed Science,* 13(2) 146-149.

Saraswat, V. N., 1989. Project Coordinator's Report (1.1.1989-31.12.1989). All India Coordinated Research Programme on Weed Control, Jabalpur, India, 218 pp.

Sarkar, P. A. and Ghosh, A. K., 1977. A study on the competitive ability of rice varieties with special reference to *Ammania baccifera.* In: *Abstracts of Papers, Annual Weed Science Conference,* Indian Society of Weed Science, India.

Siddiqi, M.Y., Glass, A. D. M., Hsiao, A.I. and Minjas, A. N., 1985. Wild oat/barley interactions : Varietal difference in competitiveness in relation to K^+ supply. *Annals of Botany*, 56 : 1-7.

Singh, R. P. and Singh, R. K., 2006. Ecological approaches in weed management. In : *Extended Summaries* (Voluntary Articles), Golden Jubilee National Symposium on "Conservation Agriculture and Environment" jointly organized by the Indian Society of Agronomy, Division of Agronomy, IARI, New Delhi and Banaras Hindu University (BHU) at the Banaras Hindu University, Varanasi, India on October 26-28, pp 301-304.

Singh, S., Malik, R.K., Balyan, R. S. and Singh, S., 1992. Weed survey of wheat in Haryana. In: *Abstracts of Papers, Annual Weed Science Conference*, Indian Society of Weed Science, March 3-4, Hisar, p 5.

Tiwari, J. P. and Bisen, C. R., 1985. Competitive impact of Phalaris weed ecosystem on sink coefficient in some rabi crops. *Indian Journal of Weed Science*, 17 (4) : 21-27.

Tull, Jethro. 1731. *Horse-hoeing Husbandry*. Berkshire, MDCC 33.

Tull, Jethro. 1733. *The Horse-hoeing Husbandry*. In : E. J. Salisbury, !961. *Weeds and Aliens*, 2nd edn., Collins, London, 384 pp.

Unger, J., 1984. *Principles and Practices of Weed Management*. College of Agriculture, Alemaya, Addis Ababa University, Addis Ababa, Ethiopia, pp 185.

Wells, G.J., 1969. Skeleton weed (*Chondrilla juncea*) in the Victorian Mallee. I, Competition with legumes. *Australian Journal of Experimental Agriculture and Animal Husbandry*, 9 : 521-527.

Williams, E. D., 1976. Components of vegetation of permanent grassland in relation to fertilizers and lime. *Annals of Applied Biology*, 83 : 342-345.

Wimschneider, W., Bachthaler, G. and Fischbeck, G., 1990. Versuche zum Konkurrenzverhalten von Avena fatua L. (Flug-Hafer) in weizen (*Triticum aestivum* L.) als Grundlage gezielter Bekampfungsma Bnahmen. *Weed Research*, 30: 43-52.

Yaduraju, N. T., 1997. Soil solarization–A novel method of weed control. Division of Agronomy, Indian Agricultural Research Institute, New Delhi–110 012, India.

Yaduraju, N. T., 2006. Herbicide resistant crops in weed management. In : *The Extended Summaries, Golden Jubilee National Symposium on Conservation Agriculture and Environment*. Octo., 26-28, Banaras Hindu University,Varanasi, pp 297-98.

Yaduraju, N. T. and Ahuja, K. N., 1992. The competition of grass weeds in relation to time of weed emergence and date of planting of wheat. *Annual Report* (1991-92). Division of Agronomy, IARI, New Delhi, p 46-47.

Yaduraju, N. T. and Ahuja, K. N., 1997. Grass weed competition in wheat in relation to its time of planting and weed emergence. *Indian Journal of Weed Science*, 29 (3&4) 167-169.

Yaduraju, N. T., Das, T. K. and Seema., 1999. Long term effect of soil solarization on weed competition and seed yield of soybean. In: *Proceedings I(A) of the 17th Asian-Pacific Weed Science Society Conference* held on November 22-27, Bangkok,Thailand, pp 242-245.

Chapter 14

Pest Forecasting for Integrated Pest Management

D.K. Das* and H.R. Sardana

National Centre for Integrated Pest Management, LBS Building, IARI Campus, New Delhi - 110 012

Introduction

The Inter society Consortium for Plant Protection (ICPP) has defined integrated pest management (IPM) as "the use of multiple tactics in a compatible manner to maintain pest population at levels below those causing economic injury while providing protection against hazards to humans, domestic animals, plants and the environment". The term 'integrated' means it is a broad interdisciplinary approach taken using scientific principles of plant protection of variety of management strategies and tactics to fuse into a single system. The terms 'pest' means all biotic agents, which adversely affect plant production. This includes insects, mites, nematodes, weeds, bacteria, fungi, viruses, parasitic seed plants and vertebrates. The term 'management' indicates to keep the pest population below some level based on economic criteria. It does not mean complete eradication of entire pest population.

Forewarning and forecasting, both deal with seeing the future scenario from present and past situation and propagating them to other people. Forecasting can be for both good (yield of crops) and bad (cyclone/earth quake/pest invasion) events or process whereas, forewarning is used for any bad, dangerous or disastrous situation. Forecasting is a broader term and in case of forewarning meaning is restricted. In case of pest and disease, any of the terminologies can be used.

---

*Present address: Division of Agricultural Physics, IARI, New Delhi – 110 012.

Why IPM?

The present day crop protection is heavily relying on chemical control of pests. The dependence on chemicals causes some serious problems to the human and its ecosystems. The problems are:

1. Insects becoming resistant to insecticides,
2. Insect resurgence after becoming resistant,
3. Biomagnifications of certain insecticides in food chain,
4. Persistence of some insecticides as residue,
5. Pollution of surface and ground water causing harm to aquatic ecosystems
6. Rejection of our agricultural products at foreign market.

The adoption of IPM strategies has received the attention to minimize the above problems.

The IPM strategy suggests:

1. Use of safe pesticides and botanical pesticides.
2. Use of pesticides in appropriate time based on pest phenology and weather information to minimize pesticide use.
3. Use of biological control agents like parasites and predators.
4. Use of resistant crop cultivars.
5. Use of cultural methods to have minimum pest problems.
6. Development of transgenic plants resistant to insect pest and disease.

Pest Forecasting

Pest forecasting models or systems are essential tools and pre-requisite of IPM as the insect population and disease incidence levels are deciding factors in pest management. The inter relationship and dependence of plants, pests, natural enemies of pest and weather and their quantification need full understanding. Therefore, forecasting system are developed pest models based on biotic and abiotic factors. These models may be deterministic, statistical, system simulation, expert system, schematic or geographical. These models make the farmers to prepare against pests and decide the timing of application of bio-agents, pesticides and other approaches in integrated pest management.

Application of chemical spray against Colorado Beetle and Potato blight in Soviet Russia based on forecasting model reduced crop losses of about 1.5 million Rubles per year. In Canada the IPM is based on the forecasting information to regulate or economize the intensity and amount of spraying. The day-degree concept is widely used in USA and. European countries for predicting the levels of attack of pests, relating degree-day with calender dates for egg hatching, forecasting the optimum moment to destroy adult moths.

Early Works on Pest Forecasting

One of the earliest forecasting work on plant pest was forecasting of late blight of potato (*Phytophthora infestans*) by Dutch rules (Van Everdingen, 1926) which considered several climatic factors together such as dew during night (at least 4 hours), minimum temperature at night (less than 10°C), mean cloudiness on following day (more than 0.8) and rainfall in the next 24 hours (minimum of 0.1 mm).

In India, one of the earliest work on pest forecasting was started by Dr S Pradhan when he gave the concept of biotic theory of locust population explosion. According to him, temperature was one of the most important factor which determines the duration of various stages in insect life cycle and number of generations. The development of biograph had seen one of the important findings to show the rate of development on different temperature regimes (Pradhan, 1967, 1982). Later, various other models were developed such as wheat stem rust rules (Nagarajan and Singh, 1975,) based on climatological parameters.

Weather parameters have been modeled by multiple regression techniques to assess the per cent incidence of 'Karnal bunt' disease of wheat by Srinivasan and Mavi (1983). The equation developed was :

$$Y = 18.475 - 0.544X_1 + 0.108X_2 - 0.059X_3 - 0.118X_4 + 1.192X_5 \quad R^2 = 0.9066$$

where,

Y: Karnal bunt disease incidence (per cent)

X_1: Maximum temperature during 11th week (°C)

X_2: Minimum temperature during 11th week (°C)

X_3: Relative humidity during 11th week (per cent)

X_4: Cumulative rainfall during 11th week(mm)

X_5: Number of rainy days during 11th week

A computer based simulation model MOTHZV has been developed by Witz et al (1985) for predicting the population dynamics of *Helicoverpa* species. The model uses early season number of eggs, larvae or adults to forecast the timing and size of later, potentially damaging population. Pheromone trap has provided the means for measuring early season number of *Helicoverpa* adults. This pheromone trap data along with climatic variable and crop phenology were used as input to the MOTHZV model to predict the timing of future *Helicoverpa* generations for a pilot test on the management of *Helicoverpa* species.

Requirement of Pest Forecasting

Weather Data

Both past and present weather data including temperature, wind, atmospheric pressure, humidity, precipitation soil moisture and radiation are useful for pest forecasting. Past data are required for forecasting model development. Present weather data are required for model operationalisation. The data should be accurate. Inaccurate data will lead to development of a inaccurate model which ultimately would lead to loss of crores of rupees.

Linking Weather Forecasting to Pest Forecasting

In all pest forecasting programme weather data is an essential component. Forecasting of future pest population is done from past/present weather. But emergence of population and size of population is determined also by the future weather in which the pests are going get exposed. In forecasting models, it is assumed that the weather of next few days or weeks would be at par with those of previous years. This may not happen in reality. Therefore, there is a need to foresee the future weather parameters. This need will be fulfilled by receiving forecasted weather data from the meteorological department.

Weather forecasting is an important branch of meteorology. Weather forecasting can be divided into 3 types.

1. Short range–up to 48 hours or 2 days.
2. Medium range–3-10 days
3. Long range–more than 10 days.

Short range weather forecasting is used to provide very short time gap to prepare for pest management practices. But it is useful for pest management purpose. The spraying operation should take into account the current rainfall, fog, mist, temperature inversion and wind speed. Medium range weather forecasting provide good scope for planning, preparation, procurement of pest management inputs, provide adequate time for pest management as well as farm management practices. Medium range forecasting of different weather parameters can be utilized for different pest management measures. Long range forecasting is also useful in IPM particularly pest forecasting as several workers have noticed the post-monsoon pest population depends on amount and distribution of monsoon rainfall. IMD makes forecast about monsoon in April. Therefore, in April itself we could have an idea about the likely pest scenario of September, October, November, December.

Apart from regular weather data some micrometeorology data are required for development of a accurate model as the environment within canopy, in which the pests thrive, differs significantly from outside environment. The micromet data which are of major significance are temperature, humidity and radiation profile within canopy and duration of wetness period. Soil temperature data also required as some stages of insects are completed within soil. This area needs more attention in future studies.

Pest Population Data

Accurate monitoring of pest population on daily basis is very much necessary for simulation models and daily to weekly basis for medium range forecasting. For monitoring the population the devices are light trap and pheromore trap. Manual counting of eggs, larvae and pupae also serves the purpose.

Crop Data

This is one of the most important data. Particularly crop phenological information is required for simulation models. With similar type of weather condition and initial population the future population can vary depending on crop area, crop phenology, and associated crops. For identification of hosts, their area coverage and phenological stage, remote sensing techniques can be suitable tools for rapid and wider area information for forecasting purpose.

Natural Enemy Data

With in the crop environment, there are some biotic agents which destroy the pest through different mechanism. These are predators and parasites. The population of natural enemies present in the crop environment determines the size of pest population and level of infestation.

Some Pest Forewarning Models Developed at NCIPM

Some of the pest forewarning models developed at NCIPM include *Helicoverpa* (Das *et al.*, 2001a) and potato aphid (Trivedi *et al.*, 2002) have been taken up by using 10-20 years historical data collected by AICRP, ICRISAT and SAUs. The linkages and coordination have been found to be effective in developing weather based forecasting models. The sensitivity and validation of these models are being tested before these are put into operation under field conditions in the localities from where the data have been collected. These simple and working models may be found suitable for field application in agriculture by using computers.

Helicoverpa armigera Forecasting Models

NCIPM has developed the group of plant protection, agrometeorology, statistics and computer scientists for the development of forewarning system for insect pests and diseases for different agro climatic zones of the country. Although, some scattered activities were carried out at different locations and forewarning systems have been developed by different agencies in few agro-ecological regions, no attempt was made to cover nationally important pests of major crops in different agro-ecological regions of India.

In India, monthly pheromone trap moth data of 1983-90 along with weather data of corresponding period collected from PAU, Ludhiana have been analysed at NCIPM by Das et al (2000) and it was found that there were major two peaks of moth catches in a year. The main peak was during March-April and the second peak was during October every year. During March-April peak, chickpea was the main crop and during October, cotton was grown almost over entire area. The peak population of *Helicoverpa* moth during March-April and October can be predicted in advance by multiple regression models using different weather parameters along with previous season's population of *Helicoverpa*. The population density during March-April (P_{M-A}) has been regressed with different weather parameters as well as pest population density of previous five months separately and cumulatively. It has been found to depend on total moth catch per trap during previous October to February (P_{O-F}), mean monthly relative humidity recorded in the afternoon during previous February (RHE_F) and mean monthly minimum temperature of previous February ($Tmin_F$) as follows:

$$P_{M-A} = -1032.65+2.06\ P_{O-F} - 34.26RHE_F + 516.74Tmin_F\ (R^2 = 0.75)$$

Apart from weather parameters, the total population cumulated from October to February is another important parameter. The population, which uses to undergo diapause from October due to severe winter is emerged again after February.

Similarly, the population density during October (P_O) can be predicted in advance with multiple regression using weather parameters as well as pest population of preceding months as follows:

$$P_O = -5.39+.12\ P_{M-A} + 1.28\ P_J - 0.51\ R_{J-J}\ (R^2 = 0.87)$$

P_{M-A} = Pest population density of March and April,

P_J = Pest population density of June,

R_{J-J} = Total amount of rainfall during June and July.

The data collected from ICRISAT, Hyderabad were analysed and a thumb rule has been found out to predict the level of attack during any year using surplus/deficit rainfall of different months and that thumb rule has justified the severity of pod borer attack in different districts of Andhra Pradesh during the winter of 1997-98 (Das *et al.*, 2001).

The surplus or deficit of monsoon rainfall (*i.e.*, the excess or less rainfall of June, July, August and September) which is referred here as 'A' and surplus or deficit rainfall during November which is referred here as 'B' of any place in this region could play the crucial role. A and B can be calculated as:

A: (Actual total rainfall during four monsoon months in a year – Long term average rainfall of that period of that place)

B: (Actual total rainfall during November month in a year–Long term average rainfall of that month of that place).

Positive (+) value of A or B indicates surplus rainfall and negative (-) value indicates rainfall deficit.

A+B-LOW	A+B+MODERATE
A-B-MODERATE	A-B+SEVERE

Validation of the Pest Forecasting Models at Different Districts of Andhra Pradesh

The constructed rule has been tested to predict the disastrous situation in crop damage at different districts of Andhra Pradesh during 1997-98 winter. The actual as well as long term average rainfall data of 1997 of Warrangal, Khammam, Karimnagar, Rangareddy and Mehboobnagar were obtained. Although, the population density data of pod borer were not available, data on the extent of damage were available. The A and B values of the above districts during the winter of 1997-98 showed A-B+ combination (Table 14.1) indicating severe attack or damage of *Helicoverpa armigera* and this was true as cotton, redgram, chilles and sunflower were devastated during this year in the above districts.

Table 14.1: A and B Parameters, Crops Affected and Level of Damage by *Helicoverpa armigera* in different districts of Andhra Pradesh

Districts	*A*	*B*	*Level of Damage*	*Crops Affected*
Warrangal	-305.5	+17.1	SEVERE	Cotton, pigeon pea, chilles, sunflower
Khammam	-142.0	+9.2	SEVERE	Cotton
Karimnagar	-321.0	+29.5	SEVERE	Cotton, pigeon pea, chilles
Rangareddy	-213.5	+17.1	SEVERE	Cotton, pigeon pea
Mehboobnagar	-250.1	+18.5	SEVERE	Cotton, pigeon pea

Refinement of Thumb Rule for Forecasting of *Helicoverpa armigera* for Gulbarga (Karnataka)

The thumb rule developed earlier to predict the level of population of *Helicoverpa armigera* at Andhra Pradesh was put into validation in chickpea-pigeon pea based ecosystem at Gulbarga, Karnataka using the historical as well as current data. After studying the historical data on monthly rainfall and percent pod damage of Gulbarga, the thumb rule was fine tuned with one modification. To compute B parameter, the rainfall values of October were taken instead of those of November (Das *et al.*, 2001b).

The thumb-rule for predicting *Helicoverpa* armigera in Deccan Region was validated by using the rainfall data. Out of 10 years, results were accurate for 9 years. Zill Parishad, Gulbarga adopted this model, published in local Kannada language and giving forecasting every year.

Table 14.2: A and B Parameters, Per cent of Pod Damage by *Helicoverpa armigera* in Different Years at Gulbarga Karnataka

Year	A	B	Predicted Level of Damage	Percent Pod Damage	Actual Level of Damage
1989	+140.9	-40.5	LOW	25.8	LOW
1990	+42.6	+88.8	MODERATE	90.7	SEVERE
1991	-232.7	-63.6	MODERATE	43.1	SEVERE
1992	-180.5	+12.6	SEVERE	100	SEVERE
1993	-156.7	+96.0	SEVERE	71.8	SEVERE
1994	-354.7	+145.4	SEVERE	65.3	SEVERE
1995	-197.5	+70.6	SEVERE	80.2	SEVERE
1996	+9.9	+36.0	MODERATE	54.6	MODERATE
1997	-126.0	+23.2	SEVERE	99.6	SEVERE
1998	+266.6	+178.4	MODERATE	32.5	MODERATE

Assumption: LOW=<30 per cent, MODERATE = 30-60 per cent and SEVERE =>60 per cent pod damage.

Prediction Rule for Pink Boll Worm (*Pectinophora gossipiella*) of Cotton

The data received from CICR Nagpur on pink bollworm (*Pectinophora gossipiella*) pheromone trap catch from 1983-84 to 1995-96 was collected. It was found from the data that the activity of this pest is confined mainly from 41st to 52nd meteorological weeks. However, the level of population fluctuated from year to year. The highest population was observed in 1995-96 and lowest population in 1988-89. The population level were categorized in three groups: Low <20, Moderate 20-60 and Severe >60 catches per trap per week. It was observed that out of 13 years, during 4 years the population was low, another 4 years, it was moderate and rest five years severe outbreak was observed (Table 14.4).

The weather and pest data were analyzed to identify weather factors responsible for out break/ severity of the pest. The earlier works done were also compared through literature. Using all there information, 4 weather based criteria and prediction rules were made. The criteria and the rules are given in Table 14.3.

Table 14.3: Weather Criteria Identified for Pink Bollworm

Sl.No.	Criteria	Condition
1.	Total Rainfall during 32-40 met. Week	<500mm
2.	Total Rainfall during 41-52 met. Week	>20mm
3.	Average Tmax during 41-52 met. Week	<33°C
4.	Average RH during 41-52 met. Week	>70 per cent

Prediction Rule

Case 1

If all criteria are satisfied- pest population during 41-52 standard meteorological weeks is expected to be Severe

Case 2

If two or three criteria are satisfied- pest population during 41-52 standard meteorological weeks is expected to be Moderate

Case 3

If none or one criteria is satisfied- pest population during 41-52 standard meteorological weeks is expected to be Low.

Based on above criteria and rules, the predicted and observed pest levels during 1983-84 to 1995-96 were computed and results are shown in Table 14.4.

Table 14.4: Criteria Table Developed for Pink Boll Worm (*Pectinophora gossipiella*) CICR, Nagpur

Years	*Developed Criteria*				*Pest Population*	
	Total Rainfall During 32-40 met. Week <500mm	*Total Rainfall During 41-52 met. Week >20mm*	*Average Tmax During 41-52 met. Week <33°C*	*Average RH During 41-52 met. Week >70 per cent*	*Prediction (41-52 met. Week)*	*Observed (41-52 met. Week)*
1983-84	590.3 (N)	16.9 (N)	28.7 (Y)	53.7 (N)	Low	5.7 Low
1984-85	167.4 (Y)	11.5 (N)	33.6 (N)	56.5 (N)	Low	12.7 Low
1985-86	406.6 (Y)	168.5 (Y)	29.5 (Y)	56.6 (N)	Moderate	19.2 Low
1986-87	439.6 (Y)	97.0 (Y)	33.7 (N)	57.6 (N)	Moderate	46.7 Moderate
1987-88	380.0 (Y)	13.0 (N)	29.8 (Y)	59.9 (N)	Moderate	45.4 Moderate
1988-89	785.8 (N)	17.4 (N)	34.8 (N)	53.4 (N)	Low	19.4 Low
1989-90	288.0 (Y)	38.1 (Y)	33.6 (N)	67.4 (N)	Moderate	44.7 Moderate
1990-91	480.0 (Y)	42.6 (Y)	29.5 (Y)	71.9 (Y)	Severe	117.1 Severe
1991-92	272.7 (Y)	5.5 (N)	33.5 (N)	49.0 (N)	Moderate	63.9 Severe
1992-93	349.6 (Y)	11.0 (N)	33.7 (N)	75.8 (Y)	Moderate	65.7 Severe
1993-94	377.5 (Y)	61.6 (Y)	29.1 (Y)	78.7 (Y)	Severe	119.4 Severe
1994-95	696.0 (N)	99.0 (Y)	30.2 (Y)	53.2 (N)	Moderate	54.6 Moderate
1995-96	448.0 (Y)	40.0 (Y)	30.3 (Y)	75.7 (Y)	Severe	187.1 Severe

Low <20, Moderate 20-60 and Severe >=60 catches per trap per week.

From Table 14.4, the predicted pest levels were same as that of observed pest levels for 10 years out of 13 years. The rules can be used to forecast pink bollworm based on forecasted weather information.

Prediction Rules for Rice Gall Midge

Data on percent infestation (Onion leaf/ silver shoot) along with weather parameters from meteorological observatory were collected at Central Rice Research Institute (CRRI), Cuttack during kharif seasons, from 1991 to 2000. From the analyzed data weather criteria identified for gall midge infestation.

Criteria	Condition
Rain (August)	> 400-500 mm
Rain (September)	< 225 mm
Bright Sunshine Hours (August)	3-3.7 hrs
Bright Sunshine Hours (September)	> 5 hrs

Forecasting Models for Sugarcane Pest

Investigations were carried out on the stalk borer infestation in relation to weather parameters in two sugarcane cultivars viz. CoJ 64 (early) and Co 1148 (late) at the Regional Centre of Sugarcane Breeding Institute, Karnal during the period 1990-91 and 1999-2000 (Sardana and Das, 2001). The following multiple regression models were found to be best to predict the number of infested canes (Y) per plot of 10mx6m size.

For CoJ 64

Y = -170.11+ 55.27 Av. Tmax (Jul-Aug) – 43.32 Av. Tmin (Jul-Aug)+0.03 Rain (Jul-Aug)+9.36Av.RHmorning(Jul-Aug)+2.43Av.RHevening(Jul-Aug) R^2=0.84

For Co 1148

Y= 496.41+ 47.35 Av. Tmax (Jul-Aug) – 54.92 Av. Tmin (Jul-Aug)+0.07 Rain (Jul-Aug)+2.31Av.RHmorning(Jul-Aug)+2.14Av.RHevening(Jul-Aug) R^2=0.86

Pest Forecasting using Remote Sensing and Geographic Information System

Although single insect can not be detected by remote sensing (RS), it can detect the initial damaged area (big enough to fit few pixels in imagery). By creating a buffer zone around it using Geographic Information System. GIS tool, the simplest pest forecast can be given. The swarms of migratory insects can be detected through radars. Addition of favourable weather and soil conditions for pest development and outbreak would enhance the accuracy. Geo-stationary satellite generated weather data grid can be used to locate the zones of pest favourable temperature, humidity, rainfall etc. Super imposition of different layers of favourable weather would pin point the level of pest infestation (polygon wise) in the study area. Crops can be identified and phenology can be detected through NDVI and other indices. Addition of crop phenological layers would enhance the accuracy. Although all the parameters required for pest forecasting can not be generated through RS, but a combination of RS, meteorology, crop based information can be integrated to give fairly accurate pest forecasting for using in IPM.

GIS is a potential tool to prepare pest affected zones. Preparation of pest maps using GIS on different rice and cotton pests at district level for different years was completed at NCIPM with collaboration of Division of Agricultural Physics. Super imposition of different years pest data along with weather and soil data would help to delineate 'hotspot', 'moderately affected' and 'less affected' zones of different pests through out India.

GIS alone can be used for forecasting insect pests and diseases. Standalone pest forecasting models can be extended to larger area. Das *et al* (2001) developed a simple thumb rule to forecast the attack of *Helicoverpa armigera* on different crops in Andhra Pradesh using rainfall data of monsoon and post-monsoon months. Later it was extended to give *taluk* wise forecast in the Gulbarga district of Karnataka using GIS.

RS and GIS combined together to developed desert locust forecasting models and decision support systems(DSS) in locust management programme. The key issues in desert locust management are:

- ☆ Bio-physical information from locust habitat using RS for better temporal sampling capability.
- ☆ Parameterization and accurate translation of RS data in relation to ground segment for model input.
- ☆ Use of Spatial variables for process modeling and subsequent field validation.
- ☆ Bio-climatic modeling for locust physiology and migration based on historical locust events and ground intelligence.
- ☆ Operationalization for forecasting of high frequency breeding areas and flight behaviour.

Regional Remote Sensing Service Centre (RRSSC), Jodhpur of ISRO has developed Geo-LIMIS (Geographically Encoded Locust Impact Minimization Information System) with the following features (Dutta *et al.*, 2006)

- ☆ Decision Support System on ARC/INFO 8.0 NT (ARC macro tools and VB)
- ☆ To provide operational decision support for habitat suitability, surveillance, prioritization of critical areas and weather conditions on spatial context to enable timely control measures at the field levels.
- ☆ Amalgamation of data from various sources:
  - RS data (optical and microwave)
  - Meteorological data from internet
  - Ground information on locust

The DSS was operationalized in Kazaktan. FAO is also issuing desert locust forewarning in African countries using similar approach.

Gaps in the Pest Forecasting Knowledge

Although monitoring of pest along with weather and crop phenology have been initiated on regular basis but still there are some gaps that need to be addressed in near future.

1. Lack of regular pest and disease monitoring data in prescribed/standard format for round the year in multicrop environment.
2. Data on natural enemies and life table studies are missing in most of the crops.
3. Lack of information on cropping system followed in specific location.
4. Results are based on location specific (point data) and it applicability over an area needs some fine-tuning.
5. The number of models for disease forecasting is less as compare to insect forecasting.

Very recently, a pilot project is launched jointly by FAO, NCIPM, CIPMC and NIC to collect pest and natural enemy data on regular basis through out the country using a hand held device developed for this purpose. The data will be mapped using GIS and every week's pest situation will be published through website.

References

Das, D. K., Trivedi, T.P., Singh Joginder and Dhandapani, A., 2000. Weather based prediction model of *Helicoverpa armigera* for Integrated Pest Management in cotton –chickpea based agro-ecosystem of Punjab. Paper presented in International Conference on Managing Natural Resources for Sustainable Agricultural Production in the 21st Century. Voluntary Papers – Natural resources, Vol.2, February 14-18, 2000, New Delhi, pp: 621-622.

Das, D. K., Yelshetty, Suhas, Trivedi, T. P. and Dhandapani, A., 2001a. Refinement and Validation of Prediction System of *Helicoverpa armigera* at Gulbarga Karnataka. Paper presented in the National Symposium on Plant Protection Strategies for Sustainable Agri-Horticulture. S K University of Agricultural Sciences and Technology. October 12-13, 2001. R S Pura Jammu.

Das, D.K., Trivedi, T.P. and Srivastava, C.P., 2001b. Simple rule to predict attack of *Helicoverpa armigera* on crops growing in Andhra Pradesh. Indian J. Agr. Sci. 71(6) 421-423.

Dutta, D.; Bhatawdekar, S.; Chandrasekharan, B.; Sharma, J.R.; Adiga, S.; Wood, Dunkan. And McCardle, Adrian, 2006. Desert Locust Monitoring System – Remote sensing and GIS Based Approach. In Satellite Remote sensing and GIS Applications in Agricultural meteorology (Eds. Sivakumar et al) Springer.

Nagarajan, S. and Singh, H., 1975. Indian stem rust rules-an epidemiological concept on the spread of stem rust of wheat. Plant Disease Reporter, 59:133-136.

Pradhan, S., 1967. Strategy of insect pest control. Indian J. Entomology. 29 : 105 -122.

Pradhan, S., 1982. Effect of temperature on the dynamics of insect development. Agricultural Entomology and pest control. ICAR. New Delhi.

Sardana, H.R. and Das, D.K., 2001. Regression modeling to predict sugarcane stalk borer *Chilo auricilius*. Indian J. Ent. 63(4): 435-438.

Srinivasan, G.S. and Mavi, H.S., 1983. *Proc. Int. Conf. on Biometeorology*, New Delhi.

Van Everdingen, E., 1926. Het verband de wears gesterdheid ende aardappelzeiekte (*Phytophthora infestans*). Tijdschr. Plantennziekten. 32:129-140.

Chapter 15

Integrated Management of Diseases of Important Spice Crops

R. Suseela Bhai and Santhosh J. Eapen
Indian Institute of Spices Research,
Marikunnu P.O. Calicut – 673 012, Kerala

Introduction

India is considered as the home of spices and a sizable area of the country is under spice cultivation. They are of considerable importance to India as these add to the economy of the country through foreign exchange. Spices are cultivated in an area of 5.98 million ha globally with a total production of 7.31 million tons (FAO, 2008). The latest statistics shows that the total cultivated area in India under spices is 2.89 million ha with a total production of 3.33 million tonnes. The major export oriented spice crops that are cultivated in our country are black pepper (*Piper nigrum* L.), cardamom (*Elettaria cardamom* Maton.), ginger (*Zingiber officinale* Rosc.), turmeric (*Curcuma longa* L.) and vanilla (*Vanilla planifolia* Andrews). Its production and marketing are the main economy stays of millions of people. However there are certain limitations in the cultivation and production of these crops due to biotic factors. The major production constraints credit due to diseases caused by fungi such as species of *Phytophthora, Pythium, Rhizoctonia, Sclerotium, Fusarium, Colletotrichum, Phyllosticta, Taphrina* etc., besides problems caused by nematodes singly or in association with fungi (disease complexes). Nematodes are known to break the resistance to fungal/bacterial pathogens in many spice crops. About 66 species of plant parasitic nematodes belonging to 27 genera are reported on ten major spices (Koshy *et al.*, 2005). Most of the diseases are both soil borne and airborne in nature and occur with the onset of south-west monsoon, besides virus diseases which occur as an all time threat to crops like black pepper, cardamom, vanilla etc. Cropping system and microclimate appear to be the deciding factors in the incidence and spread of these diseases. The diseases can be controlled only with the adoption of integrated management strategies rather than adopting individual disease control measures. However, if a specific disease constitutes as a major threat to the crop, then the integrated control strategy is directed towards the disease (Agrios, 2005). The strategies adopted for managing the various diseases of black pepper, cardamom ginger, turmeric and vanilla are dealt here.

Disease of Black Pepper (*Piper nigrum* L.)

Black pepper is susceptible to a number of diseases of fungal, viral and nematode origin. These diseases will lead to huge crop loss if left unnoticed or if no timely and proper plant protection measures are adopted. On a global scale, an annual crop loss of around 4.5-7.5 million has been reported due to foot rot alone (De waard, 1979). Even in Malaysia a crop loss of 5-10 per cent due to this disease has been reported and up to 95 per cent loss in individual gardens (Kueh and Sim, 1992). Diseases are occurring in nurseries as well as in plantations.

Diseases in Nursery

Phytophthora Rot

This disease can occur from air borne inoculum or through soil borne inoculum from contaminated soil.

Symptoms

The symptom appears on the leaves as characteristic black spots with fimbriate margins which enlarge and the leaves fall off. This may serve as primary source of inoculum for root or collar infection. Infection from contaminated soil leads to collar infection as blackening which extend both upwards and downwards. The root infection also culminates in collar infection leading to death of the plants (Anandaraj *et al.*, 1994).

Pathogen

The disease is caused by *Phytophthora capsici.*

Anthracnose

The disease appears during the end of south west monsoon.

Symptoms

The disease is characterized by the appearance of yellowish brown to dark brown irregular sunken leaf spots with a chlorotic halo on the leaves. In advanced stages of infection, the lesion coalesces and the leaves dry up. The disease is caused by *Colletotrichum gloeosporioides.* The fungus is airborne and spreads through conidia formed in pin head like pycnidial structures formed on the infected lesions (Sebastian, 1982).

Leaf Rot and Leaf Blight

Leaf blight is a serious disease in the nursery during April-May when warm humid condition prevails. Infection occurs on both leaves and stem.

Symptoms

Grayish sunken spots and mycelial threads appear on the leaves and the infected leaves are attached to one another with the mycelial threads. On stems, the infection occurs as dark brown lesions which spread both upwards and downwards. In advanced stages of infection, new flushes subtending the points of infection gradually dry up and drop off (Anandaraj, 2000).

Pathogen

The disease is caused by *Rhizoctonia solani* Kuhn. *Colletotrichum* sp. is found associated with leaf spots (Mammootty *et al.*, 1992)

Basal Wilt

The disease is mainly noticed in the nurseries during June-September months.

Symptoms

Symptoms appear as grayish lesions on stems and leaves. On the leaves white mycelium are seen at the advancing edge of the lesions. The mycelial threads later girdle the stem resulting in drooping of leaves beyond the point of infection. In advanced stages the rooted cuttings dry up. Small whitish to cream coloured grain like sclerotial beads appear on the mature lesions (Anandaraj, 2000).

Pathogen

Basal wilt is caused by the fungus *Sclerotium rolfsii Sacc.*

Plant Parasitic Nematodes

It is serious problem and widespread in occurrence. Root-knot nematode *M. incognita* and lesion nematode *R. similes* are associated with black pepper.

Symptoms

M. incognita causes typical galls or knots on the roots and elongated swelling on thick primary roots whereas *R. similis* causes necrotic lesions on white feeder roots.

Integrated Management of Nursery Diseases

Since the disease affecting the nursery occur in a short span of 3-4 months, it is better to follow an integrated management strategy to protect the plants from infection caused by various types of pathogens rather than individual treatmens. The infection may occur from contaminated soil or potting mixture besides the air borne inoculum like *Colletotrichum* sp. or other foliar pathogens. The management strategies that can control the nursery diseases are

1. *Use of certified planting material* free from nematode or virus is the foremost requirement for the production of disease free planting material. Good agricultural practices and phytosanitation helps to reduce nematode spread to a greater extent. Manual removal of nematode infested roots in black pepper rooted cuttings can also help to bring down the nematode inoculums.
2. *Disinfestation of nursery mixture* either through solarization (Mammootty *et al.*, 2008) or fumigation (Pattenshetty *et al.*, 1973)with chemicals and fortification with biological control agents is highly effective in reducing the initial nematode load and production of healthy rooted cuttings (Ramana *et al.*, 1994).

   Disinfestation of potting mixture may be done by either fumigation using formaldehyde 0.4 per cent or through solarization under direct sun for 44-50 days or by steam sterilization. Promising isolates of biocontrol agents like vesicular arbuscular mycorrhizae *viz.*, *Glomus mossae, G. fasciculatum, Acaulospora laevis, Gigaspora margarita,* which are suppressive of nematodes on black pepper (Anandaraj *et al.*, 1991, 1996) (VAM @100cc/kg of mixture and *Trichoderma* @ 5-10g⁻kg and *Pseudomonas fluorescence* @10-15g–kg of soil (Anandaraj and Sarma, 1995) can be incorporated with the nursery soil mixture.

3. *Chemical control*: As a preventive measure, black pepper planting materials can be treated with granular nematicides like phorate or carbofuran @ 0.1g a.i./plant. Application has to be made once in two months (Mohandas and Ramana, 1987). This will ensure eradication of nematodes and production of healthy roots. After the establishment of plants in the polybags, spray 1 per cent Bordeaux mixture or 0.2 per cent carbendazim on the above soil portion and drench the soil with 0.2 per cent copper oxychloride or 0.125 per cent metalaxyl–mancozeb or 0.3 per cent potassium phosphonate at monthly intervals. Once the infection is noticed, cuttings along with the defoliated leaves if any should be removed and destroyed (preferably by burning) from the nursery.

Diseases in the Plantations

Foot Rot Disease

Foot rot is the most dreadful disease of black pepper. The disease makes its appearance mainly during the south-west monsoon. All parts of the vine including leaf, stem, root and spikes are vulnerable to infection.

Symptoms

The disease is characterized by the appearance of black spots on the leaves which have typical fimbriations at the advancing margins which rapidly enlarge. The tender leaves and succulent shoot tips of freshly emerging runner shoots trailing on the soil turn black when infected. The disease spreads to the entire vine during the intermittent showers due to rain splash. When the main stem at the ground level or at the collar region is infected, the entire vine wilts followed by shedding of leaves and spikes with or without black spots. The branches breakup at nodes and the entire vine collapses in a month. When the damage is confined to the feeder roots the expression of the symptom is delayed till the cessation of rain and the vine starts showing declining symptoms such as yellowing, wilting, defoliation and drying up. This may occur during October-November months. These vines may recover after the rain and survive for more than two seasons till the root infection culminates in collar rot and death of the vine (Anandaraj and Sarma, 1995).

Pathogen

The disease is due to infection by *Phytophthroa capsici.*

Disease Management

The disease can be controlled by adopting integrated disease management practices which include phytosanitation, cultural practices, chemical and biological controls.

Phytosanitation

Removal and destroyal of dead vines along with root system from the garden helps to reduces the buildup of pathogen inoculum. It is also important that the planting materials should be collected from disease free gardens and nurseries.

Cultural Practices

Providing adequate drainage and proper shade shade regulation helps in reducing the primary inoculum source. It is better to avoid digging the plant base which may damage the roots which in turn give way to the entry of the pathogens. Prune the freshly emerging runner shoots or tie them to the standard with the onset of monsoon to avoid build up of inoculum.

Chemical Control

Spraying 1 per cent Bordeaux mixture to all the vines with the onset of South West monsoon and drenching the plant basins at a radius of 45-50cm with 0.2 per cent copper oxychloride @ 5-8 lit/vine as a precautionary measure helps to reduce the disease incidence. A second time drenching and spraying with 0.3 per cent potassium phosphonate during August-September is to be done if the monsoon is prolonged. Finally a third round of drenching during October is preferable to check the proliferation of the soil borne inoculum. Alternatively, after the receipt of a few monsoon showers all the vines can be drenched and sprayed with 0.125 per cent metalaxyl-mancozeb @ 5-10 l/vine (Sarma *et al.*, 2001).

Biological Control

With the onset of monsoon (May-June), it is preferable to apply *Trichoderma harzianum* around the base of the vine @ 50g/vine along with organic manure such as neem cake, farmyard manure (FYM), decomposed coffee pulp or coir pith. A second application of *Trichoderma* is also to be given during August-September.This has to be repeated for 2-3 consecutive years to check the pathogen spread.

Host Resistance

IISR Shakti is a moderately resistant *Phytophthora* line released from Indian Institute of Spices Research, Calicut (Anandaraj, 2006).

Slow Decline Disease

Slow decline disease of black pepper was first reported by Van der Vecht (1950) as 'pepper yellows' in the islands of Bangka, Indonesia and *R. similis* was reported to be responsible for the disease. Later Christie (1957) held responsible this nematode to the death of 20 million pepper vines. Association of *R. similis* with black pepper in India was first reported by D'Souza *et al.* (1970) and subsequently it's wide spread occurrence in black pepper plantation in South India was reported by several workers.

Symptoms

Mild to moderate foliar yellowing at different regions of the affected vine is the initial aerial symptom of this disease. With the advancement of the disease, the infected vines show defoliation and die back leading to loss of vigour, yield and finally death of the vine. Vines in the early stages of the disease look apparently normal with the onset of monsoon, but foliar yellowing becomes more pronounced with the depletion of soil moisture in summer season (Mohandas and Ramana, 1987).

Gradual decline and foliar yellowing are the predominant symptom of root knot nematode *Meloidogyne* sp. infestation. Such plants exhibit dense inter veinal chlorosis (Ramana, 1992). Typical galls or knots are seen on secondary and fibrous roots and as elongated swellings on thick primary roots. Adult females with egg masses are generally embedded within the root tissue.

Foliar yellowing and defoliation are the main above ground symptoms expressed by the vines infected with *R. similis*. Root penetration by this nematode causes necrotic lesions on white feeder roots. Subsequently the lesions merge and encircle the root cortex leading to disintegration of distal portion of the roots. Damage caused to the roots directly affects vigour and productivity of the vines. A synergistic effect of *Meloidogyne* sp., *R. similis* and *Fusarium* sp. on reduction in growth of pepper vines and increase and severity of foliar yellowing was observed (Lopes and Lordello, 1979; Sheela and Venkitesan, 1990; Mustika, 1992).

Disease Management

Preventive

Uprooting and destruction of diseased vines along with root mass and replanting after a period of 9-12 months have to be undertaken in all nematode damaged plantations. This will reduce the source of inoculum and the spread of the pathogens.

Cropping Practices

Exclusion of nematode susceptible intercrops, supports or standards can minimise the nematode infestation. Coconut and arecanut are good hosts of *R. similes. Garuga pinnata* Roxb. and *Macaranga indica* Wight are not susceptible to root knot nematodes while *Erythrina indica* Lank. and *Gliricidia sepium* (Jacq.) Walp. are less susceptible. Several inter crops like banana, ginger and turmeric and a large number of weeds are also susceptible to these nematodes (Koshy *et al.*, 1977). Large numbers of weeds that are found in pepper gardens have been recorded as hosts of the root-knot nematodes.

Soil Amendments

There is substantial evidence that the addition of organic matter in the form of compost or manure will decrease nematode populations and associated damage to crops. Some sources of organic matter known to be nematode-suppressive include oilcakes, sawdust, bone meal, manures, compost, and certain green manures. Application of neem cake @ 2 kg/vine was highly effective against *M. incognita* than *R. similis* (Ramana *et al.*, 1992). Mulching and bio-fumigation ameliorate symptoms of slow-decline/pepper yellows. Addition of chopped leaves of *Gliricidia maculata* (10g/kg soil) as green manure reduced populations of *R. similis* and increased the plant growth (Jasy and Koshy, 1992). Mulching the basins with *Gliricidia* leaves reduced root knot nematodes in Sri Lanka (Ratnasoma *et al.*, 1991).

Resistant Lines

Accessions like HP 39 and C.820 (IISR collections) were found to be resistant to *R. similis* (Eapen, 2006). One variety PW14 was reported as totally immune to *R. similis* in Sri Lanka (Gnanapragasam, 1989). *P. colubrinum* is now widely used as a root-stock to graft cultivated pepper plants (Ramana *et al.*, 1987*b*; Ramana, 1992). One cultivar, CLT-P-812 was found resistant to *M. incognita* and was later released as 'Pournami' for cultivation in root knot infested areas (Ravindran *et al.*, 1992)

Chemical Control

Carbofuran/phorate at 3 g a.i./vine applied in May/June and again in September/October resulted in the remission of foliar yellowing and reduction in nematode populations in plantations. Various nematicides like phenamiphos @ 20 g/vine (Nambiar and Sarma, 1977), aldicarb sulphone @ 8 kg/ha (Venkitesan and Setty, 1979), phorate, carbofuran and aldicarb @ 3 g a.i./vine (Venkitesan and Charles, 1979; Venkitesan and Jacob, 1985) are effective in controlling nematodes infecting black pepper in India. Mustika and others (1984) stated that the severity of pepper yellows was reduced by application of fertilizer (NPK 15:15:15) at 250 g/plant/year and either with aldicarb (50g/plant) of mancozeb (12 g/plant) or both since the disease can be caused by a multiple attack by nematodes (*M. incognita* and *R. similis*) and *Fusarium oxysporum*.The nematicide may be applied after removing the top soil without causing damage to the roots, followed by replacement of the soil.

Biological Control

Ramana (1994) and Sosamma and Koshy (1997) found that *Paecilomyces lilacinus* and *Pasteuria penetrans* suppressed nematode infestation in black pepper and increased the production of root mass. Freire and Bridge (1985) found high infestation of eggs of *M. incognita* by *P.lilacinus* and *Verticillium chlamydosporium* under artificial inoculation but to a lesser extent in the egg mass from the roots of black pepper inoculated with the fungi. *Bacillus pumilis, B. macerans* and *B. circulans* significantly reduced *M. incognita* population and improved the growth of black pepper plants (Sheela *et al.*, 1993). Two promising endophytic bacteria *viz. Bacillus megaterium* and *Curtobacterium luteum* suppressing *R. similis* have been identified recently (Eapen, 2007)

Anthracnose/Pollu Disease

This disease can be distinguished from Pollu (hollow berry) caused by pollu beetle (*Longitarsus nigripennis*) by the presence of characteristic cracks on the infected berries. The disease appears towards the end of the monsoon season.

Symptoms

The affected berries show brown sunken patches during early stages and their further development is affected. In later stages the discoloration gradually increases and the berries show the characteristic cross splitting. Finally the berries turn black and dry. The fungus also causes angular to irregular brown lesions with a chlorotic halo on the leaves. The disease is more prevalent under higher elevations

Pathogen

The disease is caused by *C. gloeosporioides.*

Spike Shedding

Spike shedding is another phenomena pronounced in the higher elevations. Predominance of female flowers, lack pf pollination in rain fed areas; heavy shade and delayed emergence of spikes were found to be the major reasons for spike shedding. The symptom of fungal infection appeared as small dark colored spots surrounded by a halo on the leaves. In serious infections, the expansion of lamina is affected resulting in crinkled appearance. Infection on spike results in spike shedding whereas infection on mature berries leads to formation of brownish split on the seeds due to unequal development (Anon, 2003).

Pathogen

The disease is found associated with *Colletotrichum gloeosporioides.*

Integrated Management of Anthracnose/Pollu Disease/Spike Shedding

Since all the three diseases occur at the same time as well as associated with one and the same organism it is better to follow an integrated approach for disease management. This includes

1. Irrigation of black pepper vines 4-5 times at an interval of 5-7 days @ 40-50 l/plant commencing from March, followed by shade regulation of support trees to provide minimum 7,500-10,000 lux light under cloudy condition (Anon, 2003).

2. Provide 40 per cent shade using shade nets
3. Plant field tolerant cultivars *viz.* Aimpiriyan and Arakulammunda (Kueh, *et al.*, 1993).
4. Spray 1 per cent Bordeaux mixture (Nair *et al.* 1986) or a combination of fungicide containing carbendazim and mancozeb or mancozeb at 0.1 per cent concentration (Sainamole, *et al.*, 2008).

Stunted Disease

Stunted disease is one of the most important viral diseases affecting black pepper in the country and the severity of the disease is high at higher altitudes (Bhat, *et al.*, 2005a)

Symptoms

The disease appears as distortion, reduction in size and mottling and mosaic on the leaves along with stunting of the whole plant, reduction in spike length and poor filling of spikes. In severe cases leaves become abnormally narrow and give a sickle shaped appearance. The internodes of the vine become short leading to stunting of plants.

Pathogen

Association of two viruses r ·mely Cucumber mosaic virus (CMV) (Sarma, *et al.*, 2001; Bhat, *et al.*, 2004 a) and Piper yellow mottle (PYMV) badna virus is reported (Bhat, *et al.*, 2003).

Transmission and Spread

Aphids are known to transmit CMV whereas mealy bugs like citrus mealy bug, *Planococcus citri* (Risso) and striped mealy bug (*Ferrisia virgata*) are known to transmit the Badna virus associated with stunted disease (Bhat, *et al.*, 2005b). The major means of spread is through infected vines.

Phyllody Disease

The Phyllody disease is characterized by varying degrees of malformation of spikes.

Symptoms

The entire spike will be transformed into leaves and branches. There is considerable increase in the length of stalk of the affected spike. The bracts and flowers were transformed to small leaf like structures which occasionally show varying degrees of fusion forming bifid or trifoliate structures. The floral buds are transformed into small branches with nodes and internodes similar to fruiting laterals. Each such lateral in turn produce leaves of different sizes. The whole deformed structure look like a tuft of leaves giving a witches broom appearance (Sarma, *et al.*, 1988 a). The disease is caused by Phytoplasma belonging to aster yellow group (Bhat *et al.*, 2006).

Integrated Disease Management

Roguiuing the infected plants as and when noticed and destroying them by burning is a prerequisite in the management of these diseases. The planting material should not be collected from diseased gardens. As the disease is of viral/phytoplasma origin, there is no chemical control measures as such. However by using biotechnological tools like PCR and DAS ELISA, the symptom less carrier plants can be detected and can be destroyed before being spread to other areas (Bhat, *et al.*, 2004b, 2006, Bhadramurthy, *et al.*, 2005). Correct and timely diagnosis help in checking the disease spread.

Diseases of Cardamom (*Elettaria cardamomum* Maton)

Cardamom (*Elettaria cardamomum* Maton) the "queen of spices" is susceptible to a number of diseases of fungal and viral origin both in the nursery and in the field.

Diseases in the Nursery

In cardamom cultivation, two types of nurseries are being maintained, primary nursery where seeds are sown and seedlings are raised up to 3-4 leaf stage. From the primary nursery the seedlings are transplanted to secondary nursery where it is grown for about 18 months and from there it is transplanted into the main field. Diseases are observed in all the stages.

Damping-off/Rhizome Rot

The disease is called damping off, if it is found in the primary nursery where as the same is called clump rot/rhizome rot in the secondary nursery. The disease appears during June-October months coinciding with the South West monsoon.

Symptoms

Damping off is characterized by the appearance of light brown discoloration at the collar region which becomes soft leading to the sudden collapse of the seedlings. In secondary nursery the symptom is expressed as dark brown discoloration of the collar portion. The affected seedlings show root and rhizome rot and leaves exhibit typical flaccidity symptoms which may later wilt and dry up and fall off with a slight disturbance.

Pathogen

The disease is caused by the soil borne fungi such as *Pythium vexans* and *R. solani*. *F. oxysporum* and *S. rolfsii* were also reported along with the disease (Siddharamaiah, 1988b). Since the disease is soil borne in nature, apparently normal seedlings from infected nurseries would serve as carriers of inoculum to the fields.

Nursery Leaf Spot

The disease is common in exposed nurseries.

Symptoms

The disease is characterized by the appearance of water soaked grayish, circular or oval shaped spots of 2-3mm which in advanced stages coalesces to form larger necrotic patches in the lower surface of tender leaves of the seedlings which later appear on the upper surface. The centre of the lesion appears sunken with dark colored pycnidia present in the tissues. In dry weather the lesion remains discrete with a definite light brown margin surrounded by chlorotic halo (Sarma *et al.*, 1994). The disease spread is through rain splash.

Pathogen

The disease is caused by the air borne fungus *Phyllosticta elettariae*. The fungus survives on infected leaves for about 12 months and can serve as the source of primary inoculum for further spread.

Plant Parasitic Nematodes

Root knot nematode is a serious problem in cardamom nurseries than in adult plantations. Widespread occurrence of root-knot nematodes, *M. incognita* and *M. javanica* has been reported in

India (Koshy *et al.*, 1976; Ali, 1986a; Raut and Pande, 1986). In primary nurseries, more than 50 per cent of the germinating seeds do not emerge due to infection of the radicle and plumule by the second stage juveniles of the root-knot nematodes. Up to 40 per cent of such seedlings do not establish in the secondary nursery.

Symptoms

The infested cardamom seedlings at the two-leaf stage show marginal yellowing and drying of leaves and severe galling on roots. On transplantation to a secondary nursery, the seedlings exhibit curling of the unopened leaves. These leaves mostly emerge after the breaking open of the pseudostem. In secondary nurseries, the infested plants are stunted and yellowed with poor tillering, drying of leaf-tips and margins, and heavy galling on roots (Ali and Koshy, 1982a; Eapen, 1995b). The leaves show varying degrees of chlorosis, narrowing and drying at leaf tips and margins. Root system of infested plants shows varying degrees of root galling. It is observed that young seedlings are more susceptible to root knot nematode attack than mature plants (Eapen, 1992).

Integrated Disease Management

Preventive

Pruning roots of cardamom seedlings is advisable to prevent the entry of nematodes to the main plantations. It is also advisable to change nursery sites every year. Disinfestation of nursery beds with methyl bromide at 500 g/10m^2 is effective in controlling root-knot infestation (Ali and Koshy, 1982b). Fumigating the nursery beds with 2 per cent formaldehyde 15 days prior to sowing (Pattenshetty *et al.*, 1973) or soil solarization and fortification of soil with VAM fungi and seed coating and soil drenching prior to sowing with biocontrol agents like *T. harzianum* also helps to suppress this disease in nurseries (Bhai and Thomas, 1988 and Thomas *et al.*, 2000b). Spraying with macozeb 0.4 per cent is also found effective in managing the rot disease (Thomas *et al.*, 1988). Application of aldicarb at 5 kg a.i./ha three times at every three months (Koshy *et al.*, 1979*a*; Jacob and Chandrasekharan, 1984; Ali, 1987b) or drenching the nursery beds with fenamiphos significantly reduced root knot nematodes (Ali, 1986b). Spraying with fungicides such as captafol (0.2 per cent), Bordeaux mixture (1 per cent) or mancozeb (0.2 per cent) at fortnightly interval (Rao and Naidu, 1974) is effective for controlling foliar infections.

Biological Control

In cardamom there are reports that *Gigaspora margarita* and *Glomus fasciculatum* suppressed *M. incognita* infestation and enhanced growth and vigour of seedlings (Thomas *et al.*, 1989). *P. lilacinus* (Eapen, 1995b) and some native isolates of T. *harzianum* and other *Trichoderma* spp. (Eapen *et al.*, 2000) are potential antagonists of root knot nematodes.

Diseases in the Plantation

Capsule Rot ('Azhukal')

It is a major threat to cardamom cultivation in the higher altitudes. The disease makes its appearance with the onset of South West monsoon, become severe during August-September and continues to prevail up to November. During years of heavy and continuous rainfall, severe disease incidence and crop losses as high as up to 40 per cent has been reported (Nambiar and Sarma, 1976).

Symptoms

Symptoms appear as water soaked lesions on young leaves and capsules which enlarge to cover the entire surface. Infected capsules show dull green discoloration which develop into rotting. Rotten capsules are shed from the panicle emitting a foul smell. Visible symptoms of rotting appear simultaneously on capsules and leaves or first on capsules followed by lesions on leaves.

Pathogen

The disease is caused by *Phytophthora meadii* Mc Rae of A2 mating type (Bhai, 1998). Presence of soil inoculum, thick shade in plantation, close spacing, high soil moisture and water logging together with favorable climatic conditions such as low temperature, high relative humidity and continuing rainfall predisposes the plant to infection.

Integrated Disease Management

As the incidence of the diseases is closely associated with monsoon showers, any plant protection operation has to be initiated prior to initiation of primary infection.

Phytosanitation

The fungus survives in the soil and infected plant parts. Trashing (removal of all dried up leaves and leaf sheath from basal portions of the plant) and removal of infected plant parts, and mulch should be done prior to the onset of monsoon.

Cultural Practices

1. Shade regulation-excess shade provides adequate microclimate for the multiplication of the pathogen. Hence shade regulation should be done in such a way that filtered light should be made available to the plant.
2. Soil drainage-Proper drainage channels should be provided to reduce the intensity of disease incidence.

Chemical Control

Two to three rounds of sprays including one round of prophylactic spray with 1 per cent Bordeaux mixture or 0.3 per cent Fosetyl–Aluminium or spraying alternate rounds of Bordeaux mixture and metalaxyl- mancozeb-72 after proper phytosanitation are effective in disease control (Anon, 1989; Thomas *et al.*, 1989; 1991a).

Biological Control

Soil application of *T. viride*, *T. harzianum* and *Bacillus subtilis* are promising in the biological control of the disease (Bhai *et al.*, 1993). Soil amendments with neem cake activate the growth of native *Aspergillus*, *Trichoderma* and actinomycetes species which are antagonistic to the pathogen (Nair, 1979). Integrated disease management is possible by one initial round foliar application of potassium phosphonate (0.3 per cent) with two subsequent rounds soil application of *Trichoderma* ($2x10^9$) (Bhai *et al.*, 1998).

Disease Resistance

Two cultivars each of Mysore and Malabar types' *viz.* MCC 60, 61, 12 and 40 (ICRI selections) were found moderately tolerant to Azhukal disease (Bhai, *et al.*, 1992b).

Rhizome Rot

The disease is also called clump rot. The disease occurs during the heavy monsoon season coinciding with the high soil moisture.

Symptoms

The disease is characterized by the yellowing of foliage and flaccidity. Rotting starts at the collar region which become brittle. The rotting extends to the rhizomes and roots and the affected tillers later breaks off with a slight disturbance emitting a foul smell.

Pathogen

The disease is caused by *P. vexans* de Bary and *R. solani* Kuhn. *F. oxysporum* is also occasionally found associated with the disease (Thomas and Vijayan, 1994)

Integrated Disease Management

Phytosanitation

Remove infected clumps and apply lime in the pits (Mayne, 1942).

Cultural Practices

1. *Shade regulation*: Excess shade provides adequate microclimate for the multiplication of the pathogen. Hence shade regulation should be done in such a way that filtered light should be made available to the plant.
2. *Soil drainage*: Proper drainage channels should be provided to reduce the intensity of disease incidence.

Chemical Control

Since different pathogens with different growth requirements are involved in the disease, combination of fungicides with different modes of action would be ideal to check the disease. Soil drenching with pentachloronitrobenzene (0.2 per cent) and captafol (0.2 per cent) at 3 l/clump, four times at monthly interval (Thomas *et al.*, 1988) or soil drenching with 1 per cent Bordeaux mixture or 0.25 per cent copper oxychloride and neem oil cake @ 500g per plant followed with one round pre-monsoon and two rounds post monsoon soil drenching with 0.25 per cent copper oxychloride (Thomas and Vijayan, 1994) were effective for controlling the disease.

Biological Control

The disease can also be controlled by the basal application *T.harzianum* alone or in combination with potassium phosphonate (0.3 per cent) as a foliar spray (Vijayan and Thomas, 2002). A consortium of biocontrol agents containing *T.harzianum* and *P. fluorescence* also is found effective in the management of the disease (Gopakumar *et al.*, 2006).

Disease Resistance

IISR-Avinash is a rhizome rot resistant line (Venugopal *et al.*, 2006).

Leaf Blight ('Chenthal' Disease)

Symptoms

The disease is characterized by the appearance of elongated water soaked lesions of varying sizes on the abaxial surface of young leaves which later turn to dark colour with a yellowish halo around. The lesions develop near leaf margin and progress towards the midrib. As the withering progresses, pseudostems wilt. The panicle in the affected plants dries up from tip downwards. The severely affected garden gives a burnt appearance.

Pathogen

The disease is caused by *C. gloeosporioides* (Govindaraju *et al.*, 1996)

Disease Management

Providing adequate shades in the plantation and mulching the plants to conserve soil moisture ensure disease control to a greater extent. Three sprays at monthly intervals with carbendazim 0.3 per cent or macozeb 0.3 per cent or copper oxychloride 0.25 per cent or Bordeaux mixture 1 per cent help to limit the spread of the disease (Govindaraju *et al.*, 1996).

Leaf Blotch

The disease was first reported as early as 1968 (Agnihotrudu, 1968). The disease appears with the advancement of monsoon. Thick shade, continuous rainfall, and high atmospheric humidity predispose the plants to infection.

Symptoms

The disease is characterized by the development of round ovoid or irregular water soaked lesions on middle leaves, usually near the leaf tips or at the mid rib areas. These areas enlarge in size, become dark brown with a necrotic centre. In moist weather, a thick grey colored fungal growth is seen on the under side of these blotched areas.

Pathogen

The leaf blotch infection is caused by *Phaeodactylium alpiniae* (Ellis, 1971).

Disease Management

Spraying of fungicides such as Bordeaux mixture 1 per cent, carbendazim 0.1 per cent, hinosan 0.1 per cent, copper oxychloride 0.2 per cent or macozeb 0.3 per cent are effective in checking the spread of the disease.

Phytophthora Leaf Blight

Leaf blight incidence is prevalent in cardamom plantations during the post-monsoon season of October-November and may even extend up to January-February. Thick shade, low night temperature and fog prevailing during the winter period pre-dispose the plants to leaf blight infection.

Symptoms

The infection starts on the young and unopened leaves as water soaked patches which soon become necrotic and dry off. A burned appearance is seen under conditions of severe infection in the plantations (Bhai and Sarma, 2005).

Pathogen

The disease is caused by *P. nicotianae var. nicotianae* (Bhai and Sarma, 2005), which can be easily identified from infected leaf using water-floating technique.

Disease Management

Leaf blight infection can rapidly spread to adjacent areas and can result in severe leaf necrosis and leaf drying unless the disease is controlled at the initial stage itself. One round of foliar spray with Bordeaux mixture 1 per cent or potassium phosphonate at 0.3 per cent is effective in limiting the spread of the disease.

Plant Parasitic Nematodes

Root knot nematode is a serious problem in plantation. Widespread occurrence of root-knot nematodes, *M. incognita* and *M. javanica* has been reported in cardamom nurseries and plantations in India (Kumar *et al.*, 1971; Koshy *et al.*, 1976; Ali and Koshy, 1982a; Ali, 1984, 1986a; Raut and Pande, 1986). The heavily shaded, hot, humid atmosphere and continuous availability of soil moisture usually prevalent in cardamom plantations are congenial conditions for the multiplication of root-knot nematodes. A yield loss of 32-47 per cent has been reported due to root knot nematode infestation (Ali, 1985, 1986b).

Symptoms

Heavy infestation with root-knot nematode in mature plants in a plantation causes stunting, reduced tillering, yellowing, premature drying of leaf-tips and margins, narrowing of leaf blades, a delay in flowering, immature fruit-drop and reduction in yield. Though galling of roots is not prominent, the infested roots exhibit a "witches broom" type of excessive branching.

Disease Management

Chemical Control

Application of phorate @ 2.5-5.0 g a.i./plant reduced the nematode population and increased the yield by more than 40 per cent (Eapen, 1995b).

Biological Control

In cardamom there are reports that *Gigaspora margarita* and *Glomus fasciculatum* suppressed *M. incognita* infestation and enhanced growth and vigour of seedlings (Thomas *et al.*, 1989). *P. lilacinus* (Eapen, 1995b) and some native isolates of *Trichoderma harzianum* and other *Trichoderma* spp. (Eapen *et al.*, 2000) are potential antagonists of root knot nematodes.

Minor Diseases of Cardamom

Cardamom is often affected by a number of fungal pathogens which causes negligible economic loss to the crop. These diseases may occur sporadically in cardamom plantations. These include various types of leaf spots and capsule spots (Table 14.1). The management strategies adopted for major diseases, if done in time is sufficient to check the spread of these diseases also.

Table 14.1: Minor Diseases of Cardamom

Disease	*Symptom*	*Pathogen*	*Disease Management*	*Reference*
Leaf rust	Yellowish rusty coloured pustules distributed on the adaxial side of the leaf in several patches which become reddish and finally dry off.	*Phakospora elettariae*	Spray Macozeb 0.2 per cent	Thirumalachar, 1943
Sphaceloma leaf spot	Scattered spherical blotches measuring a few mm in diameter and coalesce to form large necrotic patches	*Sphaceloma cardamomi* Muthappa	Bordeaux mixture 1 per cent or 0.25 per cent	Naidu 1978
Cercospora leaf spot	Rectangular water-soaked lesions arranged parallel alongside the veins which in advanced stages become grayish-brown and later dry off.	*Cercospora zingiberi* Togashi Katsaki	Macozeb	Naidu, 1978

Contd...

Table 14.1–Contd...

Disease	*Symptom*	*Pathogen*	*Disease Management*	*Reference*
Glomerella leaf spot	Circular-ovoid dark brown concentric spots on the middle leaves	*Glomerella cingulata*		Nair, 1979
Phaeotri-choconis leaf spot	Irregular papery white spots with brown margins on leaf blade. Under moist conditions the lesions enlarge and coalesce	*Phaeotrichoconis crotalariae*		Dhanalakshmy and Leelavathy, 1976
Ceriospora leaf spot	Numerous circular or oval spots of 8 mm diameter on the foliage which coalesce to form larger patches. The centre of the lesion is dirty white surrounded by light brown, circular necrotic areas.	*Ceriospora elettariae*		Ponnappa and Shaw, 1978
Sooty mould	Minute scattered dark mycelial growth on the upper leaf surface which spreads rapidly and cover the entire lamina. In severe cases the growth extends to the petioles and leafy pseudostems which are later covered with black mycelial growth and finally the leaves tear off at margin along the veins and dry prematurely.	*Trichosporiopsis* sp.		Nair, 1979
Stem rot/ Stem loadging	Brown necrotic lesions on the pseudostem, affected tillers break and fall off at the point of infection leading to lodging of pseudostem and drying of leaves	*F. oxysporum*	Three round application of carbendazim 0.2 per cent.	Dhanapal and Thomas, 2000
Anthracnose	Reddish brown sunken spots of 1-2mm diameter with soft depressed centre. The lesion varies in number and size and some times coalesces to form large lesions.	*C. gloeosporioides*	Spraying fungicides such as cuman L, captafol or carbendazim at 0.3 per cent thrice at 15 days interval	Bhai *et al.*, 1988
Capsule tip rot	Small water soaked lesions at the distal end of the capsule which later spread downwards. In advanced stages the infected portions exhibit decaying symptoms from tip downwards and extend to the whole capsule	*Rhizoctonia solani*	Praying panicles with Foltaf, Bavistin or copper oxychloride at 0.2 per cent	Unpublished

Viral Diseases of Cardamom

Cardamom is found susceptible to three important viral diseases *viz.* mosaic (Katte), 'Kokke kandu'(cardamom vein clearing virus) and Nilgiri necrosis (Cardamom necrosis virus disease). All the diseases are found economically important because of the heavy crop loss caused.

Cardamom Mosaic (Katte Disease) Virus (CAR- MV)

Mosaic is also known as katte or marble disease.

Symptoms

The symptoms appears on the youngest leaf as slender chlorotic fleks measuring 2-3 mm in length which later develop into pale green discontinuous stripes running parallel are clearer on the emerging leaves. Often mosaic like mottling is seen on the leaf sheath and young leaf shoots. In advanced stages the affected plants produce shorter and slender tillers with only a few short panicles and the plants degenerate rapidly. The disease is systemic in nature and caused by the Maclura virus belonging to the Poty viride.

Transmission and Spread

The disease is transmitted through banana aphid (*Pentalonia nigronervosa* Coq.) and also through the infected rhizomes (Uppal *et al.*, 1945).

Cardamom Vein Clearing Disease or 'Kokke kandu' (CAR- VCV)

The disease was reported in cardamom during 1993 (Venugopal and Govindaraju, 1993). Disease incidence of 0.1–80 per cent was observed in plantations and nurseries in Karnataka, India.

Symptoms

The leaves show characteristic continuous or discontinuous intraveinal clearing, stunting, resetting, loosening of leaf sheath and shredding of leaves. Leafy stems exhibit clear mottling. Clear light green patches with three shallow grooves are seen on immature capsules. Cracking of fruits and partial sterility of seeds are other associated symptoms. In summer the newly infected plants reveal only faint discontinuous vein clearing symptoms. Plants of all stages show the symptoms. New leaves get entangled in the older leaves and form hook like tiller (Venugopal, 2002).

Transmission and Spread

The disease is transmitted through banana aphid *Pentalonia nigronervosa* f. *caladii* in a semi persistent or persistent manner (Anad *et al.*, 1998). Incubation period ranges from 22-128 days and a single viruliferous aphid can transmit virus to plants of all stages. Planting symptomless virus carrying seedlings and clones was the main mode of introduction into plantations (Anon, 2003).

Cardamom Necrosis Disease (Nilgiri Necrosis Disease)

The disease was first observed in severe form in Nilgiris in Tamil Nadu and hence the name. Further the disease was observed in certain pockets in Kerala, Tamil Nadu and Karnataka (Venugopal 2002).

Symptoms

Young leaves show white, yellowish, continuous or broken streaks, proceeding from mid rib to leaf margin. In advanced stages of infection these streaks turn reddish brown. Often leaf shredding is noticed along these streaks. Leaves are reduced in size with distorted margins. Plants infected early, produce only a few panicle and capsule, and in advanced stages of infection, tillers are highly stunned and fail to bear panicles (Venugopal, 2002). All the types of cultivars are susceptible to the disease (Sridhar, 1988). Flexuous particles of 570-700nm long and 10-12nm broad belonging Carla virus group was found associated with the disease. (Naidu and Thomas, 1994).

Transmission and Spread

Spread of the disease is mainly internal and new infections occur in a centrifugal pattern from the source of inoculum. The rate of spread of the disease is very slow. No insect transmission of the disease from infected to healthy plants was recorded so far.

Infectious Variegation Virus

The disease was first noticed in Vandiperiyar area in Kerala in a severe form. Later it was also noticed in Coorg, Hassan and North Canara in Karnataka (Venugopal, 2002).

Symptoms

The infected plants show typical variegated symptoms on leaf with characteristic slender to broad radiating stripes of light and dark green on the lamina. Distortion of leaves, tillers and stunting are other common symptoms. Infected plants become unproductive with in he same year of infection (Venugopal, 2002).

Transmission and Spread

Only 2 per cent transmission was obtained through the aphid *Pentalonia nigronervosa* f. *caladii.*

Integrated Management of Viral Diseases

1. Use of virus-free planting material is the primary requirement to check the spread of the virus. Volunteers that sprout from remnants of infected plants are the potential primary source of spread (Naidu and Venugopal, 1987). Removal of infected volunteers in replanted area and total avoidance of volunteers for nursery activity in hotspots are most important for producing virus free planting material (Anon, 2003).
2. Restrict the movement of planting materials from infected regions to disease free regions.
3. Weeds and alternate hosts such as species of *Coloccasia*, *Caladium* etc. which might act as reservoir for the virus also needs to be removed from cardamom plantation to check the multiplication of the aphid.
5. Disease Resistance–IISR Vijetha which is a Katte tolerant cardamom line.

Diseases of Ginger (*Zingiber officinale* Rosc.)

Ginger (*Zingiber officinale* Rosc.) is an economically important rhizomatous spice crop susceptible to a number of diseases caused by fungi, bacteria and virus. Thirty nine micro organisms were reported to be associated with ginger causing one or other infections including storage rot. These diseases are prevalent in almost all ginger growing tracts of India. These diseases affect the crop at one or the other developmental stage resulting in huge losses to the growers and ginger industry as well. Mixed infection in rhizomes of ginger by one or the other pathogen has also been reported. Many of the diseases are either rhizome or soil borne or both

Soft Rot/Rhizome Rot

The disease has been reported from all the ginger growing states of India. Soft rot reduces the potential yield to a greater extent in the field and storage and market and may cause losses of even more than 50 per cent (Joshi and Sharma, 1980). During heavy infections, a crop loss around 90 per cent has been reported from Kerala (Rajan and Agnihotri, 1989). The disease is predisposed by heavy and continuous rain followed by water logging conditions existing in the ginger fields during the monsoon or post-monsoon season.

Symptoms

The disease is characterized by the appearance of water-soaked patches at the collar region of the pseudostem which extends to both sides and the collar becomes soft and watery leading to rotting of the pseudostem which fall off with a slight disturbance. All the under ground parts like roots, stem and newly emerging sprouts are affected. In grown up plants, the symptom appears as yellowing of the lower leaves which starts from the tip and spread downwards along the margin. The dead leaves droop and hang down the pseudo stem until the entire shoot becomes dry. Affected rhizomes turn brown and emit a foul smell characteristic of the infection. Rotten rhizomes harbor insect larvae (*Mimigralla coeruleifrons* larvae) which will be mistaken as the cause of the disease many times (Dohroo, 2005).

Pathogen

The disease is mainly caused by *Pythium* species of which *P. myriotylum* Drechsler is the predominant species (Anon, 2003). Other species of *Pythium* found associated with soft rot are *P. aphanidermatum* (Edson) Fitz., *P. butleri* Subram., *P. complectens, P. pleroticum* T.Ito, *P. vexans* deBary, *P. zingiberum, P.ultimum*, etc. (Dohroo, 2005). The primary source of infection is the oospore present in the diseased rhizomes or in the soil which germinates and infects the rhizomes or roots under favorable conditions. High soil water, high relative humidity and relatively low temperature favor the disease development and spread. Ginger planting mostly coincides with monsson rains, and during this time, the soil water and ambient temperature (25-30) become conducive to the onset of disease (Sarma, 1994).

Disease Management

Phytosanitation

Since infected rhizomes serve as the primary source of inoculum, selection of disease free healthy seed rhizomes collected from disease free ginger tracts is a prerequisite in the management of the disease (Dohroo, 1993).

Cultural Practices

Narrow ridge cultivation (Kim *et al.*, 1998) and mulching the ginger plots with neem leaves (*Melia azadirechta*) was recommended for soft rot (Das, 1999). Soil amendments using seed oil cakes of *Azadirachta indica, Calophyllum inophyllum* or *Pongamia glabra* are reported effective in reducing the rot incidence (Thakore *et al.*, 1987). Soil solarization helps in killing the pathogenic microflora present in the field. This can be done by covering the moist soil with transparent polythene film (300μ thick) during the months of March to May for about 45-60 days (Dohroo, 2005).

Chemical Control

As the disease is both seed- borne and soil borne in nature, the use of disinfested rhizomes followed by drenching the soil with a chemical is effective. Soil drenching with metalaxyll+ captafol three month after planting (Rathaiah, 1987) and seed treatment and soil drenching with metalaxyl formulations (Ramachandran *et al.*, 1989) are effective for the control of the disease. Planting of ginger under shade after treating with 1 per cent formaldehyde is another practice (Nath, 1993). Drenching the soil with zineb or mancozeb followed by rhizome treatment with carbendazim and incorporating

Thiodan dust into the soil also was found effective in managing the disease (Srivastava, 1994). In Himachal Pradesh rhizome treatment with mancozeb + carbendazim (0.25 per cent + 0.1 per cent) and soil application of phorate (12k/h) was found effective (Dohroo, 2005).Seed rhizomes should be treated with fungicides such as macozeb 0. 3 per cent or carbendazim 0.3 per cent before storing and also before sowing to prevent the seed borne inoculum reaching the soil or its multiplication inside the stored rhizomes. While planting care should be taken to avoid water logging by providing proper drainage channels since water logging predisposes the plant to infection by *Pythium* species.

Biological Control

Bio-agents *viz. T. harzianum, T. hamatum, T. virens* and bacterial isolates such as *Bacillus lentus* and *Pseudomonas fluorescens* along with VAM fungi glomus species have been found effective against the soft rot of ginger caused by *P.myriotylum* (Bhai *et al.*, 2005).

Bacterial Wilt

This is one of the most destructive diseases of ginger leading to heavy crop loss in all the ginger growing countries. Under conducive conditions a crop loss up to 100 per cent was reported in many ginger growing states in India (Dake, 1995).

Symptoms

Wilting and yellowing of the lower leaves which extends upward until all the leaves appear golden yellow is the first visible symptom of bacterial wilt. As the disease progresses, the pseudostem becomes water soaked and readily breaks off with a slight disturbance. The vascular tissue of the stem darkens and symptom progresses very rapidly until the plant collapses (Pegg *et al.*, 1974). When diseased rhizomes are cut, white milky exudates flows freely from the cut end indicating the presence of bacteria inside the vascular tissues. The Infected ginger plant become stunted and chlorotic and the lower leaves dry out gradually before the plant is killed.

Pathogen

Bacterial wilt of ginger is caused by the bacteria *Ralstonia solanacearum* biovar III (Smith) Yabuuchi. It is a vascular pathogen. The bacteria can survive in the soil for long periods in the absence of host plants and also on seed rhizomes and serve as the primary source of inoculum for field infection (Kumar and Hayward, 2005).

Integrated Disease Management

The disease management strategies (Kumar and Hayward, 2005) to be adopted for controlling bacterial wilt are as follows

1. Selection of healthy rhizomes material from disease free areas. The use of seed rhizomes collected from previously bacterial wilt infected areas will result in severe disease incidence.
2. Selection of field with no previous history of bacterial wilt.
3. Heat treatment of soil by solarization.
4. Treatment of rhizomes with streptomycin or oxytetracycline are found effective (Hartati and Supriadi, 1994). But large quantities of antibiotics are required for field application which may not be economical.
5. Clean cultivation and minimum tillage.
6. Crop rotation with non-host plants like cereals, paddy, maize etc.

7. Control of insect pest and nematode in the field
8. Soil amendments including biological control agents like *T. harzianum*.
9. Heat induction by hot air, hot water or rhizome solarization. Disinfection of seed rhizomes can be achieved by exposure to hot air for 30 min at 75 per cent RH until their centre temperature reaches 49°C (112 F). This resulted in minimal injury to the host without any adverse effect either on germination or subsequent growth. Such plants were also free from bacterial wilt under challenge inoculated conditions (Tsang and Shintaku, 1998). Soaking of ginger rhizomes in hot water (50°C) for 10 minutes is the usual preplant preparation in Hawaii for bacterial wilt management (Nishina *et al.*, 1992). Hot water disinfection using solar radiation is another economical method effective in killing the bacterial population residing inside the rhizomes (Kumar *et al.* 2003).

Ginger Yellows

The disease was first reported from Queensland and later from Hawaii and India (Haware and Joshi, 1973). It is a very serious stem rot disease that can devastate the ginger crop totally under severe infected conditions.

Symptoms

The disease appears as yellowing of the two margins of the lower leaves which gradually spreads over the entire lamina. The older leaves will dry up first followed by younger leaves. Plants may show premature drooping, wilting, yellowing and drying in patches or in the whole bed. The basal portion of the affected plants become soft and watery and fall of with a soft touch. Plants may show stunting also. Affected rhizomes show a cream/brown discoloration followed by shriveling. Central rot and root rot are also prominent. In final stages only fibrous tissue remains within the rhizome. A white cottony fungal growth also develops on the surface of the rhizomes.

Pathogen

F. oxysporum f. sp *zingiberi* was reported as the predominant pathogen though various other species were also reported. A temperature range of 15-30°C accompanied by very high humidity and the continuous presence of a free film of water is favorable for the disease development (Sharma and Jain, 1978b). The disease spread is through infected rhizomes and soil. The fungus survives in the soil as chlamydospores which may remain viable for many years in the field.

Disease Management

The main control measures are selection of healthy rhizomes, intercropping with non-host plants, followed by chemical application. Benzimidazole-type fungicides, soil fumigants and arasan as seed treatment were found effective in reducing the disease incidence. Seed treatment with 0.1 per cent $HgCl_2$ followed by soil drenching and rhizome dip with either tetramethylthiuram disulfide 0.5 per cent, ceresan wet 0.5 per cent or zineb (Kothari, 1966), macozeb and carbendazim along with biocontrol agents like *T. harzianum, T. hamatum* and *Gliocladium virens* as seed treatment and soil application are also effective (Dohroo, 1995). The disease can be managed to a greater extent using resistant or less susceptible cultivars like SG66 and Kerala Local (Dohroo, 1989a).

Phyllosticta Leaf Spot

The disease begins to appear toward the end of June when the plants are at the most susceptible stage of three-to four-leaf and have received high cumulative rainfall that is condusive for the disease spread and the temperature remains between 23.4° and 29.6 ° C and relative humidity between 80 and 90 per cent.

Symptoms

Small oval to elongated spots, measuring 1-10 mm x 0.5–4mm appear on younger leaves. The spots have white papery centres and dark brown margins surrounded by yellowish halos (Ramakrishnan, 1942). The spots later increase in size and coalesce to from larger lesions. The affected leaves become shredded and disfigured and may suffer extensive desiccation. As the plants putforth fresh leaves, they subsequently become infected. The crop attains a gray disheveled look as result of infection (Shukla and Haware, 1972).

Pathogen

The disease is caused by *Phyllosticta zingiberi* T.S Ramakr. The fungus forms amphigenus subglobose, dark brown ostiolate pycnidia measuring 78-150µm in diameter on the host (Ramakrishnan, 1942). The diseased leaves serve as a primary inoculum in the next season. Pycnidia survive in the leaf debris at the temperature range of 30 to 35°C which remains viable in soil even at 25 cm depth for 6 months. A higher intensity of rain accompanied by wind seems to exert a greater impact on leaves, as a result pycindia are splashed on more leaves and to greater distances, resulting in liberation of a greater amount of spores and spreading the disease incidence (Brahma and Nambiar, 1982, 1984).

Integrated Disease Management

Phytosanitation

The destruction of infected crop debris is an important practice to reduce the source of primary inoculum. Providing shade or ginger grown under shade is effective reducing the disease incidence.

Chemical Control

The disease is controlled by one or two applications of Bordeaux mixture 1 per cent (Ramakrishnan 1942), one spray of carbendazim (0.15 per cent) and two sprays of mancozeb (0.25 per cent) (Verma and Vyas, 1981), captan (0.3 per cent) and macozeb (0.35 per cent) + thiophanate methyl (0.01 per cent) were also found effective in the control of the disease (Das and Senapati, 1998).

Resistant Cultivars

Cultivars such as Narasapatom, Tura, Nadia, Tetraploid, and Thingpuri were moderately resistant (Setty *et al.* 1995b).

Helminthosporium Leaf Spot

Symptoms

The disease appears during the rainy season in the form of small oval spots measuring 4–5 x 2–3 mm in diameter on the leaves. The spots remain scattered on both surfaces of the leaves. The spots later enlarge in size. The mature spots have a centrally dead straw-colored area surrounded by a brown ring with a yellowish zone on the outside.

Pathogen

The disease is caused by *Helminthosporium maydi*.

Colletotrichum Leaf Spot

Symptoms

Symptoms manifest as small round to oval, light yellow spots on leaves and leaf sheaths. These spots gradually increase in size and often coalesce to form large discolored areas. Such infected areas often dry up at the center, forming holes. In the case of a severe attack, the entire leaf dries up.

Pathogen

The disease is caused by *C. zingiberi* (Sundar) (Butler and Bisby, 1931).The disease makes rapid progress during the rains. Continued wet weather with high temperature favors the disease development. The disease spread is less under shaded conditions.

Disease Management

Removal and burning of affected plants reduce the disease spread. Two sprays with Bordeaux mixture 1 per cent at an interval of 6 weeks effectively control the disease (Iyer, 1987). Growing ginger crop under shade is also effective in reducing the disease incidence (Briton-Jones, 1933).

Pyricularia Leaf Spot

Symptoms

Small pale brown spots develop on the leaves and on the collar portion which coalesce to form black blotches. Black blotches are also seen on the rhizomes, which may extend into the cortex. The pathogen also causes leaf blast symptom. The blast lesions occur during July to September (Kotani and Kurata, 1992).

Pathogen

The disease is caused by *Pyricularia zingiberi*. The fungus over winters with sclerotium-like structures on the residues of diseased ginger plants which forms the source of primary infection for ginger plants (Kotani, 1994).

Disease Management

Chemical Control

Soil fumigation with methyl bromideis effective to kill the over wintering sclerotium-like structure (Kotani, 1994). Foliar applications of mancozeb in combination with benomyl (Anon, 1977) or thiophanate-methyl (Anon, 1980) are also effective in reducing the infection caused by *P. zingiberi.*

Thread Blight

Symptoms

Symptom appeared as small water-soaked lesions on the leaf margins or other parts of the leaf during the initial stage of this disease. Later on, the infected leaves lose their turgidity, wilt and may get detached from the sheath. Fine hyphal threads spread over the infected parts with small brown sclerotia formation on the lower surface. The infected portion turns white and papery upon drying (Dohroo, 2005).

Pathogen

The causal organism is *Pellicularia filamentosa* (Pet.) Rogars (Sundram, 1954). The disease spreads rapidly in wet weather and is limited by low relative humidity.

Disease Management

Protective spraying with Bordeaux mixture 1 per cent or carbendazim 0.2 per cent before the start of heavy rains check the incidence of the disease.

Storage Rots

Ginger under storage is found affected by a number of pathogenic fungi such as *Fusarium oxysporum, P. deliense, P. myriotylum. Macrophomina phaseolina etc.* that result in rotting and decaying of the rhizomes (Dohroo, 1993).

Dry Rot

Dry rot is characterized by the black fungal growth inside the rhizomes which initiates from the cut end of the rhizome and extend to inner portions. Of late the rhizomes are invaded by mites resulting in total spoilage of the rhizomes. The disease is caused by *Macrophomina phaseolina* (Anon, 2003).

Eye Rot

Eye rot is characterized by the rotting of emerging sprouts of ginger. *The disease is incited by Fusarium oxysporum.* (Anon, 2003).

Disease Management

Seed treatment before storage for 30 minutes with a combination of mancozeb 0.3 and quinalphos 0.075 per cent is the recommended practice. A combination of Macozeb and Carbendazim treatment to rhizomes also controlled storage rot of ginger (*Dohroo et al.*, 1986a; Dohroo and Malhotra, 1995; Dohroo, 2000).

Viral Diseases

Mosaic Disease

Symptoms

The symptoms appear as a yellow and dark green mosaic pattern on leaves. The affected plants show stunting. The virus has spherical particles with a diameter of 23 to 38 nm. The virus is known to be transmitted by sap to different plants known to be hosts of CMV (So, 1980).

Chlorotic Fleck Virus

This disease was first described by Thomas (1986). The virus is detected in ginger imported into Australia from a number of countries. The geographical distribution of the virus is uncertain, but is thought to include India, Malaysia and Mauritius. The ginger chlorotic fleck virus (GCFV) has isometric particles approximately 30 nm in diameter.The virus is mechanically transmitted by *Myzus persicae, pentalonia nigronervosa, Rhopalosiphum maidis*, or *R. padi* (Thomas, 1986).

Big Bud

This disease is caused by the tomato big bud organism. Pegg *et al.* (1974) reported the disease in Queensland and suspected it to be a *Phytoplasma* disease. The affected plants cease to grow and leaves become bunched at the top of the stem. As the disease advances, plants turn yellow and die. The pathogen has a wide host range, and the disease is transmitted by leaf hoppers.

Disease Management

There are no chemical control measures for the viral or phytoplasma disease. However, controlling the aphid population by spraying insecticides is effective in controlling the spread of the diseae. Integrated disease management strategies described elsewhere for virus diseases is applicable in the case of ginger also.

Minor Diseases

Banded Leaf Blight

Rhizoctonia (*Corticium*) *solani* is associated with ginger leaves and pseudostems and rhizomes that are left too long in the ground. This fungus is also found prevalent on ginger in Himachal Pradesh. The fungus causes white spots on leaves and causes the above ground parts of plants to die prematurely (Deighton, 1936).

Disease Management

Rathaiah and Gogoi (2000) reported the effect of pigeon pea shade on the reduction of this disease. Under heavy shade, the disease intensity was about five times less than that in the open.

Basal Rot

Basal rot is caused by *Sclerotium rolfsii* (*C. rolfsii*) and is reported to occur in India (Haware and Joshi, 1973).

Violet Rot

Ito (1949) reported that the violet rot pathogen, *Helicobasidium mompa* Tanaka, infected ginger.

Black Rot

A dry rot of ginger rhizomes is of common occurrence in the Philippines and Jamaica. The causal organisms associated with the black rot disease have been reported to be *Rosellinia zingiberi* Stevens and *Atienza* sp. in the Philippines and *R. bunodes* in Jamaica (Smith, 1929).

Disease Management

The minor diseases occur very sporadically and will not cause much damage to the crop. However if infection is noticed, spray fungicides such as Bordeaux mixture 1 per cent or Mancozeb 0.2 per cent at 15 day interval twice to care of the diseases.

Plant Parasitic Nematodes

The most important nematode pests of ginger are *Meloidogyne* spp., *R. similis* and *P. coffeae*. Occurrence of *R. similis* along with *M. incognita, Pratylenchus* sp. and *Helicotylenchus* sp. has also been reported from India (Charles, 1978; Charles and Kuriyan, 1980).

Symptoms

The root-knot nematodes cause typical galling of roots. The fibrous roots were very much reduced. In addition, nematodes parasitizing the rhizomes produce galls in the outer tissues. Infested rhizomes have brown, water soaked areas in the outer tissues (Huang, 1966; Cheng and Tu, 1979). *R. similis* produce small, shallow, sunken, water soaked lesions on roots (Vilsoni *et al.*, 1976; Sundararaju *et al.*, 1979c). Heavily infested plants show external symptoms like stunting, poor tillering and chlorotic leaves with marginal necrosis. The affected ginger plants mature, dry faster and die prematurely than healthy ones, leaving a poor crop stand at harvest. Infested rhizomes serve as a source of infection and means of dissemination. *P. coffeae* is reported to cause 'ginger yellows' disease, prevalent in Himachal Pradesh and some North-Eastern States (Kaur and Sharma, 1990). The nematode infestation caused yellowing of leaves and dry rot like symptoms with dark, brown necrotic lesions on rhizomes. The dry rot symptoms caused by *Pratylenchus* spp. are indistinguishable from those caused by some fungi.

M. incognita is widely distributed in ginger fields in Kerala and causes a loss of upto 46.4 per cent (Charles, 1978). Incidence of rhizome rot of ginger caused by *Pythium aphanidermatum* is reported to be severe when rhizomes are infested with *M. incognita* and/or *P. coffeae* (Dohroo *et al.*, 1987). Recent studies have shown that, ginger plants inoculated with root knot nematodes developed disease symptoms early on inoculating with *P. aphanidermatum* (Ramana *et al.*, 1998).

Management

Preventive

Excellent reduction in populations of plant parasitic nematodes can be achieved through use of various heat treatments like steam, hot water and sunlight. Seed materials collected from healthy fields and treating the rhizomes in hot water at 50-55°C for 10 minutes were found to reduce the nematode incidence (Colbran and Davis, 1969; Anonymous, 1971).

Cropping Practices

In ginger, the control schedule for *M. javanica* involving the use of clean seed and a ginger-taro-fallow rotation has been recommended in Fiji (Haynes *et al.*, 1973).

Soil Amendments

Reduced nematode damage from increased organic matter in soil is likely a combination of improved soil structure and fertility, alteration of the level of plant resistance, release of nemato-toxins, or increased populations of fungal and bacterial parasites and other nematode-antagonistic agents. Mulching or applying well decomposed cattle manure or poultry manure or compost or neem oil cake reduced nematode build up in ginger (Colbran, 1974; Kaur, 1987; Mohanty *et al.*, 1992, 1995; Dohroo *et al.*, 1994; Vadhera *et al.*, 1998b).

Resistant Lines

IISR Mahima, a variety released from Indian Institute of Spices Research, Calicut is found resistant to *M. incognita* (Sasikumar *et al.*, 2003).

Chemical Control

Soil fumigation or application of granular pesticides like fenamiphos/carbofuran or dip treatment with fenamiphos is all recommended for control of nematodes of ginger (Colbran, 1972; Willers, 1985; Kaur, 1987; Mohanty *et al.*, 1995). Integrated management of the plant parasitic nematodes has been successful with combined application of carbofuran and neem oil cake (Makhnotra *et al.*, 1997; Mohanty *et al.*, 1992; Mohanty *et al.*, 1995; Ray *et al.*, 1995; Sheela *et al.*, 1995).

Biological Control

Among the ten fungi evaluated, *P. chlamydosporia* was superior in suppressing root-knot nematodes and increasing ginger yield (Eapen *et al.* 2008).

Diseases of Turmeric (*Curcuma longa* L.)

Turmeric (*Curcuma longa* L.), a zingiberacious member, is one of the most important commercial crop known for its aromatic and medicinal properties. The crop is susceptible to a number of fungal diseases. Bacterial and viral diseases are not so common and are of minor importance. The important fungal diseases affecting the crop are Rhizome rot, leaf blotch and leaf spots (Nageshwar Rao, 1995).

Rhizome Rot

It is the most destructive of all the diseases affecting turmeric.The disease occurs during the South West monsoon period when the crop is in the active growing stage. Water logged conditions predisposes the plant to infection.

Symptoms

The disease is characterized by the appearance of water soaked lesions at the base of the pseudo stem and yellowing of the lower leaves. The root infection is manifested as browning and rotting of the roots which later advances into the rhizomes turning them soft and rotten. The color of the rhizome changes from bright orange to different shades of brown. The infection gradually spreads to all the fingers and mother rhizomes and eventually the plants die.

Pathogen

Different species of *Pythium viz. P. myriotylum, P. graminicolum, P. aphanidermatum etc* of which *P. aphanidermatum* are found as the predominant pathogen (Anoop *et al.*, unpublished). The pathogen is soil and seed borne in nature. Infected rhizomes can harbor the maggots of *Mimegrella coerulifrons* which in certain cases is suspected to be the cause of the disease (Premkumar *et al.*, 1982).

Disease Management

Dipping rhizomes in a combination of fungicide and pesticide solution greatly prevent the incidence of seed borne infection. This is usually done just before storage of the rhizomes after harvest and just before sowing of the rhizomes. Treated rhizomes may be air dried before storage. Rhizome treatment with a combination of macozeb (0.25 per cent) and quinalphos 0.075 per cent for 15 minutes is the recommended practice. Soil drenching with metalaxyl–mancozeb (0.2 per cent) or macozeb (0.25 per cent) at 15-20 day interval twice with the first appearance of the symptom is effective in managing the disease (Sharma and Dohroo, 1982). Crop rotation is also recommended to reduce the incidence of rhizome rot.

Disease Resistance

PCT-13 and PCT-14 from Andhra Pradesh and Ca 69 and Shillong from Assam were found to be tolerant to rhizome rot. (Rathaiah, 1982).

Taphrina Leaf Spot/Leaf Blotch

Symptoms

The disease is characterized by the appearance of small scattered oily looking translucent spots on the lower leaves when the plants are in 3-4 leaf stage. The leaf spots later turn dirty yellow and deepen to colour of gold and some times to bay shade. The adjacent individual leaf spots of 1-2mm in diameter coalesce forming reddish brown blotches leading to varying degrees of leaf blight. In severe case of attack, hundreds of spots appear on both the sides of leaves.

Pathogen

The disease is caused by *Taphrina maculans*. Moist cloudy weather with temperature of 25-30°C during August-September months predisposes the plant to infection. The disease incidence is influenced by soil borne inoculum. Primary infection can also occur on the lower leaves during October-November months when the temperature falls to 21-23°C with RH80 per cent. The secondary infection is by ascospores discharged from successively maturing asci which grow into octosporus microcolonies and infect fresh leaves with out any dormancy.

The disease perpetuates from one season to other through viable ascogènous cells borne on the infected leaf debris in the field after harvest as well as through desiccated ascospores and blastospores ejected from mature asci during the crops season and over summering in the soil and leaf trash. *Curcuma amada, C. angustifolia, Zingiber cassumnar, Z. zerumbet* and *Hedychium* sp. are alternate hosts of *T.maculans.*

Disease Management

Crop Rotation

Since the infected plant material serve as the perpetuating media for the fungus, crop rotation is important to reduce the inoculum build up in the soil.

Phytosanitation

Cutting and burning of diseased leaves would greatly help in preventing the spread of the disease.

Chemical Control

Foliar spray using fungicides such as zineb (0. 2 per cent), mancozeb, blitox 50, bavistin and cuman L etc. are effective in controlling the disease (Srivastava and Gupta, 1977).

Disease Resitance

All short duration cultivars are found resistant to *T. maculans.* The varieties such as CLL 324, Amalapuram, Mydukur, Karhadi local, CLL 326, Ochira 24 and Alleppey among *C. longa* group and ca 68, ca 67, Dahgi and Kasthuri among *C. aromatica* types were resistant to infection (Nambiar *et al.*, 1977)

Colletotrichum Leaf Spot

The disease is prevalent in turmeric growing areas of South Gujarat. The disease appears usually during South West monsoon period in the month of September when there is high and continuous humidity in the atmosphere.

Symptoms

The disease appears in the form of elliptic or oblong spots of variable size. In the initial stages of infection the spots appear small and measure 1.5-2.0 inches in length and 1.0-1.5 inches in breadth and later they may increase in size. In advanced stages, two or more spots coalesce and develop into irregular patches occupying a major portion of the leaf which eventually dries up. The centre of the spot is grayish white and thin with numerous black dot like acervuli springing on both surfaces. These are arranged in concentric rings.

Pathogen

Leaf spot disease of turmeric is caused by *C. capsici*. The pathogen survives on infected leaf debris for one year, which forms the source of primary infection.

Integrated Disease Management

Cultural Practice

Application of chemical fertilizers (Nitrogen @120kg/ha, Potash @ 70-120/kg/ha) is found to be effective in reducing the disease incidence irrespective of the split doses applied.

Chemical Control

The disease can be managed by spraying 1 per cent Bordeaux mixture during August, before the appearance of the disease. Two sprays of carbendazim (0.2 per cent) at disease initiation and at 15 days interval twice, or four sprays of Captan (0.2 per cent) or zineb at monthly intervals or six sprays of mancozeb @ 0.25 per cent at 15 days interval was also recommended for the control of the disease (Thamburaj, 1991).

Disease Resistance

Long duration types are highly susceptible to the disease. The varieties Nallakatla, Sugandham, Duvvur and Gandikota were resistant to leaf spots. 'Mannuthy local' showed high degree of field tolerance.

Leaf Blight

The disease occurs during the warm and humid weather.

Symptoms

The disease is manifested as water soaked spots of varying shapes and sizes on the lower leaves which gradually increase in size resulting in blightening of entire leaf area. The blighted leaf area is divided or banded into various sectors which are characteristic symptoms for early diagnosis. The disease is caused by *Corticium sasaki*. The fungus thrives well in a wide range of temperature varying from 15- 30°C.

Leaf Spots

Turmeric is also susceptible to leaf spots caused by various other species of fungi *viz. Phaeodactylium alpiniae, Thirumalacharia curcumae*(Rathaiah, 1980a)., *Phyllosticta zingiberii, Phaeorobillarda curcumae, Cercospora curcumae-longae, Myrothecium roridum Toxde Ex.Fr., Pyricularia curcumae* and *Pestalotiopsis* sp. These diseases are of sporadic occurrence and noticed only in few areas and have not of much importance.

Disease Management

In general, the leaf spots of turmeric can be controlled by foliar sprays with fungicides such as Macozeb (0.2 per cent), Captan, Zineb or Bavistin at 15 days interval twice with the first appearance of the symptoms.

Storage Rot

Storage rot is serious problem in turmeric. Improper storage as well as heaping harvested rhizomes under sun will lead to rotting of the rhizomes. Many species of fungi are found associated with rotting of the rhizomes. Favourable incubation temp. and relative humidity at 60 per cent leads to maximum spoilage. But no rotting occurs at 15°C even when the RH varied from 30 to 90 per cent. Storage rot of rhizome of turmeric is more than 60 per cent in some disrict of Andhra Pradesh.

Pathogen

The most virulent and widespread of the fungi causing storage rot of *C. longa* are *Macrophomina phaseolina* and *Cladosporium cladosporioides*. Species of Aspergillus, *Drechsleza rostrata, Fusarium, Rhizoctonia* and *S. rolfsii* were also reported to be associated with storage rot.

Disease Management

Storage rot can be controlled by treating the rhizomes with ceresan wet or macozeb at 0.25 per cent concentration. After seed treatment with fungicide, the rhizomes should be well air dried before storage.

Plant Parasitic Nematode

Among the 35 nematode species reported infesting turmeric, *Meloidogyne* spp., *R. similis* and *Pratylenchus coffeae* are of economic importance. Two species of root-knot nematodes, *M. incognita* and *M. javanica* have also been reported to be prevalent in India.

Symptoms

Turmeric plants infested with *M. incognita* have stunted growth, yellowing, marginal and tip drying of leaves and reduced tillering with galling and rotting of roots. In the field, higher populations of *M. incognita* cause yellowing and severe stunting and withering in large patches. Plants die prematurely leaving a poor crop stand at harvest. Infested rhizomes tend to lose their bright yellow colour (Mani *et al.*, 1987). The population density of *M. incognita* increases with crop age and decreased with crop senescence (Poornima and Sivagami, 1999). An avoidable yield loss of 33.6 per cent was reported in turmeric due to root-knot nematode infestation (Ray *et al.*, 1995).

Roots of turmeric damaged by *R. similis* become rotted and most of these decayed roots retain only the epidermis devoid of cortex and stelar portions. The infested plants show a tendency to age and dry faster than healthy plants. Infested rhizomes are of a yolk yellow color compared with the golden yellow color of healthy rhizomes and have shallow water-soaked brownish areas on the surface. The scale leaves also harbour *R. similis* (Sosamma *et al.*, 1979).

Brown Rot

The disease was noticed in the collection of *C. aromatica* and *C. longa*. *Fusarium sp.* and *Pratylenchus* sp.are reported as the causal organisms.

Symptoms

The disease is characterized by discoloration and rotting of mature rhizome. During the early stages of infection, rhizomes are dull in color. In the advanced stage, the rhizome became deep grey to dark brown in color, less turgid, weight loss, wrinkled and exhibit dry rot symptoms. The fungus is more severely affected the mother rhizomes. On cutting open affected rhizomes show dark brown necrotic lesion starting from the margin in to the internal tissues initial stages, there necrotic lesions are localized, remain discontinuous and extend to a depth of 2-5 mm. Al later stages, the lesions coalesce to form larger necrotic areas and progressively extend over a major portion of the rhizome.

Integrated Management

The management practices recommended for ginger *viz.* preventive treatments, cropping practices, amendment of the soil with organic amendments and oil cakes and chemical control measures are as such applicable for turmeric also, because all the pathogens and pests are similar in the crops as well as their similar rhizomatous nature and belonging to the same family.

Chemical Control

In turmeric, significant yield increase had been noticed on applying granular nematicides like aldicarb, carbofuran, phenamiphos or phorate.

Biological Control

Application of antagonistic fungi increased the yield by 12.5 to 40.9 per cent. Several rhizobacteria were also evaluated for nematode suppression in turmeric fields (Eapen *et al.* unpublished)

Diseases of Vanilla (*Vanilla planifolia* Andrews)

Vanilla (*Vanilla planifolia* Andrews syn. *V. fragrans*), a native to the humid tropical rain forests, is now widely cultivated in India for its aromatic vanillin. It is grown as an intercrop along with arecanut, coconut, coffee and black pepper, in homestead gardens and as a pure crop in standards such as *Glyricidia, Plumeria, Dadaps* and concrete or stone poles. During initial years of cultivation diseases are not common. But as the extent of area under cultivation is increased, the crop become susceptible to a number of fungal pathogens, some of which resulted in total destruction of vines.

The pathogens affect almost all the plant parts like roots, stem, leaves and beans and occasionally inflorescence also. Fungal infections often lead to rotting of affected plant parts or wilting of the entire vine as in the case of stem rot or root diseases. Species of *Phytophthora, Fusarium, Sclerotium, Calospora* and *Colletotrichum* are the commonly occurring pathogens which cause serious damages. Generally, fungal diseases are wide spread in closely planted gardens having thick shade, intensive management with manures, frequent irrigation and where no phytosanitory measures are adopted. Viral diseases are more common. While some of the viral infections result in crop loss, others express only mild symptoms or remain as symptomless in the plant. No bacterial disease has so far been reported from any of the producing countries.

Stem Rot

Stem rot incidence is noticed mostly during monsoon seasons

Symptoms

The disease appears in the form of water soaked lesions extending to both sides of the stem giving a brown coloured appearance. Late, elongated patches develop on the stem which result in rotting of tissues. In advanced stages, the leaves turn yellowish and later dry off. When the stem in the basal or middle portions of the vines are infected and shriveled, the remaining distal portions show wilting symptoms. If sufficient number of aerial roots are not fixed in the soil, such vines perish off. In certain cases, the stem infection is followed by rotting of beans.

Pathogen

F. oxysporum f.sp. *vanillae* is the causal organisim (Tombe, *et al.*, 1992).

Disease Management

The disease can be controlled by spraying 0.2 per cent macozeb together with carbendazim (Tomb and Sitepu, 1986).

Stem Blight

Symptoms

Infection on the stem is often observed in the form of brown coloured blighted portions, sometimes extending several centimeters along the stem. The affected portion gradually shrinks, the leaves become yellowish and the vines dry off.

Pathogen

The disease is caused by P. *meadii* (Bhai and Thomas, 2000). *P. palmivora, P.capsici,* and *P.parasitica* were also reported from French Polynesia (Tsao and Mu, 1987).

Disease Management

The spread of the disease can be controlled by spraying with 1 per cent Bordeaux mixture or Potassium phosphonate (0.4 per cent).

Root Rot

Symptoms

The infection starts in the form of browning at the root tips, which later on decay and result in the death of the underground roots. The infection gradually spreads to the aerial roots also. The affected plants produce numerous aerial roots, but most of these die before they reach the soil. As a result of infection, stem and leaves become flaccid, stem shrivels and the vines give a drooping appearance. The symptoms become more conspicuous during dry periods especially in gardens with excess sunlight. During this period the vines become less capable of withstanding the infection due to lack of adequate uptake of water and minerals.

Pathogen

Root rot is caused by F. batatis Wollen var. vanillae. Tsao and Mu (1987) reported the incidence of root rot of vanilla along with leaf blight in French Polynesia where *Phytophthora* sp. was identified as the causal organism. Another type of root rot and vine death caused by *Sclerotium rolfsii* was also observed in certain vanilla gardens of India (Thomas and Bhai, 2002).

Disease Management

Cultural Practices

Providing adequate drainage, shade regulation to avoid excessive exposure, mulching the plant base, irrigation during drought period etc. are recommended for the management of this disease. Excess bearing of fruits due to over pollination was found to aggravate the disease.

Chemical Control

Soil drenching with 0.2 per cent carbendazim or Carbendazim+ Mancozeb(0.2 per cent),. Carbendazim+ Mancozeb(0.2 per cent)+ Neem oil 0.5 per cent,Thiophanate emthyl 0.2 per cent were found effective in reducing the disease incidence (Anon, 2005).

Biological Control

A bioconsortia containing bacterial antagonists suchs as *Pseudomonas fluorescens* and *Bacillus* species were, and fungal antagonist *Paecilomyces lilacinus* were found effective against root rot infection caused by *Fusarium oxysporum* f.sp. *vanillae* (Fov) (Bhai *et al.*, 2008–in press).

Bean Rot

Symptoms

Initially, rotting symptoms develop at the tips of beans, spreads to the whole bunch. Affected beans are soft, brown coloured and show abundant mycelial growth. In later stages, the disease advances to the stem, leaves, aerial roots and extend towards the basal portions. The whole vine shows decaying symptoms and the entire vine perishes.

Pathogen

The disease is caused by *P. meadii* (Bhai andThomas, 2000). Excess shade in the plantations, continuous rains, crowding of vines, water logged conditions and presence of *Phytophthora* inoculum in the field are the pre-disposing factors for the advancement of this disease.

Disease Management

Phytosanitation and spraying 1 per cent Bordeaux mixture or foliar spray with 0.4 per cent potassium phosphonate is effective in preventing the spread of the disease.

Sclerotium Bean Rot

Sclerotium bean rot is reported for the first time in India from some isolated vanilla gardens (Thomas and Bhai, 2000). This is observed in a few cases where the bunch of beans hang and come into contact with infected soil.

Symptoms

The symptoms appear as rotting of bean tips, which proceed toward pedicel regions. Affected beans show thick white coloured mats of fungal mycelium forming a mantle around the fruit bunch. Protruding hyphal strands grow abundantly on bean surface. On the leaves, stem and beans, white threads of fungal hyphae run longitudinally. Occasionally reddish brown sunken lesions are also seen on affected plant parts.

Pathogen

The disease is caused by *Sclerotium rolfsii.*

Disease Management

The disease can be managed by soil drenching and spraying with 0.2 per cent carbendazim.

Shoot Tip Rot

The disease is usually observed during post-monsoon periods of September to December.

Symptoms

The disease start as rotting in the form of brown patches on the petiole and lower portions of the youngest leaf, which is just unfolding. At this stage, the leaf is usually a funnel shaped receptacle like structure which can hold rain water or dew making congenial condition for rotting. Within a few days, the rotting extends to the affected leaves and the shoot tip becomes soft and changes to brown color and later fall off. In dry weather, as the moisture is a limiting factor, instead of soft decay, the younger most leaf and the shoot tip turn brownish black leading to die back.

Pathogen

The disease is mostly caused by *F. oxysporum. C. gloeosporioides* and *P. meadii* were also found associated with shoot tip rot disease (Bhai unpublished).

Disease Management

The disease can be managed by spraying with 0.2 per cent carbendazim.

Leaf Spots/Leaf Rot

Occurrence of leaf spot is common in almost all vanilla gardens but is considered as a minor disease as this does not cause appreciable economic loss.

Symptoms

The mature leaves show round to oval shaped spots of varying sizes. These lesions vary from one to several millimeters in diameter; appear leathery, hard and often reddish brown in color. During rainy weather, adjacent lesions coalesce together to form large areas of decay.

Pathogen

The disease is caused by *C.vanillae.*

Disease Management

The leaf spot/rot can be controlled by spraying with 0.2 per cent mancozeb or 0.2 per cent carbendazim or a mixture of both.

Inflorescence Rot

Symptoms

The symptoms start in the form of development of brown color on the flower stalks during the elongation period of the inflorescence. This discoloration gradually spreads to the flower buds and results in the shedding of the affected flower buds.

Pathogen

The casual organism was found to be *Colletotrichum* sp.

Disease Management

Spraying the plants with 0.2 per cent carbendazim is effective in checking the disease.

Bean Yellowing

The disease appears on immature beans of about five to seven months age, often during August to October months.

Symptoms

The symptoms appear in the form of development of yellow colour at the tips of the beans which slowly extends towards the pedicel portions. As this proceeds, splitting of beans starts at the tip followed by brown colouration of the affected portions. Fully affected beans fall off from the peduncle before reaching maturity. Some times the brown colouration develops into rotting of bean tip, basal portion or even the entire bean. The disease is not observed on mature beans.

Pathogen

The disease is due to *Fusarium oxysporum.*

Disease Management

The disease can be controlled by spraying 0.2 per cent thiophanate methyl at 15 day intervals for two to three times. Spraying a mixture of carbendazim and macozeb at 0.2 per cent is also effective in controlling the disease.

Premature Yellowing and Bean Shedding

Symptoms

The disease is characterized by dropping off of dried corolla from the tip of immature beans which otherwise will be attached to the beans till half way through maturity. As the corolla drop off, the exudate from the beans accumulates at the tip, and the beans turn yellow and fall off. The bean yellowing is followed by dark brown discoloration of the beans from the tip upwards. Fungal mycelium is found growing at the tip of the beans. The disease is found associated with high temperature (32.57-35.57°C) and very low relative humidity (45-63.57 per cent) prevailed during the months of February-April The yellowing and sudden dropping of beans were severe with the sudden fall in of temperature with the first summer showers. The disease was found negligible in high altitudes where temperature and humidity are maintained under forest cover (Bhai *et al.*, 2006).

Pathogen

The disease is found associated with *C.vanillae* (Bhai *et al.*, 2006).

Disease Management

Intensity of high temperature and low humidity can be controlled by providing 50 per cent shade in the plantation followed by mist irrigation for at least 4-6 h during pollination till the onset of pre monsoon showers in order to maintain a relative humidity of more than 70 per cent. Overcrowding of beans could be avoided by restricting the pollination to 15-18 flowers/inflorescence. Spraying rogor or Quinalphos @ 0.05 per cent during flowering period thrice at 15-20 days interval prevents the insect attack. Spraying fungicides such as thiophanate methyl 0.2 per cent or Carbendazim-mancozeb (0.25 per cent) at 15-20 days interval with the initiation of flowering prevents fungal infection (*Bhai et al.*, 2006).

Horse Hair Blight

Symptoms

A fungal infection is sometimes noticed on leaves in the form of hyphal strands spreading along the leaf surface. It starts near the petiole portion as whitish to gray coloured fungal strands forming a network advancing towards the tip portion covering the entire lamina. In later stages, the affected leaves start drying and give a blighted appearance.

Pathogen

The horse hair fungus, *Marasmius* sp. was associated with the disease.

Anthracnose

Symptoms

The first symptoms are seen on tender parts such as immature beans and the tip of aerial roots. On the beans the symptoms appear as extremely minute dull red or amber coloured pustules springing in small groups from discoloured patches. They turn black at the tip and the middle portion and later wilt and fall off within two or three days. As the disease progresses the leaves start yellowing and drying resulting in their death. The pathogen infects almost all the plant parts like young shoot tips, leaves, roots and beans Excess soil moisture, poor drainage, and prolonged rainy weather and over crowding of vines are the major predisposing factors (Purseglove *et al.*, 1988).

Pathogen

Anthracnose disease is caused by the fungal pathogen *Calospora vanillae* Massae which belong to Perenosporales.

Disease Management

The diseased and dead leaves, beans and aerial roots lying on the ground are the source of fresh inoculum for the repeated infection. Therefore proper phytosanitation by collection and burning of affected parts is necessary to prevent further spread of the disease.

Black Crust Disease

The occurrence of this disease is reported only in very rare cases from vanilla gardens.

Symptoms

The disease is characterized by the formation of black lesions on the stem, leaves and fruits.

Pathogen

It is caused by *Mycoleptodescus indicus* (Bezerra and Ram, 1986).

Disease Management

Treating mould affected beans with 95 percent ethyl alcohol was recommended to remove the fungus (Chalot and Bernard, 1920). However, the treatment causes fading of the original chocolate color of the beans. In severe infection, the affected portions are cut away and discarded.

Viral Diseases

Viral diseases are of common occurrence in all countries including India where vanilla is grown on a commercial scale. The crop is affected by five distinct types of viruses (Bhat *et al.*, 2004 c). In India though the commercial cultivation of vanilla started only very recently, the occurrence of viral diseases are on the higher side.

Vanilla Mosaic Virus (VMV)

Symptoms

Mosaic disease is characterized by the appearance of mosaic or mild mottling with chlorotic specks or streaks mainly on the leaves and rarely on stems followed by stunting, sterility and leaf distortions. This has been reported by Wisler *et al*,. (1987) from French Polynesia.

The viral particles are flexuous rods of 767 nm x 11 nm size. Electron microscopic study (Anon, 2003) revealed the presence of four kinds of virus particles with mosaic affected plants. Three kinds are of flexuous particles resembled *Potexvirus* (particle length of 500nm), *Potyvirus* (particle length of about 660nm) and *closterovirus* (particle length of about 1000nm) and fourth one is an isometric particle(of about 28nm).

Transmission and Spread

The virus is sap transmissible and produces local lesions in inoculated leaves of *Nicotianae tabacum*. VMV is sap transmissible and spreads through stem cuttings used for propagation. Insect vectors such as *Myzus persicae* also transmit the disease in a non-persistent manner.

Host Range

Host range studies have shown that Vanilla Mosaic Virus (VMV) is restricted to vanilla genus only (Wang and Pearson, 1992). The disease is caused by *Vanilla Mosaic Virus* (VMV) of the Potyvirus group.

Cucumber Mosaic Virus (CMV)

Symptoms

The disease is expressed as mosaic, leaf distortion and stunting of vanilla. The infection on vanilla can be detected through Double antibody sandwich-enzyme linked immunosorbent assay (DAS-ELISA). The virus is infective to members of Cucurbitaceae, Chenopodiaceae, Fabaceae and Solanaceae(Madhubala *et al.*, 2005).

Cymbidium Mosaic Virus (CyMV)

Symptoms

CyMV is characterized by mild mottle or mild chlorotic streaks in the leaves of V.*fragrans* and V. *tahitensis.* In most cases the vanilla plants infected with this virus appear symptom less (Pearson *et al.*, 1993). The virus consists of flexuous rods of size 480 nm X 13 nm. This virus belongs to 'Potex' group (Bhat *et al.*, 2006a).

Transmission and Spread

The virus is sap transmissible and also spreads through stem cuttings.

Vanilla Necrosis Poty Virus (VNPV)

Symptoms

The disease is characterized by the development of symptoms in young leaves in the form of diffuse chlorotic patches followed by leaf distortion. Necrotic lesions are seen on older leaves and stems. In advanced stages, necrosis extends resulting in the death of affected vines.

Transmission and Spread

VNPV is sap transmissible and can also spread through stem cuttings used for propagation. The virus can be easily detected using a virus specific polyclonal antiserum or a Poty virus group monoclonal antibody.

Stem Necrosis

Symptoms

The disease is characterized by the appearance of brown shriveled necrotic patches on the stem having a length ranging from a few to several millimeters. The disease initiates as a necrotic spot which enlarge and encircles the stem leading to the death of tissues which in turn affects the movement of water and other nutrients and result in the death of the vines. This is an important and fast spreading viral disease posing serious threat to vanilla cultivation. A flexuous particle belonging to Potyvirus genus was found predominant with stem necrosis. Bean common mosaic virus (bean yellow mosaic virus) is one of the virus associated with stem necrosis in vanilla(Bhat *et al.*, 2004 a, Anon, 2006).

Odentoglossum Ring Spot Virus (ORSV)

Symptoms

Vanilla plants infected with ORSV are usually symptomless. Occasionally mild mottling was observed on leaves of V.*fragrans* and V.*tahitensis*. The virus particle consists of straight rods of about 300 nm x 18 nm size. ORSV belong to tobamovirus group. The virus is sap transmissible. Spread of the virus is through stem cuttings used as planting material.

Uncharacterized Poty Virus

Symptoms

The viral infection is manifested as mosaic with sunken chlorotic patches on the foliage together with distortions of leaf margins. The virus particles consist of flexuous filaments of about 760-800nm long. This virus belongs to Poty virus group and can be indexed through ELISA. The mode of transmission is not yet known. Pearson (1990) has reported the occurrence of a viral disease in vanilla from Fiji and Vanuatu.

Integrated Disease Management

There are no chemical control measures as such for any viral infections on plants. Spread can be prevented by adopting plant protection measures as described below.

1. Use of virus-free planting material is the first and foremost requirement to check the spread of the virus.
2. If tissue culture raised plants are used, it is important to check for the presence of virus in the mother plant.
3. Regular disease surveys should be carried out to identify the plants showing symptoms of viral infection. Such infected vines should be removed and destroyed in order to avoid further disease spread.
4. Weed and crop hosts which might act, as reservoir for the virus also needs to be removed. The removed plants may be burnt or buried deep in the soil.
5. Physical contact of infected vines with healthy ones should be avoided to eliminate mechanical transmission of the disease.
6. Insects such as aphids and others may act as vectors for the different viruses. These insects whenever noticed on vanilla plants may be controlled with insecticide spray.
7. Movement of planting materials from infected regions to disease free regions should be avoided.

Guidelines were already formulated for the collection, transportation and introduction of new accessions, germplasm or planting materials from one locality to another for avoiding viral disease (Pearson *et al.*, 1991). *In vitro* germination of vanilla seeds helps in eliminating all types of pathogens including viruses. It is important that the original plantlets and the first generation subcultures should be tested since *in vitro* cultures their presence may not be detected as some viruses occur in very low concentrations. Germplasm should be tested for viral infections either in the country of origin, in quarantine station, or in the recipient country before it is taken to the planting site.

Storage Infections

Mould infections are of common occurrence in cured vanilla beans. Normally these appear on improperly cured or stored beans. Bouriquet (1954) reports infection of cured beans by species of *Penicillium, Aspergillus* and *Rhizopus* during the conditioning and storage period. Infection begins at the pedicel end of the beans. The beans become wrinkled and develop disagreeable odors. Harvesting immature beans, improper killing, lack of uniform drying and excessive moisture content are the predisposing factors for the incidence of molds in cured beans.

Conclusion

Integrated disease program is aimed against all diseases affecting a crop. However, if a specific disease constitutes the major threat to the crop, then an integrated control programme is directed against that disease. The control of the disease is most successful and economical when all available pertinent information regarding the crop, its pathogen, the history of disease in previous years, the environmental conditions expected to prevail, locality, availability of materials, land labor and costs is taken into account in developing the control programme. The main goals of an integrated disease management programme are to 1) eliminate or reduce the initial inoculum, 2) reduce the effectiveness of the initial inoculum 3) increase the resistance of the host 4) delay the onset of the disease and 5) slow the secondary cycles.

If the disease occur regularly year after year the grower in addition to using resistant varieties, should start spraying with the appropriate fungicides as soon as the disease appears or even before, and should continue the sprays. Insecticide sprays control insects and many reduce the spread of some persistently transmitted viruses. Thus in an integrated control programme, several control methods are employed, including regulatory inspections for healthy seed or nursery crop production, cultural practices, biological control, and chemical control.

Use of planting materials without symptoms of nematode infestation from known and nematode-free sources is very important aspect of successful management of the nematodes. Exclusion of nematode susceptible intercrops, shade trees, supports or standards can also minimise the nematode infestation. Soil amendments like application of organic cakes, mulching etc. are ideal eco-friendly measures to manage these pests. Efforts to identify nematode resistance have helped to recognise a few resistant lines in some spices. Chemical control of nematodes is a less preferred option now a days and as a result intensive research is on at several laboratories to isolate efficient biocontrol agents against these nematodes. Tropical agro-ecosystems are complex, more resilient and therefore, sustainable ways of nematode management have to be evolved by blending different management options.

References

Agnihothrudu, V., 1968. Description of the fungus *Phaeodactylium venketesanum* on cardamom. *Proc. Indian Acad. Sci. Sect.* B., 68, 206-209.

Ali, S.S., 1984. Occurrence of root knot nematodes in cardamom plantations of Tamil Nadu. Pp. 615-620. In: *Proc. PLACROSYM-V 1982.* Indian Society for Plantation Crops, Kasaragod, India.

Ali, S. S., 1985. Preliminary observations on the effect of some systemic nematicides and neem oil cakes in cardamom field infested with root-knot nematodes. Pp. 215-223. In: *Proc. PLACROSYM–VI 1984.* (Ed. M.R. Sethuraj). Indian Society for Plantation Crops, Kasaragod, Kerala, India.

Ali, S.S., 1986a. Occurrence of root knot nematodes in cardamom plantations of Karnataka. *Indian J. Nematol.* 16: 269-270.

Ali, S.S., 1986b. Evaluation of nemacur against *Meloidogyne incognita* in a cardamom nursery. *Indian J. Nematol.* 16: 48-50.

Ali, S.S., 1987a. Preliminary observations on the effect of some systemic nematicides and neem oil cakes in a cardamom field infested with root knot nematodes. Pp. 215-223. In: *Proc. PLACROSYM-VI 1984*, Indian Society for Plantation Crops, Kasaragod, India.

Ali, S.S., 1987b. Effect of three systemic nematicides against root knot nematodes in a cardamom nursery. *Nematol. medit.* 15: 155-158.

Ali, S.S. and Koshy, P.K., 1982a. Occurrence of root knot nematodes in cardamom plantations of Kerala. *Nematol. medit.* 10: 107-110.

Ali, S.S. and Koshy, P.K., 1982b. A note on use of Methyl Bromide for control of root-knot nematodes in cardamom nurseries. *Indian J. Nematol.* 12: 147-150.

Anand, T., Govindaraju, C., Sudharshan, M. R., and Srinivasulu, P., 1998. Epidemiology of vein clearing virus of small cardamom. *Indian J. Virology*, 14; 105-109,.

Anandaraj, M., 2000. Diseases of Black Pepper. In: P. N. Ravindran (ed.) *Black Pepper (Piper nigrum)* Harwood Academic Publishers, The Netherlands.

Anandaraj, M., 2006. Disease Management in *phytophthora* foot rot (quick wilt) affected black pepper plantations. Final report 1983. pp 55.

Anandaraj, M., Sarma, Y. R. and Ramachandran, N., 1994. *Phytophthora* foot rot of black pepper in relation to age of the host and its culmination in foot rot. *Indian Phytopath* 47: 203-206.

Anandaraj, M. and Sarma, Y. R., 1995. Diseases of black pepper (*Piper nigrum* L.) and their management. *Journal of Spices and Aromatic Crops* 4: 17-23.

Anandaraj, M., Ramana, K.V. and Sarma, Y.R., 1991. Interaction between vesicular arbuscular mycorrhizal fungi and *Meloidogyne incognita* in black pepper. Pp. 110-112. In: D.J. Bagyraj and A. Manjunath (Eds). *Mycorrhizal Symbiosis and Plant Growth.* Univ. of Agri. Sci., Bangalore, India.

Anandaraj, M., Ramana, K.V. and Sarma, Y.R., 1996c. Suppressive effect of VAM on root damage caused by *Phytophthora capsici*, *Radopholus similis* and *Meloidogyne incognita* in black pepper. Pp. 232-238. In: *Impact of Diseases and Pests in Tropical Forests.* (Eds. K.S.S. Nair, J.K. Sharma and R.V. Varma). Kerala Forest Research Institute, Peechi, India.

Anonymous, 1971. *Fiji Department of Agriculture–Report for the year 1970.* Fiji Parliament. P. 34.

Anonymous, 1977. *Annual Report of Min. Agric. Nat. Res.*, 1974. Port Louis, Mauritius p.129.

Anonymous, 1980. *Annual Report of Min. Agric. Nat. Res.*, 1978. Port Louis, Mauritius p.126.

Anonymous, 1989. *Bi-annual Report 1987-88 and 1988-1989*, ICRI, Myladumpara. p.41-47.

Anonymous,1991. *Annual Report for 1990–91.* National Research Centre for Spices, Calicut, India. 75 pp.

Anonymous, 1993. *Annual Report for 1992-93.* National Research Centre for Spices, Calicut, India. 61pp.

Anonymous, 1994. *Annual Report for 1993-94.* National Research Centre for Spices, Calicut, India. 65 pp.

Anonymous, 2003 *Annual Report 2002-03.* Indian Institute of Spices Research, Calicut, Kerala.

Anonymous, 2005 *Annual Report 2002-03.* Indian Institute of Spices Research, Calicut, Kerala 61pp.

Anonymous, 2006. *Annual Report 2006-2007. Indian* institute of Spices Research, Calicut.

Aravind, R., Dinu Antony, Santhosh J. Eapen, Kumar, A. and Ramana, K.V., 2004. Isolation and in vitro evaluation of endophytic bacteria against root-knot nematodes infesting black pepper (*Piper nigrum* L.). *National Nematology Symposium*, 17-19 November 2004, University of Agricultural Sciences,.

Beena, B., Eapen, S.J. and Ramana, K.V., 2001. Antagonistic rhizobacteria of root knot nematodes infesting black pepper (*Piper nigrum* L.). P.143. In: *Abstracts–National Congress on Centenary of Nematology in India*. Indian Agricultural Research Institute, New Delhi, India.

Beena, B., Eapen, S.J. and Ramana, K.V., 2003. Native rhizobacteria for the biological suppression of *Radopholus similis* infesting black pepper (*Piper nigrum* L.). In: *Abstracts and Short Papers–Sixth International PGPR Workshop*, Calicut, India.

Bezerra, J. L., and Ram, A., 1986. The black crust of vanilla vine (*Vanilla fragrans*) caused by *Mycoleptodiscus indicus* (Moniliales, Hyphomycetes) *Fitopatologia-brasileira* (Brazil). 11: 712-72.

Bhadramurthy, V., Retheesh, S. T., Bhat, A. I., Madhubala, R., Hareesh, P. S. and Pant, R. P., 2005. Development of ELISA based technique for the detection of a putative badna virus infecting black pepper (*Piper nigrum* L.). *Indian Phytopathology* 58: 314-318.

Bhai, R S., Thomas J. and Naidu, R., 1988. Anthracnose, a new disease of small Cardamom. *Cur. Sci.* 57: 1346-1347.

Bhai, R S., Thomas J. and Naidu, R., 1993. Biological control of Azhukal disease of cardamom caused by *Phytophthora* meadii Mc Rae. *J. Plantn. Crops.* 21 (suppln.): 134-139.

Bhai, R S., Thomas J., and Naidu, R., 1994. Evaluation of carrier media for field application of *Trichoderma* sp. in cardamom soils. *J. Plantn. Crops* 22: 50-52.

Bhai, R S., 1998 Studies on Capsule rot disease of Cardamom (*Elettaria cardamomum* Maton.) *Ph.D. Thesis*. University of Calicut.

Bhai, R. S. and Thomas J., 1998. *Trichoderma* sp. an effective bioagent for the control of seedling rot disease in cardamom nurseries. *J. Spices and Aromatic Crops* 7: 85-88.

Bhai, R S., and Kumar, A., 2005 Screening of rhizobacteria for growth promotion and pathogen suppression in vanilla (*Vanilla planifolia* Andrews.). *J. of Biological Control.*

Bhai, R. S, Sarma, Y. R., Thomas J., 2000. Integrated disease management strategy for capsule rot of cardamom. In (N. Muraleedharan and R. Raj Kumar Eds.) *Recent Advances in Plantation Crops Research*. Pp. 378-381.

Bhai, R. S. and Sarma, Y. R., 2005. Characterization of *Phytophthora* de Bary inciting Capsule Rot/ Azhukal and leaf blight disease of cardamom (*Elettaria cardamomum* Maton). *J. Plantation Crops*, 33: 193-203.

Bhai, R. S., Kishore, V. K., Kumar, A., Anandaraj, M. and Eapen S.J., 2005. Screening of rhizobacterial isolates against soft rot diseas of ginger (*Zingiber officinale* Rosc.). *J. Spices and Aromatic Crops*, 14:130-136.

Bhai, R. S., Anandaraj, M., Sarma, Y. R., Veena, S. S. and Saji, K. V., 2007 Screening of black pepper (*Piper nigrum* L.) germplasm for resistance to foot rot disease caused by *Phytophthora capsici* Leonian: *J. Spices and Aromatic Crops* 16:115-117.

Bhai, R. S., Thomas, J. and Naidu, R., 1993. Biological control of 'Azhukal' disease of small cardamom caused by *P. meadii* Mc Rae. *J. Plant. Crops* 21 (supplement) : 134-139.

Bhai, R. S., Thomas, J., 1998. Evaluation of *Trichoderma* sp. For the control of seedling rot disese of cardamom (*Elettaria cardamomum* Maton. *J. Spices and Aromatic Crops* 7: 85-88.

Bhai, R S., Thomas, J. and Naidu, R., 1992b. Evaluation of promising selections of small cardamom against Azhukal Disease. *J. Plant. Crops.* 20 (supplement): 90-91.

Bhai, R. S. and Thomas, J., 2000. *Phytophthora* rot-a new disease of vanilla (*Vanilla planifolia* Andrews) in India. *Journal of Spices and. Aromatic Crops*. 9: 73-75.

Bhai, R. S., Thomas,J. and Naidu, R., 1988. Anthracnose- A new disease of cardamom. *Curr. Sci.* 57: 1346- 1347.

Bhai. R. S, Thomas J., Potty S, N., Geetha and Solomon J.J 2002 Mosaic disease of Vanilla (*Vanilla planifolia*, Andrews)-the first report from India *J. Spices and Aromatic Crops. 12(1)*: 80-82.

Bhai. R. S, Bhat, A. I., and Anandaraj, M., 2006. Yellowing and Premature bean dropping in Vanilla (*Vanilla planifolia* Andrews). *Journal of Plantation Crops* 34: 115-117.

Bhat, A. I., Devasahayam,S., Sarma, Y. R. and Pant, R. P., 2003. Association of a badna virus in black pepper (*Piper nigrum* L.) transmitted by mealy bug (*Ferrisia virgata*) in India. *Curr. Sci.* 84: 1547-1550.

Bhat, A. I., Sarma, Y. R., Sreenivasulu, P. and Pant, R. P., 2004a. Occurrence and identification of a *Cucumber mosaic virus* isolate infecting Indian long pepper (*Piper longum*). *Journal of Medicinal and Aromatic Plant Sciences* 26: 279-284.

Bhat, A. I., Faisal, T. H., Madhubala, R., Hareesh, P. S., and Pant, R. P., 2004b. Purification, production of antiserum and development of enzyme linked immunosorbent assay–based diagnosis for *Cucumber mosaic virus* infecting black pepper. *Journal of Spices and Aromatic Crops* 13:16-21.

Bhat, A. I., Venugopal, M. N., Pant, R. P. and Bhai, R.S., 2004c Occurrence and distribution of virus diseases on vanilla (*Vanilla planifolia* Andrews) in india. *Journal of Spices and Aromatic Crops* 13: 143-148.

Bhat, A. I., Devasahayam, S., Venugopal, M. N. and Suseela Bhai, R., 2005a. Distribution and incidence of viral diseases of black pepper in Karnataka and Kerala, India. *Journal of Plantation Crops* 33: 59-64.

Bhat, A. I., Devasahayam, S., Hareesh, P. S., Preethi, N. and Thomas, T., 2005b. *Planococcus citri* (Risso)-an additional mealy bug vector of *Badna virus* infecting black pepper in India. *Entomon* 30: 85-90.

Bhat, A. I., Madhubala, R., Hareesh, P. S. and Anandaraj, M., 2006. Detection and characterization of the phytoplasma associated with a phyllody disease of black pepper (*Piper nigrum* L.) in India. *Scientia Horticulturae*, 107: 200-204.

Bhat, A.I., Bhadramurthy, V., Siju, S. and Hareesh, P.S., 2006a. Detection and identification of *Cymbidium mosaic virus* infecting vanilla (*Vanilla planifolia* Andrews) in India based on coat protein gene sequence relationships. *Journal of plant Biochemistry and Biotechnology*, 15 :33-37.

Brahma, P.. N. And Nambiar, K. K. N., 1982. Survival of Phyllosticta zingibberi Ramakr.causal agent of leaf spot of ginger. In : Nair, M. K., Premkumar T., Ravindran, P. N. and Sarma, Y. R. (eds.) *Ginger and Turmeric*, CPCRI, Kasaragod, pp. 123-125.

Brahma, R. N. And Nambiar, K. K. N., 1984. Spore release and dispersal in ginger leaf spot pathogen *Phyllosticta* zingiberi. In : Bavappa, K.V.A (ed.)Proc. Natl. PLACOSYM-V, 1982, Kasaragod, India.

Bouriquest, G. (ed) 1954. *Le Vanillier et la vanille dans le Monde Paris Editions*. Paul Lechavalier.

Briton-Jones, H. R., 1933. The control of scab disease (*Sporotrichum citri* Butler) in the British West *Indies Trop.Agric.* 10 : 40-42.

Butler, E. J., 1907. An account of genus Pythium snd some Chytridiaceae. *Mem. Dep. Agric. India* (Bot. Ser.) 1 : 70s.

Chalot, C. and Bernard, U., 1920. *Culture et preparation de la vanillae*. Paris E. L. arose.

Charles, J.S., 1978. Studies on the Nematode Diseases of Ginger. *M.Sc. Thesis*. Kerala Agric. Univ., Vellanikara, Kerala, India.

Charles, J. S. and Kuriyan, K.J., 1980. Studies on nematode incidence in ginger. Pp. 50–57. In: *Proc. PLACROSYM–II 1979*. Indian Society for Plantation Crops, Kasaragod, India.

Cheng, Y.H. and Tu, C.C., 1979. [Pathogenesis of *Meloidogyne incognita* to edible ginger]. *J. Agric. Res. China* 28: 91-99.

Christie, J.R., 1957. The yellows disease of pepper (*Piper*) and spreading decline of citrus. *Pl. Dis. Reptr* 41: 267–268.

Colbran, R.C., 1972. Studies of root knot nematode control in ginger with non volatile nematicides applied at and after planting. *Qd. J. Agric. Animal Sci.* 29: 275–280.

Colbran, R.C., 1974. Nematode control in ginger with nematicide, selection of planting materials and saw dust mulch. *Qd. J. Agric. Animal Sci.* 31: 231-235.

Colbran, R.C. and Davis, J.J., 1969. Studies of hot water treatment and soil fumigation for control of root knot in ginger. *Qd. J. Agric. Sci.* 26: 439-445.

Dake, G. N., 1995. Diseases of Ginger(*Zingiber officinale* Rosc.) and their management. *Journal of Spices and Aromatic Crops* 4: 40-48s.

Das, N., 1999. Effect of organic mulching on root-knot nematode population, rhizome rot incidence and yield of ginger. *Ann. Plant Protect. Sci.*, 7 : 112-114.

Das, N. and Senapathi, A. K., 1998. Chemical control of leaf spot of ginger. *Ann. of Plant Protect. Sci.*, 6 (6) 207-208.

Deighton, F. C., 1936. Mycological work. Rep. Dept. agric. Sierra Leone, 1935, pp.22-26.

Devasahayam, S., Anandaraj M., Thankamoni, C. K, Saji, K. V. and Jayasree, E., 2006. Black Pepper. In VA Parthasarathy and P Rajeev (Eds.) *Major Spices-Production and Processing*, Indian Institute of Spices Research 15-61.

Dewaard, P. W. F., 1979. Yellow diseases complex in black pepper on the island of Bangka,(Indonesia) *J. of Plantation Crops*. 7, 42-49.

Dhanalakshmy, C. and Leelavathy, K. M., 1976. Leaf spot of cardamom caused by *Phaeotrichoconis crotalariae. Plant Dis. Reporter*, 60, 188.

Dhanapal, K., and Thomas, J., 2000. Stem rot–A new fungal disease of small of cardamom. In (N. Muraleedharan and R. Raj Kumar Eds.) *Recent Advances in Plantation Crops Research*. Pp. 408-411.

Dhar, V., Srivastava, L.S. and Sahambi, H.S., 1981. Leptosphaeira leaf spot and blight of ginger in India. *Curr.Sci.* 50: 540.

Dohroo, N. P., 1989a. Seed transmission of pre-emergence rot and yellows of ginger. *Plant Dis. Res.*, 4 (1) 73-74.

Dohroo, N. P., 1993. Final ICAR report on Multilocational Project on Rhizome Rot of Ginger. UHF, Sola, p. 38.

Dohroo, N. P., 2005. Diseases of Ginger. In: Haywaard P N Ravindran and Nirmal Babu, K. (eds)., The *Genus Zingiber*. CRC Press.

Dohroo, N. P., Sharma, O., Sharma, M. and Sarlach, R. S., 1994. Effect of organic amendments of soil on rhizome rot, nematodes and rhizosphere mycoflora of ginger (*Zingiber officinale* Roscoe). *Ann. Biol. Ludhiana* 10: 208-210.

Dohroo, N.P., Shyam, K.R. and Bhardwaj, S.S., 1987. Distribution, diagnosis and incidence of rhizome rot complex of ginger in Himachal Pradesh. *Indian J. Plant Pathol.* 5: 24-25.

Dohroo, N. P., 2000. Etiology and management of storage rot of ginger in Himachal Pradesh. *Indian Phytopathol.* 54: 49-54.

Dohroo, N. P. and Malhotra, R., 1995. Control of storage rot of ginger in HP. In: Gupta, V. K., and Sharma R.C. (eds.) *Disease management and plant growth*, Scientific Publishers. Jodhpur, India. pp., 199-202.

Dohroo, N. P., Bharadwaj, S.S. and Shyam, K.R., 1986b. Management of Rhizome rot of ginger in storage through fungitoxicants. *Pestology*. 10: 24-25.

D'Souza, G.I., Viswanathan, P.R.K. and Shamanna, H.V., 1970. Relative distribution and prevalence of plant parasitic nematodes in coffee tracts of South Western India. *Indian Coffee* 34: 330, 342.

Eapen, S.J., 1990. A methodology for evaluation of resistance in cardamom to root knot nematode (*Meloidogyne incognita*). *Indian J. Nematol.* 20: 197–201.

Eapen, S.J., 1992. Influence of plant age on root knot nematode development in cardamom. *Nematol. Medit.* 20: 193-195.

Eapen, S.J., 1995b. Investigations on Plant Parasitic Nematodes Associated with Cardamom. Final Report, Indian Institute of Spices Research, Calicut, Kerala, India.

Eapen, S.J., 2006. Screening Spice Germplasm for Reaction to Nematodes. Final Report. Indian Institute of Spices Research, Calicut, India. 69pp.

Eapen, S.J., 2007. Endophytic Bacteria for Biological System Management of *Radopholus similis*, the Key Nematode Pest of Black Pepper (*Piper nigrum* L.)–Final Report of the DBT-sponsored Ad-hoc Project. Indian Institute of Spices Research, Calicut, India.

Eapen, S.J. andVenugopal, M.N., 1995. Field evaluation of *Trichoderma* spp. and *Paecilomyces lilacinus* for control of root knot nematodes and fungal diseases in cardamom nurseries. *Indian J. Nematol.* 25: 15-16 (Abstr.).

Eapen, S.J., Ramana, K.V. and Sarma, Y.R., 1997. Evaluation of *Pseudomonas fluorescens* isolates for control of *Meloidogyne incognita* in black pepper (*Piper nigrum* L.). Pp.129-133. In: *Biotechnology of Spices, Medicinal and Aromatic Plant–. Proceedings of the National Seminar on Biotechnology of Spices and Aromatic Plants.* (Eds. S. Edison, K.V. Ramana, B. Sasikumar, K. Nirmal Babu and S.J. Eapen) Indian Society for Spices, Calicut, India.

Eapen, S.J., Venugopal, M.N., Ramana, K.V. andSarma, Y.R., 2000. *Trichoderma* spp. for the management of root knot nematode and rhizome rot disease in cardamom nurseries. pp. 382-386. In: *Recent Advances in Plantation Crops Research.* (Eds. N. Muraleedharan and R. Raj Kumar). Allied Publishers Ltd., Mumbai, India.

Eapen, S.J., Beena, B. and Ramana, K.V., 2008. Evaluation of fungal bioagents for management of root-knot nematodes in ginger and turmeric fields. *J. Spices Aromatic Crops* (In press).

Ellis, M. B., 1971. *Demataceous hyphomycetes.* Commonwealth Mycological Institute. Kew, Survey, England, p. 608.

Food and Agricultural Organization (FAO) 2008. FAOSTAT (http://faostat.fao.org/) Accessed on 25 January 2008.

Freire, F.C.O. and Bridge, J., 1985. Parasitism of eggs, females and juveniles of *Meloidogyne incognita* by *Paecilomyces lilacinus* and *Verticillium chlamydosporium.Fitopatologia Brasileira* 10: 577-596.

George N. Agrios 2005. *Plant pathology*, Fifth Editon Elsevier, Academic Press, pp 348.

Gnanapragasam, N.C., 1989. Varietal response of pepper to infestation by the burrowing nematode, *Radopholus similis. Sri Lanka J. Tea Sci.* 58: 5-8.

Govindaraju, C., Thomas, J., and Sudharshan, M. R., 1996. 'Chenthal' disease of cardamom caused by *Colletotrichum gloeosporioides* Penz and its management. In Mathew, N. M., and Jacob, C. K. (Eds.) *Developments in Plantation Crop Research,* Allied pub., New Delhi, 255-259.

Gopakumar, B, Dhanapal, K. Thomas, J and J. Thomas 2006. Potential of a consortium of biocontrol agents for disease management in small cardamom (*Elettaria cardamomum* Maton). *J. of Plantation Crops* 34: 476-479.

Hamada, M., Hirakata, K. and Uchida, T., 1985. [Influence of southern root knot nematode, *Meloidogyne incognita* on the occurrence of root rot of pepper (*Piper nigrum* L.) caused by *Fusarium solani* f. sp. *piperis*]. *Proc. Kanto-Tosan Plant Prot. Soc.* 32: 236-237.

Hartati. S.Y. and Supriadi 1994. systemic action of bactericide containing oxytetracycline and streptomycin sulphate in treated ginger rhizomes. *J. Spices. Med. Crops*, 3: 7-11.

Haware, M. P and Joshi, L. K., 1973. Basal rot of ginger (*Zingiber officinale*) caused by *Sclerotium rolfsii* from Madhya Pradesh. *Indian Phytopath.* 26: 575-576.

Haynes, P.H., Partridge, I.J. and Sivan, P., 1973. Ginger production in Fiji. *Fiji Agric. J.* 35: 51-56.

Hegde, L., Krishnappa, K. and Bojappa, K.M., 1993. Reaction of different cultivars of cardamom (*Elettaria cardamomum* Maton) to root knot nematode (*Meloidogyne incognita*). *Indian J. Nematol.* 23: 127-128.

Huang, C.S., 1966. Host parasitic relationship of root knot nematodes in edible ginger. *Phytopathol.* 56: 755-759.

Ichinohe, M., 1980. Studies on the root knot nematode of black pepper plantation inAmazon. *Annual report of the society of Plant Protection of North Japan*, No.31. pp. 1-8.

Ito, K., 1949. Studies on morasaki–monpa disease caused by *Helicobasidium monpa* Tanaka. *Bull. For. Exp. Stat.* Tokyo 43: 126.

Iyer, R., 1987. Diseases of ginger. In: Raychaudhuri, S.P. and Verma, J.P (eds.) *Review of Tropical Plant Pathology* Vol.4. Today and Tomorrow Printers and Publishers, New Delhi.

Jacob, C.K. and Chandrasekharan, L.R., 1984. Evaluation of certain granular pesticides for the control of root knot nematodes in cardamom. Pp. 633–637. In: *Proc.PLACROSYM-V 1982.* Indian Society for Plantation Crops, Kasaragod, India.

Jasy, T. and Koshy, P.K., 1992. Effect of certain leaf extracts and leaves of *Glyricidia maculata* (H.B. and K.) Steud. as green manure on *Radopholus similis. Indian J. Nematol.* 22: 117–121.

Joshi, L. K. and Sharma, N. D., 1980. Disease of ginger and turmeric. In: Nair, M. K. Premkumar, T., Ravindran, P. N. and Sharma, Y. R. (eds) *Ginger and Turmeric*. CPCRI, Kasaragod, India. pp. 104-119.

Kaur, K.J., 1987. Studies on Nematodes Associated with Ginger (*Zingiber officinale Rosc.*) in Himachal Pradesh. *Ph.D. Thesis*, Himachal Pradesh Univ., Shimla, India., 198 pp.

Kaur, D. J. and Sharma, N.K., 1990. A new report on *Pratylenchus coffeae*–a cause of ginger yellows. *Int. Nematol. Network Newsl.* 7: 15-16.

Kim, C.H., Yang, S.S., and Hahn, K., 1998. Effects of soil disinfection, fungicide application and narrow ridge cultivation on development of ginger rhizome rot caused by *Pythium myriotylum* in fields. *Korean J. Plant Pathol.* 14: 253-259.

Koshy, P.K., Eapen, S.J. and Rakesh Pandey 2005. Nematode Parasites of Spices, Condiments and Medicinal Plants. Pp.751-791. In: *Plant Parasitic Nematodes in Tropical and Subtropical Agriculture–2 Edition*. (Eds. M. Luc, R.S. Sikora and J. Bridge). CAB International, Wallingford, U.K.

Koshy, P.K., Nair, C.P.R., Sosamma, V.K. and Sundararaju, P., 1976. On the incidence of root knot nematode in cardamom nurseries. *Indian J. Nematol.* 6: 174–175.

Koshy, P.K., Sosamma, V.K. and Sundararaju, P., 1977. Screening of plants used as pepper standards against root knot nematode. *Indian Phytopath.* 30: 128-129.

Koshy, P.K. andSundararaju, P., 1979. Response of seven black pepper cultivars to *Meloidogyne incognita*. *Nematol. Medit.* 7: 123-125.

Koshy, P.K., Sundararaju, P. andWilson, K.I., 1979a. Efficacy of certain systematic nematicides against *Meloidogyne incognita* in a cardamom nursery. *Nematol. Medit.* 7: 195–198.

Kotani, S., 1994. Disease cycle of blast of ginger and mioga caused by *Pyricularia zingiberi* Nishikado. *Ann. Phytopathol. Soc. Jpn.*, 60: 644-647.

Kotani, S. and Kurata, M., 1992. Black blotch of ginger rhizome by *Pyricularia zingiberi* Nishikado. *Ann. Phytopath. Soc. Jpn.*, 58: 469-472.

Kotani, K. L., 1966. Effect of fungicides on the germination and growth of ginger plants. *Sci. Cult.* 32: 48-49.

Kueh, T.K., 1979. *Pests, Diseases and Disorders of Black Pepper in Sarawak*. Semongok Agricultural Research Centre, Department of Agriculture, Sarawak, Malaysia. 68 pp.

Kueh, T. K. and Sim, S. L., 1992. Etiology and control of *Phytophthora* foot rot of black pepper in Sarawak, Malaysia. In P.Wahid, D. Sitepu, S. Deciyanto and U. Superman (Eds.). *Proc. The International Workshop on Black Pepper Diseases*, Bander Lampung Indonesia, Institute for Spice and medicinal crops, Bogor, Indonesia, pp, 155-162.

Kueh, T. K., Fatimah, O. and Lim, J. L., 1993. Management of black berry disease of black pepper. In M.Y. Ibrahim, C.F.J. Bong and I.B. Ipor (Eds.), *The pepper Industry: Problems and Prospects*, Universiti Pertanian Malaysia, Sarawak, Malaysia pp 162-168.

Kumar, A.C., Viswanathan, P.R.K. and D'Souza, G.I., 1971. A study of plant parasitic nematodes of certain commercial crops in coffee tracts of South India. *Indian Coffee* 35: 222–224.

Kumar, A., Anandaraj, M. and Sarma, Y. R., 2003. Rhizome solarization and microwave treatment: Ecofriendly methods for disinfecting ginger seed rhizomes. In: Allen, C., Hayward, A. C. and

Prior, P. (eds) *Bacterial Wilt and the Ralstonia solanacearum* Species Complex. American Phytopathological Society Press (in press).

Kumar, A., and Hayward, A.C., 2005. Bacterial diseases of Ginger and their control. In (Eds. P.N. Ravindran and K. Nirmal Babu) ginger, *The Genus Zingiber*. Pp 341-365.

Leong, C.T.S., 1986. Pepper nematodes. Pp. 74-78. In: *Annual Report for the Year 1984*. Department of Agriculture, Sarawak, Borneo, Malaysia.

Lopes, E.B. and Lordello, L.G.E., 1979. *Meloidogyne incognita* and *Fusarium solani* f.*piperis* associated with wilting of black pepper]. *Revista de Agric.* 49: 165-166.

Madhubala, R., Bhadramurthy, V., Bhat, A.I., Hareesh, P.S., Retheesh, S.T. and Bhai, R.S., 2005. Occurrence of Cucumber mosaic virus on vanilla (*Vanilla planifolia* Andrews)in India. *Journal of Biosciences*, 30: 339-350.

Mammooty, K.P., Sasikumaran, S. and Pillay, V. S., 1992. Management of nursery diseases of black pepper. *Indian Cocoa Arecanut and Spices J.* 15: 65-67.

Mammooty, K.P., and Koshy Abraham 2007. Effects of solarization of potting mixture on the incidence of *Phytophthora* rot in black pepper. *Focus on Pepper* 3: 15-22.

Mayne, W., 1942. Report on cardamom cultivation in South India. Misc. Bull. No. 50, *Ind. Coun. of Agric. Res.* India, p. 67.

Mohandas, C. and Ramana, K.V., 1987. Slow wilt disease of black pepper and its control. *Indian Cocoa, Arecanut Spices J.* 11: 10-11.

Mohandas, C. and Ramana, K.V., 1991. Pathogenicity of *Meloidogyne incognita* and *Radopholus similis* on black pepper (*Piper nigrum* L.). *J. Plant. Crops* 19: 41-43.

Makhnotra, A. K. and Khan, L., 1997. Assessment of yield losses in ginger due to *Meloidogyne incognita*. *Indian J. Nematol.* 27: 259-260.

Mohanty, K.C., Mahapatra, S.M. and Patnaik, P.R., 1992. Integrated management of root knot nematode (*Meloidogyne incognita*) infecting ginger. *Indian J. Nematol.* 22: 70-71.

Mohanty, K.C., Ray, S., Mohapatra, S.N., Patnaik, P.R. and Ray, P., 1995. Integrated management of root knot nematode in ginger (*Zingiber officinale* Rosc.). *J. Spices Aromatic Crops* 4: 70-73.

Mustika, I., 1990. Studies on the Interaction of *Meloidogyne incognita*, *Radopholus similis* and *Fusarium solani* on Black Pepper (*Piper nigrum L.*). Wageningen Agric. Univ., Wageningen, The Netherlands. 127 pp.

Mustika, I., 1992. Effects of *Meloidogyne incognita* and *Fusarium solani* on black pepper (*Piper nigrum* L.). *Industrial Crops Res. J.* 4: 7-13.

Mustika, I., Sudradjat, D. and Wikanda, A., 1984. Control of pepper yellow disease with fertilizer and pesticides. *Pemberitaan LPTI* 9: 37-43.

Nageshwar, Rao, T. G., 1995 Diseases of turmeric (*Curcuma longa* L.) and their management. *Journal of Spices and Aromatic Crops* 4: 49-56.

Naidu, R., 1978 Screening of cardamom varieties against *Sphaceloma* and *Cercospora* leaf spot diseases. *J. Plantation Crops*, 6 : 48.

Naidu, R., and Venugopal, M. N., 1987. Epidemiology of *Katte* virus of small cardamom. 1 disease incubation period and role of different host parts as a source of inoculum in relation to disease spread. In *Proc. PCROSYM* VI 1984, Indian Society for Plantation Crops, Kasaragod, India pp 121-127.

Naidu, R., and Thomas, J., 1994. Viral diseases of cardamom. I K.L. Chadha and P. Rethinam (Eds.) *Advances in Horticulture,* Vol 10-*Plantation and Spice Crops*, Part 2. Malhotra Publishing house, New Delhi, pp 1101-1111.

Nair, R.R., 1979. Investigations on fungal diseases of cardamom (*Elettaria cardamomum* Maton) *Ph.D. Thesis* pp.161. Kerala Agricultural University, Vellanikkara, Trichur, Kerala.

Nambiar, K.K.N., and Sarma, Y.R., 1976. Capsule rot of cardamom: *Pythium vexans* de Bary. .

Nambiar, K.K.N and Sarma, Y.R., 1977. Wilt diseases of black pepper. *J. Plant.Crops* 5: 92–103.

Nambiar, K. K. N., Sarma, Y.R., Brahma, R. N., 1977. Field reaction of turmeric types to leaf blotch disease. *J. Plant. Crops* 5: 124-125.

Nath, P. D., 1993. Effect of shade and treatment for rhizome rot of ginger. *Ann. Agric. Res.*, 14 : 327-328.

Nishina, M. S., Sato, D. M., Nishijima, W. T. and Manu, R. F. L. (1992) Ginger root production in Hawali. Commodity Fact Sheet Gin-3 (a) Manoca. Hawali Cooperative Extension Service, University of Hawali at Manoa.

Ostrec, L.J and Grubisic, D., 2003. Effects of soil solarization on nematodes in Croatia. *J. Pesticide Sci.* 76: 139-144.

Pattenshetty, H. V., Deshpande, R. S., and Sivappa, T. G., 1973. Cardamom seedlings can be protected against damping off disease by the treatment with formaldehyde. *Curr. Res.* 2: 20-21.

Pearson, M. N., 1990. Virus diseases of Vanilla in South Pacific. Working Paper 16. Sixteenth Regional Technical Meeting on Plant Protection. Auckland, New Zealand, Feb: 1990. South pacific Commission, Noumea.

Pearson, M. N., Brunt, A.A., and Pone, S.P., 1990. Some hosts and Properties of a poty virus infecting *Vanilla fragrans* (Orchidaceae) in the Kingdom of Tonga. *J. Phytopathol.* 128: 46-54.

Pearson, M. N., Jackson, G.V. H., Zettler, F.W. and Frison, E. A., 1991. FAO/IBPGR Technical guidelines for safe movement of vanilla germplasm 17 pp.

Pearson, M. N., Jackson, G. V. H., Pone, S. P. and Howitt, R. L. J., 1993. Vanilla viruses in the South Pacific. *Plant Pathology*, 1993 42:1-127-131.

Pegg, K. G., Moffett, M. L. and Colbran, R. C., 1974. Diseases of Ginger in Queensland. *Queensland Agric. J.* 100: 611-618.

Ponnappa, K. M. and Shaw, G. G., 1978. Notes on the genus cereospora in India. *Mycologia* 70: 859-862.

Premkumar, T., Sarma, Y. R., and Goutam, S. S., 1982. Association of dipteran maggots in Rhizome rot of ginger. In *Proc. National Seminar on Ginger and turmeric.* Nair MK Premkumar T, Ravindran P N and Sarma YR (Eds.) 8-9 April 1980 Calicut (pp 128-130 CPCRI, Kerala, India.

Purseglove, J.W., Brown, E. G., Green, C.L., and Robbins, S. R. J., 1988. *Spices* Vol.2. 813 pp. Tropical Agriculture Series. Longman Scientific and Technical, England.

Rajan, K. M. and Agnihotrri, V. P., 1989. *Pythium* induced rhizome rot of ginger, problems and progress. In: Agnihotrri, V. P, Sing, N., Chaube, H. S., Sing, V. S. and Dwivedi, T. S. (eds) *Perspective in Phytopathology*; Today and Tomorrow's Principles and Publishers, New Delhi, pp 189-198.

Ramachandran, N., Dake, G. N. and Sarma, Y. R., 1989. Effect of systemic fungicides on *in vitro* growth of *Pythium aphanidermatum*, the rhizome rot pathogen of ginger. *Indian Phytopathol.*, 42 (3) 463-465.

Ramakrishnan, T. S., 1942. A leaf spot disease of Zingiber officinale caused by *Phyllosticta zingiberi* n.sp. Proc. *Indian Acad. Sci. Sect.* B., 20 (4) 167-171.

Ramana, K. V., 1986. Slow-wilt disease of black pepper and the role of plant parasitic nematodes in its etiology. Pp. 17-21. In: *Proceedings of the Second Group Discussion on Nematological Problems of Plantation Crops*, Central Coffee Research Station, Balehonnur, Karnataka, India.

Ramana, K.V., 1992. *Role of Nematodes in the Incidence of Slow Decline (Slow Wilt Disease) of Black Pepper and Screening Pepper Germplasm Against Nematodes.* Final Report. National Research Centre for Spices, Calicut, Kerala, India. 149pp.

Ramana, K.V., 1994. Efficacy of *Paecilomyces lilacinus* (Thom.) Samson in suppressing nematode infestations in black pepper (*Piper nigrum* L.). *J. Spices Aromatic Crops* 3: 130-134.

Ramana, K. V., Mohandas, C and Ravindran, P. N., 1987b. Reaction of black pepper germplasm to the burrowing nematode (*Radopholus similis*). *J. Plantn. Crops* 15: 65-66.

Ramana, K.V., Sarma, Y.R and Mohandas, C., 1992. Slow decline disease of black pepper (*Piper nigrum* L.) and role of plant parasitic nematodes and *Phytophthora capsici* in the disease complex. *J. Plant. Crops* 20 (suppl.): 65-68.

Ramana, K. V., Mohandas, C. and Eapen, S.J., 1994. *Plant Parasitic Nematodes and Slow Decline Disease of Black Pepper*. Technical Bulletin, National Research Centre for Spices, Calicut, Kerala, India. 14pp.

Ramana, K. V. and Eapen, S. J., 1998. Plant parasitic nematodes associated with spices and condiments. Pp. 217-251. In: *Nematode Diseases in Plants*. (Ed. P.C. Trivedi). CBS Publishers and Distributors, New Delhi.

Rao, D. G. and Naidu, R., 1974. Chemical control of nursery leaf spot disease of cardamom caused by *Phyllosticta ellettariae*. *J. Plantation Crops*, 2: 14-16.

Rathaiah, Y., 1987. Control of soft rot of ginger with Ridomil. *Pesticides*, 21: 29-30.

Rathaiah, Y., 1980. A New leaf spot of turmeric. *Mycologia*, 72: 1210-1213.

Rathaiah, Y., 1980. Leaf blast of turmeric. *Plant Disease*. 64: 104-105.

Rathaiah, Y., 1981. A new species of Cercosptoria causing a leaf spot of ginger. *Mycologia* 73: 774-777.

Rathaiah, Y., 1982. Ridomil for control of rhizome rot of turmeric. *Indian Phytopath*. 35: 297-299.

Rathaiah, Y. and Gagoi, R., 2000. Banded leaf blight of ginger Assam. *J. Mycol. Plant Pathol.* 30: 245-246.

Ratnasoma, H. A., Gowen, S. R andWijekoon, S., 1991. Evaluation of some potential methods for control of root knotnematodes in black pepper (*Piper nigrum*). *Trop. Agriculturist* 147: 135-142.

Raut, S.P and Pande, V.S., 1986. A first report of *Meloidogyne incognita* causing root-knot disease of cardamom in Konkan. *Indian J. Mycol. Plant Pathol.* 16: 167-168.

Ravindran,P.N., Ramana, K.V., Nair, M.K., Babu, K.N. and Mohandas, C., 1992. Pournami–a high yielding black pepper selection tolerant to root-knot nematode (*Meloidogyne incognita*). *J. Spices Aromatic Crops* 1: 136-141.

Sainamole Kurian, P., Sivakumar, G., Josephrajkumar, A., Backiyarani, S., Murugan, M. and Shiva, K. N., 2008 Management of anthracnose disease (*Colletotrichum gloeosporioides* (Penz) Penz and Sac.) of black pepper (*Piper nigrum* L.) in the high ranges of Idukki District, Kerala. *J. Spices Aromatic Crops* 17: 21-23.

Sampat, N., Sonali, PandTrivedi, P. C., 2003. Interaction of arbuscular mycorrhizal fungi and different levels of root-knot nematode on ginger. *Indian Phytopathol.* 56: 297-299.

Sasikumar, B., Saji, K.V., Alice Antony, Johnson K. George, T. John Zachariah and Eapen, S.J., 2003. IISR Mahima and IISR Rejatha–two high yielding and high quality ginger (*Zingiber officinale* Rosc.) cultivars. *J. Spices Aromatic Plants* 12: 34–37.

Sarma, Y. R., Premkumar T., Ramana, K. V. and Ramachandran, N., 1988 a. Disease and pest management in black pepper nurseries. *Indian Cocoa Arecanut and Spices J.* 11: 45-49.

Sarma, Y. R., Solomon, J. J., Ramachandran, N and Anandaraj, M., 1988 b. Phyllody disease of black pepper (*Piper nigrum* L.). *J. Plant. Crops*, 16: 69-72.

Sarma, Y. R., 1994.Rhizome rot Diseases of Ginger and Turmeric. In: *Advances in Horticulture* Vol. 10: *Plantation and Spice Crops* Part 2 (Eds. K L Chadha and P. Rethinam) Malhotra Publishing House, New Delhi.

Sarma, Y. R., Anandaraj, M. and Venugopal, M. N., 1994. Diseases of spice crops. In: *Advances in Horticulture* Vol.10: *Plantation and Spice Crops* Part 2(1994) Eds. KL Chadha and P Rethinam), Malhotra publishing House, New Delhi, pp. 1015-1057.

Sarma, Y. R., Anandaraj, M. and Venugopal, M. N., 1996. Biological control of diseases of spices. In (Eds.) M. Anandaraj and K V Peter, *Biological Control in Spices*. Indian Institute of Spices Research, Kerala, India. Pl-19.

Sarma, Y. R., Ramana, K. V., Devasahayam, S and Rema, J., 2001. *The Saga of Spices Research*. Indian Institute of Spices Research (ICAR), Calicut, Kerala. pp29.

Setty, T. A. S., Guruprasad, T. R., Mohan, E. and Reddy, M. N. N., 1995a. Susceptibility of ginger cultivars to rhizome rot at west coast conditions of Karnataka. *Environ. Ecol.*, 13 (1) 242-244.

Setty, T. A. S., Guruprasad, T. R., Mohan, E. and Reddy, M. N. N., 1995b. Susceptibility of ginger cultivars to Phyllosticta leaf spot at west coast conditions. *Environ. Ecol.*, 13: 443-444.

Sharma, N. D. and Jain A.C., 1978b. Studies on biocontrol of *Fusarium oxysporum* f. sp. *zingiberi*, the causal organism of yellows disease of ginger. *Indian Phytopathol.* 31: 260-261.

Sharma, S. L. and Dohroo, N. P., 1982. Efficacy of chemicals in controlling rhizome rot of ginger (*Zingiber officinale* Rosc.). *Proceedings of the national seminar on ginger and turmeric.* Pp 120-122.

Siddharamaiah, A. L., 1988a. Seed rot wilt–a new disease of cardamom. *Curr. Res.*, 17 : 34-35.

Siddharamaiah, A. L., 1988b Stem, leaf sheath and leaf rot disease of cardamom caused by *Sclerotium rolfsii* from India. *Curr. Res.*, 17, 51.

Smith, F. E.V., 1929. Plant Diseases in Jamaica in 1928. Report of Government Mycologist. *Ann.rep.Dept. Agric.*, Jamaica 1928:pp 17-20.

Srivastava, V. P. and Gupta, J. H., 1977. Fungicidal control of turmeric leaf spot incited by *Taphrina maculans*. *Indian Journal of Mycology and Plant Pathology*, 7: 76-77.

Sarma, Y. R. and Anandaraj, M., 1988. *Phytophthora* foot rot of black pepper. In: Agnihotri, V. P., Sarbhoy, K., and Singh, D. V. (Eds.) *Management of Threatening Plant Diseases of National Importance* (pp.237-248). Malhotra Publishing House, New Delhi.

Sebastian 1982. Study on the etiology and ecology of fungal 'pollu, in pepper. *M.Sc. (Ag) Thesis*, Kerala Agriculture University, Trichur, Kerala.

Sheela, M. S. and Venkitesan, T. S., 1990. Interaction between *Meloidogyne incognita* and the fungus *Fusarium* sp. On black pepper vine (*Piper nigrum* L.). *Indian J. Nematol.* 20: 184-188.

Sheela, M.S., Venkitesan, T.S. and Mohandas, N., 1993. Status of *Bacillus* spp. as biocontrol agents of root knot nematode (*Meloidogyne incognita*) infesting black pepper (*Piper nigrum* L.). *J. Plantn Crops* 21(suppl.): 218-222.

Sheela, M. S., Bao, H., Jiji, T. and Kuriyan, K. J., 1995. Nematode associated with ginger rhizosphere and their management in Kerala. *Pest Management Hort. Ecosystems* 1 : 43-48.

Shukla, B. N. and Haware, M. P., 1972 *Phyllosticta* leaf spot of ginger (*Zingiber officinale*) in Madhya Pradesh. *Indian J. Mycol. Plant. Pathol.* 2: 93.

Sinha, S. K., Singh, V., Nimaichand, S. and Sinha, R. K. P., 1987. *Curvularia lunata* on ginger ((*Zingiber officinale* Rosc.), a new host record from Bihar. Indian. J. Plant. Pathol. 5: 213.

So, I.Y., 1980 Studies on Ginger mosaic virus. *Korean J. of Plant Prot.*, 19: 67-72.

Sosamma, V.K. and Koshy, P. K., 1997. Biological control of *Meloidogyne incognita* on black pepper by *Pasteuria penetrans* and *Paecilomyces lilacinus*. *J. Plantn Crops* 25:72-76.

Sreeja, T.P., Eapen, S.J. and Ramana, K.V., 1996. Occurrence of *Verticillium chlamydosporium* Goddard in a black pepper (*Piper nigrum* L.) garden in Kerala, India. *J. Spices Aromatic Crops* 5: 143-147.

Sridhar, V. V., 1988. Studies on Nilgiri Necrosis, a new virus disease of cardamom. *Ph.D. Thesis*, Tamil Nadu Agricultural University, Coimbatore.

Srivastava, V. P., 1994 Management of soft rot of ginger in Sikkim. *Plant Dis. Res.* 9: 146-149.

Srivastava, V. P. and Gupta, J. H., 1977. Fungicidal control of turmeric leaf spot incited by *Taphrina maculans*. *Indian Journal of Mycology and Plant Pathology*, 7: 1, 76-77.

Srivastava, L.S., Gupta, S. R., Basnet, C. P., Mahato, U. P. and Neopani, B., 1998. Microorganisms associated with ginger in Sikkim. *J. Hill. Res.*, 11 : 120-122.

Sundram, S.V., 1954. Thread blight of ginger. *Indian Phytopathol.* 6: 80-85.

Sundararaju, P., Sosamma, V. K. and Koshy, P. K., 1979c. Pathogenicity of *Radopholus similis* on ginger. *Indian J. Nematol.* 9: 91-94.

Thakore, B. B. L., Mathur, S., Singh, R. B. and Chakravarti, B. P., 1987. Soil amendment with oil cakes in ginger field for rhizome rot control. *Korean J. Plant Prot.*, 26: 267-268.

Thamburaj, S., 1991. Research in Spice Crops at TNG. *DNAU Spice India* 4: 17-18.

Thirumalachar, M. J., 1943. A new rust disease of cardamom. *Curr. Sci.* 12: 231-232.

Thomas, J. E., 1986. Purification and properties of ginger chlorotic fleck virus. *Ann. Appl. Boil.* 108: 43-50.

Thomas J, Bhai, R S. and Naidu, R., 1988. Capsule rot disease of small cardamom (*Elettaria cardamomum* Maton) and its control. *J. Plant. Crops.* 18 (supplement): 264-268.

Thomas, J., Bhai, R. S. and Naidu, R., 1989. Comparative efficacy of fungicides against *Phytophthora* rot of small cardamom. *Pesticides*. September p. 40-42.

Thomas, J. Bhai, R. S., 2000a. Sclerotium rot-a new disease of vanilla (*Vanilla planifolia* Andrews). *Journal of Spices and Aromatic Crops* 9 : 175-176.

Thomas, J. Bhai, R S. and Potty, S. N., 2000b. VA Mycorrhizae-A growth promoter for cardamom seedlings. *Spices India* 13 : 5-8.

Thomas, J. Bhai, R. S., 2002. *Sclerotium rolfsi*, a new root rot pathogen of Vanilla (*Vanilla planifolia* Andrews) in India. *Plantation Crops Research and Development in the new millennium* 541-543.

Thomas, J, Bhai, R. S. and Naidu, R., 1991a. Capsule rot disease of cardamom *Elettaria cardamomum* (Maton) and its control. *J. Plantation Crops*, 18 (suppl.), pp. 264-268.

Thomas, J. Bhai, R. S., Vijayan, A. K. and Naidu, R., 1991b. Management of rot diseases of cardamom through bio-agents. In *National Seminar on Biological Control in Plantation Crops*. (Abs.), June 27-28, RRII, Kottayam, Kerala, p.21.

Thomas, J. and Vijayan, A. K., 1994. Occurrence, severity, etiology and control of rhizome disease of small cardamom.

Thomas, G.V., Sundararaju, P., Ali, S.S and Ghai, S.K., 1989. Individual and interactive effects of VA mycorrhizal fungi and root knot nematodes, *Meloidogyne incognita* on cardamom. *Tropical Agric.* 66: 21-24.

Tombe, M., Tsuchiya, K., Nurawan, A., Nazurudin, S. B., Oniki, M. and Matsumoto, K., 1992. Experiments on the introduction of biological and cultural control of stem rot disease of vanilla. *Industrial Crop Research Journal* 4(2), 20-26.

Tombe and Sitepu 1986. The control of stem rot of vanilla with fungicides. *Edisi-Khusus-penelitian Tanaman Rempah-dan-obat* 2: 43-47.

Tsao, P. H., and Mu. L., 1987. *Phytophthora* blight and root rot of vanilla in French Polynesia: occurrence and casual species, Manila Philippines.

Tsang, M. M. C. and Shintaku, U. M., 1998. Hot air treatment for control of bacterial wilt in ginger root. *Appl. Eng. Agric.*, 14 : 159-163.

Uppal, B. N., Varma, P. M. and Capoor S. P., 1945. A mosaic disease of cardamom. Curr.Sci.14: 208-209.

Van der Vecht, J., 1950. Plant Parasitic Nematodes. Pp.16-15. In: [*Diseases ofCultivated Plants in Indonesia Colonies*]. (Eds.) L.G.E. Karshoven and J.V.d.

Venkitesan, T.S. and Charles, J.S., 1980. A note on the chemical control of nematodes infesting pepper vines in Kerala. Pp. 27-30. In: *Proc. PLACROSYM-II 1979*. Indian Society for Plantation Crops, Kasaragod, India.

Venkitesan, T.S and Jacob, A., 1985 Integrated control of root-knot nematode, infecting pepper vines in Kerala. *Indian J. Nematol.* 15: 261-262.

Venkitesan, T.Sand Setty, K.G.H., 1979. Control of the burrowing nematode *Radopholus similis* (Cobb) on black pepper. *Pesticides* 13: 40-43.

Venugopal, M. N., and Govindaraju, C., 1993. Cardamom vein clearing virus-a new threat to cardamom. *Spice India* 5: 6-7.

Venugopal, M. N., 2002. Viral Disease of Cardamom. In (Eds. P N Ravindran and K J Madhusoodanan) Cardamom, The genus *Elettaria. Medicinal and Aromatic Plants-Industrial Profiles,* Taylor and Francis New York.

Venugopal, M,N, Prasath, D. and Ravindra Mulge 2006. IISR Avinash–a rhizome rot resistant and high yielding variety of Cardamom (*Elettaria cardamomum* Maton). *Journal of Spices and Aromatic Crops* Vol. 15 : 14-18.

Verma, R. K. and Vyas, S. C., 1981. Persistance and protective activity of some fungicides in relation to *Phyllosticta* leaf spot of ginger. *Indian J. Mycol. Plant Pathol.*, 11 : 14-16.

Vijayan, A. K. and Thomas, J., 2002. *Fusarium oxysporum*, a new threat to cardamom cultivation. In (Eds. K Sreedharan, P K Vinod Kumar, Jayarama and Chulaki BM) *Proceedings of Placrosym XV*, Mysore India 10-13.

Vadhera, I., Tiwari, S. P. and Dave, G. S., 1998b. Integrated management of root knot nematode, *Meloidogyne incognita* in ginger. *Indian Phytopathol.*, 51: 161-163.

Vilsoni, F.; McClure, M.A. and Butler, L.D., 1976. Occurrence, host range and histopathology of *Radopholus similis* in ginger (*Zingiber officinale*). *Plant Dis. Reptr.* 60: 417-420.

Wang, Y.Y. and Pearson, M. N., 1992. Some characteristics of Potyvirus isolate from *Vanilla tahitensis* in French Polynesia and Cooks islands. *J. Phytopathol* (Germany.F.R) 1992. 135: 71-76.

Wilson, K. I., Sasi, P. S. and James Mathew, J., 1979a. *Fusarium* capsule disease of cardamom. *Curr. Sci.*, 48, 1005.

Willers, P., 1985. *Efficient Nematode Control a Step Closer to Ginger Producers. Information Bulletin, Citrus and Subtropical Fruit Research Institute.* No.160: 4-6.

Wisler, G. C., Zettler, F.W. and Mu. L., 1987. Virus infections of vanilla and other orchids in French Polynesia. *Plant Dis.*, 71: 1125-1129.

Chapter 16

Pea Rust (*Uromyces fabae*) and its Inheritance of Resistance

Ramesh Chand[1], Chanda Kushwaha[1] and Anil Kumar Singh[2]
[1]Department of Mycology and Plant Pathology,
[2]Department of Genetics and Plant Breeding,
Institute of Agricultural Sciences,
Banaras Hindu University, Varanasi – 221 005, U.P.

Introduction

Pea Origin and Distribution

Pea (*Pisum sativum* L., 2n = 14) is an important agricultural plant with a rich history of genetic research. Pea is one of the six major pulse crops cultivated globally and is second highest yielding legume in the world after Common bean (*Phaseolus vulgaris* L.). Pea is a member of the cultivated legume crops belonging to the order *Fabales*, family *Fabaceae* and tribe *Viciae* (Stebbins 1974; Griga and Novak 1990). The chromosome number (2n =14) of pea was established early in the last century (Cannon 1903; Casey and Davies 1993). *Vicia* and *Lathyrus* are the genera that are most closely related to *Pisum* (Blixt 1974). Pea was among the earliest known cultivated plants. Archaeological evidences indicated that it was originated in Near Eastern and Greek Neolithic settlements as early as 6000 B.C (Zohary and Hopf 1973). Pea belongs to the genus *Pisum*, a member of family papilionaceae tribe *viciae* and is composed of species *Pisum sativum* L (*Ps*) and *Pisum fulvum* (Sibth and and Sn). *Pisum sativum* L had been further divided to include several subspecies *Ps. ssp sativum, Ps ssp. elatuis, Ps ssp humile, Ps ssp arvense* and *Ps. ssp hortense*. In literature *ssp elatius* and *ssp humile* are the progenitors of *Ps ssp sativum*.

Importance of Pea

Pea is a cool season crop and is widely grown in the cooler temperate zones and in the highlands of tropical regions of the world. The nutritive value of dry pea seed is similar to other grain legumes and contains 18–30 per cent protein, 35–50 per cent starch and 4–7 per cent fiber. Furthermore, peas have better quality protein and all the essential amino acids such as lysine, methionine, tryptophan, threonine, valine, phenylanine, leucine and isoleucine. Since pea is high in protein quantity and quality, it is an important protein resource for human consumption, especially in developing countries. Consumption occurs in many forms, such as split, whole and flour. Apart from an important vegetable crop, the fresh green pods are also consumed as frozen or canned peas. The peas are also used as forage, hay, silage and green manuring. The peas used in these ways are characterized by a high degree of active lypotropic antisclerotic substances, especially choline, an important nutritional requirement which is usually deficient in daily diets (Matthews and Arthur 1985). The inclusion of peas in crop rotation is agronomically very significant. Pea is a good predecessor to other crops as it enriches the soil with the nodule bacteria- rhizobia, which fixes atmospheric nitrogen (Makasheva 1984). Moreover, peas have a higher capacity to utilize minerals, which are practically difficult to assimilate and are not available to the cereals. The root system of peas penetrates to a depth of 1-1.5m and as a result peas can extract mineral nutrients from the deeper soil layers unlike grain crops (Khvostova, 1975).

Pea ranks second to dry bean among grain legumes for worldwide production and fourth in harvested area. Dry pea is produced in more than 87 countries. India ranks fifth in production and third in harvested area (FAO, 2005). In India, pea occupies a land area of 0.64 m ha with a production of 0.52 million tons and productivity of 890 kg/ha (Anonymous, 2005). However, this productivity is quite low when compared to the highest productivity of 58.0 Q/ha in the Netherlands. Major constraints in pea production include disease, pests, frost, drought and excessive heat. Major diseases affecting pea production are Fusarium root rot; Aphanomyces root rot, Fusarial wilt, pea enation mosaic virus, powdery mildew, downy mildew, Aschochyta blight and rust. Many other minor diseases also affect the yield of this important crop. In India, problem of powdery mildew has been overcome by the use of powdery mildew resistance gene and it is mandatory to incorporate only powdery mildew resistant entries in All India Coordinated varietals Improvement trials. Rust of pea caused by *Uromyces fabae* (Pers) de Bary is considered as the most important pathogenic threat under warm and humid conditions (Chand *et al.*, 2004b). It becomes a major problem under late sown conditions in India. Our group is working since one decade on the various aspect of U. *fabae* and most of the information is presented in this chapter.

Distribution

Uromyces fabae has been reported from different part of Europe (Mendegen 199); Africa (Negussie *et al.*, 2005); Canada (Conner and Bernier, 1982); Australia (Stoddard and Herath,2001). In India it is reported from Eastern India (Chand *et al.*, 1997; Gupta *et al.*, 1990), Central India (Narsinghani *et al.*, 1980), Southern parts of India (Sokhi *et al.*, 1974; Kumar *et al.*, 1994) and Himalayan region of Himachal Pradesh (Chauhan *et al.*, 1995).

Losses

Uromyces fabae completely destroyed the crop in sub-mountainous region of North-India (Singh, 1999). Yield losses in pea due to rust was also reported by Upadhyay and Gupta (1998). Singh (1999) reported the losses in pea yield from Tarai region. On an average of about 56.81 per cent yield loss was reported in the year 1986-1988. They further classified the losses and reported 22.21 per cent of loss occurs due to reduction in seed weight of the pea. Losses also depend on different growth stages of plant. Early infections completely destroy the crop. Occurrence of disease at pod formation stage reduced the grain yield. Main attribute of yield loss is mainly due to the reduction in seed-weight (Chand *et al.*, 2006).

Symptoms

Pea rust is characterized by the appearance of two types of symptoms in India. Early symptoms develop on abaxial side of older leaves and form round to oval aecidia. Initially aecidia form creamy white to light yellow to bright orange colored pustules on the leaf and stem. Under favorable environment these pustules further developed and spread to other parts of the plants. An aecidia is a cluster of several small cups like structure on the plant. Aeciospores released from the aecial cups are deposited as yellow powder. Small aecidial pustules are mostly confined to the leaf. However it can be seen on stem also. In tendril genotypes of pea it can also be seen on the stipules and tendrils.

Uredial pustules developed on both the surface of leaf but mostly confined to the stem. They appear as powdery light brown pustules. The ruptured epidermis on infected portions of the host exposes black to brown powdery mass.

Telial symptoms appear after aecial/uredial infection late in the same season or on the part of plant leading to senescence. Teliospores are formed in the aecial or uredial pustules. Sometimes it is also formed independently Telia are mostly formed on the stem and tendril. Grain size is significantly reduced in badly infected genotype and colour of the grain becomes dull.

Causal Organism

The casual organism of pea rust in most parts of India is *Uromyces fabae* (Pers.) de Bary. It is a biotrophic, macrocyclic, autocious rust. It belongs to the Division: *Basidiomycotina*, Class: *Basidiomycetes*, Subclass: *Teliomycetidae*, Order: *Uredinales*, Family: *Pucciniaceae* (Alexopolus *et al.*, 1996).

Host Range and Occurrence of the Pathogen

Uppal (1933) and Prasada and Verma (1948) found several species of *Vicia, Lathyrus, Pisum* and *Lentil* susceptible to *U. fabae* in India and abroad. In India species of *Vicia, Lathyrus* and *Pisum* are described as host plant for *U. fabae* (Pers. de Bary) (Kapooria and Sinha, 1966; Kapooria and Sinha, 1971). They observed natural infection on *Vicia sativa* L. and *V. hirsuta* Gray, a common weed found in the fields of lentil in India also. *Vicia faba* L., *V. biennes* L., *V. hirsuta* L., *and V. arborensis* L. were described as highly susceptible to *U. fabae* and *Vicia sativa* and *Lathyrus aphaca* were found to be disease free. Conner and Bernier (1982) reported a total of 52 species of *Vicia faba* and 22 species of *Lathyrus* infected by *Uromyces viciae-fabae*. They also found this pathogen on pea, lentil and faba bean. Butler (1912) reported the occurrence of rust pathogen *U. fabae* on pea and other leguminous crops from India. Sydow and Butler (1912) reported this fungus from state of Maharastra. Pea rust (*U. fabae*) is of worldwide occurrence and attacks number of host species belonging to different genera of the family leguminoceae in the Indo-Gangetic Plains (Butler, 1918). Prasada and Verma (1948) also reported the occurrence of *U. fabae* on lentil crop from Delhi. Roy (1949) in his list of fungi of Bengal recorded the prevalence of *U. fabae* on the leaves and stems of *Pisum sativum*. Mitter and Tandon (1930); Patel (1934); Pavgi and Upadhyay (1966) and Kapooria and Sinha (1966) reported the distribution of this pathogen in the regions of Uttar Pradesh, respectively. Bilgrami *et al.* (1979) reported the occurrence of this pathogen on various host species of Pea, Lentil and *Lathyrus*. Baruah *et al.* (1980) reported that rust infection on the pea plants is caused by both *U. fabae* and *U. pisi*. Of which *U. pisi* is of rare occurrence in India. Choudhary *et al.* (1998). reported host non specificity in *Uromyces fabae* and found few legume genus infected. by the pathogen Occurrence of *U. fabae* have also been reported from Canada, Europe, Ethiopia and Australia in mild to severe forms on pea, lentil and faba bean (Conner and Bernier, 1982; Xue and Warkentin, 2002).

Pathogenic Variability

Pathogenic variability has been reported in field collection of *Uromyces fabae* (Singh and Sokhi, 1980; Conner and Bernier, 1982). The urediospores of *U. fabae* was the only infective spore in temperate climate and is used in various resistance-screening programme in pea (Xue and Warkentin, 2002) and sweet pea (Sokhi, 1984). The existence of pea specific strains from NEPZ of India has also been reported (Chand *et al.*, 1997). Aeciospores acts as as repeating spore in case of *U. fabae* and play important role in the out break of disease under warm humid conditions.

Aecidial State

The aecidium is small, whitish and cup shaped structure and bear aeciospores. The aeciospores are round to angular or elliptical, yellow in colour and posses fine warts. They measures 14- 22 µm in diameter. Aeciospores are sessile and formed in chain of 7–8 spores. Aecidia are deep seated in the spongy mesophyll cells.Aeciospores released after rupturing peridium of aecial cups. Large amount of aeciospores are deposited on the plant surface look bright yellow.

Uredial State

The Urediospores are round to ovate, light brown, spiny with 3-4 germ pores, measure 20- 30 × 18- 26 µm. Uredia developed infrequently. Urediospores are stalked, round to elliptical. Urediospores come out in group after rupturing host epidermis.

Teleospores

Teleospores are sub-globose ovate or elliptical with round or flattened apex which is considerably thicked and appears papilate the spore are smooth measures 25–30 × 18- 27 µm. The stalks are generally presistant on the detached spores and pale yellowish brown, thick and up to 90 µm.

Basidiospore

The basidiospore of U. *fabae* germinated within 3 days of rehydration and it was not influenced by the photoperiod (Freytag *et al.*, 1988). The germination percent of basidiospores was high on the substrates except glass slide. On water agar (2 per cent) basidiospores produced long germtube 50–100 µm. The differentiation in the germtube starts after 3 days. Infection structure resoled better when germination was studied on 5 per cent water agar. The short germtube swell apical and formed small appressorium. Subsequently penetration peg formed and oblong vesicle expanded in agar.

On the glass slide appressorium formation often occurs however, vesicle formation rarely observed (Freytag 1988). Basidiospore differentiation was most effectively induced on nitrocellulose membrane. Less than 5 per cent basidiospore produced long germtube and majority of basidiospores reached to the appressorium stage.

Two nuclei are always seen in basidiospore. During early infection structure development both the nuclei migrated together from cytoplasm in to the vesicle. Subsequently penetration peg was plugged and vesicle grew apically, forming a primary hypha which was some time delimited by a septum. Up to primary hypha stage no more than two nuclei was observed.

Haustoria

Biotrophic plant pathogenic fungi differentiate specialized infection structures called haustorium within the living cells of their host plants. These haustoria have been linked to nutrient uptake ever since their discovery. The flow of sugars from the host *Vicia faba* to the rust fungus *Uromyces fabae* seems to occur largely through the haustorial complex. One of the most abundantly expressed genes in rust haustoria, the expression of which is negligible in other fungal structures, codes for a hexose transporter. Functional expression of the gene termed HXT1 in *Saccharomyces cerevisiae* and *Xenopus laevis* oocytes assigned a substrate specificity for D-glucose and D-fructose and indicated a proton symport mechanism. Abs against HXT1p exclusively labeled haustoria in immunofluorescence microscopy and the haustorial plasma membrane in electron microscopy. These results suggest that the fungus concentrates this transporter in haustoria to take advantage of a specialized compartment of the haustorial complex. The extrahaustorial matrix, delimited by the plasma membranes of both host and parasite, constitutes a newly formed apoplastic compartment with qualities distinct from those of the bulk apoplast. This organization might facilitate the competition of the parasite with natural sink organs of the host.

Mode of Survival of the Pathogen

The aeciospores and urediospores of *Uromyces fabae* did not survive at a temperature more than 30 °C for one week therefore, they were not supposed to survive the high temperature of the intervening crop season. Teliospores survives during the intervening season and germinates to produce basidiospore that produced pycnia, and subsequently causes infection in pea (Singh, 1999). The debris of pea plant carrying telelia inoculated to the pea plant and no spermagonia were found on the plants. However, aecial infection was observed on the few plants (Kushwaha *et al.*, 2006). This result needs to be verified at different places in order to confirm the role of teleospore in the initiation of primary infection. Being a wide host range pathogen Conner and Bernier (1982) have suggested the role of collateral host in the occurrence of disease on pea. They reported *Vicia* and *Lathyrus species* as a collateral host to *U. fabae* and pathogen survive on these hosts during the absence of the main crop. There is no information about the migration of pathogen form collateral hosts from India and other countries. Pea is grown through out the year in the different part of the India but its role in the multiplication and further spread of the *U fabae* is not known. In the recently concluded A.I.C.P.I.P experiment (2005) incidence of rust was noticed earlier in Pant Nagar than the Varanasi and Dholi. There is need of extensive survey of collateral hosts and off season grown pea crop temperate, sub-tropical and tropical part of India to confirm the role of collateral hosts in perpetuation of the pathogen from year to year.

Germination of Different Spores

Germination of spores is independent of the host. Water agar (0.2 per cent) was found to induce 46.67 per cent germination in urediospore and 9.67 per cent germination in aeciospores. Occasional germination was observed in teliospores that were stored for three years after the incubation at 18 °C for 15 days. The germination of teliospores was less than the 1 per cent. Aeciospore, urediospores and teliospores germinated by a single germ tube. Germination in the aeciospore was initiated after 8 hrs on 0.2 per cent water agar at 25 °C. The urediospores were also found to germinate on 0.2 per cent water agar at 15 °C after 4 hours of incubation.

Effect of Relative Humidities and Temperatures on the Aeciospore and Urediospore Germination

The optimum condition for aeciospore germination was 25 °C in combination with 100 per cent relative humidity. The per cent germination of the aeciospores decreased gradually with the decreasing temperature below 25 °C. None of the aeciospores germinated at relative humidity of 88.5 per cent at a temperature below 25 °C (Kushwaha *et al.*, 2006). Aeciospores germination was minimum (0.17 per cent) at 5 °C in combination with 100 per cent relative humidity. Relative humidity of 98 per cent in combination with 15 °C favored maximum (3.5 per cent) urediospore germination. Longer leaf wet duration increased the rust severity (Chauhan and Singh, 1994)

Relationship Between Temperature and Disease Severity

The mean temperature during the initiation of the disease ranged between 15–20 °C (Chanda *et al.*, 2006). The incidence of the disease was delayed when occurrence optimum temperature (25–30 °C) was late in the season due to delayed onset of cool weather. Urediospores of *U. viciae-fabae* (broad bean rust) germinated well in the range 5-26 °C, with fastest germination at 20 °C. Exposure to 30 °C gave poor germination and damaged the spores. Infection of *Vicia faba* leaves depended on a moisture film. At 20 °C some infection occurred with only 4 h leaf wetness, but longer wet periods up to 24 h gave increased infection. At lower temperatures, the infection process was slower and final pustule numbers were also less. Spore germination was delayed by daylight and by all artificial light sources that contained far-red (700-800 nm) wavelengths. The delay was increased at higher light intensities. When spores were subjected to alternating periods of light and darkness, it was found that 40 min of darkness was sufficient for the irreversible induction of germination at 20 °C (Joseph and Hering 1997)

Penetration by Uredia

The Pre-Penetration Phase

The uredospores of the rust fungus *Uromyces fabae* form an adhesion pad and release a cutinase and two specific esterases after contacting the host cuticle. Apparently, adhesion of the pads is improved by these enzymes. The spores have reduced ability to attach to the leaf surface when these enzymes are inactivated (Deising *et al.*, 1992).

Host Surface Perception

Appressorium formation by urediospore germ tubes of the bean rust *Uromyces appendiculatus* is induced by physical differences in the topography of the leaf surface, such as stomatal lips of guard cells, or by defined ridges of 0.5 µm height formed on an artificial surface (Hoch *et al.*, 1987; Kwon and Hoch, 1991). In addition, it has been shown that many rust fungi exhibit species-specific responses on membranes with defined topographies (Allen *et al.*, 1991).

Cell Wall Degradation (By Enzymatic Action)

Nevertheless, penetration by obligately biotrophic parasites, such as rust fungi and powdery mildews or some hemibiotrophs, requires only minor damage of the cell wall. Degradation of the cell wall is limited to the site of penetration as shown by Xu and Mendgen (1997). Secretion of cellulytic enzymes of these pathogens is either developmentally regulated or triggered by environmental signals (Mendgen *et al.*, 1996).

For example, cellulase activity of *Uromyces fabae* germlings has been shown to be strictly regulated by differentiation. It increases during appressorium formation and reaches a maximum during development of infection hyphae and haustorial mother cells (Heiler *et al.*, 1993). Also, the production of the pectic enzymes pectin methylesterase and polyglacturonate lyase (Deising *et al.*, 1995) and extracellular proteases (Rauscher *et al.*, 1995) of this rust fungus depends on the differentiation of infection structures. Apparently, the concerted action of cell wall degrading enzymes enables the hyphal growth through the leaf tissue but prevents extensive cell wall maceration and cell death which would interfere with the biotrophic lifestyle of the fungus also.

The fungus penetrates through leaf stomata and forms a fusiform substomatal vesicle (Emeron *et al.*, 2005) Haustorial mother cells elongate and attempt to penetrate the leaf mesophyll cells. If penetration succeeds, a nutrient absorbing haustorium develop in the mesophyll cell and enables fungal growth of subsequent hyphae (Parlevliet and Kievit 1986). The rust infection can be hampered at very early stages of fungal development from spore deposition to stomata recognition, resulting in reduced infection. Nevertheless most germinating spores proceed and develop normal haustorial mother cells. However, haustorium formation may be prevented by papilla formation within mesophyll cells. Papillae have callose matrix and contain various inorganic and inorganic constituents including antimicrobial proteins and auto fluorescence phenolic compounds. Papillae are deposited on the inner surface of the mesophyll cell walls where fungus attempt to penetrate and act as a physical or chemical barrier Skalamera and Heath 1996, Zeyen *et al.*, 2002. when penetration resistance fails and haustoria develop within the host cells.

Post Penetration Development

Aeciospores germinate to give rise germ tube and later on approsorium this process found to take about 48 to 72 hours in host tissue later there is formation of infection peg which grows intercellularly inside the host tissue after entering through stomatal opening then substomatal vesicle is formed which give rise to haustorial mother cell which give rise to First haustorium formed inside host cell. Post-penetration process consist of spread of mycelium in the inter celluar spaces and and later on thy replace some of the host spongy mesophyll cells. A perdial layaer of single cell was formed aroud the mycelial aggregation). These structure expend further and attained a size (μM).Through out the development of aecium mycelium was dikaryotic. Dikaryotic mycelium later on which forms sporogenous cells of the aecium or aeciosporophores its form at the base of the aecial primordium each cell contains two nuclei that divide conjugately during the formation of an aeciospore initial. Two of the daughter nuclei remain in the sporogenous cell and the other two moves towards aeciospores initials. After the initial is delimited from the mother cell by septum, the nuclei in the initial divide again and a transverse septum separates the initial into a binucleate aeciospores and small wedge shaped, binucleate sterile, intercalary or disjunctor cell. The entire process repeated several times resulting in the formation of a chain of aeciospores a disjunctor cell. Arrangement of aeciospores is younger towards the base while older towards the apex.initial arised in the aecial cavity. Peripheral cells of the aecial base undergo successive division to produce a wall that surrounds the spore chains. This wall is the peridium. In a young aecium that has not broken the host epidermis the peridium surrounds the spore chains on all sides, forming a complete dome over them. As the aecium mature the spore chains push through the roof of the peridium forms a lip around the aecial cup. As an aecium develops, the disjunctor cells disintegrate and spores separate from each other. When aecia develop in a leaf they generally are located in the lower portion and break through the lower epidermis. In case of extra sensitive variety of pea generally large number of aecial initials start and form the aecidiophoere but the development of aeciospores are hampered and resulted in the non brusting aecidia. In these pea genoypes it has also been observed that number of aecidiosporophores are significantly less than the other genotypes. But the number of aeciospores in chain was almost 3- 6 per aeciosporophore. The reduction of number of aecisporephore comes mostly from the reduction of aecisporophoes. In *U. fabae* number of aecidiospores varied from 10–40. The aecidiosporophores grow up right in the cavity towards the epidermis. The perdial layer replace the mesophyll cell reached below epidermis. After some time pressure exesrted by the mature aeciospores rupture the epidermis

Nuclear Conditions

Basidiospores were binucleate. Vesicle contains two nuclei and septum seprate two nuclei in infection structure (Fretag *et al.*, 1988). Nuclear staining of aeciospore and urediospore clearly revealed the binucleate structure. Both the nuclei they were together during the septum formation in the germ tube (Kushwaha, 2005).

Effect of Host Factor on the Outbreak of Disease

Effect of Temperatures and Plant Growth Stages on Spore Production

The aeciospores were produced abundantly at both the temperature regimes *i.e.* 10–15 °C and 20–25 °C at all the growth stages of crop. At higher temperature regime (20–25 °C), during pod formation the number of aecidia/pustules/leaf was highest. Higher temperature promoted more numbers of aecial pustules than the lower temperature. Teliospores were produced, when plants entered into the senescence stage (110-130 days after sowing).

Resistance Components in Host

There are several resistance components that individually or in tandem with other components reinforce the resistance. Most of these components are influenced by the environmental factors *i.e.* temperature and relative humidity. Therefore a multi year and multi locations testing of these components are most essential in order to capitalize these traits in the crop improvement. Most of the cases these traits are scattered among the germplasm lines of pea, lentil and vicia. Individual effects of these components are very less therefore difficult to demonstrate under field conditions. However, using molecular techniques these can be detected. The protection efficiency of these traits can be verified by comparing the disease severity, AUDPC, test weight under protected and unprotected conditions (Chand *et al.*, 2006). Yield gain under fungicidal protection will be always higher in the susceptible genotype than slow rusting.

Effect of Incubation and Latent Period on Rust Severity

Pea

Incubation period of aeciospore ranged from 7.17–17.84 days on different pea genotypes. The correlation between incubation period and AUDPC were negative–0.68. Highest incubation period was shown by Pant P 13 and lowest incubation period was represented by HUVP 1 (Kushwaha, 2006).

Latent Period

varied from 8.23 to 17.83 days among the pea genotypes. Latent period affects the disease severity and its spread. Genotype having higher latent period delayed the out break of disease. The correlation (r) between latent period and AUDPC were negative–0.58. There are many genotypes having higher latent period (Chand *et al.*, 2006). Latent period has great significance in delaying the on set of disease and its further spread. Most of resistance pea genotypes screened at rust hot spots like Bangalore, Pune and Varanasi are generally possess higher latent period.

Lentil

Significant difference were reported among the lentil cultivars for the latent period for urediospores in the susceptible cultivars it was noticed 8 days while same amount of pustule was noticed after 15 days in the resistant cultivars (Negussie *et al.*, 2005).

Vicia

there was significant difference in the latent period in the Vicia germplasm lines. Most of the slow rusting lines had longer latent period than the susceptible one. Latent period was negatively correlated with the disease severity, AUDPC and rate of spread (Sellaro *et al.*, 2000). The LP was negatively correlated with the CS, AUDPC, DS, and r.

Colony Size

Pea

Significant difference are reported for the aecial colony size that varied from 1.9 to 4.9 mm^2 (Chand *et al.*, 2006). Small colony size are often associated with the slow rusting genotypes. However, in some cases effect of small colony size is nullified by the high frequency of colony/unit area.

Lentil

The difference in the pustules size were reported in lentil that varied from 0.096 0.56 mm^2 (Negussie 2005).

Vicia

Colony measured after inoculation were larger on seedlings than the adult plant. In the susceptible control, the colony size was almost double in size than on the adult plant. The hypersensitive lines showed the smallest colonies at both plant maturity stages. Few non hypersensitive lines also showed smaller colonies than the susceptible check. (Scillaro *et al.*, 2000)

Infection Frequency

Pea

under field condition some of the pea line showed less number of colony on the leaf when spores were blowing on all the lines. But the infection frequency increased when free water prolonged for longer period on plant surface with a temperature 25- 30 °C.

Lentil

lentil genotypes displayed enough variation for the infection frequency. Most of the lines with the low infection frequency retard the disease development and identified as slow rusting types.

Vicia

A significant difference was recorded among the vicia genotypes for infection frequency when inoculated with the equal amount of urediospre under indentical conditions of disease development. Most of the slow rusting lines showed low infection as compare to the susceptible one. This trend was also noticed under field situation.

Genetics of Pea Rust Resistance

Breeding for resistance has been suggested as an efficient means to reduce the economic loss caused by pea rust (Singh and Srivastava, 1985; Kumar *et al.*, 1994; Singh and Ram, 2001). For a successful resistance breeding programme, availability of genetic resources that contain different genes for resistance

Inheritance of Pea Rust Resistance

Inheritance studies on rust resistance in pea are limited and still not well established. There were reports of existence of both monogenic as well as polygenic forms of resistance towards rust in pea (Pal *et al.*, 1979, Katiyar and Ram, 1987; Singh and Ram, 2001 and Kumar *et al.*, 1994). Lack of hypersensitive reaction in pea against *U. fabae* suggests the absence of race specific monogenic forms of resistance. Clear information about inheritance of rust resistance in pea would help to design a suitable strategy to enhance existing resistance in pea and could be helpful in any pea improvement programme. It would also allow for the development of suitable long-term disease management strategies.

Resistance breeding would be greatly facilitated if information on the biology of this pathogen and its interaction with environment were available. Plant growth stages also affect the infection and influence the total yield loss. Only specific yield attributing characters are affected under such conditions. Since disease appears during late vegetative stages it is only the seed weight that is most affected in pea (Chand *et al.*, 2004). Therefore, it becomes necessary to establish a relationship between the disease and reduction in seed weight incurred by it. This will also enable the workers to predict the losses in the yield by knowing the stage for disease incidence and its severity.

Breeding for Rust Resistance

Breeding for resistance to pea rust has been initiated only recently. Because knowledge of the genetic variation of the pathogen and genetics of host resistance is very limited, well-designed breeding programmes are not yet possible. However, mapping of the resistance gene and possible use in marker assisted selection (Vijayalkshmi *et al.*, 2005) has provided the basis to design a breeding programme aimed at transferring and combining resistant genes in useful background *i.e.*, agronomically superior pea cultivars. With the advent of novel techniques and the increasing understanding of the host-pathogen system, more efficient breeding methods will be applicable in breeding for resistance to pea rust, while considering the following points.

Gene Pyramiding

Incorporation of more than two resistance genes into a single cultivar against different pathotypes of the same pathogen is known as gene pyramiding. By combining genes conferring resistance to different pathotypes, the cultivar can be used in more diversified environments where different pathotypes are likely to be dominant. In addition, multiple resistance genes may have additive effects and the presence of more genes implies that pathotypes have to be more virulent to all the genes before a resistance cultivar will lose resistance (Crute, 1988). Such procedure has been suggested for breeding for durable resistance in many crops *e.g.* incorporation of *Ur-1* to *Ur-11* genes has been done in common bean against bean rust (*U. appendiculatus*). This method can also be used in breeding for pea rust resistance with the identification of different pathotypes and pathotype specific resistance.

New Resistant Sources

The progress in the development of resistant varieties has been slow due to lack of well characterized resistant sources in pea. So, large scale screening of germplasm for resistance is required. Further, the screening of germplasm lines should be done keeping in view the different slow rusting components. As, against a variable pathogen population only slow rusting lines are known to display durable resistance (Wilcoxson, 1981). Slow rusting is a form of quantitative resistance and quantitative resistance is greatly affected by growth stages of plants and environmental conditions, which masks the actual performance of resistance. Therefore, it is required to ascertain the actual performance of the various genotypes and separate the environmental influence on slow rusting components.

Slow Rusting and/or Partial Resistance

One of the best possible ways to stabilize the productivity of pea crop is to grow rust resistant varieties of pea. Therefore, enhancement of resistance to rust in pea is a major challenge. There are several strategies of developing varieties with durable resistance. These include multilines (Marshall, 1977), partial resistance/slow rusting (Wilcoxson *et al.*, 1975) and gene pyramiding (Green, 1975; Pederson and Leath, 1988). Slow rusting has been observed in food legumes for *Uromyces* rusts (Habtu and Zadoks, 1995; Conner and Bernier, 1982a). The phenomenon of slow rusting to pea rust has not been explicitly reported in literature and little is known about its durability and inheritance (Kumar *et al.*, 1994; Chand *et al.*, 2006).

Incomplete resistance makes the problem tricky to the pea breeders (Chand *et al.*, 2004a; Gupta, 1990; Singh and Srivastava, 1985). The available resistant sources are of slow rusting type (Kumar *et al.*, 1994) and it retards disease develpoment, resulting in intermediate to low disease levels against prevalent pathogen. So, Selection for slow rusting components along with the yield traits are likely to be done when the distribution of rust in the screening field would be normal with 90 per cent rust severity on susceptible check and <20 per cent on the resistance check. (Kushwaha *et al.*, 2007). Selection gain of these lines can be verified in term of less disease severity, low AUDPC, higher seed test weight than the susceptible checks. The gain in increased yield and test weight would be less in resistant genotypes when grown under fungicide protected and unprotected condition (Chand *et al.*, 2006).

Host-Pathogen Interaction System

To identify and incorporate new sources of resistance genes into the breeding programme, it is necessary to have a good understanding of the pathogenic variability of *U. fabae* and host-pathogen interaction. The obligate nature of *U. fabae* makes it difficult to maintain the pathogen in culture and apply to screen segregating population under controlled growth condition. Complication is likely to increase when both the uredial and aecidial spores create disease. Since, infection process of both the spore state is different. Infection by uredia is mostly confined to the epidermal cells and few layers of mesophyll cell. Whereas in the case of aecidial state the infection are deep in the mesophyll and spongy tissues in order to form the aecial cups. Therefore, the level of resistance of same genotype may vary with the different spore and need to be worked out. It can be expected that a better understanding of the pathogenic variability of *U. fabae* will soon be obtained by using molecular markers- based techniques.

Molecular Markers

In recent years, DNA based markers have shown great promises in expediting plant breeding procedures. Identification of molecular markers for resistance genes can efficiently facilitate pyramiding major genes into a valuable background in less time and make it more cost effective. In such special cases of disease resistance breeding, marker assisted selection (MAS) takes on special roles, whereby pyramiding several major resistance genes into a valuable genetic background is simplified (Tanksley *et al.*, 1989).

Using bulked Segregant analysis (BSA), Vijayalkshmi *et al.*, (2005) identified two RAPD markers *viz.* SC10-82_{360} (primer, GCCGTGAAGT) and SCRI-71_{1000} (primer, GTGGCGTAGT), flanking the pea rust resistance gene (*Ruf*) with the distance of 10.8 cM (0.097 rf and LOD of 5.05) and 24.5 cM (0.196 rf and LOD of 2.72). These RAPD markers are not close enough of *Ruf* to allow a dependable marker-assisted selection for rust resistance. However, if the two markers flanking *Ruf* are used together, the effectiveness of MAS would be improved considerably.

Disease Management

Although, fungicides provide an effective means to rust management (Singh, and, Musymi, 1979; Singh, 1984; Singh, *et al.*, 1985). However, it is not cost effective and associated with several environmental and health related hazards. Therefore, cultivation of disease resistant varieties of the crop provides a cheap, effective and safe means of disease control. Resistant varieties as a component in integrated management are also recommended for this disease (Bond, 1994).

Optimum population and timely sowing of crop help in escaping of rust (Rai and Gupta 2003). Incidence of rust in the North India mostly appeared from second week of January on wards. October sown crop generally affected at later stage after grain filling. Late sown crop mostly coincide with most vulnerable stage.

Future Thrust

Incidence of rust in different part of the country especially in the northern and southern hill should be investigated. Host range of rust population in the pea growing area should be worked out. Region wise involvement of uredial and acecial in out break of rust will provide better insight into rust biology. Metrological factors that influence the on set of rust should be studied in detail. Various features of the host which influence the disease is yet to be studied. Studies related to the inheritance of the rust resistance components are essential in order to pyramid the different components of resistance in agronomically suitable background. Latent periods which influence the disease has been characterized in few donors and recombinant inbreed lines raised in susceptible back ground need to be mapped by molecular tools Along with the biochemical changes that delayed the on the set of disease needs to be investigated thoroughly.

References

Alexopolus, C.J., Mims, C.J. and Blackwell, M., 1996. *Introductory Mycology*. John Wiley Sons Inc., 869 pp.

Allen, E.A., Hazen, B.E., Hoch, H.C., Kwon, Y., Leinhos, G. M. E., Staples, R. C., Stumpf, M. A. and Trehune, B.T., 1991. Appressorium formation in response to topographical signals by 27 rust species. *Phytopathology*, 81: 323-331.

Anonymous, 2005. *Annual Report (Rabi)*. All India Coordinated Research Project on MULLaRP, Indian Institute of Pulses Research, Kanpur (India).

Baruah, H K., 1980. *Textbook of Plant Pathology*. Oxford and IBH- New delhi.

Beckett, A. and Woods, A.M., 1987. The pattern of the formation of the uredium of *Uromyces viciae–fabae* in *Vicia faba*. *Canadian J. of Bot.*, 65: 1998-2006.

Beckett, A., Tatnell, J.A. and Taylor, N., 1990. Adhesion and Pre-invasion behaviour of urediospores of *Uromyces viciae-fabae* during germination on host and synthetic surfaces. *Mycological Research*, 94: 865-875.

Bhalla, M.K. and Bernier, C.C., 1984. Evaluation of components of rate reducing resistance in *Vicia faba* to *Uromyces viciae-fabae*. *Phytopathology*, 74: 819.

Bilgrami, K., Jamaluddin, S. and Rizvi, M.A., 1979. *List of Fungi: Part-I. List of References*. Today and Tomorrow's Printers and Publishers, New Delhi, 264 pp.

Blixt, S., 1974. The pea. In: King RC (ed) *Handbook of Genetics*. Plenum Press, New York, Vol 2, pp. 181–221.

Bond, D.A., Gellis, G.S., Rowland, G.G., Le Guen, J., Robertson, L.D., Kholid, S.A. and Le-Juan L., 1994. Present status and future strategy in breeding Faba bean (*Vicia fabae* L) for resistance to biotic stress. *Euphytica*, 73: 151-166.

Butler, E.J., 1918. *Fungi and Diseases in Plants*. Thatcher Spink Co., Calcutta, 547 pp.

Cannon, W.A., 1903. Studies in plant hybrids: the spermatogenesis of hybrid peas. *Bull. Torrey Bot. Club*, 30: 519-543.

Casey, R. and Davies, D.R., 1993. Peas: genetics, molecular biology and biotechnology. *Biotechnol. Agric. Ser.* No. 10, CAB International.

Chand, R., Srivastava, C.P. and Kushwaha, C., 2004. Screening technique for pea (*Pisum sativum* L.) genotypes against rust disease (*Uromyces fabae* Pers de Bary). *Indian Journal of Agricultural Sciences*, 74: 166-167.

Chand, R., Srivastava, C.P., Singh, B.D. and Sarode, S.B., 2006. Identification and characterization of slow rusting components in pea (*Pisum sativum* L.). *Genetic Resources and Crop Evolution*, 53: 219–224.

Chand, R., Srivastava, C.P., Singh, R.M. and Singh, R B., 1997. Pea specific strains in the *Uromyces fabae*. *Indian Journal of Pulses Research*. 10: 127-128.

Chaudhary, R.G., Lal, S.S., Kumar, R. 2000. Lentil rust epiphytotics of Mokama 'Tal' of Bihar. *IIPR Newsletter*, 11(1): 1-16.

Chaudhary, R.G., Naimuddin and Dhar, V., 1998. Host non-specificity of Kanpur strain of pea rust pathogen *Uromyces fabae* (Pers.) de Bary. *Indian Journal of Pulses Research*, 11: 156-158.

Chauhan, M.P., Singh, I.S. and Singh, R.S., 1996. Genetics of rust resistance in lentils (*Lens culinaris*). *Indian Phytopathology*, 49: 387-388.

Chauhan, R.S. and Singh, B.M., 1994. Effect of different durations of leaf wetness on pea rust. *Plant Dis. Res.*, 9: 200-201.

Conner, R.L. and Bernier, C.C., 1982. Host range of *Uromyces viciae-fabae*. *Phytopathology*, 72: 687-689.

Conner, R.L. and Bernier, C.C., 1982a. Slow rusting resistance in *Vicia fabae*. *Can. J. Plant Pathol.*, 4: 263–265.

Conner, R.L. and Bernier, C.C., 1982b. Race identification in *Uromyces vicia fabae*. *Phytopathol.*, 72: 687-689.

Crute, I.R., 1988. The elucidation and exploitation of gene-for-gene recognition. *Plant Pathol.*, 47: 107-113.

Das, M.K., Rajaram, S., Mundtt, C.C. and Singh, R.P., 1993. Associations and genetics of three components of slow rusting in leaf rust of wheat. *Euphytica*, 68: 99-109.

Deising, H., Jungblut, P.R. and Mendgen, K., 1991. Differentiation-related proteins of the broad bean rust fungus *Uromyces viciae-fabae*, as revealed by high resolution two-dimensional polyacrylamide gel electrophoresis. *Archives of Microbiology*, 155: 191-198.

Ellison, P.J., Cullis, B.R. and Kable, P.F., 1992. The effect of light and temperature on *in vitro* germination and germ tube growth of urediniospores of Tranzschelia discolor *Australian Journal of Agricultural Research*, 43: 451-464.

Emeran, A.A., Scillero, J.C., Niks, R.E. and Rubiales, D., 2005. Infection structures of host- specialized isolates of *Uromyces viciae fabae* and of other species of *Uromyces* infecting leguminous crops. *Plant Disease*, 89: 17-22.

FAO, 2005. http://faostat.fao.org/faostat. Last updated February 2005.

Freytag, S., Bruscaglioni, L., Gold, R.E. and Mendgen, K., 1988. Basidiospores of rust fungi (*Uromyces* species) differentiate infection structures *in vitro*. *Experimental Mycology*, 12: 275-283.

Griga, M. and Novak, F.J., 1990. Pea (*Pisum sativum* L.). In: Bajaj, Y.P.S. (ed) *Biotechnology in Agriculture and Forestry 10, Legumes and Oilseed Crops*, pp. 65-99.

Gupta, R.P., 1990. Evaluation of pea germplasm for their reaction to powdery mildew and rust. *Indian Journal of Pulses Research*, 3: 186-188.

Hahn, M. and Mendgen, K., 1997. Characterization of in planta-induced rust genes isolated from a haustorium-specific cDNA library. *Molecular Plant Microbe Interactions*, 10: 427-437.

Hahn, M. and Mendgen, K., 1992. Isolation of ConA binding haustoria from different rust fungi and comparison of their surface qualities. *Protoplasma*, 170: 95-103.

Hahn, M., Neef, U., Struck, C., Gottfert, M. and Mendgen, K., 1997. A putative amino acid transporter is specifically expressed in haustoria of the rust fungus *Uromyces fabae*. *Molecular Plant Microbe Interactions*, 10: 438-445.

Hall, J.L. and Williams, L.E., 2000. Assimilate transport and partitioning in fungal biotrophic interactions. *Australian Journal of Plant Physiology*, 27(6): 549-560.

Heath, M.C., 1976. Ultrastructural and functional similarity of the haustorial neckband of rust fungi and the Casparian strip of vascular plants. *Canadian Journal of Botany*, 54: 2484-2489

Heath, M.C., Skalamera, D., 1997. *Adv. Bot. Res.*, 24: 195-225.

Heisteruber, D., Schulte, P. and Moerschbacherm, B.M., 1994. Soluble carbohydrates and invertase activity in stem rust-infected, resistant and susceptible near-isogenic wheat leaves. *Physiological and Molecular Plant Pathology*, 45: 111-123.

Hiratsuka, Y., 1968. Morphology and cytology of aeciospores and aeciospore germ tube of host alternating and pine-to-pine races of *Cronartium flaccidum* in northern Europe. *Canadian Journal of Botany*, 46: 1119-1122.

Joseph, M.E. and Hering, T.F., 1997. Effects of environment on spore germination and infection by broad bean rust (*Uromyces viciae-fabae*). *Journal of Agricultural Science*, 128: 73-78

Kapooria, R.G. and Sinha, S., 1966. Studies on host range of *Uromyces fabae* (Persoon) de Bary. *Indian Phytopathology*, 19: 224-230.

Kapooria, R.G. and Sinha, S., 1971. Further studies on the host spectrum of *Uromyces fabae* (Persoon) de Bary in India. *Indian Phytopathology*, 24: 170-171.

Katiyar, R.P. and Ram, R.S., 1987. Genetics of rust resistance in pea. *Indian Journal of Genetics*, 47: 46-48

Khvostova, V. V., 1975. *Genetics and Breeding of Peas*. Oxonian Press Pvt. Ltd.

Kumar, T.B.R., Rangaswamy, K.T. and Ravi, K., 1994. Assessment of tall field pea genotypes for slow rusting resistance. *Legume Research*, 17: 79-82.

Kushwaha, C., Srivastava, C.P., Chand, R. and Singh, B.D., 2007. Identification and evaluation critical time for assessment of slow rusting in pea against *Uromyces fabae*. *Field Crop Research*, 103: 1-4.

Kushwaha, C., Chand, R. and Srivastava, Chandra Prakash, 2006. Role of aeciospores in outbreak of pea (*Pisum sativum* L) rust (*Uromyces fabae*). *European J. Plant Patholgy*, 115: 323–330.

Long, D.E., Fung, A.K., McGee, E.E.M., Cooke, R.C.and Lewis, D.H., 1975. The activity of invertase and its relevance to the accumulation of storage polysaccharides in leaves infected by biotrophic fungi. *New Phytologist*, 74: 173-182.

Makasheva, R.K., 1984. *The Pea*. Russian translation series. Gorokh, Kolos Publishers, Leningrad.

Matthews, P. and Arthur, E., 1985. Genetic and environmental components of variation in protein content of peas. In: Hebblethwaite PD, Heath MC, Dawkins TCK (ed) *The Pea Crop*. Butterworths.

Mc Phee, K., 2003. Dry pea production and breeding: A mini review. *Food, Agriculture and Environment*, 1: 64-69.

Mendgen, K. and Hahn, M., 2002. Plant infection and the establishment of fungal biotrophy *Trends in Plant Science*, 7: 352-356.

Mendgen, K., Struck, C., Voegele, R.T. and Hahn, M., 2000. Biotrophy and rust haustoria. *Physiological and Molecular Plant Pathology*, 56(4): 141-145.

Mims, C.W., Rodriguez-Lother, C. and Richardson, E.A., 2001. Ultrastructure of the host-parasite interaction in leaves of *Duchesnea indica* infected by the rust fungus *Frommeela mexicana var. indicae* as revealed by high pressure freezing. *Canadian Journal of Botany*, 79: 49-57.

Mitter, J.H. and Tandon, R.N., 1930. Fungi flora of Allahabad, India. *Journal of Indian Botanical Society*, 9: 190-196.

Narsinghani, V.G., Singh, S.P. and Pal, B.S., 1980. Note on Rust resistance pea varieties. *Indian J. Agricultural Sciences*, 50: 453.

Negussie, T., Pretorius, Z.A. and Bender, C.M., 2005. Components of rust resistance in lentil. *Euphytica*, 142: 55-64.

Pal, A.B., Brahmappa, B., Sohi, H.S. and Rawal, R.D., 1979. Studies on inheritance of resistance to rust (*Uromyces fabae* Pers. de Bary) on pea. *SABRAO J.*, 11: 101-103.

Parlevliet, J.E. and Kievit, C., 1986. Development of barley leaf rust *Puccinia hordei*, infection in barley 1. Effect of partial resistance and plant stage. *Euphytica*, 119: 953- 959

Patel, M.K., 1934. *Indian Bulletin of Plant Protection*, (8): M199-200.

Pavgi, M.S. and Upadhyay, H.P., 1966. Parasitic Fungi from North India IV Mycopathology et. *Mycological Applications*, 30: 257-260.

Persoon, D.C.H., 1801. *Synopsis methodica fungorum*, 1: 224.

Pfender, W.F., 2001. A Temperature-Based Model for Latent-Period Duration in Stem Rust of Perennial Ryegrass and Tall Fescue. *Phytopathology*, 9: 111-116.

Prasada, R. and Verma, U.N., 1948. Studies on lentil rust, *Uromyces fabae* (Pers) de Bary in India. *Indian Phytopathology*, 1: 142-146.

Rai, O.P. and Gupta, R.P., 2003. Effect of sowing date and population density on yield and rust in dwarf pea. *Indian Journal of Pulses Research*, 16: 34-35.

Rashid, K.Y. and Bernier, C.C., 1991. The effect of rust on yield of Faba bean cultivars and slow-rusting populations. *Canadian Journal Plant Science*, 71: 967-972.

Roy, T.C., 1949. *Fungi of Bengal*. Directorate of Agriculture. Govt. of West Bengal.

Rubiales, D., Emeran, A.A. and Sillero, J.C., 2002. Rusts on legumes in Europe and North Africa. *Grain Legume*, 37: 8-9.

Sillero, J.C., Moreno, M.T. and Rubiales, D., 2000. Characterization of new sources of resistance to *Uromyces viciae-fabae* in a germplasm collection of *Vicia faba*. *Plant Pathology*, 49: 389-395.

Sillero, J.C. and Rubiales, D., 2002. Histological characterization of resistance to *Uromyces viciae-fabae* in faba bean. *Phytopathlogy*, 92: 294-299.

Singh, K., Rawal, H.S., Singh, S. and Gumber, R.K., 1995. Physiological specialization in *Uromyces fabae* causing rust of lentil. *Lens Newsl.*, 16: 41.

Singh, R. and Ram, H., 2001. Inheritance of days to flowering and rust resistance in peas. *Researches on Crops*, 2: 414-418.

Singh, R.M. and Srivastava, C.P., 1985. Evaluation classification and usefulness of pea germplasm lines for quantitative characters. *Legume Research*, 8: 68-73.

Singh, R.S., 1999. *Plant Diseases*. Oxford IBH Pub., New Delhi, pp. 686.

Singh, S.J. and Sokhi, S.S., 1980. Pathogenic variability in *Uromyces viciae fabae*. *Plant Disease*, 64: 671-672.

Singh, S.J. and Sokhi, S.S., 1984. Evaluation of pea cultivars to *Uromyces viciae fabae*. *Indian Phytopathology*, 21: 85.

Singh, J.P. and Musymi, A.B.K., 1979. Control of rust and powdery mildew by a new systemic fungicide, Bayleton. *Pesticides*, 13: 51-53.

Singh, K., 1984. Field evaluation of fungicides against lentil rust caused by *Uromyces viciae fabae*. *Indian Phytopathology*, 37: 396.

Singh, K., Jhooty, J.S. and Cheema, H.S., 1986. Assessment of losses in lentil yield due to rust caused by *Uromyces viciae-fabae*. *LENS Newsletter*, 13(1): 28.

Singh, R.R., Singh, M. and Singh, M., 1997. Chemical control of pea rust. *Annals of Plant Protection Sciences*, 5: 118-119.

Singh, S.N., Srivastava, S.K. and Agrawal, S.C., 1985. Relative efficacy of fungicides in the control of seedling blight and rust of lentil. *Indian Journal of Agricultural Sciences*, 55: 284-286.

Singh, S.N., Srivastava, S.K., Agrawal, S.C., 1985. Relative efficacy of fungicides in the control of seedling blight and rust of lentil *Indian Journal of Agricultural Sciences*, 55: 284-286.

Sinha, R.P. and Yadav, B.P., 1989. Inheritance of resistance to rust in lentil. *Lens Newsl.*, 22: 46-47.

Sohn, J., Voegele, R.T., Mendgen, K. and Hahn, M., 2000. High level activation of vitamin B_1 biosynthesis genes in haustoria of the rust fungus Uromyces fabae. *Molecular Plant Microbe Interactions*, 13: 629-636.

Sokhi, H.S., Sokhi, S.S. and Rawal, R.D., 1974. Varietal reaction of pea to powdery mildew (*Erysiphe polygoni*) and rust (*Uromyces vicia fabae*). *Mysore J. Agricultural Sciences*, 8: 529-532.

Sokhi, S.S., 1984. A distinct physiologic race of *Uromyces viciae fabae* on sweet pea. *Indian Phytopathology*, 37: 123-125.

Staples, R.C., 2000. Research on the rust fungi during the twentieth century. *Annual Review of Phytopathology*, 38: 49-69.

Stebbins, G.L., 1974. *Flowering Plants: Evolution Above the Species Level*. Belknap Press of Harvard University Press, Cambridge, Massachusetts.

Stoddard, F.L. and Herath, I.H.M.H.B., 2001. Genetic analysis of partial rust resistance in faba beans. *Australian Journal of Agricultural Research*, 52: 73–84.

Sydow, H. and Butler, E.J., 1912. Fungi Indiae Orientalis pars IV. *Annals of Mycology*, 9: 372-421.

Tanksley, S., 1983. Molecular markers in plant breeding. *Plant Mol. Biol. Rep.*, 1: 3-8.

Upadhyay, A.L. and Singh, V.K., 1994. Performance of pea varities/lines against powdery Mildew and Rust. *Indian Journal of Pulses Research*, 7: 92-93.

Uppal, B.M., 1933. *International Bulletin of Crop Protection*, 7: M103 &M746.

Vijayalakshmi, S., Yadav, K., Kushwaha, C., Sarode, S.B., Srivastava, C.P., Chand, R. and Singh, B.D., 2005. Identification of RAPD markers linked to rust resistance in pea (*Pisum sativum* L). *Euphytica*, 144: 265-274.

Voegele, R.T. and Mendgen, K., 2003. Rust haustoria: Nutrient uptake and beyond. *New Phytologist*, 159: 93-100.

Voegele, R.T., Struck, C., Hahn, M. and Mendgen, K., 2001.The role of haustoria in sugar supply during infection of broad bean by the rust fungus *Uromyces fabae*. *Proceedings of the National Academy of Sciences of the United States of America*, 98 : 8133-8138 .

Voegele, R.T., Hahn, M., Lohaus, G., Link, T., Heiser, I. and Mendgen, K., 2005. Possible roles for mannitol and mannitol dehydrogenase in the biotrophic plant pathogen *Uromyces fabae*. *Plant Physiology*, 137: 190-198.

Wilcoxson, R.D., 1981. Genetics of slow rusting in cereals. *Phytopathology*, 71: 989-993.

Wynn, W.K., 1976. Appressorium formation over stomata by the bean rust fungus response to a surface contact stimulus. *Phytopathology*, 66: 136-146.

Xue, A.G. and Warkentin, T.D., 2002. Reactions of field pea varieties to three isolates of *Uromyces fabae*. *Canadian Journal of Plant Science*, 82: 253–255.

Zohary, D. and Horpf, M., 1973. Domestication of pulses in the old world. *Science*, 182: 887-894.

Chapter 17

Management of Fungal and Bacterial Diseases of Tomato

M.K. Naik, G. Ramegowda, S. Sudha and Y.S. Amaresh
University of Agricultural Sciences, College of Agriculture, Raichur – 584 101, Karnataka

Introduction

Tomato (*Lycopersicon esculentum* Mill.) is one of the most popular and widely grown vegetables in the world ranking second in importance to potato in many countries. Tomato is known for its outstanding nutritive value, and is popular also because it supplies vitamin-C. On the basis of fresh weight, vitamin-C content averages about 25 mg/100g, however the values vary with cultivars. Tomato also contains folic acid, pantothenic acid, riboflavin (0.01 mg) and thiamine (0.07 mg). It has 3.6 per cent carbohydrates, 1.9 per cent protein, 0.1 per cent fat, 0.7 per cent fiber, 900 I.U. of Vitamin A and 0.5 per cent minerals. The global production area is around 4,397,873 ha, which is second only to potato production among vegetable crops. In India tomato is produced in an area of 497000 ha, with a production 8637000 million tons. The prevalence of high humidity and warm temperature during growing season makes the crop vulnerable to the attack of various fungal and bacterial pathogens causing significant yield losses. The present status of knowledge of fungal and bacterial diseases regarding their distribution, economic importance, symptomatology, etiology, epidemiology and management practices has been reviewed in this chapter.

Fungal Diseases

Damping Off

The disease is common in poorly managed nursery beds, incited often by the soil borne pathogens. It is responsible not only for the poor seed germination and stand of seedlings but also for carry over of the pathogens to the field where transplanting is done. The symptoms of disease occur in two phases *i.e.*, pre-emergence and post- emergence damping off. There is failure of seedling emergence from the soil either due to seed rots or killing of young seedlings before their emergence from the soil, hence resulting in patchy appearance of seedlings stands in the nursery in early stages of growth. In case of post-emergence damping off, the disease outbreak is characterized by toppling over of infected seedlings at any time after their emergence from the soil. The infected tissue initially appears to be water soaked and soft and subsequently, the stem at the infection points get constricted resulting in toppling over and mortality of the seedlings. In the nurseries the disease usually radiates from initial infection points and thus gives the appearance of Islands where almost all seedlings are killed.

Etiology

The commonly associated pathogens with this disease are *Pythium aphanidermatum* (Edson) Fitzp., *P. butleri* Subram., *P. inflatum* Mathews, *Phytophthora nicotianae* var. *parasitica* (Dastur) Waterhouse, *Fusarium oxysporum* Schlechtend ex Fr., *F. solani* (Mart.) Sacco f.sp. minus, *Rhizoctonia solani* Kuhn and *Sclerotium rolfsii* Sacco (Tripathi and Grover, 1978; Chandra *et al.* 1983; Verma, 1987 and Chima-Wokocha and Wokocha, 1987). Among these fungi, the species of *Pythium* are more common than others. The mycelium of *Pythium* is intracellular in the host. It consists of slender, coenocytic and much branched hyphae. Asexual reproduction is mostly by means of zoospores which are always differentiated in a vesicle extruded by the zoosporangium. The sporangium forms a beak, the tip of which softens and out of which flows the protoplasmic mass of zoosporangium to form a spherical vesicle. In the vesicle, protoplasm cleaves to form zoospores which are kidney shaped and biflagellate. They rupture the wall of the vesicle and escape in the surrounding film of water. After swimming for a while, the zoospores encyst and then germinate by germtube. In sexual reproduction, the oogonium is plurinucleate, arises terminally less often intercalary on long or short unmodified hyphae with homogeneous protoplasm and only one nucleus migrate into the periplasm. The antheridium is also binucleate at first but only one nucleus functions in the fertilization. The branch bearing the terminal antheridium may arise at a distance from the oogonium on a separate branch or from the same branch. Through a conjugation tube, one male nucleus passes into the egg which then forms a thick wall and becomes oospore. *Pythium aphanidermatum* and *P. butleri* are closely related species. The mycelium of *P. aphanidermatum* is white, well developed and woolly but not as copious or robust as of *P. butleri*. The zoosporangia are lobed, toruloid, branched but smaller and less profusely branched than those of *P. butleri*. The swellings sometimes only form bud like outgrowths from the main sporangia. Encysted zoospores measure 7.6 to 121µm. Oogonia are spherical, smooth walled, terminal on lateral hyphae and 19 to 29 µm in diameter. Antheridia are monoclinous, intercalary or terminal. One or two antheridia attach to each oogonium. They are usually barrel or dome shaped and measure 9-11 x 1 0–14 µm in size. Oospores are aplerotic, single, moderately thick walled and 17-19 µm in diameter. *P. butleri* differs from *P. aphanidermatum* in terms of more copious robust aerial mycelium, larger, more swollen and more branched sporangia, larger zoospores, antheridia, oogonia and oospores.

Epidemiology

The pathogens associated with this disease are either seed or soil borne or both. Species of *Pythium* attack the plants when they are in juvenile stage. Soil moisture near saturation stimulates mycelial growth as well as asexual reproduction. Propagules capable of causing infection of tomato include vegetative mycelium, germinating sporangia, zoospores and oospores. Some propagules, particularly zoospores may come into contact with tomato plants by chemotaxis. Infective propagules are capable of direct penetration, but wounds enhance penetration and infection. High soil moisture and relatively higher soil temperature favour the rapid development of the disease. The disease is further aggravated in ill-aerated soils with poor drainage having thick stand of the seedlings. Soils low in percentage clay content were more conducive to the disease than those with high clay contents (Chima Wokocha, 1987). All salt levels increased infection of tomatoes by damping-off caused by *Rhizoctonia solani, Fusarium solani* in field soil and *Pythium aphanidermatum* (Kassim *et al.*, 1989).

Management

Cultural Practices

Cultural practices to reduce damping off include avoidance of nursery sowing in same bed year after year and frequent heavy irrigations. Soils with low clay content were more conducive to the disease particularly in case of *Sclerotium rolfsii* than with high clay content (Chima- Wokocha and Wokocha, 1987). Aqueous leaf extracts of *Hyptis suaveolens, Murraya koenigii and Ocimum canum [O. americanum]* completely inhibited the growth of *Pythium aphanidermatum* and *P. debaryanum in vitro.* Soil amendments with leaves of *H. suaveolens, M. koenigii and O. americanum* controlled damping-off disease of tomato by up to 100.0, 55.6 and 33.3 per cent, respectively, in soil infested with *P. aphanidermatum* and by up to 100.0, 55.3 and 77.8 per cent, respectively in soil infested with *P. debaryanum* (Pandey and Dubey, 1997).

Damping-off of tomato caused by *Rhizoctonia solani* was controlled in a pot test using the biological agent, *Bacillus subtilis* RB14-C, and the chemical pesticide, Flutolanil. The co-utilization of *B. subtilis* RB14-C, and Flutolanil decreased the amount of Flutolanil used from 375 micro g/pot when Flutolanil was used alone to 94 µg/pot, while exerting the same effect of reducing disease occurrence Kondoh *et al.* (2000).

Chemical Control

Soil sterilization and seed treatment with fungicides are commonly used for the management of this disease. Soil sterilization and soil application of mercuric chloride (1:1500) was recommended for the control of this disease. Seed treatment with cuprous oxide, copper sulphate or phenyl mercury acetate, soaking in copper sulphate solution, copper oxychloride, S- 7250 and RH-893 (2-n octyl-4-isothiozoline-3-one), carbendazim, captan and thiram is recommended (Tripathi and Grover, 1978; Vyas *et al.*, 1981). The best control was given by copper oxychoride and/or Emisan Brassicol [quintozene] followed by Topsin M [thiophanate-methyl] were most effective at controlling damping off of tomato due to *R. solani*. Thiophanate-methyl and Bavistin [carbendazim] were highly effective at controlling damping off caused by *M. phaseolina* on tomato and chilli, respectively (Satija and Indra-Hooda, 1987). Efficacy of soil fumigation with metham sodium and methyl bromide in reducing soil borne fungi has been well documented (Mc Carter *et al.*, 1978). Pre-sowing drench with S- 7250 and RH-893 and a post-sowing soil drench with captan, captafol, dichlozoline or S- 7258 gave better control of the disease (Tripathi and Grover, 1978). The best control of damping-off of tomatoes was noticed by 2-Methoxyethylmercury chloride at 1 g/kg seed or in combination with copper oxychloride (1+1 g/kg seed) was the best fungicide in controlling the disease on both the crops. However, a combination of thiophanate-methyl with 2-methoxyethylmercury chloride and copper oxychloride was also very effective (Dharam-Vir and Indra-Hooda, 1989). Ridomil (metalaxyl), captan (captan) and dithane M-45 (mancozeb), each at 3 different concentrations applied as soil drench, effectively controlled the disease. Captan was effective only at higher concentration. The number of leaves, and root and shoot lengths were significantly reduced by all the fungicides (Ayub *et al.*, 1998).

Biocontrol and Integrated Management

Krishnamoorthy and Bhaskaran (1990) reported that the soil inoculation with *Trichoderma viride*, *T. harzianum*, and *Laetisaria arvalis* gave good control of *P. indicum*, and there was a manifold increase in seed germination in comparison to control. The seed treatment with these bioagents was comparable with soil drenching with copper oxychloride or seed treatment with captan or thiram. Although soil and seed applications of *Gliocladium virens* and *Trichoderma longibrachiatum* against *Rhizoctonia solani* and *Pythium aphanidermatum* were also efficacious yet, in general soil application of these bioagents conferred greater protection than the seed dressings (Sreenivasaprasad and Manibhushanrao, 1990). Inoculation of mycorrhizal fungus *Glomus fasciculatum* either simultaneously or 2 weeks before inoculation with *Pythium aphanidermatum* reduced damping off and increased plant height, and shoot and root weight (Hegde and Rai, 1984). However, *Gomus leptotichum* and *Acaulospora morrowae* a mycorrhizal fungi showed greater disease resistance than non-mycorrhizal ones (Cassiolata and Melo, 1991). Mycorrhizal inoculations not only reduced the percentage of disease incidence but also increased the fruit yield and fruit weight. AM association also significantly increased the host plant height, total biomass and dry matter by enhancing nutrient status (Reddy *et al.*, 2006). The integration of *T. viride* and sawdust (1.0 per cent) was more effective in the reduction of the incidence of damping off (15 per cent) than *T. viride* (44 per cent) or wheat bran (59 per cent) in *P. indicum* inoculated unsterilized soil (Neelamegam, 2006). Mycorrhizal inoculations not only reduced the percentage of disease incidence but also increased the fruit yield and fruit weight. AM association also significantly increased the host plant height, total biomass and dry matter. It is clear from the study that inoculation of plants with AM fungi will improve plant health by enhancing nutrient status, yield and also by protecting the host plants against pathogens. (Reddy *et al.*, 2006). Treatments comprised metalaxyl seed treatment at 3 g/kg seed (ST), neem cake at 900 g/bed (NC), *Trichoderma harzianum* at 320 g/bed (BA), and/or soil drenching with 1 per cent Bordeaux mixture (SD). Among the treatments, pre-emergence damping off was less in treatments having soil application of *T. harzianum* as evidenced by good percent of seed germination. Of them, the lowest disease incidence was observed in ST+NC+BA+SD, followed by ST+NC+BA, showing a seed germination of 60.71 per cent and 59.57 per cent, respectively. (Kumar and Rao 2006). Seed treatment with *T. harzianum* formulations reduced the incidence of damping-off disease of tomato by up to 74 per cent and enhanced plant biomass under greenhouse and field (Tamil Nadu, India) conditions. (Jayaraj *et al.*, 2006).

Soil solarization with white polythene sheet for 30 days during the peak of summer resulted in the lowest incidence of damping-off in tomato (3.9 per cent), aubergine (1.5 per cent) and chilli (1.3 per cent) caused by *Pythium*, *Fusarium* and *Phytophthora*, followed by seed treatment with captan. Among the fungal antagonists, *Trichoderma viride* recorded the lowest incidence of damping-off (5.7 per cent in tomato, 4.7 per cent in aubergine and 4.7 per cent in chilli) followed by seed treatment with *Pseudomonas fluorescens*, *T. harzianum* and *Azotobacter chroococcum* which were at par with each other. (Rahman *et al.*, 2003). *T. viride* along with soil solarization effectively reduced the pre- and post-emergence damping off by 75.23 and 88.34 per cent, respectively, over the control with concomitant increase in seedling vigour index and biomass of tomato (Gnanavel and Jayaraj, 2003). *Pseudomonas fluorescens* isolate Pf1 was effective in reducing the damping-off incidence in tomato and hot pepper in greenhouse and field conditions. Isolate Pf1 was further tested for its ability to induce production of defence-related enzymes and chemicals in plants. Earlier and increased activities of phenylalanine ammonia-lyase (PAL), peroxidase (PO) and polyphenol oxidase [catechol oxidase] (PPO) were observed in *P. fluorescens* Pf1 pretreated tomato and hot pepper plants challenged with *Pythium aphanidermatum*. (Ramamoorthy *et al.*, 2002). Damping-off of tomato caused by *Rhizoctonia solani* was controlled in a pot test using the biological agent, *Bacillus subtilis* RB14-C, and the chemical pesticide, Flutolanil. The co-utilization of *B. subtilis* RB14-C and Flutolanil decreased the amount of Flutolanil used from 375 µg/pot when Flutolanil was used alone to 94 micro g/pot, while exerting the same effect of reducing disease occurrence. (Kondoh *et al.*, 2000). The effects of *Trichoderma viride* and farmyard manure (FYM), singly or in combination, on damping-off (*Pythium indicum* [*P. deliense*]) incidence and growth of tomato cv. Co-1 was studied in a pot culture experiment. *T. viride* singly or in combination with FYM reduced the disease incidence in *P. indicum*-inoculated unsterilized (44 and 14 per cent) and sterilized (56 and 41 per cent) soils. *P. indicum* reduced seedling length and biomass; *T. viride* with FYM enhanced these parameters (Neelamegam and Govindarajalu, 2002).

Integration of soil solarization of raised nursery bed with polythene sheet for six weeks during summer and followed by seed treatment with metalaxyl at rate of 5g/kg seeds and then seed coating with bioagent, *Trichoderma viride* resulted in management of damping of tomato under field condition (Naik *et al.*, 2001; Naik, 2003). The population of pathogen *Pythium, Phytophthora, Rhizoctonia* and *Fusarium* was reduced by 90 per cent and incidence of damping of came down drastically. Further the stand and vigour of the crop improved substantially (Naik *et al.*, 2001).

Buckeye Rot

In India, this disease has been reported for the first time from Solan, Himachal Pradesh (Jain *et al.*, 1961), subsequently from Karnataka, Haryana and Punjab. Losses from this disease ranged from 35 to 40 per cent (Jain *et al.*, 1961). The fruit rot incidence may go upto 90 per cent under high humidity and good rainfall conditions while the annual loss has been estimated to be about rupees 3 million (Sharma, 1971).

Buckeye rot (*Phytophthora nicotianae* var. *parasitica*) is a major constraint in tomato cultivation and has been reported from all tomato growing areas of the world, where high relative humidity coupled with warm weather conditions prevail. In India, losses from this disease range between 35 and 40 per cent, which may rise with the prevalence and severity of disease depending upon the favourable weather conditions. (Gupta *et al.*, 2005). The pathogen has been reported to infect the plants at all the growth stages causing damping off, collar rot and stem canker, blossom blight and buckeye fruit rot (Sharma, 1971) of which buckeye fruit rot is very conspicuous especially in those areas where tomatoes are grown during monsoon season. Infection is generally noticed first on any region of the lower immature green fruits which are either in contact with soil or near to soil level, as water soaked light brown discolouration. These spots increase readily showing concentric dark brown rings slightly resembling the markings on a buckeye nut. Subsequently, the disease spread to the upper fruits. The lesions rapidly enlarge and in 3 to 4 days entire fruit surface turn dark brown and become soft. White flocculent superficial growth of the fungus consisting of sporangia develops profusely on the diseased fruits in warm and humid weather. Later, these fruits may drop down in the field. Besides, the zonated fruit spots, the spots without zonations are also observed.

Etiology

The disease is caused by *Phytophthora nicotianae* Breda de Hann var. *parasitica* (Dastur) Waterhouse. The hyphae are thin, sparsely branched and aseptate. The sporangiophore arise from anyone of the hyphal threads and produce sporangia generally at the tip in nature while in culture they are tenninal as well as intercalary. The sporangia are broadly ovoid to globose in shape or sometimes elongated to barrel shaped which are borne terminally on the sporangiophore. The dislodged sporangia have sometimes a very short pedicel and usually have one hemispherical papilla at the tip but occasionally two papillae are also found. The sporangial size vary from 25.5–58.0 x 20.0-52.0 µm on the host while in culture are 24.0-52.5 x 19.5–48.5 µm. The sporangia usually germinate by zoospores, though germination by germtube also takes place at high temperature (27°C and above). The chlamydospores are abundantly produced on oat meal agar in cultures. They are smooth, globose, slightly yellowish with thick brown walls and measure 20-60 µm in diameter. They germinate both by zoospores and germtubes. Antheridia are amphigynous, spherical or oval and l0x 12-16 µm in size. Oogonia are usually produced in single culture though often very sparsely or not until after some weeks. They are readily produced when opposite strains are grown together in culture. Oogonia measure less than 28 µm, rarely 31 µm. On maturity, oogonia become rough, thick walled and yellow brown. Oospores are aplerotic, 18 to 20 µm in diameter with a 2 µm thick wall.

Epidemiology

The pathogen survives in the form oospores in soil and in the infected seed (Sharma *et al.*, 1976a). The fungus remains viable for more than two years under ordinary field conditions in the soil and a depth of 10 cm is more favourable for the penetration of the fungus and it can remain viable for a longer period under wet condition as compared to dry condition. Rainfall has been found to be the most important factor for the development of the disease. Sharma *et al.* (1979) reported that disease is not expected to occur at or below 20°C and 22 to 25°C supplemented with slight rainfall (10mm) will result in disease appearance after 4 days. The disease is not serious at temperature exceeding 25°C. Higher doses of nitrogen gave more yield but significantly higher disease incidence whereas, potash levels generally had little effect on yield or incidence (Shanna and Sohi, 1983). Cultural and pathogenic variation among different isolates of *P. nicotianae* var. *parasitica* from different host plants have been reported (Sharma and Sharma, 1987).

Gupta *et al.* (2005) observed that the pathogen has a wide host range infecting both cultivated as well as wild plant species. This fungus survives in the form of chlamydospores or oospores in the soil and infected seed. Infection takes place through zoospores which is favoured by temperature ranging from 20 to 25 °C, relative humidity above 60 per cent and high rainfall.

Management

Cultural Practices

Williams *et al.* (1943) recommended the destruction of diseased fruits and trimming of lower leaves to avoid moist and stagnant air conditions. Plant should be staked or tied to frames to prevent contact with the soil (Welch, 1949). Wider spacing 1.2-1.8 m x 0.3 m should be adopted in unstaked commercial plantings (Loest, 1948). Foliage and fruits should be removed from ground level to control the disease (Sharma *et al.*, 1976b).

Chemical Control

Various workers have recommended the spray applications of a wide array of systemic and non-systemic fungicides *viz.* Bordeaux mixture, tribasic copper sulphate, mancozeb, basamid, Solasan, Vapam and ditrapex (Raicu and Stan, 1973, Raicu *et al.*, 1974), Difolatan and copper oxychloride (Sokhi and Sohi, 1974), Ridomil MZ (0:25 per cent), Curzate (0.03 per cent) + Indofil M-45 (0.25 per cent) and Indofil M-45 (0.25) (Shyam and Gupta, 1996b). Pine needle mulch with Ridomil, MZ (0.25 per cent) spray on mulch effectively reduced the disease (Shyam and Gupta, 1996a). An integrated approach consisting of spray of mancozeb or Bordeaux mixture or captafol along with staking and removal of foliage and fruits upto 30cm was found to be most effective and economical (Sharma *et al.*, 1976c, Sharma, 1992). Further, the sequential sprays of captafol, mancozeb and copper oxychloride at

40, 55 and 70 days after transplanting increased yield by 50.5 per cent by reducing the disease incidence (Bhardwaj, 1991). Integration of fungicide (Indofil M-45) sprays, cultural practices like periodical clipping of lower leaves (15-20 cm), weeding, mulching with white polythene sheet and removal of affected fruits not only reduce disease incidence but also increased fruit yield (Dodan *et al.*, 1994).

A cultural and chemical management trial on buck eye rot disease was conducted at the Agriculture Research Station, Lumle, Nepal during July-December 2000. The effects of straw mulching, cultivar, fungicide {metalaxyl} and {copper oxychloride} and their interactions were evaluated. Mulching showed a significant difference in marketable and unmarketable yields. (Malla and Sah, 2002). Further, the interaction between fungicide and mulching showed a significant difference in marketable yield. The combination of Ridomil and mulching produced the highest marketable yield (4.07 t/ha). Similarly, the interaction between fungicide and cultivar had a significant difference in marketable yield. The combination of Ridomil and Pusa Ruby produced the highest yield (4.18 t/ha). (Malla and Sah, 2002).

Grote and Bucsi (2004) noticed that preventative and curative leaf applications of fosetyl were effective for the control of *P. nicotianae* on tomatoes grown in hydroponic culture. Two curative applications of 800 mg fosetyl/litre controlled disease for 21 days, and one preventative application controlled disease for 50 days.

Although various methods have been adopted to control this disease, which include host resistance, cultural practices, use of chemicals and integrated disease management, information regarding management of this disease through biological methods, soil amendments, soil solarization, biopesticides and other eco-friendly methods is lacking.

Host Resistance

Several varieties/hybrids such as hybrid N-64 (STEP 390 x HES 6578), Rossal, EC 129171/PI–2, EC 128-969, EC 129-166, EC 72901, EC 122062- 1 and hybrid I 0 (Thareja and Srivastava, 1989) were resistant to the disease. Dodan *et al.* (1995) screened 110 cvs/lines and only one line *viz., Lycopersicon esculentum* var. *ceresiformae* was found resistant under natural epiphytotic conditions. The factors affecting the buckeye rot resistance are plant height, ascorbic acid content, total acidity and total nitrogen content of green fruits (Sharma *et al.*, 1976). Polyphenol oxidase activity is also higher in resistant varieties than in susceptible ones (Sharma and Bhardwaj, 1977).

Joshi *et al.* (2005) evaluated 73 genotypes under Indian conditions. None of them was completely free from the disease. Seven genotypes (LO 701, EC 113809, EC 392693, EC 129602, EC 129603, LO 1003 and EC 191538) were moderately resistant, whereas 47 were moderately susceptible. Some of these moderately resistant or moderately susceptible genotypes possessed desirable horticultural traits. EC 129602 gave the highest yield (985.70 g per plant), average fruit weight (45 g) and ascorbic acid content (47.5 mg/100 g), along with the lowest (6.15 per cent) level of disease incidence.

Late Blight

This disease is widespread in all the tomato growing areas of the world where the temperature and moisture permit the pathogen to thrive. On tomato, it was first recorded in India in 1900. Late blight affects leaves, stems and fruits. On leaves, it appears as pale, green, irregular spots on the tips and margins which in moist weather enlarge rapidly with central tissue turning necrotic and dark brown or black. On the ventral surface of the leaves, a white downy growth of the fungus appears around the dead areas. In dry weather, the growth of the fungus is checked and the water soaked areas dry up and turn brown. Brown streaks develop along the stems. On green fruits olivaceous greasy spots appear which gradually cover the entire fruit. Initially, the tissue remain firm with varying depths of discolored tissue below the skin but when blight is followed by soft rot, the fruit soon disintegrates.

Etiology

The disease is caused by *Phytophthora infestans* (Mont.) de Bary. The fungus is identified by its characteristic lemon shaped, hyaline sporangia which emerge through leaf stomata. Sporangia (21-38 x 12- 23 µm) may germinate directly through germtube or under cool moist conditions through zoospores which are upto eight in number. Each biflagellate zoospore swim in film of water for sometime, shed flagella, become encysted and initiate new infections. Sexual reproduction in *P. infestans* is not very common in nature. The antheridium is amphigynous. The oogonium is spherical. The germination of oospore has not been seen and is presumed that it germinates by germtube. Vartanian and Endo (1985) identified two races *viz.*, T_0 and T_1 by the reaction on tomato cultivars.

Epidemiology

The fungus is soil borne and mostly survives in infected potato seed tubers, cull piles, neighboring potato fields, infected plant debris and other host plants. For development of the disease, low temperature and high humidity are of fundamental importance. The most favorable temperature for successful germination of sporangia either via zoospores or directly by germtubes is 12 and 24°C, respectively. Temperature above 30°C are considered unfavorable for disease development.

Garcia *et al.* (2001), analysed late blight severity which showed recurring patterns, such as the earliest onset of the disease in the area where both potato and tomato were grown, strengthening the hypothesis that infected potato tubers are the main source of primary inoculum.

Management

Cultural Practices

Avoidance of direct contact of the fruits and foliage with soil is recommended. In case of indeterminate varieties like Solan Gola and hybrids like Naveen and BSS-99 staking is recommended. Mulching with pine needles and dried grass can also be helpful in reducing the fruit rot phase. Diseased fruits should be collected regularly and dumped in the pits and while spraying the fields with fungicides, pits should also be sprayed. Improve drainage of the field and keep field free from weeds.

Tumwine *et al.* (2002) evaluated three cultural practices in repeated experiments for their effectiveness in alleviating the adverse effects of sanitation. Tomatoes grown within plastic shelters early in the production cycle were taller, and had more healthy leaves than those grown late. Fungicide-treated plants retained the highest number of flowers and attached fruits and gave the highest yields.

Chemical Control

All the fungicides used for the control of buckeye fruit rot can be safely applied against this disease also. In recent years some systemic fungicides like metalaxyl and oxadixyl have been found effective. However, there are chances of development of resistant strains. It will always be better to apply such systemic fungicides either in combination with non-systemic fungicides like mancozeb or copper oxychloride or alone once or twice to reduce the build-up of inoculum in the field. Mancozeb has poor curative activity. (Samoucha *et al.*, 1987) but when it was applied in protective spray programme along with one spray of Ridomil MZ at a critical stage of disease appearance followed by again a spray of mancozeb provided excellent control of the disease and increased fruit yields. Besides, prophylactic application of aqueous and ethanolic leaf extracts of *Reynoutria sachalinensis* has also been claimed to provide 45 to 70 per cent control of the disease (Herger *et al.*, 1988).

Metalaxyl at 50 g/kg was successfully applied to control late blight of tomatoes. However, treatments using metalaxyl + chlorothalonil (80 + 400 g/kg) and metalaxyl + mancozeb (80 + 640 g/kg) were more effective in controlling late blight, and increasing tomato yields. (Rodrigues *et al.*, 2001).

Chander mohan and Thind (2001) evaluated the performance of cymoxanil+mancozeb (Curzate M-8, 0.25 per cent). The treatments were given as preventive and curative applications of fungicides at 7 and 10 day intervals for contact and systemic fungicides, respectively. Among the various prophylactic treatments, Ridomil MZ alone, Ridomil MZ followed by Indofil M-45 and Indofil M-45 followed by Ridomil MZ were the most effective control. The highest fruit yield was obtained with first spray of Ridomil MZ followed by 3 sprays of Indofil M-45.

Host Resistance

Rubin and Cohen, (2006) developed an improved, soil-free laboratory method to induce late blight infection in detached tomato leaves or tomato seedlings with in *vivo*-produced oospores of *Phytophthora infestans*. Oospores were produced on detached tomato leaves infected with an inoculum mixture of A1 and A2 isolates. The infected leaves were homogenized in distilled water. The homogenates were exposed to two cycles of drying and wetting to kill sporangia, zoospores, and mycelia and, thereafter, mixed with perlite and water. Tomato leaves or seed were floated on the perlite mixture in petri dishes and incubated for up to 30 days. Results from 13 experiments showed that, in dishes containing oospores, 96 of 820 leaflets (11.7 per cent) developed late blight lesions within 6 to 30 days; whereas, in 3 experiments with seedlings, 12 of 1,400 plants (0.9 per cent) were blighted within 17 to 26 days. This improved method may facilitate our understanding of the role of in vivo-produced oospores in the epidemiology of late blight disease progress. Tomato cultivars/lines *viz.*, Heinemannas Jubilaum (Czyzewska and Bialkowska, 1973), Rossol VFN, Piacenza 0164 and Roma (Erinle and Quinn, 1980), GP 12, GGP 13 and GP 18 (Petrovskaya and Stolpnikova, 1989) Ottawa-30, Ottawa-31, PI 270407, W-Va- 700, Pierline New Yorker, Nova, FI 814, Hossiline and Hybrid TH 2312 behaved as resistant. Some of the varieties like New Hampshire, Sure Crop, New Yorker, Nova and Virginia known as resistant elsewhere (Yang, 1979) behaved as susceptible under Punjab conditions (Kaur *et al.*, 1988).

Singh, *et al.* (1995) evaluated 205 tomato varieties and breeding lines screened for resistance to *Phytophthora infestans*. None of them was immune. Ottawa 30, Ottawa 31, PI 112835 and PI 207407 were resistant and the rest were susceptible.

Early Blight of Tomato

It is a commonly occurring disease of tomato and occasionally assuming destructive status at any stage of plant growth and may cause losses upto 78 per cent. Though, *Alternaria solani* has been reported to be the most common pathogen causing early blight disease, yet there are reports that the pathogens, *A. alternata* and *A. alternata* f.sp. *lycopersici* do cause leaf spots in tomato, (Grogan *et al.*, 1975; Malatharakis, 1983; Sahi and Shyam, 1993). *Alternaria solani* Sorauer causes this disease in main season crop grown in plains. The fungus can affect seedlings but generally is a problem of older plants. Lowest leaves are attacked first and then the disease progresses upwards. Dark brown spots with concentric rings develop on the leaves which give target board effect, the most characteristic symptom of the disease. In humid weather, the affected areas coalesce and big rotting patches appear. In severe attacks, affected leaves shrivel and fall down prematurely resulting in early defoliation. On leaves, *A. alternata* (Fr. ex Fr.) Keissl. f. sp. *lycopersici* Grogan *et al.* produces small angular scattered, light brown spots, progressing between veins towards the leaflet margin. The severely infected leaflets curl and dry, out prematurely. Subsequently, the pathogen invades the adjoining healthy leaflets and gradually progresses on the foliage upward from the lower most infected leaves. Symptoms also appear on stems and branches as water soaked Spots which soon turn light to dark brown in colour. The spots are circular to irregular often elongated and enlarge rapidly when the humidity is high coupled with frequent rains or continuous dew formation. The older lesions become dark grey in colour often encircling the branches.

On the lower most leaves, *Alternaria alternata* Fr. ex Fr. (Keissl.) spots start appearing about one month after transplanting as minute water soaked specks which soon turn into small, circular scattered, dark brown Spots. Older spots are surrounded by yellow halo and the leaflet become chlorotic. The affected leaves dry and fall prematurely.

Etiology

The disease is caused by three pathogens *viz.*, *Alternaria solani* Sorauer, *A. alternata* f.sp. *lycopersici* Grogan *et al.* and *A. alternata* (Fr. ex Fr.) Keissl. The morphological characteristics of each pathogen are described below: *Alternaria solani*: The mycelium consists of septate, branched, light brown hyphae which become darker with age. The hyphae in the host are at first intercellular, later penetrating into the cells of the invaded tissues. Conidiophores emerge through the stomata from the dead centers of the spot. They are relatively short, 50 to 90 µm and dark colored. Conidia (120-296 x 12-20 µm) are beaked, muriform, dark and borne singly or in chains of two. Five to ten transverse septa and a few longitudinal septa are present in each conidium. Germination of conidia occurs within 35 to 45 minutes in water at 28 to 30°C and produces 5 to 10 germtubes.

Alternaria alternata f.sp. *lycopersici*: The young colonies after 48 hours of inoculation on S-medium (sucrose + $CaCO3$ + agar), are fluffy at the margins and off white greenish in the center, subsequently turn dusky, neutral greenish grey within 96 hours of growth. Later these become appressed and nearly black in colour. The conidia occurring on leaflet and stem lesions under field conditions are 1 to 5 septate measuring 17.0–50.0 x 10.0–13.0 µ and beak length ranged from 3.0–7.0 µm.

Alternaria alternata: The young colonies possessed much fluffy margins with off white centers which turned into dusky neutral grey within 96 hours. Later these colonies become appressed and nearly greyish black in color. The conidia occurring on lesions are 1 to 6 septate measuring 13.0–46.0 x 7.0–10.0 µm. The size of conidial beak ranges from 3.0 to 10.0 µm.

Epidemiology

These fungi overwinter in diseased plants debris and persist for one to two years. Seed borne nature of *A. solani* has also been established by Khulbe and Sati (1987) and can be introduced through infected seed. The plants receiving normal doses of either N or P showed more susceptibility to the disease but the effect was reversed with increased doses of nitrogen and phosphorus (Bedi and Dhiman, 1983) while increased doses of K (double then normal) increased susceptibility to the disease. High soil moisture, humidity, dew and rainfall favour disease development. The optimum temperature for *A. solani* is 28 to 30°C (Singh, 1987) while for *A. alternata* f. sp. *lycopersici* and *A. alternata* range between 25 to 30 and 20 to 25°C, respectively (Sahi, 1990). Once infection has occurred, conidial dispersal continues throughout the growing season of the crop. Maximum dispersal of conidia occurs between 9 a. m and 12 noon (Datar and Mayee, 1982).

A field trial was conducted (Paul *et al.*, 2004) in order to study the development of tomato early blight caused by *Alternaria solani* using four different spraying treatments: (1) weekly sprays; (2) application according to the TOMCAST (VSD20); (3) CUFAST; and (4) FAST forecasting system. The fungicide used was Vicosa Mixture, a nutrient amended Bordeaux mixture. Weekly sprays were begun in the second week after planting while the other spray treatments were made according to each respective forecasting model. Disease severity was assessed weekly with the aid of a diagrammatic scale. Linear regression analysis was used.

Subsequently, Batista *et al.* (2006) developed decision support systems which are important tools in reducing the large amount of fungicides applied to suppress disease intensity. Systems developed for early or late blight were validated in two cropping seasons under Brazilian conditions. For tomato early blight, FAST, CUFAST, and TOMCAST systems were compared in the spring-summer, 2006. Based on that four sprays of fungicide are recommended.

Management

Cultural Practices

Destruction of infected plant material, use of healthy seed and crop rotation are desirable to reduce the disease. Periodical clipping of lower leaves upto 20 cm height from the ground level has been reported not only to reduce the disease severity but also increase the fruit yield (Sahi, 1990).

Chemical Control

Fungicidal sprays with captafol, mancozeb, maneb, benomyl, carbendazim, metiram, copper oxychloride plus dichlofluanid and copper oxychloride plus folpet are recommended for the control of this disease (Ramakrishnan and Kandaswamy, 1978; Alexanderi *et al.*, 1981; Rajgopal and Vidyasekharan, 1982, 1983). Soil application of captafol plus folpet prevented the stem infection caused by *A. alternata* f. sp. *lycopersici* for more than 25 days (Malathrakis, 1983). Sequential application of captafol, mancozeb and copper oxychloride, 40, 55 and 70 days after transplanting increased yield and reduced the incidence of *A. solani* and *A. alternata* and captafol sprays gave highest net income (Bharadwaj, 1991). Shyam and Gupta (1996a) reported the efficacy of metalaxyl plus mancozeb in managing the leaf spots and the highest cost-benefit ratio was in Indofil M-45 application. *A. alternata* infection was reduced by 100 per cent when extract of *Clerodendrum aculeatum* was applied as a foliar spray one day before inoculation (Sharma, 1992).

Tofoli, *et al.* (2003) evaluated the effectiveness of various groups of fungicides for controlling early blight as well as their effect on tomato fruit yield. Three experiments were carried out in commercial fields. The highest levels of disease control, quality and increase on fruit yields were obtained with pyraclostrobin + metiram, fenamidone + chlorothalonil, famoxadone + cymoxanil + mancozeb, kresoxim-methyl, azoxystrobin, difenoconazole, tebuconazole, pyrimethanil, cyprodinil, famoxadone + mancozeb followed by prochloraz, fluazinam, procymidone and iprodione. Mancozeb and chlorothalonil also resulted in considerable reduction in disease.

Prasad and Naik (2003) Studied the impact of pre-harvest spray of fungicide on post harvest incidence of fruit rot of tomato. The maximum control of fruit rot was noticed in iprodione and mancozeb sprays. In post harvest treatments, iprodione, mancozeb and wax coating were found to be better in controlling *Alternaria* fruit rot of tomato.

A glasshouse study was conducted (Wickramaarachchi, *et al.*, 2003) to evaluate the potential use of non-conventional chemicals to control early blight by inducing resistance in tomato. Foliar spray of aqueous solutions of 9 non-conventional chemicals, *i.e.* sodium dihydrogen orthophosphate at 50 mM, sodium bicarbonate at 2000 ppm, potassium dihydrogen orthophosphate at 1000 ppm, dipotassium hydrogen orthophosphate at 1000 ppm, potassium bicarbonate at 1000 ppm, potassium permanganate at 20 ppm, potassium nitrate at 1000 ppm, calcium nitrate (hydrated) at 0.5 per cent and calcium chloride (fused) at 1000 ppm, were tested against early blight of tomato along with mancozeb at 2000 ppm as the standard fungicide. All chemicals, except potassium bicarbonate, reduced the disease. The percent disease index was lowest (9.26) in the fungicide control. Among the test chemicals, spray application of potassium hydrogen orthophosphate gave the highest yield.

In the field, among the selected effective plant extracts, plant oils and antagonists, spray application of palmarosa oil (0.1 per cent), at the initial appearance of the disease symptoms significantly reduced the leaf blight disease incidence besides increasing tomato yield. (Vadivel and Ebenezer *et al.*, 2006).

Integrated disease management (IDM) package was developed comprising of biological, cultural, physical and need based use of chemicals for targeting above four diseases. Overall disease control performance of IDM module was calculated in terms of yield. Average yield on OP variety (DVRT-1) for IDM was 290.92 q/ha as compared to control (149.32 q/ha). Yield in IDM block was almost double than control and statistically highly significant. (Pandey, *et al.*, 2005)

Host Resistance

Varieties 6718 VF, Rutgers, Marvel, Marglobe, Pritchard, Ace-55, Improved Garden State and Victor were resistant to *Alternaria alternata.* f.sp. *lycopersici* (Grogan *et al.*, 1975). Cultivar L-15 was found resistant to *A. solani* while TI 244 had good field tolerance (Erinle and Eroctor, 1992). Sahi (1990) reported four germplasm lines *viz.*, EC-174069, EC-174076, EC-174025, EC-174057 and a cultivar Marglobe resistant to *A. alternata* and *A alternata* f.sp. *lycopersici*. Biezen *et al.* (1995) reported two wild species of tomato *viz. Lycopersicon pennellii* and *L. peruvianum* highly insensitive to AAL-toxins as well as resistant to *A.a.* f.sp. *lycopersici*. Genetic analysis showed that high insensitivity to *A.alternata* f.sp. *lycopersici*-toxins from *L. pennellfii* is inherited as a single complete dominant locus. Subsequent classical genetics, RFLP mapping and allelic testing indicated that high insensitivity to AAL-toxin from *L. pennellii* is conferred by a new allele of the ASC locus.

Among 63 genotypes, none of them showed immune reaction. However, genotypes Arka Alok, Arka Abha, Arka Meghali, Arka Sourabh, IIHR-305, IIHR-308, IIHR-2266, IIHR-2285 and IIHR-2288 showed resistant reaction. (Prasad and Naik 2003).

The development of cultivars with high levels of resistance has been hampered by the lack of sources of strong resistance in the cultivated tomato and by the quantitative expression and polygenic inheritance of the resistance. In some accessions of wild species, high levels of early blight resistance have been found, but breeding lines still have unfavorable horticultural traits from the donor parent. Recently, the first linkage maps with loci controlling early blight resistance have been developed based on interspecific crosses. These maps may facilitate marker-assisted selection. This overview presents the current knowledge about the *A. solani tomato* complex with respect to its biology, genetics, and breeding. (Chaerani, and Voorrips, 2006)

Septoria Leaf Spot

It was first reported from Argentina by Spegazzini (1882). In India, its occurrence was first reported from Coonoor, South India (Ramakrishnan and Sundaram, 1954). Now it is widely distributed throughout the world and causes considerable loss by premature defoliation. The plant may be attacked at any stage of its growth. This fungus produces spots on the leaves, stems, the persistent calyx and the fruits. The initial symptoms appear on the lower leaves as a water soaked spots on the under surface of the leaves. As the spot grow larger, these become more or less circular in outline and show definite brown coloured margins with grey centers. A few black glistening pin-head size pycnidia may be seen within the center of spots. The diseases on the stems appear as small, slightly elongated, dark spots containing numerous black pycnidia. Small elongated specks appear on calyx which are intermediate in size as those found on the leaves and stems. Black circular spots may also be seen on young as well as mature fruits.

Etiology

The disease is incited by *Septoria lycopersici* Speg. On the host, young mycelium is hyaline, thin walled and sparingly septate. The protoplasmic contents are densely granular, vacuolated, contain numerous small oil globules and the diameter ranged from 1.3–5.6μm while the older mycelium is brown, infrequently branched and septate, ranging from 2.5-5.9 μm in diameter. In the initial stage of pycnidial formation, several hyphae aggregate at certain point and after intertwining, form a mass of hyphal tissue that assume the appearance of pseudoparenchyma. Pycnidia are subglobose and composed of 2 to 3 layers of brown cells ranging between 53-165 μm in width. Pycnidiospores are filiform, slightly curved, hyaline and septate with pointed or rounded ends and ranged from 34.5-69.0 x 2.2-4.8 μm. Sexual stage has not been reported so far. At least two physiologic races of the pathogen have been identified.

Epidemiology

The pathogen perpetuates in diseased plant debris left in the field and also, on or in the seed. The fungus has wide host range including various solanaceous crops and weeds growing in and around tomato fields. Temperatures of 20 to 25°C with 75 to 92 per cent relative humidity are congenial for disease development.

In order to determine the effects of spore density, leaf age, dew period and temp. on the susceptibility of tomato to S. *lycopersici*, growth chamber studies were conducted. Depending on the leaf area, 3-5 ml of water containing 10^2, 10^3, 10^4, 10^5 and 10^6 spores/ml were brushed onto the third true leaf up to the seventh true leaf. The log of the disease severity (lesions/cm^2 leaf) was linearly correlated with the log of the inoculum density (spores/cm^2 leaf). Older leaves were more susceptible than younger leaves to Septoria leaf spot when inoculum densities were high (10^4 spores/cm^2 leaf), but younger and older leaves were uniformly susceptible when inoculum densities were low (10-10^2 spores/cm^2 leaf). (Elmer and Ferrandino, 1995).

The study conducted by Kumar, and Sugha, (1999) revealed that spores remained viable for up to 6 months after harvest on plant debris stored under laboratory and screenhouse conditions. Plant debris buried near the soil surface and disease infested soil helped in the over seasoning of the pathogen and the successful initiation of disease, whereas seed did not play any role in the perpetuation of disease.

Sugha, and Suman Kumar (2000) noted that an increase in inoculum accelerated the progress of disease development. An inoculum load of 525-1050 spores/ml and leaf wetness in the range of 48-60 h were optimum for the normal development of disease under artificial inoculation tests.

The Tom-Cast disease forecasting model, based on leaf wetness and temperature, predicted that in a dry season fungicide sprays were unnecessary for control of early blight in processing tomato crops (Minchinton *et al.*, 2006).

Management

Cultural Practices

Destruction of infected plant material and use of clean seed is useful in checking this disease (Singh, 1966). Various cultural practices such as soil solarization, early sowing and dry season planting, crop rotation, choice of open airy location, trimming and staking of plants, avoiding walking among plants and ploughing under all plant material (Buchhlotz, 1944) have also been recommended for its management.

Chemical Control

Sprays of various fungicides, such as benomyl, mancozeb, captafol, metiram, copper oxychloride plus dichlofluanid, copper oxychloride plus folpet, captan, thiram, Mycodil (folpet+captafol), cuprous oxide, copper oxychloride, Ridomil MZ, chlorothalonil, Polihom (metiram zinc + copper oxychloride), Ridomil plus 48 (metalaxyl + copper oxychloride), Temoal (cymoxanil + metalaxyl+copper oxychloride) have been recommended (Rajagopal and Vidhyasekaran, 1982; Tedla, 1985; Szekely *et al.*, 1990; Bharadwaj, 1991).

Host Resistance

Several varieties/lines/wild species such as EC2750, EC7785, HP2453, KT- 4 and Natural Cross, Sweet 72, Sel. 152, S-12, Marglobe and Gambde (Chauhan, 1987), L. chilense, *L. parviflorum, L. racimigerum, L. peruvianum* vars. *humifusum* and *glandulosum, L. pimpinellifolium* forms PI126444 and PI 126936, *L. esculentum* accessions 85602009 and 85602015 and *L. esculentum* var. *ceraciforme* (Sotirova and Rodeva, 1989), L-15 (Madalageri *et al.*, 1988), *L. esculentum* var. *cerasiforme, L. pimpinellifolium* acc. PI 1269205, *L. hirsutum f. typicum* accession PI-127826 and *L. hirsutum f. glabratum* accession PI 19931 (Sotirova *et al.*, 1992), TI 420 (Erinje and Eroctor, 1992) have been reported resistant to this disease. IPA-5 and Rio Fuego were resistant while Caribe and Contessa revealed intemiediate resistance (Moretto *et al.*, 1993). Line IPA-5 had higher level of horizontal resistance than Caribe and Contessa (Moretto and Barreto, 1993).

Sotirova and Rodeva, (1989) studied 5 botanical varieties and 45 accessions of *Lycopersicon esculentum* and 20 wild forms including the *Solanum* species *S.* [*L.*] *pennellii*, S. *lycopersicoides* and *S. sisymbriifolium*, the following showed a resistance reaction under artificial infection: the wild species *L. chilense, L. parviflorum, L. racemigerum, L. peruvianum vars. humifusum* and *glandulosum* and *L. pimpinellifolium* forms PI126444 and PI126936; the *L. esculentum* accessions 85602009 and 85602015; and *L. esculentum* var. cerasiforme. Pot reactions to the disease were also recorded under artificial epiphytotic conditions during 1994. Marglobe, BWR1 and EC5889 were resistant under natural conditions, while EC50366, EC50523, EC129156, EC179909, EC191530, Pusa Early Dwarf and VFN4 were moderately resistant. Sugha, and Kumar,(1998)

Among the screened lines, one (LE-415) was resistant, whereas 6 (EC-163605, EC-52961, EC-57724, EC-35520, EC-326104 and West Virginia 700) were moderately resistant. (Pandey and Pandey, 2002). Joshi *et al.* (2004) identified the lines EC 401927 and CLN 5915-206-D4 2-2-0 showing moderate degree of resistance to Septoria leaf spot and early blight (13.00 and 20.67 per cent, respectively).

Powdery Mildew

This disease is widely prevalent in dry areas around the Mediterranean, Central Europe, Palestine and Near East. It causes about 40 per cent reduction in fruit yield (Jones and Thompson, 1987).

Symptoms

The fungus produces a white talcum like covering on the lower surface of the leaves while the corresponding upper surface turns yellow. The infected leaves shed prematurely.

Etiology

The fungus responsible for this disease is *Leveillula taurica* (Lev.) Arnaud. The mycelium of the causal agent is endophytic, produces long and often branched conidiophores and conidia are borne singly or in short chains. Conidia are dimorphic (pyriform and cylindrical) and isolates vary considerably in conidial size. The ranges of mean conidial measurements are 49.7–71.4 x 16.6–24.1 μm and 44.6–65.2 x 16.2–22.7 μm for the pyriform and cylindrical conidia, respectively. Formation of perithecia is rare but within the stroma, embedded in the mycelium and are 135–250 μm in diameter. Numerous hyphae like short and branched appendages are attached to the perithecia. Usually, there are 20 to 35 asci, 70-110 x 25-40 μm in size, within each perithecium. Each ascus contains two ascospores. The ascospores are large cylindrical to pear-shaped, slightly curved and measure 25-40 x 12-22 μm.

Epidemiology

The fungus is an obligate parasite and perennates on weed hosts. Wherever perithecia are produced they explain the mode of perennation. Dry weather conditions coupled with low humidity are favourable for disease initiation and spread. Temperature of 18 to 24°C and relative humidity 70 to 100 per cent, respectively are favourable for infection.

Management

Cultural Practices

Disease incidence has been reported to be 30 to 40 per cent less in sprinkle irrigated tomato crop as compared to furrow irrigated crop (Dixon, 1981).

Chemical Control

Disease can be effectively managed by spraying dinocap (0.05 per cent) and wettable sulphur (0.2 per cent) but ergosterol biosynthesis inhibitor fungicides like triadimefon, propiconazole, triforine are more effective in reducing the disease intensity (Jones and Thompson, 1987; Correll *et al.*, 1988; Mahajan *et al.*, 1991).

Fusarium Wilt

Fusarium wilt of tomato crop is emerging as a major disease in those parts of world where soil temperature remains high during the period of cultivation. The disease is characterized by clearing of vein lets and drooping of petioles of young plants. Lower leaves show yellowing and die prematurely. Symptoms continue to appear on successively younger leaves. Later, the whole plant wilts and dies prematurely. After the disease has advanced for a few weeks, browning of the vascular system can be seen in a cross section of the lower stem or by removing stem tissue near the collar region with a Knife.

Etiology

The disease is caused by *Fusarium oxysporum* Schlechtend ex Fr. *f. sp. lycopersici* (Sacc.) Snyder and Hans. The mycelium is white to pink, often with a purple tinge. Microconidia are borne on simple phialides arising laterally, oval-ellipsoid, straight to curved, 5-12 x 2.2-3.5 μm, and non-septate. Macroconidia are sparse to abundant, borne on branched conidiophores or on the surface of sporodochia and are thin walled, three to five s~ptate, fusoid subulate and pointed at the ends, have a pedicellate base and measure 27-46 x 3-5 μm if three septate or 35-60 x 3-5 μm if five septate. Three septate spores are more common. Chlamydospores, both smooth and rough walled, are abundant and form terminally or on an intercalary basis. They are generally solitary but occasionally form in pairs or chains. Three physiological races *viz.*, race 1, 2 and 3 have been reported world over (Jones *et al.*, 1991).

Hirano (2006) used the polymerase chain reaction technique for differentiating the pathogenic type (form and race) of *Fusarium oxysporum*, which generates wilt symptoms on tomato. A combination of amplifications with these primer sets effectively differentiated the pathogenic types of *F. oxysporum* in tomato.

A total of 62 isolates of *F. oxysporum*, obtained from symptomless tomato plants, were characterized using IGS typing and pathogenicity tests on tomato plants. All *Fusarium* isolates were highly pathogenic on tomato. *Fusarium oxysporum f. sp. radicis-Fusarium oxysporum f.sp. lycopersici* isolates were separated into five IGS types. From the 53 *F. oxysporum f. sp. radicis-Fusarium oxysporum f.sp. lycopersici* isolates, 34 isolates have the same IGS types (IGS type 25), and the remaining 19 isolates were distributed into four IGS types. However, only nine isolates of *F. oxysporum* f. sp. *Fusarium oxysporum f.sp. lycopersici* have six different IGS types. This difference of diversity between the two formae speciales suggests that F. *oxysporum* f. sp. radicis-*Fusarium oxysporum f.sp. lycopersici* isolates have a foreign origin and may have been accidentally introduced into Tunisia (Hibar *et al.*, 2007).

Epidemiology

The fungus is soil borne and can persist in soil for many years. The favourable temperature for disease development is 26 to 28°C. The disease is more severe in acidic soils than alkaline. Low nitrogen and high potassium levels in soil predispose the plant to this disease. Application of potassium 100 kg per ha or in combination with nitrogen (100 kg N + 100 kg K/ha) decreased the disease incidence to 62 and 67 per cent, respectively, while nitrogen alone 150 kg per hectare resulted in 83.33 per cent disease incidence (Raj and Kapoor, 1995).

Management

Akram and Khan (2006) conducted field trials in Uttar Pradesh, India to evaluate the effects of *Meloidogyne incognita* and *Fusarium oxysporum f.sp. lycopersici*, applied via sequential and concomitant inoculation, on the growth and yield of tomato. The nematode and fungus acting alone caused root galls and foliage wilt, respectively, and reduced plant growth and yield. Fusarium wilt became considerably severe in the presence of root-knot nematode. Plants inoculated with the nematode before the fungus (pre-inoculation) exhibited 41.8 per cent reduction in yield, followed by concomitant treatment (33.3 per cent). Gall formation and egg mass production of the nematode was adversely affected in the presence of *Fusarium oxysporum f.sp. lycopersici* and their lowest number were noticed.

Cultural Practices

At soil pH 6.2 the disease index was 70.05 but on liming to pH 8.1 it reduced to 42.4 per cent (Sarhan, 1982). Wilt symptoms were less severe in plants supplied with nitrate N and more so in those given ammonium N. Application of combination of high NO, low NH_4 and lime reduced disease development actively. In liquid culture also the fungus was more virulent wit NH_4 as the sole N source than with NO (Woltz and Jones, 1973). Application of copper, boron, manganese and zinc increased the resistance of tomato plants to this pathogen (Ahmed *et al.*, 1987).

Chemical Control

Disease incidence was reduced with the application of carbendazim (0.1 per cent) and benomyl (0.1 per cent) as soil drench (Sen and Kapoor, 1974). Etebarian (1992) reported the efficacy of benomyl, iprodione plus carbendazim and carbendazim alone both *in vitro* and *in vivo*. Treatment of tomato seedlings with herbicide, hloracetarnide, induced resistance in tomato cultivar Rehovot 13 seedlings to Fusarium wilt (Cohen *et al.*, 1992).

Host Resistance

Nath (1976) reported Marglobe and Rutgers as commercially acceptable and resistant varieties against this disease. Mohamed, and Badr,(2004) conducted studies on inheritance of resistance to Fusarium wilt of tomato the parental, F1, F2, Bc1 and Bc2 generations of three tomato inter specific crosses Super-Marmande x LA1375, Edkawi x LA1375 and UC.97-3 x LA1375 were evaluated during 2000-02 in field and greenhouse tests in Egypt to study the inheritance and nature of resistance to Fusarium wilt disease. Resistance to Fusarium wilt was found to be controlled by one dominant gene. The ratios of resistance to susceptibility in the F2 and Bc1 (from a backcross to the susceptible parent) populations were 3:1 and 1:1, respectively.

Biological Control

Trichoderma harzianum did not parasitize *Fusarium* but it excreted beta- 1, 3-glucanase and chitinase into the medium when grown in liquid media containing laminar in or chitin as a sole carbon source (Sivan and Chet, 1986). Seed coating or application of wheat bran-peat preparation of *T. harzianum* decreased the incidence of wilt and increased yield. Phae *et al.* (1992) reported the efficacy of *Bacillus subtilis* NB 22 against wilt pathogen. A binary microbe system established by using an antagonistic rhizoplane bacterium, *Serratia marcescens* a chitin-degrading organism and *Streptomyces anulatus* effectively controlled wilt disease (Toyoda *et al.*, 1993). *Fusarium* propagules were consistently lower in the presence of VAM fungus *Glomus* and tomato root necrosis decreased (Caron *et al.*, 1986).

Clasing and Herrera (2006) conducted greenhouse tests to control *Fusarium oxysporum f.sp. lycopersici* with *Trichoderma harzianum* strain Thv (Thv) and *Paenibacillus lentimorbus* strain 629 (PI 629) in tomato cv. Bonny Best plants. These strains of T. *harzianum* and *P. lentimorbus* were previously selected in *in vitro* tests as effective biological control agents of *Fusarium oxysporum f.sp. lycopersici.* Tomatoes were planted in a soil previously fumigated with CH3Br (75.5 g/m2) or solarized during 40 days. Height, aerial, root and total fresh and dry weights of the plants were evaluated. The best results were obtained with CH3Br and solarization.

Bacterial Diseases

Bacterial Wilt

The disease has been reported throughout the world wherever tomato is grown. In India, the disease is known to occur in Punjab, Himachal Pradesh Madhya Pradesh, Uttar Pradesh, Bihar, West Bengal, Assam, Orissa, Kerala, Karnataka and Tamil Nadu on potato, eggplant, tomato, chilli etc. (Rao, 1977; Khan *et al.*, 1980). Loss in tomato yield due to the disease has been reported upto 90.62 per cent (Kishun, 1987).It attacks more than 200 species of cultivated plants and weeds in 33 plant families. Characteristic symptoms of bacterial wilt on most of the hosts are rapid and complete wilting of normal grown up plants. During wet and warm climatic conditions, the most conspicuous symptom seen on tomato plants is sudden drooping of the leaves, without yellowing often accompanied by rotting of the stem. The roots appear healthy and often are well developed. Pathogen is mostly confined to vascular region. In advance cases, it may invade the cortex and pith and cause yellow brown discolouration of tissues. Bacterial polysaccharides mechanically block the vascular system (Balasubramanian and Nflakantan, 1976) which checks the translocation of water and minerals resulting in wilting of plants.

Etiology

Ralstonia solanacearum (Smith) Yabuuchi *et al.* is gram negative rod shaped bacterium frequently occurs in pairs, motile with 1 to 4 polar flagella and measures 0.5-0.7 x 1.5-2.5 µm. It is aerobic and catalase and oxidase-positive and forms nitrites from nitrates. It, like some other nonfluorescent pseudomonads, produces intracellular, refractile, sudanophilic inclusions consisting of polyhydroxy-butyric acid. Some strains produce a brown diffusable pigment on complex media and a black pigment on autoclaved potato slants. The pathogen is negative for levan production and starch hydrolysis. It causes weak hydrolysis of gelatin. Among the three races known, race 1 and 3 occur in India (Rao, 1977).

Epidemiology

The pathogen is soil borne and survives atleast for 2 years even in the absence of any host (Shekhawat *et al.*, 1979). The bacteria from tomato and brinjal survive up to 3 years in fallow and upto unlimited period in cultivated land. Bacterium is found up to 45 cm depth in cultivated and fallow lands but its population decreases with the increase in depth (Kishun and Sohi, 1982). The pathogen is apparently not seed borne and seeds from affected tomato plants produced neither diseased plant nor served as symptom less carrier (Shekhawat *et al.*, 1979). However, Devi and Menon (1980) were able to produce bacterial wilt in tomato plants grown from seeds collected from partially wilted plants. Disease spreads through infected plant material, irrigation, water, soil and implements (Kishun, 1986). The disease is more serious in sandy loam soil than red loam soil (Yuan *et al.*, 1992). Infection always occurs through wounds caused by transplanting, cultural operations or nematode invasion. The organism first moves into the large xylem vessels.

When a single lateral bundle is invaded, drooping of leaves is common but if all bundles are invaded, the entire plant wilts. Infection and disease development is favoured by high temperature (optimum 30-35°C) and high soil moisture.

Management

Cultural Practices

Crop rotations such as cowpea-maize-cabbage, okra- cowpea-maize, maize-cowpea-maize and finger millet-eggplant-french bean are effective in reducing the disease (Sohi *et al.*, 1981). Disease was absent or the incidence reduced in fields with deep ditches and high beds for drainage and less or shallow tillage or no tillage to avoid root injury (Yuan *et al.*, 1992). Michel *et al.* (1997) found that the effect of the soil amendment was not consistent when applied to soil from 3 other sites in Taiwan; in soil from 2 sites no reduction of the pathogen population occurred. At these sites, tomato bacterial wilt in the field was not reduced significantly after amending. In comparison with a non-amended control, the addition of CaO alone significantly reduced the *R. solanacearum* population in AVRDC soil ($P<0.05$) but the reduction was significantly greater when the complete soil amendment was added., but in the same treatment tomato yield was significantly reduced by 48 per cent ($P<0.05$) compared with non-amended treatments.

Chemical Control

Applications of bleaching powder 15 kg per ha (Kishun and Chand, 1988; Ho, 1988) and groundnut cake before planting have been found effective against this disease. Seedling dip with streptocycline (lg/40 L water) for 30 minutes (Duttaand Verma, 1969) and penicillin and agrimycin-100 in Hoagland's solution (Sitaramiah and Sinha, 1983) before planting is effective. Bacterinol-100, when applied as dry seed dresser, a nursery spray and field applications resulted in 11.5 per cent yield increase over untreated control (Rao, 1990). Streptomycin sulphate or oxytetracycline when sprayed at 200 ppm at 4 and 7 days intervals provided good control (Gunawan, 1988; 1989). Terlai, a bactericide also gave good control of this disease (Hanudin and Machmud, 1994).

Host Resistance

Tomato lines/cultivars *viz.*, VC 9-lUG and VC II-lUG (Bedekar, 1977), VC 4(Graham *et al.*, 1977), Scorpio (Peterson *et al.*, 1983), Musashikyo-ryoku 560, Matsurao, root stocks LD89 and BFM-R (Yoneyama, 1985), Redlands Summertaste and Redlander (Mofett, 1986), Arka Abha (BWR-,J1), Arka Alok (BWR-5) (Sinha *et al.*, 1988), Sonali (Sel.2) (Patil *et al.*, I 1990), Zayou 1 (Yuanetal., 1992) are resistant while accessions PI 126408, PI 212441, PI263722, PI 365930, Hawaii 7981, 7997 and CRA66 (Sel. A,D and E) have been reported highly tolerant (Sonoda and Augustine, 1 1978).

Graft of susceptible cultivars of tomato on resistant rootstocks belonging to Solanum spp. was reported to be effective in management of this disease. Rajput *et al.* (1982) controlled this disease by grafting susceptible tomato cultivars onto resistant eggplant (Pusa Purple Cluster) rootstocks. When plants of cultivars Roma and Floradel scions grafted onto Solanum forum rootstock, plant survival was 66.6 and 50.5 per cent, respectively (Peregrine and Ahmad, 1982).

Tomato cv. Healani somaclones were evaluated in a field infected with bacterial wilt. Survival percentages of selected somaclonal lines ranged from 40 to 100 per cent, while the original Healani had a survival rate of 0 per cent and resistant cv. Kewalo had 30 per cent survival (Bobisud, *et al.*, 1996). Osiru, *et al.* (2001) studied the inheritance of resistance to bacterial wilt of tomato. Two bacterial wilt-resistant (MT 74 and MT 55), two susceptible (MT 164 and MT 15) and two mildly resistant (MT 86 and MT 29) tomato cultivars were crossed in a half diallel fashion in a screenhouse at Kabanyolo, Uganda. Parents and F1's were root inoculated and data were collected on bacterial wilt intensity at 3-days intervals from 6 to 21 days after inoculation and analysed for bacterial wilt reactions. The results further revealed that resistance to bacterial wilt is controlled by two genes.

Eight tomato parental lines and 28 F1 crosses developed at the station were tested in bacterial sick plot, in Jharkhand, The mean of four years data revealed that three parental lines *viz.* CHDT-4 (EC 339074 released as Swarna Lalima), CH-180 (BT-17) and CHDT-5 (EC-369060-A released as Swarna Naveen); and three F1 crosses CHDT-4 x CHDT-1 x CHDT-1 x CH-180 and CH-195 x CH-180 showed resistant reaction to bacterial wilt. (Sharma, *et al.*, 2006).

Biological Control

Aspiras and Cruz (1985) reported that *Bacillus Polymyxa* F46 and *Pseudomonas fluorescens* soil applications were highly effective in reducing.the disease. *B. polymyxa* may also be conveniently applied as a seed or seedling dip during sowing or planting. Seedling root dip in selected strains of *Pseudomonas glumae* not only provided protection against wilt disease but also induced resistance in host tissue (Wakimoto, 1987; Furuya and Matsuyama, 1992).

Pseudomonas fluorescens treated tomato plants were protected from wilt upto 95 and 36 per cent under greenhouse and field, respectively (Anurath and Gnanamanickam, 1990). Pouring of culture suspension of *Bacillus subtilis* NB 22 in infested soils has also been recommended to control the disease (Phae *et al.*, 1992). Misaghi *et al.* (1992) reported that isolates of *Bacillus* sp., *Pseudomonas fluorescens* and *P. putida* with mobility (ability of the bacterium to follow developing roots in the absence of percolating water) effectively controlled this disease.

Garlic bulb (35 g in 77 ml sterile water) suspension (10 ml) or ground garlic bulb (6 g) when added to rhizosphere not only reduced the disease but also resulted in normal plant growth and increased fruit weight (Hutagalung, 1988).

Pradeep Kumar and Sood (2002) conducted pot experiments on seedlings of tomato cultivars EC191536 (moderately resistant) and Solan Gola (susceptible) to determine the efficacy of bacterial antagonists *Pseudomonas fluorescens* and *Bacillus cereus* and vesicular arbuscular mycorrhiza *Glomus mosseae* against tomato bacterial wilt (*Ralstonia solanacearum* race 1 biovar III). The combined use of *P. fluorescens* and *B. cereus* was more effective than the individual applications. Further, they studied the population of *R. solanacearum* which was least in treatments where the antagonists were combined with VAM. The synergistic effect in the reduction of wilt incidence was observed when *G. mosseae* was combined with antagonists.

However, an analysis on the colonization of endophytic bacteria B47 and its control on tomato bacterial wilt was made by Li-Qi Qin *et al*(2006). The endophytic *Bacillus subtilis* strain B47, isolated from tomato stem, was an efficient biological control agent of tomato bacterial wilt. A strain resistant to the streptomycin derivative of the wild strain B47 was used to study the colonizing ability, the fluctuation of bacterial population in the soil and plant organs, and bacterial wilt control. Results indicated that B47 could colonize both in the soil and in tomato organs. After inoculation in soil, the population of strain B47 increased.

The interactions of Karanje cake, bleaching powder and seedling root dip in Streptocycline was the most effective treatment as significantly minimum wilt incidence was recorded at 60, 90 and 130 days after transplanting along with the highest fruit yield during both years. The next effective treatment was Karanj cake+bleaching powder. The seedling root dip in Streptocycline was the least effective treatment. Organic amendments increased the Ca, K and Na, and reduced Al contents in the soil. Soil solarization reduced the number of weed seeds in the soil, mainly the monocotyledons. Methyl bromide treatments and chicken litter amendments significantly reduced bacterial wilt incidence and disease progress (Baptista, *et al.*, 2007).

Bacterial Speck

The disease appears on the leaves, stems and petioles as well as fruits. Lesions on leaflets are circular and dark brown to black. A halo is lacking in the early stages but develops with time. The lesion extends throughout the leaf and most prominent on the abaxial surface. Spots may coalesce, killing large areas of tissue. Lesions on stems and petioles are oval to elongate. On fruits, minute lesions or specks develop that are dark and rarely larger than 1 mm in diameter. The tissue around each speck may be a more intense green than unaffected areas. The specks are dark brown, slightly raised and have definite margins. A dark green halo around these spots may be associated.

Etiology

This disease is caused by *Pseudomonas syringae pv.* Tomato (Okabe) Young *et al.* The bacterium is a strictly aerobic, gram negative rod with rounded ends, measuring 1.8-2.8 x 0.69-0.97 µm arranged in pairs and occasionally in chains, motile with 1-3 polar flagella. The bacterium produces a diffusible green pigment which fluoresces when exposed to ultraviolet light on King *et al.* medium B. It is negative for oxidase reaction and arginine dihydrolase. The organism causes pitting of sodium polypectate at pH 4, is positive for levan production and glucosidase activity, utilizes D (-) tart rate but not erythritol or DL-lactate as sole carbon sources, and produces a hypersensitive reaction when infiltrated into tobacco at high concentrations.

Epidemiology

The bacterium over winters in infected seed and plant debris. Chambers and Merriman (1975) reported that the bacterium survived in plant debris either one or 2 cm below the soil surface for 30 weeks and thermal death point green and ripe fruits were 48 and 49°C, respectively. The bacterium is disseminated by splashing rain. The pathogen under artificial inoculation conditions affects only *Capsicum annuum* (Rangaswami and Gowda, 1963). The disease development and spread is only at temperature between 13 and 28°C and at high relative humidity when there is free water on the leaves (Yunis *et al.*, 1980).

Management

Cultural Practices

Use of healthy seed and proper hygiene during seedling growth is effective in reducing the disease (Naumann, 1981).Seed should be extracted by the fermentation process and pulp containing natural and lor added populations of *Pseudomonas syringae* pv. tomato were non-infective at I and 48 hours after the acid and fermentation process began, respectively (Chambers and Merriman, 1975; Naumann and Griesbach, 1990).

Chemical Control

Pyke *et al.* (1984) reported complete control of the disease by seed treatment with streptomycin sulphate slurry. Under *in vitro* conditions, agrimycin- 100 (100 ppm) followed by Sandox Seed Dressing 6334 (200 ppm) and streptocycline (100 ppm) are the best. Spraying of copper hydroxide or copper oxychloride and streptomycin sulphate provided significant control (Jardine and Stephens, 1987). Copper oxychloride (0.3 per cent), Zn ion maneb plus fixed copper, copper oxychloride (0.3 per cent) plus mancozeb (O.02 per cent), cupric hydroxide and streptomycin sprays effectively controlled the disease (Issa *et al.*, 1985; Jardine and Stephens, 1987; Ramos *et al.*, 1989).

Host Resistance

Tomato lines/cultivars ON T 7710, Oregon Cherry and Early Cherry and Droplet (Pitblado and Mac Neill, 1983) and Macedonia, ACE 55 FV tomatoes and BOG-AT 69 (Panagopoulos *et al.*, 1988) have been reported resistant.

Biological Control

Weekly applications of two antagonistic *Pseudomonads* (strains 208 and 38 r) gave a level of protection equal to that of copper compounds (Colin and Chaflk, 1986). A non-pathogenic Tn 5 mutant of *P. syringe* pv. tomato retained its ability to survive epiphytically on tomato leaves and inoculations of the mutant with a pathogenic strain, significantly reduced the incidence of bacterial speck. When a derivative of the non pathogen with tolerance of high levels of copper was co-inoculated with a copper- sensitive pathogenic strain to plants treated with a copper bactericide significantly greater control was achieved than with either the non- pathogenic strain or the copper treatment alone (Cooksey, 1988).

Bacterial Canker

This is one of the important bacterial diseases of tomato and occasionally become very destructive, if weather conditions are favourable. It was first observed in USA. During 1909 In India, the disease was recorded during 1949 from Nagpur (Jain, 1951). On young plants, the first symptom is a drooping of the petioles and if one of the top leaflets is broken off next to the stem, a brown discolouration is found in the sap ducts. The same discolouration extends downward into the stem to its base and if the disease is severe, yellowish-white streaks, becoming grey brown, appear on the stems, especially when infection is close to the surface. These streaks may crack open and form cankers. Leaflets on one side of the stem show wilting while those on the opposite side appear healthy. The spots on fruits have been referred to as birds-eye spot, are superficial and 3 to 6 mm in diameter. In the early stages, the spots appear as white to light brown slightly raised areas. As the spots enlarge, the brown center becomes more prominent and is surrounded by a characteristic white halo. Internally, fruit may contain yellow vascular areas leading to the seed.

Etiology

Tomato bacterial canker is caused by *Clavibactor michiganensis* subsp. *michiganensis* (Smith) Davis *et al.* It is non spore forming, non acid fast, gram positive and aerobic. Reports on motility and encapsulation are varied, but in general the bacterium is considered negative for these characteristics. Cells may be pleomorphic, minute, coccoid, or club shaped, depending upon growth conditions, but those from plant material are characteristically rod shaped. Reproduction is characterized by "snapping" division, resulting in V–and Y–shaped arrangement of cells. Colonies on nutrient agar are characteristically yellow and reach a diameter of 2 to 3 mm in 5 days. They are smooth and have a entire margins and butyrous consistency.

Epidemiology

The bacterium perennates in soil, plant debris, weed hosts, volunteer plants and infected seed and upto 70 per cent seed borne infection has been reported (Nedumeran and Vidhyasekaran, 1982). Secondary spread occurs by means of splashing water, contaminated equipment and workers hands. The density of bacterium required to cause disease is more than 10^6 cfu/g tomato leaves (Shirakawa *et al.*, 1991). It may develop at temperature as low as 16°C and as high as 36°C while 28°C is optimum. The disease is more severe in wet weather.

Management

Cultural Practices

Use of disease free seed, destruction of crop residue and proper sanitary measures reduces the disease (Panagopoulos *et al.*, 1988). Tomato should not be grown in infested fields for atleast 3 years (Jindal, 1992). Soil solarization was better than soil fumigation with methyl bromide (Antonio *et al.*, 1995). Fermentation of fruit pulp from field grown tomatoes for 24 or 48 hours at 20 or 25°C successfully disinfected the seeds while seed from greenhouse tomatoes required fermentation for 96 hours at 30°C and 120 hours at 20 or 25°C (Dhanvantari, 1989).

• *Chemical Control*

In an attempt to eradicate seed borne inoculum, seed treatment with hot water, chemicals and soaking of seed in antibiotics has been advocated by various workers. Seed soaking in 0.6 M HCl for 1 hour, 0.05 per cent O-hydroxydiphenyl (2-phenylphenol) for 15 minutes were highly effective treatments in reducing the disease (Dhanvantari, 1989). Bacterium was eradicated from naturally infected seeds when soaked in 0.6 M HCl for 5 hours, in 0.25 or 0.50 per cent acidified cupric acetate (ACA) for 20 minutes, or in hot water at 52°C for 20 minutes or at 56°C for 30 minutes but the seed treated with hot water (52°C for 20 minutes) germinated at a greater rate than untreated (Fatmi *et al.*, 1991). Dhanvantari and Brown,(1993) observed that when seed soaked in 0.1 M HCl or 0.05 per cent (W/V) 0- hydroxydiphenyl for hour germinated well, seed infection reduced to <I per cent, field incidence of primary canker to <2 per cent compared with 24 per cent in untreated check. There was no infection in harvested fruits from' these treatments as compared to 97 per cent in infected check plots. Kritzman (1993) also eradicated the bacterium from infected seed by immersing in a solution containing cupric acetate, acetic acid, pentachloro–nitrobenzene, e 5-ethoxy-3 (trichloromethyl)–1,2,4 thiabendazole and Triton x-IOO for 30 minutes at 25°C.

Soaking of seeds for 6 hours at 25°C in Terramycin (I per cent) was 58 to 76 per cent effective and dipping the seedling roots in streptomycin (1000 ppm) for 3 hours at 20°C was 72 to 75 per cent effective (U1ukus and karaca, 1985). Soaking tomato seeds in plant extract Arenarin (extracted from Helichrysum arenarium) at 1: 1000 ratio gave good control of this disease (Bychenko, 1974). Marinescu (1978) reported cryptonal (Chinosol) at 0.05 per cent and 0.5 per cent Miceram (copper oxychloride + zineb) as effective treatments against bacterial canker.

Host Resistance

Tomato cultivars lines H 2990 (Emmatty and John, 1973), Okitsu Material; selected from the F8 of Okitsu 3X *Lycopersicon hirsutum* var. *glabratum* PI 134418 as breeding material (Kuriyama and Kuniyasu, 1974), line L-59 and *Solanum multiflorum* (Docea and Marinescu, 1977) and BOG-AT 69 (Panagopoulos *et al.*, 1988) are resistant to bacterial canker.

Bacterial Spot

This disease is widely prevalent throughout the world on tomato and pepper. It was probably flrst reported in Texas on chillies in 1912 and on tomato in 1970. Under field conditions, it is found only on tomatoes, peppers and one of the nightshades, although some weed hosts like *Physalis angulata, P. minima, P. peruviana, Amaranthus dub ius, Acalypha viriginica, Solanum nigrum, S. americanum* and *Lagascea mollis* have also been reported (Srinivasan *et al.*, 1962; Kishun *et al.*, 1977; Munoz and Granada, 1990). Crop losses result from the yield reduction occurring due to defoliation as well as severely spotted fruits which are unsalable.

Symptoms

The bacterium affects all above ground plant parts. On the leaves and stems, the spots are generally brown circular and upto 3 mm in diameter. During rainy periods or in the presence of dew, the spots are watersoaked which are darker in color, less uniformly distributed and not surrounded by halo. In favourable weather conditions, spots coalesce together and a general yellowing occur on leaflets with many lesions. Later blightening of foliage takes place which gives a scorched appearence. On young fruits, the lesions appear as small, dark, slightly raised dots, sometimes with a narrow water soaked borders. On mature green fruits, the spots appear as brownish-black elevated scab like areas 2 to 6 mm in diameter with feathered or irregular margins. For this reason, the disease is also known as scab. At harvest time, they are generally sunken and appear grey or bleached, owing to the dry paper like conditions of the affected skin.

Etiology

The disease is incited by *Xanthomonas vesicatoria* (ex. Doidge) Vauterin *et al.* The bacterium is a gram negative, short rod, strictly aerobic, motile by a single polar flagellum and measures 0.7-1.0 x 2.0-2.4 ~m. On nutrient agar it grows relatively slowly and the colonies are circular, wet, shinning and naples yellow. The bacterium produces acid but no gas from arabinose, glucose, sucrose, galactose, trehalose, cellobiose and fructose. It also produces xanthomonadins pigments.

Epidemiology

The pathogen is mainly seed borne and also perennates in plant debris. The disease spreads through wind splashed rains, infected seed and planting material. It penetrates through stomata and wounds created by wind driven sand, insect punctures or mechanical 'means. Disease is favoured by temperature of 24 to 30°C and high precipitation. A higher rate of nitrogen and potassium application reduces the epiphytic populations of bacterium while defoliation increases with the increasing concentration of both magnesium and calcium (Mc Guire *et al.*, 1991).

Management

Cultural Practices

Use of disease free seed and three years crop rotation has been suggested to reduce the disease (Jindal, 1992).

Chemical Control

Root dip in 0.05 per cent BF-3 (di-methyl- (2 oxy-4-phenyl butone-3yl-2) phosphon) for 30 minutes or 1 hour before planting reduces the disease (Basymbekov and Firsov, 1975). Three sprays of Agrimycin-100(100ppm) at 10 days interval have been found effective (Kishun and Sohi, 1982). Weekly applications of maneb, zineb and copper oxychloride mixture reduced the disease incidence (Maringoni *et al.*, 1986). Jones and Jones,(1987) reported that twice a week application of copper-mancozeb combination gave better control than individual fungicide once but premixing of copper compounds and mancozeb for upto 4 hours had no effect on disease control.

Host Resistance

Roma VF group (Roma VF, Ronita VFN, Rossol VFN, Napoli VF, Heinz 324-1 and IPA-2) was less susceptible to this pathogen than cultivars belonging to the UC 134 group (Barbosa and de Silva, 1982). Cultivar Hawaii 7998 had hypersensitive resistance to this disease (Somodi *et al.*, 1989).

References

Ahmed B.Y., Sharif, F. M., and Sarhan A. R.T., 1987. Effect of certain micronutrients on Fusarium wilt of tomato. *I. J. Agric. Water Resources Res. Plant Prod.* 6: 13-28.

Akram, M., Khan, M. R., 2006. Interaction of *Meloidogyne incognita* and *Fusarium oxysporum f.sp. lycopersici* on tomato. *Annals of Plant Protection Sciences.* 14(2): 448-451.

Alexandri A. A., Eliade E., and Nagler M., 1981. Some aspects of the effect of some fungicides on tomato plants. *Analele Institutului de Cercetari pentruProtectia-Plantelor.* 16: 331- 350.

Anonymous, 1996. State Department of Agriculture Estimate, H.P. Govt., Directorate of Antoniou P. P, Tjmos, E. C. and Panagopoulos, C. G., 1995, Use of soil solarization for controlling bacterial canker of tomato in plastic houses in Greece. *Plant Pathol.* 44: 438-447.

Anuratha C. S., and Gnanamanickam S. S., 1990. Biological control of bacterial wilt caused by *Pseudomonas solanacearum* in India with antagonistic bacteria. *Plant Soil* 124: 109.

Arun and Joshi., Gupta, S.K., Kohli, U.K., 2004. Resistance to Septoria leaf spot (*Septoria lycopersici*) and buckeye fruit rot (*Phytophthora nicotianae* var. *parasitica*) of tomato (*Lycopersicon esculentum*). *Indian J. of Agri. Sci.* 74(4): 194-199.

Arun and Joshi., Kohli, U.K., Sanjeev and Sharma., 2005. Screening tomato (*Lycopersicon esculentum*) genotypes for buckeye fruit rot (*Phytophthora nicotianae*) and quality traits of resistant donors. *Indian J. Agri. Sci.* 75(10): 663-666.

Aspiras R.B., and de la Cruz A.R., 1985. Potential biological control of bacterial wilt in tomato and potato with Bacillus polymyxa F46 and *Pseudomonas fluorescens*. ACIAR Proc. 13: 13.

Ayub, M., Khan, M., Khan, A., Amin, M., 1998, Efficacy of some of fungicides for managing pre-emergence damping-off in tomato and their effect on seedlings vigor. *Sarhad Journal of Agriculture.* 14(5): 475-481.

Balasubramanian A. and Nilakantan G., 1976. Effect of aldicarb on a non-target soil micro-organism. *Madras Agric. J.* 63: 509-514.

Baptista, M. J., Reis., Junior F. B. dos., Xavier, G. R., Alcantara, C. de., Oliveira, A. R.de., Souza, R.B., Lopes, C.A., 2007. Efficiency of solarization and biofumigation on tomato bacterial wilt control in the field. *Pesquisa Agropecuaria Brasileira.* 42(7): 933-938.

Barbosa V., and de Silva Neto J. M., 1982. Incidence of bacterial spot (*Xanthomonas vesicatoria*) in processing tomotoes in Sao Paulo state. *Proc. Trop. Reg.,Am. Soc. Hortic. Sci.* 25: 461-464.

Basymbekov, M. B. and Firsov, V. F., 1975. Trials of some new preparations in the control of bacterial spot of tomato. *Vestn. S. Kh. Nauki Kazakhstana,* 18: 32-35.

Batista, D. C., Lima, M. A., Haddad, F., Maffia, L. A., Mizubuti, E. S. G., 2006. Validation of decision support systems for tomato early blight and potato late blight under Brazilian conditions. *Crop Protec.,* 25(7): 664-670.

Bedekar H.S., 1977. Reaction of four tomato cultivars of different mixtures of isolates of *Pseudomonas solanacearum*. *Veg. Sci.* 4: 1-5.

Bedi P. S., and Dhiman J. S., 1983. Influence of nitrogen, phosphorus and potassium on the development of early blight of tomato. *Indian Phytopath.* 36: 546-548.

Bharadwaj C. L., 1991. Efficacy and economics of fungicides spray schedules for the management of foliar and fruit diseases in tomato. *Indian. J. Mycol. Plant Pathol.* 21: 56-59.

Biezen E. A., Van der Glagostskaya T., Overduin B., Nijkamp HJJ., Hille J., and Vander BEA., 1995. Inheritance and genetic mapping of resistance to *Alternaria alternata* f. sp. *lycopersici* in *Lycopersicon pennellii*. Mol. Gen. *Genet*. 247: 453-461.

Bobisud, C.A., Martin, S.P., Sekioka, T.T., 1996. Field testing bacterial wilt-resistant tomato somaclones. *Journal of the American Society for Horticultural Science,* 121(3): 384-387.

Buchholtz W. F., 1944. Tomato leaf spot diseases in South Dakota. *Circ. Agric. Ext. Servo* 408: 7.

Bychenko N.I., 1974. The effectiveness of certain antibiotics for disease control in glasshouse vegetables. *Mikol. Fitopatol.,* 8: 419-420.

Caron M., Fortin J.A., and Richard C., 1986. Effect of Glomus intraradices on infection by *Fusarium oxysporum* f. sp. *radicis-lycopersici*. *Can. J. Bot.* 64: 552-556.

Cassiolata, A. M. R., Melo, I. S., de 1991. Interaction between *Rhizoctonia solani* and vesicular arbuscular mycorrhizal fungi in tomato. *Summa Phytopathological*. 17(3-4): 95-200

Chaerani, R., Voorrips, R.E., 2006. Tomato early blight (*Alternaria solani*): the pathogen, genetics and breeding for resistance. *Journal of General Plant Pathology*. 72(6): 335–347

Chambers S.C., and Merriman Pro., 1975. Perennation and control of Pseudomonas tomato in Victoria. *Aust. J. Agric. Res.* 26: 657-663.

Chauhan M.S., 1987. Reaction of tomato varieties to leaf spots and fruit rots in Haryana. J: *Mycol. Plant Pathol.* 17: 355.

Chima Wokocha, R., 1987. Effect of soil type on the damping-off of tomato seedlings caused by *Sclerotium rolfsii* in the Nigerian savanna. *Plant and Soil*. 98(3): 443-444.

Clasing, C., Herrera, R., Montealegre, J. R., 2006. Use of bio antagonists and soil solarization to control *Fusarium oxysporum* f. sp. *lycopersii* in tomato.

Cohen R, Blair B., and Katan J., 1992. Chloroacetamide herbicides reduce incidence of Fusarium wilt in melons. *Crop Prot.* 11: 181-185.

Correll J. C., Gordon TR and Elliott VI, 1988. Powdery mildew of tomato: the effect of planting date and triadimefon on disease onset, progress, incidence and severity. *Phytopathology* 78: 512-519.

Czyzewska S., and Bialkowska H., 1973. Observations on the soundness of fruits in a collection of field tomatoes. Biul. Warzywniczy 15: 223-248.

Datar V. V., and Mayce C. D., 1982. Conidial dispersal of *Alternaria solani* in tomato. *Indian Phytopath.* 35: 68-70.

Devi L.R., and Menon M.R., 1980. Transmission of *Pseudomonas solanacearum* through tomato seeds. Agric. Res. J: Kerala 18: 120-122.

Dhanvantari B.N., 1989. Effect of seed extraction methods and seed treatments on control of tomato bacterial canker. *Can. J. Plant Pathol.* 11: 400-408.

Dixon G. R., 1981. *Vegetable Crop Diseases*. A VI Publishing Co., Westport, Connecticut, 404 pp.

Docea E., and Marinescu G., 1977. Contribution to the control of the bacterial canker of tomato caused by Coryne bacterium michiganense (E.F. Smith) Jensen. *Acta Hortic.* 58: 469-474.

Dodan D. S., Shyam K. R., and Bharadwaj S. S., 1995. Relative response of tomato cultivars lines against buckeye rot under field conditions. *Plant Dis. Res.* 10: 135-136.

Dodan D. S., Shyam K. R., and Gupta S. K., 1994. Integrated management of buckeye rot of tomato. *Plant Dis.* Res. 9: 141-144.

Dutta A.K. and Verrna S.P., 1969. Control of bacterial wilt of eggplant with Streptocycline. *Hind. Anti Biot. Bull.* 11: 260-261.

Elmer, W. H., Ferrandino, F. J., 1995. Influence of spore density, leaf age, temperature, and dew periods on Septoria leaf spot of tomato. *Plant Disease*. 79(3): 287-290

Ennle I. D., and Quinn J. G., 1980, an epiphytotic of late blight of tomatoes in Nigeria. *Plant Dis.* 64: 701-702.

Erinle I. D., and Eroctor P. G. E., 1992. Testing tomato varieties for resistance to Septoria leaf. Spot in Nigeria. Samaru *J: Agric.* Res. 9: 63-67.

Etebarian H.R., 1992. Studies on Fusarium wilt of tomato and its chemical control in Varamin area. Iran. J. Agric. Sci. 23: 1-14. FAa. 1997. *Quarterly Bulletin of Statistics* 10: 79-80.

Furuya N., and Matsuyama N., 1992. Biological control of the bacterial wilt of tomato with antibiotic producing strains of *Pseudomonas glumae. J. Fac. Agric. Kyushu Univ.* 37:159-171.

Gnanavel, I., Jayaraj, J., 2003. Integration of soil solarization and Trichoderma viride Pers. Fr. to control damping-off of tomato (*Lycopersicum esculentum* Mill.). *Journal of Biological Control.* 17(1): 99-101.

Graham K.M., Tan H., Chong K.Y., Yap T.C., and Vythilingam S., 1977. Breeding tomatoes for low lands of Malaysia. *Malaysian App. Bioi.* 1: 34.

Grogan R. G., Kimble K. A., and Misaghi II., 1975. A stem canker disease of tomato caused by *Alternaria alternata* f. sp.lycopersici. *Phytopathology* 65: 880-886.

Grote, D., Gabler, J., 1999. Quantification of Phytophthora nicotianae in tomato plants. *Zeitschrift-fur-Pflanzenkrankheiten-und-Pflanzenschutz.* 106(5): 445-454.

Gunawan as., 1988. The effect of spraying intervals of streptomycin sulfate/ox tetracycline 15/1,5 WP on bacterial wilt of tomato, *Pseudomonas solanacearum. Bull. Penalties Hortic.* 16: 114-116.

Gunawan as., 1989. Control of bacterial wilt (Pseudomonas solanacearum E.F. Smith) with Agrimycin 15/1.5 WP on tomato inpago, Bandung. *Bull. Penelitian Hortic.* 17(3):41-44.

Gupta, S. K., Sachin., Upmanyu., Sharma, R.C., 2005. Biology, epidemiology and management of buckeye rot of tomato. *Challenging Problems in Horticultural and Forest Pathology.* 183-199.

Hanudin and Machmud M., 1994. Effec to f bactericide Terlai' and *Pseudomonas fluorescens* on bacterial wilt of tomato. *ACIAR Bacterial Wilt Newsl.* 10: 10-12.

Herger G., Klingauf F., Mangold D., Pommer E. H., and Scherer M., 1988. Efficacy of extracts of Reynoutria sachalinensis (F. Schmidt) Nakai against fungal diseases, especially powdery mildews. Nachrichtenbl. *Dtsch Pflanzenschutz-dienst.* 40: 56-60.

Hibar, K., Edel., Herman, V., Steinberg, C., Gautheron, N., Daami., Remadi, M., Alabouvette, C., El., Mahjoub, M., 2007. Genetic diversity of Fusarium oxysporum populations isolated from tomato plants in Tunisia. *Journal of Phytopathology.* 155(3):136-142.

Hirano, Y. and Arie, T., 2006. PCR-based differentiation of *Fusarium oxysporum* f. sp. *Fusarium oxysporum f.sp. lycopersici* and radicis-*Fusarium oxysporum f.sp. lycopersici* and races of *F. oxysporum* f. sp. *Fusarium oxysporum f.sp. lycopersici. Journal of General Plant Pathology.* 72(5): 273-283

Ho, B.L., 1988. Bacterial wilt control in tomato using household disinfectants. *MARDI Res. J.* 16: 73-76.

Hutagalung L., 1988. Garlic bulb as a material for suppressing the incidence of bacterial wilt (*Pseudomonas solanacearum* E.F. Smith) on tomatoes. *Bull. Penelitian Hortic.* 16: 84-93.

Issa E.R., Ramos da. S., Sinigaglia C., Rodrigues I.N., and Chib S., 1985. Observations on the control of small bacterial spot due to *Pseudomonas syringae* pV. tomato (Okabe) Young *et al.*, on tomato (*Lycopersicon esculentum* Milt.) *Rev. Agric.* 60: 241-247.

Jaime and Garcia, R., Orum, T.V., Felix and Gastelum, R., Trinidad and Correa, R., VanEtten, H.D., Nelson, M.R., 2001. Spatial analysis of *Phytophthora infestans* genotypes and late blight severity on tomato and potato in the Del Fuerte Valley using geostatistics and geographic information systems. *Phytopathology*, 91(12): 1156-1165.

Jain S. S., Sharma, S. L., and Juneja S.L., 1961. Studies on buckeye rot of tomato. New to Himachal Pradesh. *Proc. Indian Sci. Congo Assoc.* 50: 351-352.

Jayaraj, J., Radhakrishnan, N. V., Velazhahan, R., 2006. Development of formulations of Trichoderma harzianum strain M1 for control of damping-off of tomato caused by *Pythium aphanidermatum*. *Archives of Phytopathology and Plant Protection*. 39(1): 1-8

Jindal K.K., 1992. The versatile bacteria how they show up on vegetable crops, the remedies thereof. *Indian Hortic.* 36: 18-21.

Jones J. B., Jones JP, Stall R. E'., and Zitter T A. (Eds.) 1991. Compendium of Tomato Diseases. *The American Phytopathological Society, St. Paul, Minnesota*, USA, 73 pp.

Jones J.B., and Jones J.P., 1987. The effect of bactericides, tank mixing time and spray schedule on bacterial leaf spot of tomato. Citrus Veg. Mag. 50: 34-35.

Jones J.B., Jones J.P., Stall RE' and Zitter T. A. (Eds.) 1991. Compendium of Tomato Diseases. *The American Phytopathological Society*, St. Paul, Minnesota, USA, 73 pp.

Jones W. B., and Thomson SV. 1987. Source of inoculum yield and quality of tomato as affected by Levei/-lula taurica. *Plant Dis.* 71 : 266-268.

Kassim, N. A., Taha, K. H., Seddyk, I. A. S., 1989, Effect of different concentrations of sodium chloride on some fungi causing damping-off of tomato seedlings. *Mesopotamia J. of Agriculture*, 21(2): 343-351.

Kaur S., Suljan Singh., Kanwar J.S., and Cheema D.S., 1988, Variability in tomato for resistance to late blight under field conditions. *Indian Phytopath.* 41: 486-487.

Khan A.N.A., Shetty K.S., and Patil R.B., 1980. Acutance of bacterial wilt of chilli in Karnataka and its relationship to the wilts of other solanaceous vegetables. *Indian Phytopath.* 32:507-512.

Khulbe R. D., and Sati M. C., 1987, Seed transmission of *Alternaria solani* in tomato. *Indian Phytopath.* 40: 106-107.

Kishun R., 1986. Occurrence of Pseudomonas solanacearum on different hosts in India–A review. *ACIAR Bacterial Wilt News lett.* 1: 7-8.

Kishun R., and Chand R., 1988. Effect of bleaching powder on *Pseudomonas solanacearum* and other soil micro flora. *J. Soil Bioi. Ecol.* 8: 90-94.

Kishun R., and Sohi H.S., 1982. Chemical control of bacterial spot of tomato. *Indian Phytopath.* 35: 323-324.

Kishun R., 1987. Loss in yield of tomato due to bacterial wilt caused by *Pseudomonas solanacearum. Indian Phytopath.* 40: 152-155.

Kondoh, M., Hirai, M. and Shoda, M., 2000. Co-utilization of Bacillus subtilis and Flutolanil in controlling damping-off of tomato caused by Rhizoctonia solani. *Biotechnology Letters.* 22(21): 1693-1697.

Krishnamoorthy A.S., and Bhaskaran R., 1990. Biological control of damping off disease of tomato caused by *Pythium indicum* Balakrishnan. *J. Biol. Control* 4: 323-324.

Kritmtan G., 1993. A Cherni-thermal treatment for control of seed borne bacterial pathogens of tomato. *Phyto Parasitica* 21: 101-109.

Kumar, K.V.K. and Rao, K.C., 2006. Efficacy of *Trichoderma harzianum* in controlling tomato damping-off disease and promoting seedling vigour. *J. Plant Prot. and Env.* 3(1): 104-109.

Kumar, Pradeep and Sood, A. K., 2002, Management of bacterial wilt of tomato with VAM and bacterial antagonists. *Indian Phytopath.* 55(4): 513-515.

Kumar, S. and Sugha, S. K., 19990 Over seasoning of *Septoria lycopersici* an incitant of leaf spot disease in tomato. *Journal of Mycology and Plant Pathology*. 29(1): 83-86

Kuriyama T., and Kuniyasu K., 1974. Studies on breeding disease resistant tomatoes by means of interspecific hybridization III. Selecting new material resistant to bacterial canker. *Bull. Veg. Orn. Crops Res. Stn.* 1: 93-107.

Li., Qi., Qin., Luo., Kuan., Lin., Wei., Lu., YanHui., Ye. and YunFeng, 2006. Analysis on the colonization of entophytic bacteria B47 and its control on tomato bacterial wilt. *Acta Phytophylacica Sinica.* 33(4): 363-368

Madalageri B. B., Dharmatti Pro., Srikant K., Madalageri M. B. and Kulkarni S., 1988, Screening of tomato genotypes for combined resistance to *Stemphylium solani, Alternaria solani* and *Septaria Iycopersici. Plant Pathol. Newsl.* 6: 18.

Mahajan, P.D., Boramanikar, P.K. and Sawant, D.M., 1991. Efficacy oftriforine in the control of powdery mildew of tomato. *Maharashtra J. Hortic.* 5: 72-74.

Malathrakis N. E., 1983. Alternaria stem canker of tomato in Greece. *Phytopathol. Mediterr.* 22: 33-38.

Malla, S. and Sah, D.N., 2002. Use of cultural and chemical techniques in managing Septoria leaf spot and buck eye rot diseases in tomato. *Working Paper Lumle Agricultural Research Centre.*

Marinescu G., 1978. Identification and control of bacterial diseases of tomato. *Prod. Veg. Hortic.* 27: 13-17.

Mc Carter S. M., Jaworski C. A., and Johnson A. W., 1978. Effect of continuous plant culture and soil fumigation on soil borne plant pathogens and on growth of tomato transplants. *Phytopathology* 68: 1475-1481.

Michel, V. V., Wang, J.F., Midmore, D. J. and Hartman, G.L., 1997. Effects of intercropping and soil amendment with urea and calcium oxide on the incidence of bacterial wilt of tomato and survival of soil-borne *Pseudomonas solanacearum* in Taiwan. *Plant Pathology*. 46(4): 600-610.

Misaghi U., Olsen M.W., Billotte, J.M. and Sonoda, R.M., 1992. The importance of rhizobacterial M mobility in biocontrol of bacterial wilt of tomato. Soil Bioi. Biochem. 24: 287-293.

Mohamed, F.G. and Badr, L.A.A., 2004, Genetic studies on fusarium wilt resistance in tomato. *Annals of Agricultural Science, Moshtohor.* 42(2): 711-728.

Moretto K. C. K. and Barreto M., 1993, Evaluation of the behaviour of seven tomato cultivars M in relation to the Septoria leaf spot. *Summa Phytopathol.* 19: 18-20.

Naik, M. K., Reddy, M. V., Poonam Sinha and D. Mc. Donald., 1997, Differential Solarization- a decision making tool for management of soilborne plant pathogens. *Indian Phytopath.* 50: 102-104.

Naik, M.K., 2003, Challenges and Opportunities for research in soil borne plant pathogens. *J. Mycol. Pl. Pathol.*, 33: 79-96.

Nath, Premo., 1976. *Vegetables for Tropical Region*. Indian Council of Agricultural Research, New Delhi 104 pp.

Nedumeran S., and Vidhyasekaran P., 1982. Seed borne infection of *Corynebacterium michiganense* in tomato. *Indian Phytopath*. 35: 510-511.

Neelamegam, R., Govindarajalu, T., 2002. Integrated application of Trichoderma viride Pers. Fr. and farmyard manure to control damping-off of tomato [*Lycopersicum esculentum* Mill.]. *Journal of Biological Control*. 16(1): 65-69.

Neelamegam, R., 2006. Efficiency of Trichoderma viride Pers: Fr. with wheat bran/sawdust on damping-off of tomato seedlings caused by *Pythium indicum* Bal. *Advances in Plant Sciences*. 19(2): 381-386.

Osiru, M.O., Rubaihayo, P.R. and Opio, A.F., 2001. Inheritance of resistance to tomato bacterial wilt and its implication for potato improvement in Uganda. *African Crop Science Journal*. 9(1): 9-16.

Panagopoulos C.G., Cavalloro R., and Pelerents C., 1988. The present status of bacterial diseases of tomatoes in Greece. Proc. CEC-IOBC Experts group meeting, Rennes, 1985, 221-222.

Pandey K.K., 2005. Development and testing of an integrated disease management package for multiple disease of tomato. *Indian Phytopath*. 58(3): 294-297.

Pandey, P.K. and Pandey, K. K., 2002. Field screening of different tomato germplasm lines against Septoria, Alternaria and bacterial disease complex at seedling stage. *Journal of Mycology and Plant Pathology*. 32(2): 234-235.

Pandey, V. N. and Dubey, N. K., 1997. Antifungal potentiality of some higher plants against Pythium species causing damping-off in tomato. *National Academy Science Letters*. 20(5/6): 68-70.

Pandey, V.N. and Dubey, N.K., 1994. Antifungal potential of leaves and essential oils from higher plants against soil phytopathogens. *Soil Biology and Biochemistry*. 26(10): 1417-1421.

Patil, B.P., Rajput, J.C., Palve S.B., Jambhale A.S. and Diwakar, M.P., 1990. New tomato 'Sonali' high yielding and wilt resistant. *Indian Hortic*. 35: 6-9.

Paul, P.A., Vale, F.X.R., Zambolim, L., Fontes, P.C.R., Coelho, R.R., Macabeu, A.J., 2004; Comparative epidemiology of tomato early blight on tomatoes under four different spraying regimes. *Fitopatologia Brasileira*, 29(5): 475-479.

Peregrine, W.T.H. and Ahmad K. Bin., 1982. Grafting a simple technique for overcoming bacterial wilt of tomato. *Trop. Pest. Management* 28: 71-76.

Peterson R.A., Inch A.J., Herrington M.E. and Saranah J., 1983. Scorpio, a tomato resistant to bacterial wilt bio var III. Aust. *Plant Pathol*. 12: 8-10.

Phae, C, G., Shoda, M., Kita, N., Nakano, M. and Ushiyama, K., 1992. Biological Control of crown and root rot and bacterial wilt of tomato by *Bacillus subtilis* NB 22. *Ann. Phytopath. Soc*. 58: 329-339.

Pitblado R.E. and McNeill B.H., 1983. Genetic basis of resistance to *Pseudomonas syringae* pv. tomato in field tomatoes. *Can. J. Plant Pathol*. 5: 251-255.

Prasad, Y., and Naik, M. K., 2004, Evolution of genotypes, fungicides and plant extract against early blight of Tomato caused by *Alternaria solani*. *Indian J. Plant Protection*, 31:49-53.

Prasad, Y., and Naik, M.K., 2004. Status of Alternaria blight of tomato in North-Eastern Karnataka. *Karnataka J. Agric. Sci.* 17(3): 607-608.

Rahman, M. A., Vijaya, M. and Chiranjeevi, C., 2003. Performance of soil solarization, captan and biocontrol agents in management of damping-off disease in solanaceous vegetable nursery. *Indian-Journal-of-Plant-Protection*. 31: 71-75.

Raicu C., and Stan G., 1973. *Phytophthora parasitica* Dast. Its symptomatology and control. *Rev. Hortic. Si. Vitic.* 22: 65-70.

Raicu C., Stan G., and Costache M., 1974. Control of *Phytophthora parasitica* Dast. *An. Inst. Cerc. Legumic. Si. Floric.* 3: 349-356.

Raj agopal R., and Vidhyasekaran P., 1983. Effect of fungicidal control of leaf spot diseases of tomato on the quality of fruits. *Indian Phytopath.* 36: 352-354.

Raj H., and Kapoor, 1995. Influence of different plant nutrients on severity of tomato wilt (*Fusarium oxysporum*). *Indian J. Mycol. Plant Pathol.* 25: 290.

Rajagopal R. and Vidhyasekaran P., 1982. Control of leafs pot diseases of tomato. *Pesticides* 16: 16.

Rajput J.C., Palve S.B., and Limaye U.P., 1982. Control of bacterial wilt of tomato. *J. Maharashtra Agric. Univ.* 7: 191.

Ramakrishnan L., and Kandaswami T. K., 1978. Efficacy of fungicide on the control of *Alternaria* leaf spot of tomato. *Madras Agric. J.* 65: 122-123.

Ramakrishnan T. S., and Sundaram N. Y., 1954. Notes on some fungi from south India. *J. Indian Phytopath.*, 7: 140-151.

Ramakrishnan T. S., and Soumini C. K., 1947. Fruit rot of tomatoes caused by *Phytophthora palmivora Butl. Proc. Lindian Acad. Sci.* Sec. B. 25: 39-42.

Ramamoorthy, V., Raguchander, T. and Samiyappan, R., 2002. Enhancing resistance of tomato and hot pepper to Pythium diseases by seed treatment with fluorescent pseudomonads. *European J. Pl. Pathol.* 108(5): 429-441.

Ramos R. S., Sinigaglia C. and Chiba S., 1989. Chemical control of bacterial speck (*Pseudomonas syringae pv. tomato*) of tomato. *Biological Sci.*, 55: 1-3.

Rangaswami G., and Gowda S.S., 1963. On some bacterial diseases of ornamentals and vegetables in Madras state. *Indian Phytopath.* 16: 74-85.

Rao, M. V. B., 1977. Races of *Pseudomonas solanacearum Sm. Indian J. Plant Protec.* 5: 38- 40.

Rao, A.N.S., 1990. Protection of tomato from bacterial wilt (*Pseudomonas solanacearum Smith.*) by Bactrinol-100. *Cur. Res. Univ.Agric. Sci. Bangalore.* 19(6): 100-101.

Reddy, B. N., Raghavendra, C. R. and Sreevani, A., 2006. Approach for enhancing mycorrhizal-mediated disease resistance of tomato damping-off. *Indian-Phytopathology* 59(3): 299-304.

Rodrigues, C., Ribeiro, L.G., Lopes, J.C., Freitas, F.S.de. and Azevedo, L.A.S.de., 2000. Efficiency of metalaxyl to control tomato late blight. *Horticultura-Brasileira*. 18(1):65-67.

Rubin, E. and Cohen, Y., 2006. An improved method for infecting tomato leaves or seedlings with oospores of *Phytophthora infestans* used to investigate F1 progeny. *Plant Dis.*, 90(6): 741-749.

S. V. and Rai P.V., 1984. Influence of *Glomus fasciculatum* on damping off of tomato. *Curr. Sci.*, 53: 588-589.

Sahi H. P. S., 1990. Epidemiology and Management of Alternaria Leaf Spots of Tomato in Himachal Pradesh. *Ph.D. Thesis*, Dr YS Parmar Univ. Hort. and Fty., Solan 97 pp.

Sahi H.P.S. and Shyam K. R., 1993. Occurrence of Alternaria leaf spots of tomato in Himachal Pradesh-a new record. *Plant Dis. Res.* 8: 140-141.

Samoucha Y., Hugelshofer U. and Gisi U., 1987. Effects of disease intensity and application type on efficacy and synergy of fungicides mixture against *Phytophthora infestans*. *Phytopathol.* 120: 44-52.

Sarhan A. R. T., 1982. The influence of soil pH on the severity of tomato Fusarium wilts infection. *Acta Phytopathol. Acad. Sci. Hung.* 17: 291-294.

Satija, D.V, Indra and Hooda, 1987. Note on fungicidal control of damping-off of tomato and chilli caused by *Rhizoctonia solani* and *R. botaticola*. *Haryana J. Hortic. Sci.* 16(3-4): 294-297.

Sen B., and Kapoor I. J., 1974. Chemical control of Fusarium wilts of tomato. *Pesticides* 8: 40-42.

Shanna S. L., Sohi H. S., Chowfla S. C., and Peshin S. N., 1976. Note on the reaction of tomato varieties in field to the causal organism of the buckeye fruit rot disease. *Indian J. Agric.* Sci. 44: 323-324.

Sharma A. K., 1992. Integrated management of buckeye rot of summer tomatoes in the hills. *Indian J. Mycol. Plant Pathol.*, 22: 287-288.

Sharma N., 1992. Control of *Alternaria alternata* by *Clerodendrum aculeatum*. *Natl. Acad. Sci. Lett.* 15: 107-109.

Sharma S. L., 1971. Buckeye Rot of Tomato Caused by *Phytophthora parasitica* Dastur. *Ph.D. Thesis*, Punjab University Chandigarh, 168pp.

Sharma S. L., and Bhardwaj S. S., 1977. Relation of polyphenol oxidase activity in tomatoes to buckeye rot [*P. n. var. parasitica* (Dast) Waterh.]. *Indian. Exp. Biol.*, 15: 688.

Sharma S. L., Chowfla S. C., and Sohi H. S., 1976b. Chemical control of buckeye rot of tomato. *Indian J. Mycol. Plant Pathol.* 6: 130-134.

Sharma S. L., Chowfla S.C., and Sohi H. S., 1976c. Control of buckeye rot of tomato by cultural practices. *Indian J. Mycol. Plant Pathol.* 6: 51-54.

Sharma S. R., and Sohi H. S., 1983. Buckeye rot of tomato as influenced by different levels of N, P, and K fertilizers. *Turk. Phytopathol.* 12: 19-26.

Sharma S.L., Chowfla S.C., and Sohi H.S., 1979. Epidemiology and forecasting of tomato buckeye rot. *Indian Phytopath.* 31: 424-429.

Sharma, J. P., Jha, A. K., Singh, A. K., Pan, R. S., Mathura Rai, Kumar, S., 2006. Evaluation of tomato against bacterial wilt (*Ralstonia solanacearum*) in Jharkhand. *Indian Phytopath.* 59(4): 405-409.

Shekhawat G.S., Kishore V., Singh D.S., Khanna R.N., Singh R., and Bahal V. K., 1979. Survival of *Pseudomonas solanacearum* under diverse agroclimates in India. *Indian J. Agric. Sci.* 49: 736-738.

Shyam K. R., and Gupta S.K., 1996b. Management of buckeye fruit rot of tomato through non-foliage application of fungicide. *Plant Dis. Res.* 11: 202 (Abst.).

Shyam, K. R., and Gupta, S.K., 1996a. Fungicidal control of buckeye fruit rot (*Phytophthora nicotianae parasitica*) and *Alternaria* leaf spots (*Alternaria alternata, A. alternata f. sp. lycopersici*) of tomato (*Lycopersicon esculentum*). *Indian J. Agric. Sci.* 66: 309- 312.

Singh R. N., 1966, You can fight diseases of tomato. *Indian Hortic.* 11: 9-14.

Singh, H.P., Kaur, S., Sokhi, S.S., Surjan and Singh., 1995. Resistant sources to late blight in tomato. *Indian Phytopath.* 48(1): 79-80.

Sinha S.K., Sinha S.N., and Jain B.P., 1988. Reaction of wilt resistant tomato variety and lines to *Pseudomonas solanacearum*. *ACIAR Bacterial Wilt News L:ett.* 4: 3.

Sitaramaiah K., and Sinha S.K., 1983. Relative efficacy of some selected antibiotics on bacterial wilt (*Pseudomonas solanacearum* biotYpe-3) ofbrinjal. *Indian J. Mycol. Plant Pathol.* 13: 277-281.

Sivan A. and Chet, I., 1986. Possible mechanisms for control of *Fusarium spp.* by *Trichoderma harzianum. Pests Dis.* 2: 865-872.

Sohi H.S., Rao M.V.B., Rawal R.D., and Kishun R., 1981. Effect of crop rotations on bacterial wilt of tomato and eggplant. *Indian J Agric. Sci.* 51: 571-573.

Somodi, G. C., Jones, J. B., Scott, J. W., Green, S. K., Griggs, T. D. and Mclean, B.T., 1989, Relationship of lesion development on resistant and susceptible tomatoes to internal populations of *Xanthomonas campestris* pv. *vesicataria*. *AVRDC Publ.* 89: 317.

Sonoda, R.,M. and Augustine J., 1978. Reaction of bacterial wilt resistant tomato lines to *Pseudomonas solanacearum* in Florida. *Plant Dis. Rep.* 62: 464-466.

Sotirova V., and Rodeva R., 1989. Genetic sources of resistance to Septaria lycopersici (Speg.) in tomato. *Genet. Sel.* 22: 480-484.

Sotirova, V., Stoimenova E., and Rodeva R., 1992. Source of combined resistance to diseases in wild species of *Lycopersicon* Mill. *Genet. Sel.* 25: 38-44.

Sreenivasaprasad, S., and Manibhushanrao, K., 1990. Biocontrol potential of fungal antagonists *Gliocladium virens* and *Trichoderma longibrachiatum*. *Z. Pflanzenkr. Pflanzenschutz* 97: 570-579.

Srinivasan, M. C., Patel, M. K., and Thirumalachar, M. J., 1962. Two bacterial leaf diseases on Physalis minima and studies on their relationship to *Xanthomonas vesicatoria* (Doidge) Dowson. *Proc. Indian. Acad. Sci.* B. 56: 193-196.

Sugha, S. K. and Suman, Kumar, 2000. Factors affecting the development of *Septoria* leaf spot of tomato. *Indian Phytopath.* 53(2): 178-180.

Sugha, S. K. and Kumar, S., 1998, Reaction of tomato (*Lycopersicon esculentum*) genotypes to septoria leaf spot (*Septoria lycopersici*). *Indian J. of Agric. Sci.*, 68(10): 709-710.

Szekely I., Baicu T., and Alexandri A..A., 1990. Investigations of the efficacy of new fungicides for control of late blight [*Phytophthora infestans* (Mont.) de Bary] and other leaf diseases of tomato. *Acad. Stin. Agric. Silv.* 23: 245-254.

Tedla T., 1985. Effect of Captafol and Ridomil (R) MZ in the control of late blight (*Phytophthora infestans*) and Septoria leaf spot (Septoria lycopersici) on tomato. *Acta Hortic.* 158: 389-399.

Thareja R.K. and Srivastava M.P., 1989, Varietal resistance of tomato against buckeye rot. *Indian Phytopath.* 42: 297 (Abst.).

Tofoli, J.G., Domingues, R.J., Garcia. Junior, O. and Kurozawa, C., 2003; Tomato early blight control by fungicides and its effects on yield. *Summa-Phytopathologica.* 29(3): 225-233.

Toyoda H. M., Morimoto Kakutani K., Morikawa M., Fukamizo T., Ooto S., Terada H., and Ouchi S., 1993. Binary microbe system for biological control of Fusarium wilt of tomato: enhanced root colonization of an antifungal rhizoplane bacterium supported by a chitin-degrading bacterium. *Ann. Phytopathol. Soc. Jpn.* 59: 375-386.

Tripathi, N. N. and Grover, R.K., 1978, Comparison of fungicides for the control of *Pythium butleri* causing damping off of tomato in nursery beds. *Indian Phytopath.*, 30: 489-496.

Tumwine, J., Frinking, H.D. and Jeger, M.J., 2002; Integrating cultural control methods for tomato late blight (Phytophthora infestans) in Uganda. *Annals of Applied Biology.* 141(3): 225-236.

Ulukus I., and Karaca I., 1985. Investigations on disease rates, causal agents and symptoms of bacterial diseases on tomatoes and capsicums in Elazig, Diyarbakir and Mardin provinces and control measures against the most important disease. *Ege Univ. Zir. Fak. Der.* 22: 61-68.

Vadival S., and Ebenezer, E.G., 2006, Eco-friendly Management of Leaf Blight of Tomato caused by *Alternaria solani. J. Mycol. Pl. Pathol.* 36(1): 76-83.

Verma B.L., 1987. A resins root disease of tomato caused by *Pythium inflatum* Mathews. *Curr. Sci.* 56: 616-617.

Wakimoto S., 1987. Biological control of bacterial Wilt of tomato by non-pathogenic strains of *Pseudomonas glumae. Korean J. Plant Pathol.*, 3: 300-303.

Welch F. J., 1949. Years record of research in Mississippi, farm problems. *Mississippi Farm Res.* 12: 3-8.

Wickramaarachchi, W.A.R.T., Reddy, P.N.N., Reddy, T.V. and Rao, P.V., 2003. Suppression of tomato early blight by spraying of animal manure based compost water extracts. *Annals of the Sri Lanka Department of Agriculture.* 5: 281-290.

Williams P. H., Sheard E., Read W. H., and Selman I. W., 1943. *Mycol. Rep. Exp. Res. SIn. Cheshunt.*, 28: 52.

Woltz S. S. and Jones J.P., 1973. Interactions in source of nitrogen fertilizer and liming procedure in the control of *Fusarium* wilt of tomato. *Hortic. Sci.* 8: 137-138.

Yang Charles Y., 1979. Bacterial and fungal diseases of tomato. *Proc. First Int. Symp. On Tropical Tomato,* 23-27 Oct., held at Shanhua Taiwan, Republic of China, pp.111- 123.

Yoneyama S., 1985. Resistance of some cultivars and rootstocks to bacterial Wilt of tomato. *Proc. Kanto Tosan Plant Proto Soc.* 32: 85.

Yuan L.H., Su Q.L., and Yi Y.H., 1992. A brief report on the technique for control of tomato bacterial Wilt disease. *Plant Protect.*, 18: 48-49.

Yunis H., Bashan Y., Okon Y., and Henis Y., 1980. Weather dependence, yield and control of bacterial speck of tomato caused by *Pseudomonas syringe* pv. tomato. *Plant Dis.*, 64: 937-939.

Chapter 18

Integrated Pest Management in Rice: A Bio-Intensive Approach

Gururaj Katti, A.P. Padmakumari,
Ch. Padmavathi and Chitra Shanker
Directorate of Rice Research, Hyderabad – 500 030, A.P.

Introduction

Rice is the staple food for 65 per cent of the population in India. It is the most important source for meeting the caloric and dietary protein needs of the people as well as for generating employment and income particularly for low income groups in rural areas. India is the second largest rice producing country in the world. The crop accounts for about 22 per cent (44.6 million ha) of the total cropped area with an output of 87million tonnes which forms approximately 46 per cent of the cereals production and 42 per cent of the total food grains. Rice cultivation in India extends from 8° to 35° N latitudes across very diverse ecosystems such as irrigated, rainfed upland, rainfed lowland, semi deep and deep water as well as coastal saline regions.

There has been a marked improvement in rice production in the last 30 years primarily due to the improved technologies such as high yielding varieties, increased fertilizer and other input use efficiency, improved irrigation facilities and adoption of improved crop and pest management practices. At the present population growth, India needs to add not less than 2.5 million tonnes of milled rice every year to sustain the present level of self sufficiency. In order to achieve this, adoption of modern and intensive agricultural practices by the farmers is inevitable. However, along with the practice of intensive agriculture, there is a constant battle to overcome biotic constraints like insect pests, diseases, nematodes and weeds.

Insect Pests

More than hundred species of insects have been recorded as pests of rice, of which about a dozen are of economic significance in India. The coordinated network trials conducted at different centres in India have indicated that by controlling insect pests alone an yield loss of around one tonne per hectare could be averted.

Major Pests

Five insect pests *viz.*, rice yellow stem borer (YSB), *Scirpophaga incertulas* (Walker), gall midge (GM),*Orseolia oryzae* (Wood-Mason), brown planthopper (BPH), *Nilaparvata lugens* (Stal), whitebacked planthopper (WBPH), *Sogatella furcifera* (Horvath), and leaf folder, *Cnaphalocrocis medinalis* (Guenee) have been considered to be of national significance as they are one of the major biotic stresses responsible for significant fluctuations or decline in rice yields across the diverse rice ecosystems.

Other Important Pests

Other pests *viz.*, whorl maggot, *Hydrellia* spp. in all major rice growing areas, hispa, *Dicladispa armigera* (Olivier) occurring in Andhra Pradesh, Himachal Pradesh, Bihar, West Bengal, Orissa and north eastern region, green leafhopper (GLH) *Nephotettix virescens* (Distant) prevalent in Bihar, West Bengal, Assam, Orissa, Madhya Pradesh, Andhra Pradesh and Tamil Nadu, gundhi bug, *Leptocorisa* spp. in Uttar Pradesh, Bihar, West Bengal, Orissa, Madhya Pradesh, Manipur and parts of Andhra Pradesh, climbing cutworm, *Mythimna separata* (Walker) in coastal upland rice growing areas, swarming caterpillar, *Spodoptera mauritia* (Boisduval) in low lying rice in Bihar, Gujarat, West Bengal, Assam and Orissa and thrips in several southern and eastern parts of India, have regional significance. Some pests such as mealy bug, termites, and case worm are pests of growing concern particularly in eastern region. Recently, leaf and panicle mites, black bugs, blue beetle have also started causing serious concern in some parts of the country.

Extent of Pest Caused Losses

Assessment of losses through analysis of 135 multi-location trials under the All India Coordinated Rice Improvement Project (AICRIP) revealed that the avoidable losses due to insect pests averaged 28.8 percent in rice crop. At the global level, Oerke *et al.* (1994) reported losses of 51.4 per cent of the attainable yields in rice of which 20.7 per cent was due to animal pests including insect pests, 15.6 per cent due to weeds and15.1 per cent due to pathogens. The equivalent monetary loss in yield for rice annually amounted to $113 billion. Madhuban Gopal *et al.* (2006) also reported heavy losses (18.6 per cent) due to infestation by insect pests and diseases accounting to monetary loss worth 55120 million rupees.

Integrated Pest Management: A Long Term Solution

Integrated Pest Management is a system that, in the context of the associated environment and the population dynamics of the pest species, utilizes all suitable techniques and methods in as compatible a manner as possible and maintains the pest populations at levels below those causing economically unacceptable damage or loss (FAO). It is an ecologically based strategy that focuses on long term solution of pests through a combination of techniques such as use of resistant varieties, biological control, modification of agronomic practices and habitat manipulation. Pesticides are carefully selected and applied to minimize the risks to the human health, beneficial and non target organisms and environment (Amerika Singh and Sharma, 2004).

Concerted efforts have been made in the last two decades to develop and implement location specific integrated pest management strategies across the diverse ecosystems in rice (Razak, 1986; Sankaran, 1987; Krishnaiah and Reddy, 1989). Rice farmers particularly in the irrigated ecosystem have shown willingness to minimise or replace chemical control with the effective and holistic IPM system for managing pest problems (Pasalu *et al.*, 2006). Though, there are very few such reports in the rainfed rice production systems (Singh and Sasmal, 2004), recently location specific IPM technologies suitable for rainfed rice production systems have been verified and demonstrated in farmers fields in Andhra Pradesh, eastern states of Orissa and West Bengal as well as northeastern states of Assam and Manipur (Gururaj Katti *et al.*, 2006). Though, the validation of effectiveness of IPM in rice has been carried out in farmers fields, there is still adequate scope for increase in use of eco-friendly components within the framework of IPM thereby possibly minimizing or avoiding the use of toxic chemicals, for long term economic, ecological and social benefits to farmers.

Biointensive IPM: An Eco-friendlier Version

Biointensive IPM incorporates ecological and economic factors into agricultural system design and decision making and addresses public concerns about environmental quality and food safety. The benefits of implementing biointensive IPM include reduced chemical input costs, reduced on-farm and off-farm environmental impacts and more effective and sustainable pest management. An ecology-based IPM has the potential of decreasing inputs of fuel, machinery, and synthetic chemicals– all of which are energy intensive and increasingly costly in terms of financial and environmental impact. Such reductions will benefit the grower and society (Dufour 2001).

An important difference between conventional and biointensive IPM is that the emphasis of the latter is on measures to restructure the agricultural ecosystem to the disadvantage of a pest and to the advantage of its parasitoid and predator complex. At the same time, biointensive IPM shares many of the same components as conventional IPM, including monitoring, use of economic thresholds, judicious application of eco-friendly insecticides etc.

Pest Monitoring

Pest monitoring is an integral part of biointensive pest management. In rice, conventionally, light traps have been in vogue for keeping watch on development of insect pest populations in the field. The technique is simple, cost effective and easy to operate, though it can provide only a general estimate of the pest population. In this regard, recent research advances in the field of semio-chemicals which play an important role in insect communication, have led to development of valuable sex pheromone based tools for monitoring pest populations. Pheromone traps provide more accurate and species specific information on pest populations compared to light traps. The pheromone traps are also simple and easy to handle and have already proved popular as tools for monitoring yellow stem borer populations in rice fields (Krishnaiah *et al.*, 2004). Installation of three sleeve traps per acre each loaded with a rubber septum impregnated with 5 mg pheromone chemical, are effective. A trap capture of 30-35 male moths/week/trap could be taken as the capture threshold for pest management decisions. The pheromone mediated monitoring is very practical and useful for the farmers as it is pest specific, is unaffected by weather factors and has no adverse effects on environment. In view of recent advances, development of pheromone traps for leaf folder also looks promising (Ganeshwar Rao, 1996; Kawazu *et al.*, 2002 and 2005).

Host Plant Resistance (HPR)

HPR for pest control is the main basis for eco-friendly nature of biointensive IPM. Resistant cultivars are the most durable, economical and practical means of tackling pest problems, being compatible with all other components of IPM (Table 18.1). In India, research work under the All India Coordinated Rice Improvement Project (AICRIP) in the last three decades has led to release of 836 improved rice varieties for commercial cultivation (Shobha Rani *et al.*, 2008). These include varieties resistant to major insect pests and diseases. Many of these cultivars being high yielders are widely adopted for cultivation in the pest prone areas. However, in future, due to changing and multiple pest complex scenario there is a need for emphasis on multiple resistant varieties. A number of donors like Velluthacheera, Banglei, Aganni, ADR 52, Pandi, Chennellu etc. with proven resistance to gall midge, BPH and WBPH have already been identified. Utilizing these donors multiple resistant varieties *viz.*, Suraksha, Vikramarya, Shaktiman, Rasmi, Daya were developed for release in states like A.P., Orissa, West Bengal, Kerala, M.P. etc.

Table 18.1: List of Varieties Resistant to Important Insect Pests of Rice

Sl.No.	*Insect Pest*	*Released Varieties*
1.	Gall midge	Sneha, Pothana, Kakatiya Erramallelu, Kavya, Rajendradhan 202, Karna, Ruchi, Samridhi, Usha, Asha, MDU 3, Bhuban, Samalei, Orugallu, Abhaya, Shakti, Suraksha, Daya, Pratap, Udaya, IR 36, Shaktiman, Tara, Kshira, Sarasa, Neela, Lalat, Phalguna, Mahaveer, Vibhava, Divya, Dhanya Lakshmi, Surekha, Vikram, Kunti, Triguna, Sita, Samleswari, Karma Mahsuri, Dhanarasi, Mahamaya, Jyothi
2.	Brown Plant hopper	Chaitanya, Krishnaveni, Vajram, Pratibha, Makom, Pavizham, Manasarovar, Co-42, Chandana, Nagarjuna, Sonasali, Rasmi, Jyothi, Bhadra, Neela Annanga, Daya, Aruna, Kanaka, Remya, Bharatidasan, Karthika, Vijeta, Cotton Dora Sannalu, KRH2, PA6201, Hybrid 6129, ADT37
3.	White backed plant hopper	Haryana Basmati, HKR120, Laatha, Narendra 2002, Jitendra, Satyaranjan
4.	Green Leaf hopper	Vikramarya, Lalat, Khaira, Nidhi. GR12, Gr103, Birsa dhan 108, Birsa Dhan 104, Prakash, CoRH1
5.	Stem borer	Ratna, Sasyasree, Vikas, Ravi, Pothana, Dhanya lakshmi, Chandram

Further, the modern versatile tool of biotechnology has enough potential to overcome problems in conventional means of resistance breeding. So far, donors with only moderate level of resistance/ tolerance against stem borer and leaf folder are available. Hence, efforts could be focused on transfer of resistance trait from the wild accessions or through incorporation of novel genes like *cry IA* (*b*) and *cry IA* (*c*) from suitable *B.t.* (*Bacillus thuringiensis*) strains into commercial rice varieties utilizing gene transfer technology for resistance to yellow stem borer (Manju Sharma *et al.*, 2003). Marker assisted breeding is another area for exploitation and DNA marker technology has helped in conferring resistance to hopper pests and gall midge (Bentur, 2004).

Cultural Control

Manipulation of cultural practices offers good scope for indirect suppression of pest populations, particularly in rainfed rice. Integrated nutrient management helps to strengthen plant health thereby enabling to withstand and overcome pest attack. Nitrogen application in splits as well as use of slow release formulations such as sulfur coated urea and urea super granules, helps to lower the incidence of insect pests such as leaf folder, planthoppers and stem borer and also realize higher yields. Similarly, balanced application of other plant nutrients like phosphorus, potassium, zinc etc. also contributes towards resisting pest damage as well as in maintaining pest population below levels that cause economic damage. Other cultural practices such as water management, leaving of alley ways and regular field sanitation are not only eco-friendly but also offer direct benefits to the farmer in pest suppression at no additional cost.

In the last few years, chemically intensive rice cropping systems have led to disastrous consequences such as soil and air pollution, problem soils, indiscriminate destruction of soil microbial fauna etc. Hence, organic rice based farming systems have come into focus for alleviation of these problems with additional commercial benefits as in the case of export quality basmati rice Studies on insect pest scenario in organic input based rice fields have revealed that vermicompost, neemcake and poultry manure application generally lowered pest incidence, however yield levels of rice in these plots could not be maintained on par with that of chemical input treatments (Gururaj Katti *et al.*, 2006). Further studies are needed to confirm whether organic rice cultivation can outweigh existing benefits from chemical based rice farming to the farmers and public.

Implementation of IPM in rice with specific thrust on in situ conservation of natural enemies can be achieved through modification of the rice environment. Habitat management is one such eco-friendly option to enable these organisms to make significant impact in rice IPM. Attempts to modify rice habitat by employing Pusa Basmati variety as 'trap crop' for the management of yellow stem borer have been successful (Padma Kumari and Pasalu, 2003). Based on the basic principles of trap cropping a strategy was developed at Directorate of Rice Research to trap the YSB insects on a susceptible rice variety (Pusa Basmati1) and minimize the damage on the main crop in stem borer endemic areas. Planting of main crop and trap crop is done at the same time by varying the sowing dates so that the trap crop flowers a week earlier than the main crop. For every 2.5- 3m of main crop, one row of trap crop (Pusa Basmati-1) should be planted preferably in east–west direction. In case of BPH endemic areas when alley ways are advocated it was proved that trap crop can be planted along the alley way (Padmakumari *et al.*, 2006).

Biological Control

Biological control is the use of living organisms–parasites, predators, or pathogens–to maintain pest populations below economically damaging levels, and may be either *natural* or *applied*. A first step in setting up a biointensive IPM program is to assess the populations of beneficials and their interactions within the local ecosystem. This will help to determine the potential role of natural enemies in the managed agricultural ecosystem.

There is a tremendous scope for exploiting the rich and diverse populations of natural enemies available *in situ* in rice fields (Beevi *et al.*, 2000) to check the multiplication of pests and consequently their adverse impact on yield (Table 18.2).

Identifying key natural enemies and vulnerable life stages are essential for careful planning and developing practical management programs of insect pests. Analysis of life table data of rice leaf folder based on natural field populations revealed various mortality factors responsible for the fatality of leaf folder at different stages. K-values indicated that the total generation mortality was 72.33 per cent in *Dry* and 54.15 per cent in *Wet* seasons, of which biotic factors contributed 34.06 per cent and 29.72 per cent mortality while other factors resulted in 38.27 per cent and 24.43 per cent mortality in both the seasons, respectively. Bio-mortality factors included seven parasitoids, one pathogen and five predators. Of these, pupal parasitoids *viz.*, *Brachymeria* sp., *Xanthopimpla flavolineata* Cameron and *Tetrastichus* sp were the key mortality factors causing 34.05 per cent mortality in *Dry* and 32.85 per cent in *Wet* seasons of the total pupal mortality. Parasitism was also high in late larval instars (20.88 per cent in *Dry* and 17.27 per cent in *Wet seasons*) due to *Macrocentrus philippinensis* Ashmead and *Goniozus* sp. In early instars, *Apanteles angustibasis* (Gahan) and *Apanteles* sp. caused 8.54 per cent (Dry season) and 8.7 per cent (*Wet season*) mortality. The lone pathogen recorded was *Beauveria bassiana* (Balsamo) Vuillemin causing 3.71 per cent and 1.89 per cent mortality of late larval instars in both dry and wet seasons, respectively. Thus it is evident that biotic factors play a major role in the population dynamics of leaf folder and conservation of late larval and pupal parasitoids in the field is crucial for efficient management of this pest in rice (Padmavathi *et al.*, 2005).

Table 18.2: Major Natural Enemies of Rice Pests

Insect	*Natural Enemies*	*Stage Attacked*
Stem borer	**Parasitoids**	
	Tetrastichus schoenobii	Egg
	Telenomus dignoides	
	Trichogramma japonicum	
Gall midge	**Parasitoid**	
	Platygaster oryzae	Egg/larva
Planthoppers and leafhoppers	**Parasitoids**	
	Anagrus spp.	Egg
	Oligosita spp.	Egg
	Gonatocerus spp.	Egg
	Gonatopus spp.	Egg/Nymph
	Predators	
	Cyrtorhinus lividipennis	Egg/Nymph/adults
	Tythus parviceps	Egg/Nymph/adults
Leaf folder	**Parasitoids**	
	Trichogramma chilonis	Egg
	Macrocentrus philipiennsis	Larva
	Brachymeria spp.	Larva/Pupa
	Tetrastichus spp.	Pupa
	Predators	
	Ophionea nigrofasciata	Larva
	Paederus fuscipes	Adult

General Predators in rice ecosystem

Spiders

1. *Pardosa pseudoannulata*
2. *Tetragnatha maxillosa*
3. *Oxyopes javanus*
4. *Argiope catenulata*

Coccinellids

1. *Harmonia octamaculata*
2. *Micraspis* sp.

Aquatic predators

1. *Microvelia douglasi atrolineata*
2. *Limnogonus fossarum*

In rice, polyphagous beneficial arthropod predators can effectively reduce the populations of some insect pests (Barrion *et al.*, 1991). The two important predatory beetles found in rice ecosystem are Staphylinid beetle, *Paederus fuscipes* Curtis and carabid beetle, *Ophionea* sp. The number of predatory beetles was significantly higher in *rabi* (4.22 to 14.16 per sq.m) compared to *kharif* season (1.66 to 7.83 per sq.m). Among the two predators, population of *P. fuscipes* was significantly more (3 to 14.16 per sq.m) than that of *Ophionea* sp (1.66 to 9.33 per sq.m) at all the stages of crop growth in both seasons (Table 1). Within the season, number of predatory beetles was significantly more at reproductive stage (6.67 to 14.16 per sq.m) than at vegetative stage (1.66 to 6.55 per sq.m). In both choice and no-choice tests, *P. fuscipes* preferred BPH nymphs > WBPH nymphs > leaf folder II instar larvae > BPH adults > GLH nymphs > WBPH adults > GLH adults > leaf folder III instar larvae while the order of preference for *Ophionea* sp was BPH nymphs > WBPH nymphs > BPH adults > WBPH adults > leaf folder larvae (second instar) > GLH nymphs > GLH adults > leaf folder larvae (third instar) (Padmavathi *et al.*, 2007).

Farmers Field Schools (FFS) approach of educating the farmers about the role of natural bio control in their fields implemented by the Directorate of Plant Protection, Quarantine and Storage has also contributed towards strengthening of biointensive IPM. Agro-ecosystem analysis, a holistic soil and plant health approach involving careful planning, surveillance and monitoring of the crop condition throughout should also be an essential feature of rice IPM in future.

Multi-location studies conducted under the All India Coordinated Rice Improvement Programme (AICRIP) have revealed that judicious application of insecticides results in maintenance of a pest population at very low levels thereby helping in build up of natural enemy populations which can be economical and a more practical way to ensure higher yields (Gururaj Katti *et al.*, 2007).

Applied biological control, also known as augmentative biocontrol, involves supplementation of beneficial organism populations, for example through periodic releases of parasitoids, predators, or pathogens. In India, inundative releases of egg parasitoids, *Trichogramma japonicum* and *T. chilonis*, against stem borer and leaf folder, respectively in different ecosystems situations along with other components of rice IPM, have resulted in higher net returns (PDBC 1995-2004, Gururaj Katti *et al.*, 2001). Development of pesticide resistant strains of natural enemies for inundative releases is essential to overcome the deleterious effects of insecticides. Project Directorate of Biological Control, Bangalore (PDBC, 1999) has developed insecticide resistant strain of *T. chilonis* with multiple tolerance to endosulfan and monocrotophos.

Biodiversity in rice IPM

Pest control has often been highlighted as an important ecosystem service provided by biodiversity and one that is threatened by modern agricultural practices. Intensification of agriculture tends to simplify agricultural systems reducing natural enemy diversity, destabilize arthropod populations, often resulting in pest outbreaks (Swift *et al.* 1996). In relation to the rice crop, the fauna and flora in rice fields include pests, their natural enemies (predators and parasitoids) and neutral forms which may be resident, migratory or aquatic species. Some of the minor insects and aquatic insects that have no direct economic impact on rice production have received little attention. But in reality many play an important role in the ecological services provided to the rice ecosystem and a fundamental link in the complex food webs. Pest management practices have a direct impact on this biodiversity. Injudicious use of chemicals causes a drastic drop in beneficial insect diversity and lead to pest outbreaks.

A study conducted at DRR during *kharif* and *rabi*, 2007 revealed that even least invasive pest management practices like soil application of granular formulations and neem spays had an impact on the abundance of insects in the ecosystem and to a lesser extent on the diversity of arthropod species. Chironomids, a neutral insect, which has been reported to be helpful in early colonization of spiders in the ecosystem, were affected by pest management practices. Similarly, the ecosystem abunds with a number of aquatic arthropods, that adds to the health of the ecosystem. These require quantification and conservation through ecofriendly pest management practices.

Use of Novel Tools like Pheromones

Semiochemicals have emerged as one of the promising and safer components in biointensive IPM of several crops including rice. Among them, sex pheromones have been well researched and also one of the best approaches as they are eco-friendly, chemically safe and offer sustainability. Pheromones at minimal doses have also showed potential for direct control of yellow stem borer. Pheromone mediated mass trapping of yellow stem borer is now a practical, cost effective and environmental friendly option for the farmers. Installation of 20 traps loaded with 5 mg impregnated lures per hectare can mass trap male moths of yellow stem borer and help in effectively lowering population of this pest in farmers fields (Krishnaiah *et al.*, 2004). Strategic integration of this technology along with release of egg parasitoids against leaf folder has put biointensive IPM on a sound footing particularly in areas endemic to stem borer and leaf folder (Gururaj Katti *et al.*, 2001)

Chemical Control

Chemical control with its curative effects and ease of application is unavoidable in case of pest outbreaks. Hence, judicious application of insecticides based on economic threshold levels (Table 18.3) can be deployed as last resort under biointensive IPM. So far, several insecticides, both granules and spray formulations have been found effective against insect pests specific to different ecosystems, under coordinated and lead research programmes of Directorate of Rice Research, Hyderabad. In the last few years, newer eco-friendly chemicals with novel modes of action and effectiveness at very low doses such as, cartap hydrochloride against stem borer and leaf folder, ethofenprox, thiomethoxam, imidacloprid and clothianidin against planthoppers fit well into rice IPM programmes.

Use of Botanical Pesticides

In view of the potential hazards arising out of indiscriminate use of synthetic chemical pesticides, non-traditional pesticides like neem and plant products are likely to play a major role in rice IPM in future. Therefore, fine-tuning of technologies with a relook at the utilisation of plant products particularly neem, in pest management is the need of the hour.

Earlier, greenhouse and field studies carried out at the Directorate of Rice Research, Hyderabad, as well as under AICRIP, revealed that neem formulations were moderately effective against stem borer, planthoppers, hispa and leaf folder at recommended concentrations (2 per cent in oil based formulations). Trials on optimization of use of neem formulations in conjunction with insecticides for pest management showed that by replacing one or two insecticide sprays with applications of neem formulations, there is a possibility of reducing environmental contamination without lowering either pest control or grain yield in rice. Commercial neem formulations such as Neemax, Rakshak, Econeem, Neemazal and Neem gold have also been found safe to major natural enemies of rice ecosystem (Krishnaiah, 2000, Gururaj Katti *et al.*, 2005).

Use of Biopesticides

Microbial pesticides such as *Bt* (*Bacillus thuringiensis*) formulations with endotoxins are specific to insect pests and quite safe to humans, natural enemies of insect pests and other non-target organisms. Evaluation of some of these formulations has revealed that they are moderately effective against leaf folder and stem borer, but need to be made commercially viable.

Some of the other promising microbials include fungal pathogens such as *Beauveria bassiana* against rice hispa (Hazarika and Puzari, 1997), *Pandora delphacis* against BPH (Narayanasamy, 1995).

Table 18.3: Eco-friendly Insecticides Recommended Against Insect Pests of Rice

Insecticide	*Group*	*Rate (g a.i./ha)*	*SB*	*GM*	*WM*	*LF*	*RH*	*BPH*	*WBPH*	*GLH*
Granular formulationns										
Carbofuran	CB	750	***	**	***		**	***	***	**
Chlorpyriphos	OP	1000	**	***	**		*	*	*	*
Cartap	NT	750	***			***		**	**	**
Fipronil	PP	75	**	***	**	**	**	**	**	**

Insecticide	*Group*	*Rate (g a.i./ha)*	*SB*	*LF*	*RH*	*BPH*	*WBPH*	*CW*	*GLH*
Spray formulations									
Quinalphos	OP	500	**	***	**		*	**	
Monocrotophos	OP	400	***	***	**	***	***	**	***
Chlorpyriphos	OP	500	***	***	**			**	
Acephate	OP	750	*	**		**	***		**
Carbaryl	CB	750	*	**	**	***	***	*	**
BPMC	CB	500	*			**	**		**
Cartap	NT	300	***	***		**	**	**	**
Ethofenprox	ED	75	*	*		***	***		***
Ethiprole	PP	50	*	*		***	***		***
Imidacloprid	NN	25	*			***	***		***
Thiamethoxam	NN	25	*			***	***		***

*: Moderately effective; **: Effective; ***: Highly effective.
NT: Neiristoxin; OP: Organophosphate; CB: Carbamate; ED: Ether derivative; NN: Neonicotinoid; PP: Phenyl pyrazole.

Role of Indigenous Technical Knowledge (ITKs)

Of late, role of ITKs looks quite promising with their low cost, suitability to local environment as well as ready availability at farmers' level for use. Surveys in north eastern parts of India have revealed indigenous practices such as incorporation of leaves and twigs of herbs, *Eupatorium odoratum*, *Ziziphus sp.*, application of goat excreta, peel of Pummelo etc. by the farmers for pest control (Pasalu *et al.*, 2004). Similarly, pest specific practices have been investigated at CRRI such as planting of *Sachharum* sp. in rice fields for checking caseworm damage. Also, fixing of snail traps @ 10 traps/ha or planting of small branches of local karada plant (*Cleistanthus collinus*–Phyllanthaceae) 3–4 m apart in rice fields can be a very good eco-friendly alternative for the management of gundhi bug, because insecticide use at milky stage of rice crop is not only difficult but also hazardous. Hence, judicious exploitation of ITKs as an integral component of IPM can result in several ecological benefits to rice farmers.

The innovative nature of indigenous communities can be seen among the farmers of Meghalaya, and is evident from the array of pest management methods they have developed to control different kinds of pests which came along with the newly introduced settled wet terrace paddy farming. Most of these methods were developed and are being developed based on locally available resources. The methods are environmentally friendly because chemical pesticides are not used, and most importantly, because they have strong foundations in wellestablished biological and ecological principles. Farmers of the West Garo Hills plant fruit bearing plants like *Bridelia retusa, Dendrophthoe falcata, Morus macroura* and *Sapium baccatum* in terraces, sometimes in home gardens and jhum fields. These plants attract predatory birds by providing shelter and food. The birds eventually keep pest populations down by feeding on the different kinds of insect pests, mainly larvae, caterpillars and nymphs. The farmers' philosophy about this method is simple: "We arrange food and shelter for the birds, in return they take care of our pests". Rats are also controlled by employing the same principle. Branches, preferably of bamboo, are put in terraces. They act as a resting place for owls at night, which will keep the rat population in check. (Bikramjit Sinha *et al.*, 2007).

The Khasis of West Khasi Hills practice another interesting method, and use cow's blood to repel birds from paddy fields. A small bird species, locally known as *phreit*, comes in flocks and damages the paddy, breaking the tillers. Usually it destroys the crop during the maturation period. To stop this, farmers put fresh cow's blood in a bamboo pipe. The pipe is tied with a stick and is placed in the paddy fields. According to the farmers, after 2 or 3 days the blood starts to emit a smelly gaseous substance, which acts as repellent to the birds.

Perhaps the most widely-used of all the traditional pest management practices is the use of decomposing crabs to control rice bugs (as the filling of paddy grain starts, locally available crabs are smashed and put on pointed bamboo sticks in terraced paddy fields). This is practised throughout the entire state by all communities. On-farm trials of the method revealed that by using a crab of 2.5 x 3 cm size as bait, 80–85 bugs can be trapped per 5 m2 in five days. The modified technique has been incorporated in the IPM recommendations for rice bugs, and is being successfully implemented in the entire state. (Bikramjit Sinha *et al.*, 2007).

References

Bentur, J.S., 2004. Insect resistance in rice:keeping pace with changing scenario. Pp. 55-68 in: *Proceedings of the National symposium on Frontier areas of Entomological Research*, November 5-7, 2003, IARI, New Delhi. p. 506

Bikramjit Sinha, Randhir Singha and Dhrupad Choudhury, 2007. Ecological pest management for emerging pest problems. *LEISA Magazine* 23 (4): 11-13

Ganeswara Rao, A., 1996. Sex pheromone systems of rice pests with special reference to leaf folder *Cnaphalocrocis medinalis* Guenee. *Ph.D. Thesis*, Andhra Pradesh Agricultural University, Hyderabad.

Guruaj Katti, R. Jotimani, Ch. Padmavathi, I.C. Pasalu, J.S. Prasad and N.V. Krishnaiah., 2006. Effect of organic manures on pest incidence in rice. Poster presented (Abstract ID: 5067) p 441 in: Abstracts– *26th International Rice Research Conference*, 2nd International Rice Congress 2006, October 9-13, New Delhi, India.

Gururaj Katti, Krishnaiah, N.V., Pasalu, I.C., Jhansilakshmi V., Padmakumari A.P. and Padmavathi, Ch., 2005. Scope of botanicals in rice pest management with special reference to neem formulations. pp. 158-163 In: *"National workshop on promotion of bio-pesticides and bio-fertilizers in agriculture"*. Proceedings of the National workshop organised by National Institute of Rural Development, Hyderabad, 20-22, September, 2005.

Gururaj Katti, I.C. Pasalu, A.P. Padmakumari, Ch. Padmavathi, V. Jhansilakshmi, N.V. Krishnaiah, J.S. Bentur, J.S. Prasad and Rao, Y. Kondala, 2007. *Biological control of insect pests of rice*. Technical Bulletin No. 22. Directorate of Rice Research, Rajendranagar, Hyderabad- 500 030, A.P. p. 22.

Gururaj Katti, I.C. Pasalu, R.C. Dani, D.K. Bora, M.P. Singh, C.R. Satpathi P.S. Reddy and Venkateswarlu, B., 2006. Integrated Pest Management in Rainfed Rice Production Systems of India–A Farmers' Participatory Study. *Oryza*, 43(4) 296-304.

Gururaj Katti, Pasalu, I.C., Varma, N.R.G. and Dhandapani, N., 2001. Quantification natural biological control in rice ecosystem for possible exploitation in rice IPM. *Indian J. Ent.*, 43(4): 439-448

Hazarika,L.K and Puzari,K.C., 1997. Field efficacy of white muscardine fungus (*Beauveria bassiana*) on rice hispa (*Dicladispa armigera*). *Indian J. Agri.Sci.* 67, 10: 463-465

Katti, G., Pasalu, I.C. and Krishnaiah, K., 2000. Quantifying the role of natural biological control in rice- a case study in a farmer's field. *J. Biol. Control*, 14(2): 15-21

Kawazu, K., Nagata, K., Zhang, Z., Sugie, H., and Tatsuki, S., 2002. Comparison of attractiveness in Japan and China of three synthetic pheromone blends based on geographic variations in the rice leaffolder, Cnaphalocrocis medinalis (Lepidoptera: Pyralidae). *Bull. Entomol. Res.* 92:295-299.

Kei Kawazu, Yoshito Suzuki, Yutaka Yoshiyasu, Efieda B. Castillon, Hiroshi Ono, Pham Thi Vuong, Feng-kuan Huang, Taro Adati, Takehiko Fukumoto and Sadahiro Tatsuki., 2005. Attraction of *Cnaphalocrocis medinalis* (Lepidoptera: Crambidae) males in Southeast Asia to female sex pheromone traps: Field tests in southernmost China, northern Vietnam and southern Philippines with three synthetic pheromone blends regarding geographic variations. *Applied Entomology and Zoology*, 40 (3): 483–488.

Krishnaiah, K. and Reddy, P.C., 1989. Operational Research Project on Integrated Control of Rice Pests in Medchal area, Ranga Reddy district, A.P. Achievements and Constraints. Annual Rice Workshop, HAU, Hisar, April 19-22, 1989, pp. 9

Krishnaiah, K. Gururaj Katti, Pasalu, I.C., Varma, N.R.G. and Syed Zainulabedin., 2004. Management of rice yellow stem borer, *Scirpophaga incertulas* Walker with sex pheromones. DRR Technical Bulletin No. 6, 2004-2005. p 32

Krishnaiah, N.V., 2000. Final report of the ICAR AP cess fund scheme on "Bioefficacy and Environmental impact of neem formulations in insect pest management of Rice". Directorate of Rice Research, Rajendranagar, Hyderabad–500 030, A.P., India. p 61

Narayanasamy, P., 1995. Mycoinsecticide : A novel biopesticide in Indian scenario. Biotechnology and Development Review, January–June,1995 pp.15-20.

Padma Kumari, A.P., and Pasalu, I.C., 2003. Influence of planting pattern of trap crops on yellow stem borer, *Scirpophaga incertulas* (Walker) damage in rice. *Indian J. Plant Prot.*, 31 (1): 78-83

Padmakumari, A.P., Gururaj Katti and I.C. Pasalu., 2006. Eco-friendly management of rice yellow stem borer, *Scirpophaga incertulas* (Walker) in farmers field. Poster presented p 422 in: Abstracts–26th International Rice Research Conference, 2nd International Rice Congress 2006, October 9-13, New Delhi, India.

Padmavathi, Ch., Gururaj Katti, A.P.Padmakumari and I.C.Pasalu (2005). Key mortality factors in life table of rice leaf folder. Paper presented in National Conference on Applied Entomology: Current status, Challenges and Opportunities", held at Rajasthan college of Agriculture, Udaipur, September 26–28. pp 38–39.

Padmavathi, Ch, Gururaj Katti, A.P.Padmakumari and I.C.Pasalu (2007). Prevalence, prey preference and predatory potential of staphylinid beetle, Paederus fuscipes and carabid beetle, Ophionea sp. in rice. Poster presented in National conference on "Organic waste utilization and eco-friendly technologies for crop protection", held at ANGRAU, Hyderabad from March 15-17, 2007. pp 90, pg.233.

Pasalu, I.C., Gururaj Katti, Dani, R.C., Bora, D.K., Singh, M.P., Satpathi, C.R. Reddy, P.S., Rao, G.G.S.N. and Venkateswarlu, B., 2004. Integrated pest management in rainfed rice production systems. Technical Bulletin No. 10. Directorate of Rice Research, Rajendranagar, Hyderabad- 500 030, A.P. p. 47.

Pasalu, I.C., Gururaj Katti, P. Krishnamurthy, L.V. Subba Rao, C.S. Reddy, S.V. Subbaiah, N.V. Krishnaiah and J.S. Prasad., 2005. Integrated pest management in rice. Technical Bulletin No. 15. Directorate of Rice Research, Rajendranagar, Hyderabad- 500 030, A.P. p. 53.

Pasalu, I.C., Muralidharan, K., Garg, D.K., Gururaj Katti, Krishnaiah, N.V., Kumari, A.P.P., Prasad, J.S., Reddy, C.S., Krishnaveni, D., and Srinivasprasad, M., 2006. Status and Prospect of Integrated Pest Management Strategies in Selected crops : Rice. In : *Integrated Pest Management Principles and Applications Volume 2: Applications*. pp 38-86. (Eds. Amerika Singh, O.P. Sharma and D.K. Garg) CBS Publishers and Distributers, New Delhi.

PDBC, 1995-2000. Annual reports (1995-2004). Project Directorate of Biological Control, Bangalore.

Rani, N. Shobha, G.S.V. Prasad, L.V. Subba Rao, I. Sudarshan, Manish K Pandey, V.R. Babu, T. Ram, G. Padmavathi, K. Muralidharan, I.C. Pasalu and B.C. Viraktamath., 2008. *High Yielding Rice Varieties of India*. Techjnical Bulletin No. 33, Directorate of Rice Research, Rajendranagar, Hyderabad 500 030, Andhra Pradesh, India. 188 pp.

Razak, R.L., 1986. Integrated pest management in rice crop. *Plant Prot. Bull.*, 38:1-4

Rex Dufour. Biointensive Integrated Pest Management. Fundamentals of sustainable agriculture. www.attra.ncat.org

Sankaran,T., 1987. Biological control in Integrated Pest control- Progress and Perspectives in India. Proceedings of the National Symposium on integrated Pest Control Progress and Perspectives, October 15-17, 1987 (Eds) Mohan Das, N and George Koshy. pp 545.

Sharma, Manju, Charak, K.S. and Ramanaiah, T.V., 2003. Agricultural biotechnology research in India:Status and policies. *Curr. Sci.*, 84(3):297-302

Singh, Amerika and Sharma, O.P., 2004. Integrated pest management for sustainable agriculture. pp 11-24 in:Proceedings 11-Integrated Pest Management in Indian Agriculture (Eds. Pratap S. Birthal and O.P. Sharma), National Centre for Agricultural Economics and Policy Research (NCAP), New Delhi, p. 265

Singh, B.N. and Sasmal, S., 2004. Integrated pest management techniques for rainfed rice ecologies. pp 51-64 in: Proceedings 11-'Integrated Pest Management in Indian Agriculture" (eds. Pratap S. Birthal and O.P. Sharma), National Centre for Agricultural Economics and Policy Research(NCAP), New Delhi, India and National Centre for Integrated Pest Management (NCIPM), New Delhi, India, p. 265.

Chapter 19

Integrated Pest Management in Groundnut

V. Nandagopal[1], S. Desai[2] and T.V. Prasad[3]
[1]Division of Entomology, CRRI, Cuttack – 753 006
[2]Principal Scientist, CRIDA, Hyderabad, A.P.
[3]National Research Centre for Groundnut (ICAR), Ivnagar Road, Junagadh – 362 001, Gujarat

Introduction

Many insect and non-insect pests are known to be associated with groundnut crop. All the pests are not economically important, instead only a few would be economically important at a given time in any situation. Advanced breeding lines possessing resistance to major insect pests are also available. Bio-control agents like NPV against *Spodoptera* and thuricides for red hairy caterpillar, have been successfully used. Intercrops such as pearl millet and soybean against thrips, leaf hoppers and leaf miner, castor against leaf hoppers and *Spodoptera* are useful as traps/repellent crops for reducing the incidence of these pests. New approaches like pheromones for gram pod borer, tobacco caterpillar and leaf miner, and use of neem products and bio-pesticides against major insect pests would go a long way in reducing the economic burden of groundnut farmers. The best plant protection options would be to integrate all the possible pest management measures, even if were to be one or two components with minimum inputs with out affecting yields and environment.

Biotic Stress

The groundnut crop suffers from a variety of insect pests, pathogens and weeds, to an extent of 2380 million rupees per annum in India (Balasubramanian, 1986), afterwards no attempt were made approximate the possible economic loss in groundnut. The relative importance of pests in different groundnut growing zones are variable depending upon the local cultivation practices, the environment and other varieties and the cropping patterns. Some pests are widely distributed and cause significant yield reduction while others are restricted in distribution and are not considered to be economically important. Timely pest surveillance is of greater importance.

Chemical pesticides have played an important role in increasing groundnut production in the past, but, their indiscriminate use for over three decades has led to several problems such as development of resistance in pest to pesticides, pesticides residues, destruction of beneficial insects and the out break of secondary pests. In Holistic Integrated Pest Management (HIPM), a series of steps are followed to keep the population level of the pests below the economic threshold level and demands a judicious use of pesticides, resistant/tolerant varieties, biological suppression, use of pheromones, timely spray of pesticides etc. to form part of the plant production umbrella. Special attention should be given to those pests which occur consistently and cause heavy losses. With economic considerations of input versus out put ratios, it would be most appropriate and feasible to develop IPM Programme for a given location and implement it for the effective management of pests in groundnut.

Major Pests of Groundnut

The Economic Threshold levels for a few of insects of groundnut have been worked out (Table 19.1).

Table 19.1: The Economic Threshold Levels for the Insects of Groundnut

Insects	*ETL*	*Remarks*
Thrips	5 adults/terminal buds	Attains severe status only on young crop during rainy season when the crop is sown in cold weather
Jassids	5 to 10 per plants up to 30DAE	Can cause damage to crop at the seedling stages
Leaf miner	5 mines/plant up to 30DAE 10 mines mines/plant up to 45DAE 15 mines mines/plant up to 50 DAE	Need to look for parasites. If 50 per cent of the larvae are parasitized and the population is close to the threshold level then there is no need to spray
Spodoptera and other defoliators	20 to 25 per cent defoliation up to 40 DAE	From 50DAE the crop can tolerate up to 50 per cent defoliation without any economic loss

Information on period of occurrence and congenial climate for development of major insect pests and diseases (Tables 19.2 and 19.3).

The foliage feeding insects which cause economic loss mostly belong to the order Lepidoptera. These include the leaf miner, *Aproaerema modicella* Dev. (Gelechidae), the red hairy caterpillars, *Amsacta albistriga* Wlk and *A. moorei* Bult. (Arctiidae); Bihar hairy caterpillar, *Spilosoma obliqua* Wlk. (Arctiidae); tobacco caterpillar, *Spodoptera litura* (F); (Notcuidae) and gram pod borer *Helicoverpa armigera* Hb (Noctuidae). Among the sucking insects, the leaf hoppers, *Balclutha hortensis* Lindb. and *Empoasca kerri* Pruthi; aphids *Aphis craccivora* Koch, and *H setaria*); mites, the two spotted white spider mite, *Tetranychus hypogaea* and two spotted red spider mite, *Tetranychus cinnabarinus* are important. Symptoms of damage caused by various insects is given by Amin (1988).

Table 19.2: Period of Occurrence and Congenial Climate for Development of Major Insects

Insect-Pests	*Period of Occurrence*	*Congenial Climate*
Leafminer	March-October	High temperature favours population buildup
White grub	August-October	Temperature coupled with humidity favours the survival of the grub
Thrips	March-October	Early drought favours the population buildup
Jassid	March-October	Dry weather, bright sun shine and occasional rains favour the population
Aphids	July-September	High water content in plants favour the population
Hairy caterpillars	June-October	Long drought period followed by rains favour the population
Tobacco caterpillar	March-October	Hot and humid climate favours the population
Gram pod borer	March-October	High rain fall followed by drought for few days favours the population
Termites	September-October	Warm humid weather favours the population mean temperature passively favours the population

Table 19.3: Period of Occurrence and Eongenial Climate for Development of Major Diseases

Diseases	*Period of Occurrence*	*Congenial Climate*
Early leafspot and Late leafspot	10-28 days after emergence	Mean min. temp. : 18-23°C Mean Max. temp. : 31-35°C RH (per cent) : above 80 per cent Mean monthly rainfall : at least 60mm Conidia show typical forenoon pattern of diurnal periodicity for the release with daily peaks recurring at 10h. For an epidemic : 90-93 per cent RH with around 20°C temp. for 6-7 days. Temperature/RH index for each of five days for prediction.
Rust	40-50 days after emergence	In summer(May-June) incubation period is 18 days. Continuous dry period, RH below 70 per cent and high temperature > 26ºC delays rust epidemic. In low altitude areas disease appears severely. For spore dissemination diurnal periodicity with peak occurrences around noon. Optimum temperature of 20º C, high humidity and moisture on the leaf surface favour infection
Collar rot	Within 30-35 days after planting	High soil (30-35°C)- and air temperatures; soil moisture levels of 13-16 per cent. In sandy and sandyloam soils more incidence. In Punjab July sown crop suffers more damage.
Stem rot	Through out the crop season	29-32°C during most of the day, and night temp. around 25°C, Soil moisture to the extent of 40 to 50 per cent of water holding capacity
Dry root rot	Mature plants are affected	26-35°C congenial. High incidence in shallow fields.
Peanut bud necrosis disease	30-40 days after planting	Vector thrips are mainly carried by the wind, temperature around 30º C and a wind speed of 10 km/ha favour migration of thrips. Major spread of disease in August and September in *Kharif* sown crop and in December, January in R/S sown crop. High plant population reduces severity. Intercropping of bajra reduces disease incidence.
Peanut clump	3-4 weeks after emergence	–
Peanut mottle	Young developing leaflets	29-32°C favour severe disease.
Pod and root-knot	Through out crop season	–
Pod and root lesion	Through out crop season	–
Kalahasti malady	Through out crop season	–

Subterranean Pests

Among the subterranean insects, *Holotrichia consanguinea* Blanch is a major pest causing severe damage to groundnut crops in Rajasthan, Gujarat, Haryana, Punjab, Uttar Pradesh, Andhra Pradesh and Bihar. Another species, *H. serrata* occurs in southern parts of the country like Karnataka and Tamil Nadu. *Odontotermes obesus* (Rambur), *Odontotermes brunneus* (Hagen)and *Trinervitermes biformis* (Wasmann) and *Microtermes* sp. are important termites affecting groundnut.

Rodents and Birds

The rodents and birds are of minor importance as pests of groundnut under field and storage conditions. Three species of rats are associated with groundnut *viz.*, *Mus booduga*, *Rattus meltoda* and *Talera indica.* Under field conditions the rats cause a clear runway with legs impression seen from channel to field or from field bunds to field. Holes with loose soil near the entry could be seen. Pieces of groundnut shells and cut branches are also seen in the corners of house, storage go-downs or near storage bins. The birds are generally not harmful to groundnut except crows (*Corvus splendens*), pigeons and black ibis (*Psaubis papillasa*). The former two birds generally pick up the sown seeds of groundnut leading to reduction in plant population and the occurrence of gaps in the fields. The latter one removes the matured pods from plant and feed on the kernel. Otherwise all these birds are helpful by way of predating the insects like white grubs, grasshopper and grown up lepidopteron larvae.

Diseases

Foliar Fungal Diseases

Foliar fungal diseases, take heavy toll on the crop in congenial environmental conditions. The magnitude of yield losses caused by these diseases is very high and ranged from 10 per cent to 70 per cent, all over the groundnut growing zones (Ghewande,1983; Ghewande *et al.*, 1983; Ghewande, 1985,1987, 1989a,1989b, 1990, 1990a; Ghewande and Desai,1992; Ghewande *et al.*, 1993; Ghewande *et al.*, 1993a; Moraes and Godoy,1985; Subramanium and Mcdonald,1983). Early leaf spot is caused by *Cercospora arachidicola* Hori (Perfect stage-*Mycosphaerella arachidis* Deighton) and late leaf spot is caused by *Phaeoisariopsis personata* (Berk and Curt) V.Arx (Perfect stage-*Mycosphaerella berkeleyi* Jenkins). Temperatures in the range of 25°–30° C and high relative humidity favor infection and disease development. The pathogens may survive on volunteer groundnut plants and infected crop debris. Long distance distribution of the pathogens may be by airborne conidia, by movement of the infected crop debris, or by movement of pods or seeds that are surface-contaminated with conidia. Early and late leaf spot pathogens are also soil-borne in nature (Moraes and Godoy,1985). Groundnut rust is caused by *Puccinia arachidis* Speg. Groundnut crop or volunteer crops are available in one or the other parts of the groundnut zones enabling the survival of uredinial stage of rust round the year. Long distance dissemination of rust may occur via air-borne urediniospores, by the movement of infected crop debris, or by the movement of pods or seeds surface contaminated with urediniospores. An optimum temperature of 20° C, the presence of moisture on the leaf surface, and high relative humidity, favour infection and subsequently disease development. Spread of thę disease within the field is facilitated by wind movement, by rain splash and also by insects (Subramanium and Mcdonald,1987).

Seed and Soil-Borne Diseases

Seed and soil-borne diseases may develop from fungi already established in the seeds or seedlings by soil inhabiting fungi. Collar rot/crown rot/seedling blight (*Aspergillus niger* Van Tieghem, *A. pulverulentus* (McAlp) Thom), stem rot/Sclerotium wilt (*Sclelrotium rolfsii* Sacc) dry wilt or dry root rot

(*Macrophomina phaseolina* (Tassi) Goid = *Rhizoctonia bataticola* (Taub.) Butler) have been recognized as the major diseases. These diseases cause severe seedling mortality resulting in patchy crop stand mostly in sandy loam soils and reduce the yields by 25-50 per cent (Ghewande,1985). Collar rot is serious in the sandy loam and medium black soils of Punjab, Tamil Nadu, Uttar Pradesh, Rajasthan and Haryana. Dry root rot and stem rot are sporadic in nature of occurrence. Dry root rot is reported to occur in Rajasthan, Uttar Pradesh, Tamil Nadu, Andhra Pradesh and Maharashtra. Stem rot is becoming serious in Tamil Nadu, Andhra Pradesh, Karnataka, Madhya Pradesh and Gujarat. The collar rot disease spreads by the spores and mycelium adhering to the seeds pods from season to season. In addition, collar rot pathogens are saprophytic, common soil dwellers in soil humus and plant debris initiate the infection. The carry over from season to season is in plant debris rotting in the soil and on the seeds harvested from a infected field. *A. niger* can tolerate low soil-moisture and optimum development at temperature between 30° C and 35° C. Dry root rot has wide host range. The pathogen. *M. phaseolina* is facultative saprophyte and a soil dweller. Infected soil, plant debris, and pods serve as sources of inoculums. It can be obtained from kernel shell surface and shell tissue. The optimum temperature for seedling infection is 29°–35° C. For pod invasion the temperature range between 26°–32° C is preferable. The dissemination of the fungus is by sclerotia via plant debris and soil, and also infected pods, shell and kernel (Garren,1964). The Sclerotium wilt has also a wide host range and spreads through infected soil, rain drops, sclerotia and in infected farm yard manure.

Virus Diseases

Economically important virus diseases are bud necrosis diseases (PBND), peanut mottle virus (PMV) and clump virus (PCV). Peanut stripe virus (PStV) disease has been reported on groundnut from India (Reddy *et al.*, 1983). Peanut Stem Necrosis diseases has been reported. Bud necrosis disease is widely distributed in India and has been recognized as one of the most important virus diseases of groundnut in Andhra Pradesh, Uttar Pradesh and Madhya Pradesh and of late in Punjab, Haryana, Rajasthan, Maharashtra, Northern Karnataka and Tamil Nadu on rainy season crop. It has also been reported on rabi/summer (post rainy season crop) crop of Mehaboob Nagar district of Andhra Pradesh and Northern and Vidarbha region of Maharashtra. The incidence of the disease was observed to be as high as 92 per cent on spanish bunch varieties of groundnut during summer, 1985 in Jamnagar district of Saurashtra region of Gujarat (Ghewande,1987). Bud necrosis disease has been reported to cause yield losses up to 50 per cent in early infected crop of groundnut in India (Ghewande,1983). The first sign of infection of PBND appears as chlorotic speckling or ring spots on the terminal leaf followed by necrosis of the terminal bud, stunted and shoots bear small leaves that are distorted and mottled leading to death of the plant. In late infections, pod size is reduced. Seeds are often shrivelled and have mottled and discolored testa (Reddy *et al.*, 1991). The virus is sap transmissible. *Thrips palmi* Karny has been shown to transmit the disease. Crops like green gram, black gram, pea, soybeans, beans, groundnut, tomatoes and pea are highly susceptible to both virus and vector apart from ornamental and weed plants. Major spread of the disease is associated with thrips builds up again during January and February and as such crops sown in December-January suffer the maximum (Amin,1988).

Peanut mottle virus (PMV) is reported to occur on rabi/summer groundnut mainly in the states of Andhra Pradesh, Maharashtra and Gujarat. In PMV, newly formed leaves show mild mottling and vein clearing, whereas older leaves show upward curling and interveinal depression with dark green islands. Seeds from infected pods are discolored. PMV is reported to be seed transmitted in a rage from 0.1 to 3.5 per cent depending on the type of cultivar. The disease can cause up to 30 per cent loss in yield (Kuhn and Demski,1984). PMV is transmitted in a non-persistent manner by aphids, *A. craccivora* and *Myzus persicae* (Kuhn *et al.*, 1978). It occurs in nature on several important legume crops and weeds.

Peanut clump virus (PCV) has been reported from Punjab, Gujarat, Andhra Pradesh and Uttar Pradesh on crops grown in sandy soils. It has also shown its presence in certain pockets of Rajasthan, particularly around Jaipur. Yield losses up to 60 per cent have been recorded in late infected plants. The virus has a wide host range and several weeds commonly occurring in groundnut fields are also infected by the virus. Indian PCV isolates are transmitted through seed (Ratna *et al.*, 1991). The virus has also been reported to be soil-borne and possibly transmitted by fungus *Polymyxa graminis*.

Peanut Stripe Virus (PStV) was observed in a few groundnuts at the fields under All India Coordinated Research Project on Groundnut (AICORPG) at NRCG in the beginning and subsequently in Andhra Pradesh, Karnataka, Tamil Nadu, Maharashtra and Gujarat, during 1987 to 1992. It was also noticed in some farmers fields around Junagadh district of Saurashtra region of Gujarat during kharif, 1991 and summer, 1992 (Vijendra Singh *et al.*, 1993). The typical symptom of the PStV is the development of discontinuous dark green banding along with lateral leaflet veins. The other variations of PStV symptoms include blotches, chlorotic rings surrounding the blotching of round leaflets, oak leaf pattern and vein clearing, a mosaic pattern of dark green islands and stunting of plants due to primary infection. PStV is reported to be transmitted through seed and several species of aphids, vectors. It has a wide host range. No authentic reports of yield losses due to PStV are available from India.

Nematode Diseases

Although more than 90 species of plant-parasitic nematodes have been reported in association with groundnut crop (Sharma *et al.*, 1990), only a few species are known to be economically important (Varaprasad and Sharma,1990) Pod and root-knot disease, caused by four species of *Meloidogyne viz.*, *M. arenaria* (Neal) Chitwood, *M. hapla* Chitwood, *M.incognita* Kofoid and White and *M. javanica* (Treub) Chiitwood is the most important disease of groundnut. It is almost distributed throughout the groundnut growing areas of the world. *M. arenaria* and *M. javanica* are more common in the warmer regions, and *M. hapla* in the cooler regions and *M. incognita* in the Mediterranean region (Sharma and Mcdonald,1990). *M. arenaria* is the most wide spread of the groundnut root-knot nematodes. Pod and root-knot nematode disease are considered to be important in India (Sharma *et al.*, 1990). Groundnut plants infected with root-knot nematode commonly develop enlarged roots and pegs which develop into galls of various sizes. *M.hapla* produces small galls but extensive root proliferation. Roots often form a dense mat in severe infection. Pods also become infected and develop knobs, protuberances, or small warts. Development of the root system is commonly much reduced. Plants infected with root-knot nematodes may show various degrees of stunting, chlorosis and wilting. Pod and root lesion disease caused by *Pratylenchus brachyrus* (Godfrey) *Filipjevs Schstekh* was recently reported in India from one of the major groundnut growing areas of Andra Pradesh (Sharma,1988), but it is distributed in other states in India is not known. *Cricoemella ornata* (Raski) de Grisse and Loof has also been reported to be associated with groundnut in India (Sharma,1988). Pod and root lesion disease commonly called as Kalahasti malady caused by *Tylechorhycus brevilieatus* Williams is also important disease which damages groundnut crop in Nellore and Chittur districts of Andhra Pradesh. The disease caused by *P. brachyrus* is characterized by the production of small brown flecks where the nematode enters and increase in size with time. *T. brevilieatus* infection causes brownish black coloration of the pod surface. Small brownish yellow lesions appear on pegs, and young developing pods. Affected plants are stunted and greener than foliage (Reddy *et al.*, 1984).

Weeds

Weeds are the regular menace in most part of groundnut growing area in India during rainy seasons. They interfere with growth and development of groundnut apart from acting as alternate host for many insects and diseases. Controlling the weeds at appropriate time is essential to realize good yield of the groundnut. The critical periods for weed competition in groundnut was between 3 to 6 weeks after planting. Hence, the weed control practice should be taken up during early part of crop phase.

Major Weeds

The weeds occurring in groundnut ecosystem represent atleast 15 families of plant species. They can be classified broadly into monocot and dicot weeds. Monocots: *Brachiaria eruciformis* (Graminae)); *Cynotis* sp. *Commelina bengalensis* L. (Commelinaceae); *Cynodon dactylon* L. Pers, *Dactloctenium aegyptium* (L.) P. Beaue; *Diotaria sanguinalis* L.Scop.; *Echnochloa colonum* (Linn) Link; *E.cruegalli* (L.) Beanv.; *Eleusine indica* (L) Gaertn; *Eragarostis tenella* (L) Stapf, *Paspalum distichum*; *Setaria glauca* (Graminae). Dicots: *Abutilon indicum* (Malvaceaea); *Acalypha indica* (Papavaeraceae); *Alvsicarpus* sp. (Papilionaeae); *Acalypha indica* (Euphorbiaceae); *Celosia argaentea* L. (Amaranthaceaea); *Argaemone mexicana* (Tiliaceaea); *Digera arvensis* Forsk (Amaranthaceaae); *Euphorbia parviflora, E. microphylla* (Eupborbiaceaea); *Leucas aspera* (Lebiatae); *Portulaca oleraceae* L. (Portulacecaea); *Vernonia cinerae* (Composite) and sedges: *Cyparus rotundus* L.; *C. defformis*; *C. compressus* (Cyperaceae) (Devidayal *et al.*, 1987).

Methods of Pests Control

In general, the pests control programme involves use of pesticides mostly and very rarely farmers go in for other alternative methods such as cultural, physical, biological etc. The losses due to pests are of high magnitude all over the world. Hence, comprehensive pest management programme is most desirable. Various pest control methods adopted in India and elsewhere are discussed below:

Insect Pests

Host Plant Resistance

A groundnut line, ICGV 50 was found to be highly resistant to leaf miner and tobacco caterpillar. Two cultivars, ICGV 86031 and ICGV 86699 developed at ICRISAST were found to be highly resistant to leaf miner. A variety Girnar 1 developed by the NRCG during 1988 was found to possess high level of resistance to leaf hopper (*B. hortensis*) (Anon., 1989). To develop a resistant variety for all the insect pests occurring in a locality is rather difficult as the feeding behaviour of various insects differs considerably.

Cultural Practices

Cultural methods involve all aspects of pest management except use of insecticides, natural enemies and resistant varieties. Considering yield as well as the low level of resistance to leaf miner, August 1st or 2nd week appeared to be the best sowing time for higher yields in TMV 7. A significant negative correlation was observed between incidence of leaf miner and maximum-minimum temperature, while positive correlation existed with relative humidity and maximum-minimum temperature (Logiswaran *et al.*, 1982). Groundnut either as pure crop with 30 x 10 cm or 20 x 10 cm spacing alone with straw as mulch gave the maximum control of leaf miner as well as increased the extent of larval parasitization (Logiswaran and Mohanasundaram,1985). As an intercrop, cowpea with groundnut turn out to be the best combination with least larval population. Interestingly, groundnut with paddy straw mulch resulted in least weed number/unit area. Nandagopal (1992) reported the low incidence of leaf miner, when bajra was intercropped with groundnut and 74 per cent reduction in egg laying by leaf miner on the foliage treated with 10 per cent concentration of bajra leaf water extract. A highly significant positive correlation was observed between three generations of leaf miner and percent leaflets attacked and larvae per plant. The population of leaf miner moth catch was high in the month of October followed by August and September (Logiswaran and Mohanasundaram,1985; Srinivasan *et al.*, 1986) reported high leaf miner population at high plant populations levels irrespective of varieties or level of nitrogen applied. This observation was further confirmed as there was highly significant positive correlation between the plant density and leaf miner incidence (r=0.588). In case of thrips and damage by BND, the incidence was reduced when crop was sown with the onset of rainy season. Closer spacing also lessens the impact of diseases (Reddy *et al.*, 1983). Intercrop or multicrop groundnut with cereals reduce the incidence by 10 per cent (Reddy and Wightman,1989).

Biological Control

Several parasites, predators and pathogens have been reported on different insect pests of groundnut. A brief account on the natural enemies reported on the major pests of the groundnut in India is presented here under:

Aphids

Many parasites and predators are reported to cause mortality in aphids (Amin,1988, Bakhetia and Siddhu,1976; Upadhyay and Vyas,1987). Among the predators, *Menochilus sexmaculatus* Fab., *Coccinella repanda* Thumbeg. var. *transversalis* Fab., *C. 8 maculata*, *C.septempunnctata*; *Brumus suturalis* Fab and syrphid, *Ischiodon scutellans* Fab., I.javan are important. Some of the bugs such as *Anthocoris* sp., and green lacewings *Chrysoperla cornea* were also reported to play a dominant role in the suppression of aphids. The peak activity of coccinellids appeared from 6th week to 10th week after sowing at Anand during 1979-1981 (Upadhyay and Vyas,1987). Among coccinellids, *M.sexmaculatus* dominated with more than 50 per cent mean population followed by *C.septumpunctata* and *B.suturalis*.

Leafhoppers

Not much work has been done on the role of parasites and predators in the control of leafhoppers in groundnut. Some lygaeid bugs and spiders were observed predating on them, but they did not appear to cause large reduction (Amin,1988). A predatory fly *Crassopalpus* sp. (Empididae: Diptera) was found predating on *E. kerri*, *B. hortensis* and *Exitianus taeniaticeps* (Kirchbaum) and it was reported feeding 27 leafhoppers in 10 h when field condition was simulated (Nandagopal,1988).

Thrips

There were only two reports on the natural enemies of thrips, Muralidharan and Ananathakrishnan (1978) wherein a bug *Carayonocoris indicus* feeding on *F. schultzei*. Three species belonging to Heteroptera and four belonging to Salticidae (Spiders) were recorded predating on *C. indicus* in Junagadh (Nandagopal, 1992).

Leaf Miner

Though 28 parasites have been recorded from South India, Maharashtra and Gujarat (Mohammad,1981; Shanower *et al.*, 1992) none was put to use under field conditions. Kothai (1974) reported a peak (23.5 per cent) in the parasitism of leaf miner during October, Khan and Raodeo (1978) reported high parasitism during August to September from Maharashtra. Yadav *et al.* (1987) reported a mean parasitism of 50 per cent during March-May by Goniozus sp. and 79.2 per cent by *Stenomesius japonicus*. Oblisami *et al.* (1969) reported a fungus *Aspergillus flavus* on the larvae of leaf miner. Occurrence of a mermithed nematode from larvae of leaf miner was also reported Kothai,1974). A series of experiments at ICRISAT centre has shown that larval parasitism can reach to 90 per cent due to this control pressure and carry over of larval population from one generation to the next was not observed (Wightman, and Amin,1988).

Chemical Control

Synthetic insecticides are inevitable in the management of insect pests in agriculture and they continue to dominate as major insect control agents. Despite increased cost, handling hazards, farmers are inclined to go in for synthetic insecticides, as the quick knock-down effect of these synthetic insecticides is convincing to them. A lot of information has been generated on the use of synthetic insecticides by various workers in India and abroad and only cost effective chemicals (Alexander Jesudasan *et al.*, 1989; Amin, 1988, Das,1988; GHorpade and Thakur,1989; Jayaraj and Rabindra, 1985; Joshi *et al.*, 1978; Kothai,1974; Mohammad,1981; Mohanasundram,1986; Muthaih and Hussain,1991; Nandagopal,1992; Upadyay and Vyas,1989; Wightman and Amin,1988) are presented here under: Following insecticides were found to be effective against various insect pests. Against aphids Metasystox 0.1 per cent; trithion 0.06 per cent; thiometon 0.08 per cent; menazone 0.03 per cent, demetan-o-methyl 0.05 per cent, 25 per cent EC 400 ml/ha, against Leaf hoppers Dimethoate 400 ml/ha monocrotophos 0.05 per cent, 0.07 per cent; 0.5 kg/ha phosalone 4 per cent D; phenthoate 50 EC 750 ml/ha; dichlorovos 76 per cent EC 65, 62 ml/ha; phosphamidon 375 ml/ha, demeton-o-methyl 0.3 per cent; thiometon 0.03 per cent., against thrips Carbofuron seed dress 4 per cent; monocrotophos 0.075 per cent, 0.05 per cent, 0.04 per cent; phosphamidon 0.02 per cent; dimethoate 0.05 per cent; demeton-o-methyl 0.03 per cent., against Leaf miner, Dimethoate 0.03 per cent, 0.05 per cent, dichlorovas 0.05 per cent, 0.1 per cent; chloropyriphos EC 0.05 per cent; carbofuron EC 0.025 and 0.05 per cent, WP 0.05 per cent; Carbaryl WP 0.2 per cent and 10 per cent D; BHC WP 0.05 per cent, 0.1 per cent monocrotophos 180 g a.i./ha, EC 1.6, 1kg/ha, 0.04 per cent, 0.05 per cent, 0.07 per cent; phosalone EC 0.25 and 0.05 per cent; quinolphos EC 0.025 per cent and 0.05 per cent; endosulfan 0.5 kg a.i./ha; ripcard 50 g a.i./ha were found to be effective.

Moult Inhibitors

The cuticular chitin in insects is degraded along with protein at the time of moulting by chitinase enzyme. If synthesis of chitin is prevented the newly formed cuticle will be devoid of chitin. The insect will be unable to withstand the muscular contraction during the moulting which may prevent the casting of excuvium and finally leading to the death of insect (Balasubramanian,1986). These moult inhibitors include diflubenzuron, penfluron, triflumuron and apart from 11 coded components.

Sex Pheromones

The sex pheromones in insects are the chemicals excreted outside which are species specific in attraction and in some species some components could attract subspecies of a species (David *et al.*, 1985). Sex pheromones have the potential role in IPM in groundnutand have been successfully used in the field for the management of *H. armigera* and *S. litura* (Anon., 1974-1991; Pawar *et al.*, 1988). The presence of sex pheromones in leaf miner was reported recently (Nandagopal and Reddy, 1990). Though these sex pheromones are under field use, no economics or its role was reported in reducing the larval population in the field. Sex pheromones for *H. armigera* and *S.litura* are being explored in groundnut through the AICRPG (Anon., 1974-1991). With regard to *H. armigera* in groundnut, a mean of 3.9 males were trapped at 1 m height of sex lure and then there was decrease with increase in height of the trap (Pawar *et al.*, 1988). In an another study when the pheromone of *S.litura* was mixed with *H. armigera*, there was 200 per cent decrease in catch of this moth as compared to trap kept without mixing. Under AICORPO the adult male moths collected using sex pheromone revealed that in Ludhiana the insects continued to be present all through the year (Anon., 1974-1991). Pest population based on three years data (1986, 1987, 1988), the maximum population occurred during April (340), May (649), June (581), September (246), and October (308) while in Jalgaon last week of July (125) and first week of August (108) were the peak periods (Anon., 1974-1991). The population of leaf miner at Hyderabad occurred through out the year recording a maximum of 1662, 1344 and 1190 males/week between March 3rd week to 4th week of February, 1989, respectively (our un–published data). Pawar and Srivastava (1988) reported that even though the sex pheromone of *S.litura* was mixed with pheromone of *H. armigera*, no reduction in the moth catch in the former was observed, however there was drastic reduction in *H. armigera* catch.

Neem Products

Efficiency of neem products in insect control was reviewed by Saxena (1989) and Schemutterer (1990). According to these reviews not much work has been done on the use of neem products in the control of insect pests in groundnut. Neem oil, neem leaf extract and neem leaf decoction at 5 per cent gave 87 to 91 per cent cumulative mortality over 48 hrs of exposure of the leaf hopper *B. hortensis* (Nandagopal *et al.*, 1991). Under the AICORPG system, the first year trial with neem based products like Margoside CK 0.1 per cent and aqueous leaf extract resulted significantly higher pod yield at Vridhachalam and Junagadh centres apart from reducing the leaf miner damage (Anon,1974-1991). Chandramohan and Sivasubramanaian (1987) reported a yield of 2383 kg pods/ha in neem oil treated plot which was on par with monocrotophos. Patel (1968) evaluated efficacy of neem seed extract, neem leaf extract at 10 per cent and neem oil 2 per cent in comparison with monocrotophos against *Amsacta albistriga*. Though monocrotophos gave the highest pod yield, crude neem oil gave 1542 kg/ha pod which was significantly higher compared to control. In an another study both neem kernel suspension (1 per cent) and neem seed extract inhibited the growth of *H. armigera* and *S. litura* larvae (Koul,1985). The neem based insecticides (Margoside OK and CK) suppressed the *S.litura* damage and also gave higher pod yield (Anon., 1974-1991). Growth regulating hormones for *S.litura* have been recorded from 43 plants species belonging to 27 families. Among them Indian neem (*Azadrirachta indica*), *Pongamia glabra* and *Parthenium hysterophorus* are important. Crude neem oil has been successfully utilised in the management of gram pod borer, hairy caterpillar and leaf miner in groundnut. Ovicidal property of crude neem oil was demonstrated in leaf miner (Nandagopal,1992).

Control of Rodents and Birds

Storage (Rodents)

Mixture of 10 g of Zinc phosphide (3 to 5 kg) + popped rice (1 kg) + groundnut oil (5 g)+ sugar (5 g)} was found effective in the management of rodents. The bramodiolon is the single doze anti-coagulant available at 0.005 per cent concentration. This is marketed in the name of "moosh moosh". Each cake can be divided into 4 pieces and each piece can be placed inside a live borrow and the mouth of the borrow should be plastered with soil.

Birds Control

The current practice of birds control includes destroying the roosting sites near crop areas, normal scaring through drumming, birds scaring through guns, use of colour balloons and bright lighting ribbons (Sivanarayan,1985).

Diseases

Host Plant Resistance

Proven stable sources have been identified both in the primary gene pool and in the wild relatives of *A. hypogaea*. Hundreds of land races, advanced breeding lines, cultivars and wild species have been screened for resistance against foliar fungal diseases by several workers, and a large number of resistant sources have been identified (Ghewande *et al.*, 1992). Sources of multiple resistance especially for late leaf spot and rust are available from ICRISAT. These are: EC 76446 (292), NCAc 17133 (RF), PI 259747, PI 350680, PI 389595 and USA 63. Recently five multiple disease resistant germplasm accessions, namely NCAc 17149, NCAc 927, NCAc 17133(RF), PI 393646 and PI 341879 have been identified to be resistant to early leaf spot, late leaf spot, rust and Alternaria leaf spot. Genetic resistance to foliar diseases caused by *P. personata* and *C. arachidicola* has usually been associated with low yields and late maturity (Reddy,1988). It is not difficult to obtain lines with moderate levels of resistance and produce reasonably high yields with minimal or no fungicidal applications (Huang Bing Chao, 1991). As a result of disease resistance breeding programmes, for the first time in the country, multiple foliar disease resistant groundnut varieties *viz.*, Girnar 1 from NRCG, ALR 1,DH 8TG 3, MH 16, ICG 37 TAG 24, ICGV 87160 and ICGV 86590 from India and Southern Runner and Carduipayne from America have been released for general cultivation. In general, it is very difficult to get promising sources of resistance to many of the seed and soil-borne pathogens such as *S. rolfsii, M. phaseolina* and *A. niger* as they are not very specialized pathogens and they have a very wide spectrum of host attacking mechanism (Shew *et al.*, 1987; Garren,1964). However, a few tolerant genotypes have been identified to these pathogens Chohan *et al.* (1970) identified one highly resistant genotype, EC 21115 (U4-47-7) to collar rot. J 11 a spanish bunch released variety and other genotypes like C No.45-23, PI 350680, NCAc 17157, B 201 and RAU 31-3 have also been reported to be resistant to collar rot (Anon., 1974-1991; Chahal *et al.*, 1974). Resistant sources like TMV 1, TMV 8 (Sharma,1982), B 30 and B 31 (Sakhuja and Sethi,1985) for root rot (*M. phaseolina*) have been reported. Genotypes, T 17, T 11-11, EC 1682, RB 4, T 25, T 9, NC 2, NC 18016, NCAc 17941 AX Florigiant have been reported to be resistant to stem rot pathogen (*S. rolfsii*) (Garren,1964;Sharma,1982; Rodrigujez-Kabna *et al.*, 1991).

Among released varieties, Kadiri 3 (R 33-1) and ICGS 11 are tolerant to bud necrosis disease. These varieties should be used in endemic areas. At ICRISAT, ICGV 86029, ICGV 86030, 86031, ICGV 86032, ICGV 86033 and ICGV 86538 were identified to possess field tolerance to bud necrosis with good agronomic traits (Reddy *et al.*, 1991). *Arachis chacoense, A. glabrata, A. skp* (PI 262848) and *A. pusilla*, were found to be immune to bud necrosis (Reddy *et al.*, 1983; 1991). while *A.chacoense* and *A.pussilla* were found to be resistant to peanut mottle virus (Reddy,1988). Two genotypes *viz.*, EC 76446 and NCAc 17133 RF have not shown any seed transmission of peanut mottle virus (Ghewande,1983) and PI 261945 and PI 261946 were tolerant to PMV (Reddy,1988). A. sp. ICG 30036 was reported to be tolerant to peanut clump virus. Genotypes, PI 468141, PI 468142, PI 468144, PI 468345, PI 468174, PI 468363, PI 468366 were reported to be resistant to PStV (Culver *et al.*, 1987). ICG Nos. 4983, 8215, 8958, 8973, 11558, 11560, 11562, and 12168 were found to be resistant to PStV. Out of these, ICG 11558 was immune (Prasada Rao *et al.*, 1988; Vijendra Singh *et al.*, 1993). Four genotypes JSP 1, ICGMS 2, NC 4X and CGC four have been reported to be resistant to root knot disease caused by *M.arenaria* (56). Sakhuja and Sethi (1985) identified G201, Japtin 220-15, local-256, local Ambali and 4018 as resistant to *M.javanica*. PI 1290606, PI 1295233 and PI 1365553, have been found resistant to pod and root lesion disease caused by *P.brachyurus* (Starr,1984; Varaprasad and Sharma,1990).

Cultural Control

Removal of volunteer groundnut plants, removal or burial of infected crop debris, crop rotation etc. are important cultural operations in reducing the leaf spots and rust diseases (Subramanium *et al.*, 1989). Early sowing of the crop reduces incidence of rust and leaf spots (Ghewande *et al.*, 1993; Ghewande and Desai,1992). Sowing date also influenced the severity of the early leaf spot (Subramanian and Hildebram, 1993) Foliar disease severity could also be reduced by altering plant population (Ghewande *et al.*, 1986; Ghewande *et al.*, 1993). Leaf spots and rust being mainly air borne diseases, spread quickly where there is continuity of host plants over large areas. Therefore, various economically profitable crop combinations were tried to check the spread of these diseases at the NRCG and elsewhere.

Muthiah *et al.* (1991) reported that the incidence of late leaf spot in groundnut + cowpea was minimum followed by groundnut + pearl millet and groundnut + black gram and the rust incidence was minimum in groundnut + sorghum. Deep ploughing of fields and rotation of groundnut with gram and wheat are helpful in reducing the collar rot disease incidence. Chohan (1974) reported control of collar rot by resorting to early sowing in June. Groundnut planted with a spacing of 30 cm between the rows had lowest root rot incidence than when planted with a row spacing of 45 cm or 60 cm (Sharma, *et al.*, 1990). The lower incidence of collar rot and stem rot in early sown crop and closer plant spacing has also been reported (Ghewande,1983). Cultural practices such as deep covering or burial of organic matter before planting, cultivation by avoiding movement of soil up around the base of plants and prevention of development of organic debris are useful in reducing the incidence of stem rot. Cotton is a suitable rotation crop for management of *S. rolfsii* (Rodriguez *et al.*, 1991).

Plant density can be increased to decrease the proportion of infected plants of bud necrosis (Reddy *et al.*, 1991). Groundnut intercropped with pearl millet and early sowing showed less incidence of bud necrosis disease as compared with sole crop at ICRISAT (Ghewande, 1983). Cultivation of groundnut away from soybean, cowpea and navy beans could reduce peanut mottle virus considerably as in these crops the virus is seed transmitted and serve as foci of inoculums (Garren, 1964). Groundnut crop grown during *rabi* escapes the clump disease (Delfosse *et al.*, 1996). PStV could be managed by adopting measures like use of virus free seed for sowing, sowing of suitable trap crops around groundnut fields, practice of suitable mixed cropping and intercropping to reduce secondary infection due to aphids, adopting crop rotations of groundnut with non-legumes, roughing of PStV infected plants and burning them, removal of weed hosts of PStV and avoiding of growing crops like soybean, cluster bean, French bean, green gram, cowpea and sesame in the nearby areas of groundnut crop since these are hosts of PStV (Vijendra Singh *et al.*, 1993). Crops like maize, sorghum, pearl millet and rice may be rotated with groundnut to reduce the incidence of root-knot namatodes. Sesamum, castor and cotton are good crops to include in rotations for reducing the incidence of *M. arenaria* (Rodriguez *et al.*, 1991). Nematode problems in groundnut can be effectively managed through deep ploughing during summer months, increase the use of F.Y.M., oilcakes and other organic amendments and soil solarization (D.J.Patel, personal communication).

Biological Control

Mycoparasites, *Dicyma pulvinata* (= *Hansfordia pulvinata*) and *Verticillum lecanii* were found to parasitize the early and late leaf spot pathogens (Ghewande *et al.*, 1991a). *Acronomium persicinum, A.salmonaum, Darluca filum, Eudarluca caricis, Tuberculina costaricana, Penicillium islandicum* and *V. lecanii* were found parasitizing groundnut rust (Ghewande *et al.*, 1983; 1991a). *P. islandicum* and *V. lecanii* and their culture filtrates were found to be effective in controlling both late leaf spot and rust diseases under field condition (Ghewande *et al.*, 1991a). Harvey and Bailey (1983) reported that *Bacillus thuringiensis* and *Pseudomonas ceparia* provided upto 70 per cent control of early and late leaf spot in small scale field tests. In 2 of 3 field tests, significant reductions in the severity of early leaf spot was observed due to *Bacillus cerus* (strain 304) on chitin amended leaves (Kokalis and Burelle *et al.*, 1992). Some soil fungi, actinomycetes and bacteria antagonistic to collar rot pathogen have been reported (Chogan, 1974). *Trichoderma harzianum, T. auareoviride, T. viride, T. polysporum, Bacillus spp.* are reported to be antagonistic to *S. rolfsii, M. phaseolina* and *Rhizoctonia solani.* For example, it is estimated that about 140 kg of Trichoderma granules are required to suppress *S. rolfsii* in 1 ha area (Beckman and Rodriguz-Kabana,1975). Both *T. viride* and *T. harzianum* were found to be capable of reducing the sclerotial population of *M. phaseolina* (Sharma,1982). Seed treatment with spores and mycelium of *T. polysporum* protects the seeds from invasion by *M. phaseolina* (Chohan,1974). Elad *et al.* (1982) have proved the beneficial effects of *T. harzianum* against *S. rolfsii* and *R. solani. T. harzianum* was found to be antagonistic to *A. niger* (Lastin *et al.*, 1989). *Gliocladium virens* was found to be effective against *S. rolfsii* (Papavizas and Cillins,1990).

Chemical Control

Several systemic and non-systemic fungicides were tried for reducing the severity early leaf spot, late leaf spot and rust in India and elsewhere (Salako and Adu,1990; Culbreath and Breneman,1992). However, no single fungicide was found to be very effective against all the foliar diseases. In India, leaf spots and rust were effectively managed when three sprays of carbendazim (0.05 per cent) +mancozeb (0.2 per cent) were given at 15 days interval commencing from the first appearance of symptoms (Anon., 1974-1991). Vidhyasekharan (1981) reported that chlorothalonil was effective against leaf spots and rust. Similarly, positive effects of chlorothalonil in controlling leafspots were reported (Breneman and Sumner,1990). Sekhawat *et al.* (1985) reported that the spray schedule consisting two sprays of mancozeb (0.2 per cent) at 35 and 70 days after germination and one spary of carbendazim (0.025 per cent) at 50 days after germination was most economical. Vyas *et al.* (1986) also recommended that carbendazim (0.075 per cent) + mancozeb (0.15 per cent) mixture should be sprayed around August 15 and second and third spray should be repeated depending on the intensity of both leaf spot and rust. The rust and leaf spots were substantially reduced due to the application of carbendazim (0.025 per cent) + tridemorph (0.4 per cent) five times at fortnightly intervals commencing 35 days after sowing during summer season (Ghewande,1983). Application of 1.06 kg a.i./ha of tridomorph-carbendazim-maneb resulted in the most effective-foliar diseases control and the highest pod yield (Shekhawat *et al.*, 1985). Gorbet *et al.* (1990) reported that five rounds of spray, of carbendazim (0.05 per cent) + mancozeb (0.2 per cent), starting from 30 days after sowing with 12 days intervals was found to have maximum control of both late leaf spot and rust diseases which resulted in 57 per cent increased in dry pod yield.

Seed treatment with either thiram @ 3g/kg seed or carbandazim @ 2g/kg seed is effective in controlling seed and soil-borne diseases (Ghewande, 1989a). Seed treatments of peanut with Vitavax-captan at 3g/kg seed and the spraying of crop with benlate at a rate of 2.5g/L caused a significant reduction in crown rot incidence (Nofal *et al.*, 1990). Good control of pre-emergence root rot caused by *M. phaseolina* was achieved by seed dressing with captafol (Shanmugam and Govindaswami,1973). Soil drenching with carboxin has been reported to be effective against *S. rolfsii* (Amma and Shanmugam,1974). Flutolanil was found effective against *S. rolfsii* and *R. solani* (Csinol, 1987). Diniconazole was effective against *R. solani* and *S. rolfsii* (Csinos *et al.*, 1987). To avoid re-infestation of thrips transmitting bud necrosis disease, an application of insecticides with strong residual effect is required (Reddy and Wightman, 1989). Granular formulations of thiofanox, aldicarb, acephate and disulfotan gave significant control of thrips when applied at the time of sowing. Subsequently the crop may be applied with systematic insecticides at regular intervals (Reddy *et al.*, 1991). Similarly soil application of 1,2-dibromide, 3-chloropropane, carbofuran and temik reduce peanut clump disease effectively, but they are not economical (Delfosse *et al.*, 1996). Aphids and other species of PStV vectors should be controlled through commonly used systemic insecticides to avoid secondary transmission of PStV (Vijendra Singh *et al.*, 1993). Application of granular nematicides like nemacur 10 g, carbofuran 3g, thimet 10g, mocap 10g etc. @ 3kg a.i./ha is useful in reducing the incidence of nematodes in groundnut. Seed treatments with 3 and 6 per cent aldicarb, sulfone and carbofuran reduce the damage by *M. arenaria* (Patel *et al.*, 1986).

Other Methods of Control

Use of chemicals in the control of diseases has had created a lot of environmental problems. Therefore, use of renewable sources of alternative chemotherapeutants is highly desirable. Green plants appear to be the reservoirs of effective chemotherapeutants and can provide renewable sources of useful biofungicides. For examples, aqueous leaf extracts (2 per cent) of neem and mehandi (*Lawsonia inermis*) found to be effective and economical in controlling leaf spots and rust diseases (Ghewande, 1989). Neem oil (1 per cent) and nerium leaf extract (10 per cent) reduced the incidence of ring mosaic, late leaf spot and rust (Ganapathi and Narayanaswami,1990). Usman *et al.* (1991) reported that foliar application of neem kernel extract (3 per cent) reduced the late leaf spot and rust diseases. Ghewande *et al.* (1993) reported that the application of 3 sprays of 5 per cent aqueous neem leaf extract reduced early leaf spot by 14 per cent and late leaf spot by 19 per cent. The aqueous neem leaf extract was found to be efficacious in controlling both late leaf spot and rust diseases and in increasing pod yield at hot spot locations in Tamil Nadu, Karnataka, Maharashtra and Orissa (Anon., 1974-1991). Seed treatment with neem seed extract can check the infection of *S. rolfsii* and *R. bataticola* pathogens (Venkatarao and Alice,1991).

Weeds

Mechanical/Cultural Control

Experiments conducted through out the country revealed that 2-3 hand weeding or harrowing upto 45 days are optimum for groundnut. First weeding should be done at about 20-25 days after sowing and second a fortnight later.

Chemical Weed Control

Many weedicides like Nitrofen, Fluchloralin, Alachlor have been recommended to control the weeds in groundnut crop throughout the country (Anon., 1974-1991). Pendimithalin has also been recommended by All India Coordinated Research Project on Oilseeds in groundnut. Alachlor is applied primarily as a pre-emergence treatment at the rate of 1.5 to 2 kg a.i./ha. This is effective in controlling annual grasses and *Acalypha, Amaranthes* and *E. species.* Dephenemid is a water soluble wettable powder applied at the planting time @ 2.25 to 3.40 kg a.i./ha for the control of annual and perennial grasses and partial control of broad leaved weeds. It persists for 3 to 6 months in soil. Trifluralin is available as both EC and granules, for pre-emergence application @ 0.5 kg a.i./ha. Fluchloralin is a selective pre-emergence weedicide controlling grasses and broad leaved weeds. Immediate irrigation is necessary after soil application @ 1.25 to 1.5 kg a.i./ha. Pendimithalin is a selective, pre-emergence weedicide controlling annual grasses and certain broad leaved weeds. Application of this weedicide @ 1 kg a.i./ha controls most of the grassy weeds in groundnut crop. Fusilade is a post-emergence weedicide formulated as EC and is very effective against annual and perennial grasses when applied @ 0.25 to 1 kg a.i./ha.

Integrated Pest Management

Various control measures need to be suitably integrated to cover a broad spectrum of pests. The approach, therefore, should be an integrated protection of the crop rather than integrated management of individual pest. Unfortunately, serious efforts were not made in the past to integrate the various pests control measures and as such research on integrated pest management in groundnut is still in its infancy. However, there are a few research efforts where the pest control measures have been integrated and it appears that there is a scope for an integrated pest management in groundnut (Ghewande and Nandagopal, 1997).

Integrated Management of Diseases

Collar rot causes very high plant mortality at the seedling stage in the groundnut crop during June-July, An integration of date of sowing (10th or 20th June), and seed treatment with thiram @ 5 g/ ha kernel reduced the plant mortality and gave higher returns at Ludhiana and Samrala in Punjab (Anon,1974-1991). At Aliyarnagar and Coimbatore in Tamil Nadu adoption of practices such as closer spacing, advancement of date of sowing by two weeks, spray of insecticides like monocrotophos and orthene, spray of antiviral principles like sorghum (JVP) and coconut (CAVP), removal of bud necrosis infected plants particularly at the early stages of crop development and removal of weeds reduced the incidence of bud necrosis disease (Anon., 1974-1991). It is reported that when low rates of insecticides are applied for the control of an insect complex in groundnut, NC 6 a resistant variety yielded 16 per cent more yields than a susceptible Florigiant variety of groundnut (Campbell *et al.*, 1977). Gorbet *et al.* (1990) have successfully used host resistance along with fungicides for the control of leaf spots. Combining Southern Runner, a moderately resistant variety to late leaf spot, *S. rolfsii* and tomato spotted wilt virus with applications of Chlorothalonil can provide late leaf spot control (Vidhyasekharan, 1981).

At NRCG, various control measures were integrated for the management of early and late leaf spot during *kharif* 1988 to 1990. Three years data showed that most effective early and late leaf spot control was achieved when groundnut was intercropped with pigeon pea and sprayed twice (55 and 70 DAS) with fungicide mixture (carbendazim 0.05 per cent + mancozeb 0.2 per cent) or groundnut intercropped and received one spray each of neem leaf extract (2 per cent), fungicide mixture, and cell-free filtrate of *P.islandicum*. These treatments and the intercrops with neem leaf extract at 55 and 70 DAS and with inoculum *P.islandicum* at 55 and 70 DAS gave the highest monetary returns over three years (Ghewande *et al.*, 1989a).

Integrated Management of Weeds

So far no attempt has been made to integrate all the feasible methods of weed control in groundnut. Any single weed control method either chemical or cultural is not completely effective in managing the weed problem. For effective weed management all the feasible methods have to be used harmoniously. The known practice such as row spacing, tillage and crop rotation are worth to include in any weed control strategy. Generally, application of mixture of two or more herbicides is effective against different species of the weeds compared to application of single weedicide. For example, mixture of Nitrofen + fluchloralin @ 0.2 kg a.i. and 0.5 kg a.i./ha respectively controls various weeds in rainfed crop. The pre-emergence application of mixture of alachlor and Nitrofen is more effective in controlling weeds. Pre-sowing application of fluchloralin followed pre- emergence application of Nitrofen @ 0.7 to 1 kg a.i./ha. control weeds effectively. So the approach should be sequential application of two different herbicides which will be more effective. Pre-emergence application of fluchloralin followed by one or two hand weeding at 25 to 30 and 50 to 60 days after sowing is the most economical. Integration of herbicide and hand weeding at 30 DAS gave consistent and effective weed control and higher pod yields over two years (Ramkrishna, 1991).

Holistic IPM Developed in Groundnut

A holistic IPM was developed in groundnut, based on 6 years independent approach from entomological, pathological and weeds management aspects, followed by three years rainy season experiments. The holistic IPM approach revealed that under varied raining pattern, the groundnut productivity can be sustained in India. It is not only productivity that is sustained but also the requirements for a farm family and his cattle. The need based use of crude neem oil with pesticides to the minimum, along with use of intercrops/trap crops such as pearl-millet, castor, soybean and pigeon-pea increases the net profit apart from acting as a security crop against the failure of groundnut crop due to some reason or the other.

The biotic stresses were managed more effectively, when the.IPM package *i.e.* (seed treatment with Carbendazim + trap crops (pearl millet, castor–around the field and soybean and pigeon pea–intercrop) + insecticide mixture (neem oil 2 per cent + Phosphomidon 0.02 per cent or Endosulfan 0.04 per cent) + pheromone trap (for *Helicoverpa armigera, Spodoptera litura* and *Aproaererma modicella* + 2 per cent neem leaf extract spray(40 DAS), Mancozeb 0.25 per cent + Carbendazim 0.05 per cent (55 DAS), culture filtrate of *Penicillium ilandicum*(70 DAS). (Weedicide: Flochloralin 1.5kg a.i./ha + one Interculturing (35 DAE) + one hand weeding (30 DAE) was fallowed. The IPM module along with pigeon pea, castor, pearl-millet and soybeans contributed for higher CBR as growing of intercrops was always profitable in groundnut. Intercrop system always less likely to fail and more stable than mono-cropping. The higher CBR of 3.13 was obtained with a net return of Rs. 17810/ha. Mixed crops always proved profitable in terms of net profit as well as catering the needs of the farm family.

Components Used in the HIPM Module

Seed Treatment

Carbendazim (Bavistin) 2g/kg of seeds for seed borne fungi.

Cultural Practices

1. Soil amendments with castor cake or mustard cake or neem cake @ 1000 kg/ha (preferably 15 days before sowing) for soil borne diseases like stem rot and collar rot.
2. Intercropping of groundnut with red gram/pearl millet/sorghum (3:1) depending upon the locally recommended intercrops for foliar fungal diseases
3. Use of trap crop such as cowpea/soybean/castor as border crop for sucking/defoliating insects
4. One hand weeding after the application of pre-emerged weedicide Flochloralin@ 1.5Kg a.i/ha.

Biological Control

Use of commercially available preparations of *Trichoderma* Sp. (for the control of stem rot and seed rot as 1) seed treatment @ 4g/kg seed 2) soil application @ 25 to 62.5 kg/ha 3) application of B.T. (*B. thuringensis* @ 300-500 g/Ha) for *Spodoptera and Helicoverpa* (between 45-60 days)

Use of Pheromone Traps for *Spodoptera* (10 per ha), *Helicoverpa* (10 per ha) and Leaf Miner (25 per ha)

Need Based Application of Chemicals

1. Foliar spray of crude neem oil @ 2 per cent in teepol (for both foliar fungal and sucking pests
2. Foliar spray of fungicides (0.05 per cent carbedazim + mancozeb 0.2 per cent) at 45 and 60 Days after sowing for major foliar fungal diseases like early and late leaf spots and rust)
3. Weedicides: Flochloralin@ 1.5Kg a.i/ha as pre-emergence weedicide.

This holistic IPM module was tested under farmers conditions for assessment and refinement through Technology Assessment and Refinement through Institution Village Linkage Programme (TAR-IVLP). Certain modifications were made to suit the needs of farmers during *kharif* 1999-2000

The intercrop especially castor could not be taken as border crop, instead it was taken as intercrop after every 3-4 rows of groundnut. Components used were: Seed treat,emt wotj carbendazim @ 2g/kg seed, soil application of Castor Cake @ 500kg/ha., soil application of *Trichoderma vir ide*@ 62.5 kg/ha, Castor as an intercrop, foliar application of neem oil @ 2 per cent in teepol for both foliar fungal diseases and sucking pests and use of pheromone traps for *Spodoptera* and *Helicoverpa* (10 traps/ha each).

This IPM technology gave control of major insect pests ranging from 24-46 per cent and diseases from 28-48 per cent and average increased in yield by 19 per cent, with highest gross monetary return of Rs. 52000/ha as against farmers' practice of Rs. 36099/ha. Considering the added advantages, farmers have now realized the importance of the IPM package.

Plant Protection Options Available for Groundnut Growing Farmers in India

Use of Resistant/Tolerance Varieties

Following varieties resistant/tolerant to various major insect pests and diseases can be used.

Insect Pests/Diseases	*Resistant/Tolerant varieties*
☆ Sucking Pests (Aphids, Jassids and Thrips)	Girnar 1
☆ Defoliators (*Spodoptera litura, Helicoverpa armigera* and leaf miner	BD 2
☆ Early leaf spot, late leaf spot and rust	ALR1,ALR2, ALR3, Girnar 1, ICGV 86590, ICGV 87160, ICGV 86325, CSMG 84-1, OG-52-1, RSHy 1, DRG 12, DRG 17, TAG 24, BSR1, VRI 5, CO 4
☆ Aflaroot and Collar rot	J 11, JCG 88 and OG-52-1
☆ Stem rot	OG-52-1, Dh 8, ICGV 86590
☆ Peanut Bud Necrosis Disease(PBND)	ICGS 11, ICGS 44, ICG 37, Kadiri 3, ICGV 86325, K 134, DRG 12, R 8808, JCG–88, CSMG 884, Chandra

Cultural Practices

☆ Deep burying of crop residues; destruction of crop debris by burning; removal of volunteer groundnut plants (self sown plants) early planting and wider row spacing (40-45cm) for managing early leafspot, late leafspot, and rust.

- Against collar rot, avoid deep sowing (not more than 2 inches) and avoid injury to seedlings and deposition of soil particles on cotyledons during intercultural operations.
- Non-dirtying cultivation in combination with minimizing defoliation due to leaf spot; deep burial of surface organic matter and crop debris by ploughing it to a depth of 8-10 inches; early sowing and close planting to manage stem rot pathogen.
- Deep tillage for management of soil-borne insect-pests and pathogens
- Soil amendantment with castor cake/neem cake/mustard cake @ 500-1000 kg/ha for termites and soil borne pathogens
- For PBND management: Early sowing (first fortnight of June) for Peninsular and central India, late sowing (first fortnight of July) for Northern India; and close spacing (20x10cm/ 30x7.5cm).

Barrier/Repellant Crops

- Pearlmillet against thrips, leafminer, peanut bud necrosis disease
- Castor: Against *Spodoptera litura* and *Achoea Janata*

Trap Crops

- Castor for *Spodoptera litura*, soybean/cowpea for leafminer

Intercropping

- Soybean as a trap crop against leafminer, thrips and jassids
- Cowpea: Against leafminer
- Pearlmillet, sorghum, maize and pigeonpea : Early leafspot, late leafspot and rust, Pearl millet and maize for PBND
- Mixed cropping with Moth (*Phaseolus aconitifolius*) in alternate rows for stem rot
- Crop Rotation with Cotton, Wheat, Maize, Onion, Garlic for Stem rot (*Sclerotium rolfsii*)

Biological Control/Biopesticides

- Nuclear polyhedrosis virus for the management of Spodoptera and Helicoverpa @ 250 LE ($6x10^9$/LE/ha).
- Spray of aqueous neem leaf extract (2-5 per cent) for the management of leafspots and rust
- Spray of neem seed kernel extract (5 per cent), or crude neem oil in teepol (2 per cent) against defoliators, mites and foliar pathogens.
- Release of *Trichogramma chelonis* @ 50000/ha, two times at 7-10 days interval followed by release of Bracon hebetor @ 5000/ha two times at 7-10 days against leafminer and defoliators.
- Pheromone traps @ 10 traps/ha for *Spodoptera* and *Helicoverpa* and 25 traps/ha for leafminer.
- Erection of Bird perches @ 10-12/ha.
- Vegetative trapping of migratory larvae of Red hairy caterpillars by keeping twigs of *Jetropha* or *Calotrpis* and neem branches for trapping adult-white grubs.
- Spray of Bt @ 300-500 kg/ha against RHC

- ☆ Seed treatment with *Trichoderma viridae/T. harzianum* against seed and soil-borne pathogens @ 4g/kg seed.
- ☆ Soil application of *Trichoderma viridae/harzianum* @ 25-62.5 kg/ha, preferably in conjunction with organic amendments such as castor cake or FYM against seed and soil borne diseases
- ☆ Two foliar sprays of water extract of Oak leaf @ 1.0 per cent with in 40 days after germination for PBND

Chemical Methods (Need Based)

Leaf miner	Endosulfan 35 EC at 500 ml to 1000ml/ha OR Bendiocarb (Garvax 80 WP) 0.5 kg ai/ha or Carbosulfan 10 G at 4 kg ai/ha OR Cartap (Padan 50 EC) at 1 kg ai/ha
Tobacco caterpillar	Diazinon 300 g ai/ha OR Endosulfan 1 Kg ai/ha
American boll worm	Endosulfan 644 ml ai/ha or Dichlorvos 350-500 g ai/ha
Red Hairy Caterpillar	Dichlorvos 300-500 g ai/ha or Dicrotophos 150-200 g ai/ha OR Monocrotophos 200-300 g ai/ha or Phosphamidon 200-300 g ai/ha
Sucking pests (leaf hopper, thrips, aphids)	Carbaryl 1.12 kg ai/ha or dichlorvos 350 g to 500 g ai/ha OR monocrotophos 150g to 200 g ai/ha OR phosphamidon 150 to 200 g ai/ha or malathion 1.12 kg ai/ha OR dimethoate at 200 to 250 ml ai/ha OR Phorate 1.12 kg ai/ha OR Phosphamidon 150 to 200 g ai/ha
Termite and white grub	Soil application of phorate granules at 1.1 and 1.6 kg ai/or 5 per cent diazinon granules at 25 kg/ha or carbofuran 3G at 10 kg/ha
Mites	Kelthan 500 ml or Endosulfan 350 to 400 ml, or 0.03 per cent Phosphamidon + 0.04 per cent endosulfan

Chemical Control Measures for Protection of Stored Groundnuts

How to Use	*Insecticides*	*Rate of Application*
Prevention of infestation		
Spray interior surfaces of infested stores before refilling	Malathion (w.p.) OR Fenitrothion (E.C.) OR Chloropyriphos-methyl (E.C.) OR Pirimophos-methyl (E.C.) OR Iodofenphos (w.p.) OR Deltamehtrin (w.p.)	250g 25 per cent a.i. in 5L water 100 m^{-2} or 200 ml 50 per cent a.i. in 5L water 100 m^{-2} 200 g 50 per cent a.i. in 5L water 100 m^{-2} or 50 g 2.5 per cent a.i. in 3.3 L water 100 m^{-2} 1 stripe 30 m^{-2}
Space treatment of empty stores before refilling	Dichlorovos (resin strips)	Apply at half the rate recommended above
Direct application of spray to pods or to stacks containing pods or kernels	Malathion (w.p.) OR Fenitrothion (E.C.) OR Chloropyriphos-methyl (E.C.) Pirimophos-methyl (E.C.) OR Iodofenphos (w.p.) Deltamehtrin + OR Piperonyl butoxide(E.C.)	" " " " "
Surface application to, or admixture of insecticidal dust with pods	Malathion (dust) OR Deltamehtrin (dust) OR Fenitrothion + carbaryl (dust) OR Bromohos (dust)	10 ml in 990 ml water t^{-1} or 500g 2 per cent a.i.t^{-1} 500g 0.2 per cent a.i.t^{-1} 500g 1.2 per cent + 0.8 per cent a.i.t^{-1} 200g 2 per cent a.i. of surface

Diseases	Spray schedule
Early leafspot, late leafspot, rust	Two sprays of Mancozeb 0.2 per cent at 35 and 70 days after germination and one spray of Carbendazim 0.025 per cent at 60 days after germination (ICBR 1:2.85) or Application of carbendazim 0.025 per cent + tridemorph 0.04 per cent five times at fortnightly intervals commencing 35 days after sowing during summer season OR Spray application of carbendazim 0.05 per cent + mancozeb 0.2 per cent at 2-3 weeks interval, 2 or 3 times starting from 4-5 weeks after planting (CBR 1:14.8 to 1:24.4) OR Foliar application of chlorothalonil (0.1 per cent)
Collar rot	Seed treatment with carbendazim 1-2g/kg seed of mancozeb 2-3 g/kg seed or chlorothalonil/ captafol 2 g/kg seed
Stem rot	Seed treatment with carbendazim 2g/kg seed or mancozeb 3g/kg seed Soil drenching with 0.2 per cent carbendazim
PBND	Foliar application of dimethoate or quinalphos 0.02 per cent (two spray) within 40 day of germination

Need based spray of crude neem oil (CNO) 2 per cent with insecticide, phosphamidon 0.02 per cent could effectively control the population build up of leafhopper and thrips. Intercropping of pearl millet reduces the insect pests when grown with groundnut (Gangwar *et al.*, 1994; Kennady *et al.*, 1990.) This is in conformity with the finding of Nandagopal (1992) and Nandagopal *et al.* (1995). Number of moths trapped were 1690 *S. litura*, 1520 *H. armigera*, and 2610 *A. modicella* during rainy season of 1993. Fungicide mixture (mancozeb 0.025 per cent + carbandazim 0.05 per cent) could effectively control the ELS and LLS of groundnut. Application of 2 per cent aqueous neem leaf extract spray at 40 DAS, followed by spray of fungicide mixture 55 DAS, and application of culture filtrate of *P. islandicum* (50 per cent) (1ml culture filtrate in 2 ml of water) at 70DAS also effectively controlled the diseases (ELS, LLS and rust). The low per cent leaf damage due to GLM in the unsprayed plot during *kharif* 95 was due to smothering of weeds over the groundnut crop. The weed intensity was significantly suppressed under control wherever Flochloralin at 1.5 kg./ha was applied along with one hand weeding and one inter culture operation.

It was obvious that, wherever groundnut was intercropped with pigeon pea, the monetary return was higher mostly because of additional returns obtained from pigeon pea. In the good year *i.e.* when rainfall was optimum, the CBR was (1:1.93) in the treatment.

It can be concluded that the biotic stresses were managed more effectively, when the IPM package *i.e.* (seed treatment with carbendazin + trap crops (pearl millet, castor–around plot and soybean and pigeonpea–middle of plot) + insecticide mixture (neem oil 2 per cent + phosphamidon 0.02 per cent or endosulfan 0.04 per cent) + pheromone trap (for *H. armigera, S. litura* and *A. modicella*) (trap crop (red gram) + 2 per cent neem leaf extract spray(40 DAS), Mancozeb 0.25 per cent + carbendezin 0.05 per cent (55 DAS), culture filtrate of *P. ilandicum* (70 DAS). (weedicide: fluchloralin 1.5kg a.i./ha + one inter-culturing (35 DAE) + one hand weeding (30 DAE) was fallowed. The IPM module along with pigeonpea, castor, pearl millet and soyabean contributed for higher CBR as growing of intercrops was always profitable in groundnut (Devi Dayal and Reddy, 1991, 1991 a). Also intercrop system always is less likely to fail and more stable than mono-cropping (Rao and Willey, 1980). In Indian situation, where failure of either beginning, mid season or end season rainfall is frequent, the inter crop such as pigeon pea and castor may support as security crop to the farmers. Lingappa (1990) reviewed the effect of intercropping on the pest management in crops in India and expressed that reduced pest damage is possible in intercropping system. For an Indian farmer to whom intercropping is the norm, and so may not be an impossible task to persuade the farmer to take up intercropping (Anon., 1997).

Future Prospects of IPM

It is evident from the literature that there is a great scope for integration of suitable and compatible pest management options and thus develop cost effective and highly efficient package fc·· the integrated pest management in groundnut. The components of IPM such as use of multiple resistant varieties mass trapping of lepidopetran pests, using sex pheromones, use of biocontrol agents like NPV, thuricides, *V. lecanii, P. islandicum*, use of moulting inhibitors, use of suitable intercrops, use of plant products of neem origin, soil solarization, alellopathy for weeds and judicious use of pesticides should be integrated suitably wherever possible keeping in view of the environmental conditions. The HIPM system is in operation on a limited scale in India it needs to be strengthened and implemented in a large scale for economically important pests.

References

Alexander Jesudasan, R.W., Arulsamaraj, D.and Priyakumari, Joyce, 1989. Field evaluation of new insecticides against the groundnut leaf miner. *Pesticides*, 22: 26-27.

Amin, P.W., 1988. Insect and mite pests and their control. In *Groundnut* (ed. P.S. Reddy), pp. 393-452.

Amma, P.A. and Shanmugam. N., 1974. Studies on the root rot of groundnut caused by *Sclerotium rolfsii* Sacc. *Madras Agric. J.* 61,973-978.

Anonymous, 1974 to 1991. Progress Report, Annual *Kharif* Oilseeds Work shop, All India Coordinated Research Project on Oilseeds Research (AICORPO), Directorate of Oilseeds Research, Hyderabad.

Anonymous, 1989. *Annual Report*, National Research Center for Groundnut (Indian Council of Agricultural Reseach), Junagadh, Gujarat, pp. 75.

Anonymous, 1997. *Annual Report*, National Research Center for Groundnut (Indian Council of Agricultural Reseach), Junagadh, Gujarat, pp. 69.

Backman, P.A. and Rodriguez-kabana, 1975. A system for the growth and delivery of bilogical control agents to the soil. *Phytopathology*, 65: 819-821.

Bakhetia, D.R.C. and Sidhu, A.S., 1976. Biology and seasonal activity of the groundnut aphid *Aphis craccivora* Koch. *J. Res. Punjab Agric. Univ.*, 14: 299-303.

Balasubramanian, M., 1986. Integrated Pest and Disease Management: The need of the day. In: *Pest and Disease Management* (Ed.)S. Jayaraj, pp. 252.

Brenneman, T.B. and Sumner, D.R., 1990. Effects of tractor traffic and chlorothalonil applied via ground sprays or Center Pivot irrigation system on peanut and pod yield. *Pl. Dis.* 74: 277-279.

Campbell, W.V., Wyne, J.C. and Emery, D.A., 1977 Control of an insect complex with a resistant variety No. 6. *Proceedings of APRES*. 9: 69.

Chahal, A.S., Dutt, S. and Chogan, J.S., 1974. Varietal resistance in groundnut (*Arachis hypogaea* L.) to collar rot (*Aspergillus niger*) in the Punjab state. *J. Res. Punjab Agric. Univ.*, 11: 200-203.

Chandramohan, N. and Sivasubramaian, P., 1987. Evaluation of neem product against leaf miner *Aproaerema modicella* (D.). *Neem Newsl.*, 4: 44.

Chohan, J.S., 1974). Recent advances in diseases of groundnut in India. In: *Current Trends in Plant Pathology*, Lucknow, Uttar Pradesh, India, University Press, pp. 171-184.

Chohan, J.S., Sandhu, G.S. and Dalal, J.L., 1970. Screening of world collection of groundnut (*Arachis hypogaea* L.) germplasm for resistance to collar rot disease caused by *Aspergillus niger* Van Tieghem in Punjab. *Indian J. Agric. Sci.*, 40: 546-551.

Csinos, A.S., Littrell, R.H. and Kvien, C.S., 1987. Activity of diniconazole on foliar and soil-borne diseases of peanut. *Appl. Agric.Res.*, 2: 113-116.

Culbreath, A. K. and Bresnneman, T.B., 1992. Combining center pivot irrigation, application of chlorothalonil with moderately resistant cultivars for control of late leaf spot in groundnut. *Plant Diseases*, 76: 29-32

Culver, J.N., Sherwood, J.L. and Melouk, H.A., 1987. Resistance to peanut stripe virus in *Arachis* germplasm. *Plant Dis.*, 71: 1080- 1082.

Das, B.B., 1988. Efficacy of some synthetic pyrethroids and conventional insecticides spray against groundnut pest complex. *Pesticides*, 22: 10-12.

David, H., Nesbitt, B.F., Easwaramoorthy, S. and Nandagopal, V., 1985. Application of sex pheromone in sugarcane pest management. *Proceedings of the Indian Academy of Sciences, Animal Sciences*, 94: 333-339

Delfosse,. P., Legreve, A., Devi, P.S., Reddy, A.S. and Maraite, H., 1999. Indian peanut clump virus is highly infectious on graminaceous plants. *Zeitschrift-fur-Pflanzenkrankheiten-und-Pflanzenschutz*, 106(4): 418-424

Demski, J.M. and Kuhn, C.W., 1983. Peanut mottle. In *Peanut Disease Compendium* (eds. D.M. Porter, D.H.Smith and R. Rodriguez-Kabana) pp. 45-46, St. Paul minnesota. *The U.S. American Phytopathological Society.*

Devidayal and Reddy, P.S., 1991. Effect of intercropping sunflower (*Helianthus annus*) in different planting pattern of groundnut (*Arachis hypogaea*) on yeild and economics in Saurashtra region of Gujarat. *Ind. J. Agric. Sci.*, 61(7): 490-492.

Devidayal, Basu, M.S. and Reddy. P.S., 1987. Weed control in groundnut Technologies for better crops (Bulletin No.32), ICAR,New Delhi, pp. 16.

Elad, Y., Hadar, Y., Chat, I. and Hemis, Y., 1982. Prevention with *Trichoderma harzianum* Rifai of reinfestation by *Sclerotium rolfsii* Sacc. and *Rhizoctonia solani* Kuhn of soil fumigated with methyl bromide and improvement of disease control in tomatoes and peanuts. *Crop Prot.*, 1: 199-211.

Ganapathy, T. and Narayanasamy, P., 1990. Effect of Plant Products on the incidence of major disease of groundnut. *Int. Arachis Newsl.*, 7: 20-21.

Gangwar, S.K., Singh, Y.P. and Patel, C.S., 1994. Influence of intercropping on infestation by insect pests of crops at medium-high altitude of Meghalaya. *Indian Journal of Agricultural Sciences*, 64(2): 137-140.

Garren, K.H., 1964. Inoculum potential and differences among peanuts in susceptibility to *Sclerotium rolfsii. Phytopathology*, 54: 279-281.

Ghewande, M.P. and Nandagopal, V., 1997. Integrated Pest Management in Groundnut (*Arachis hypogaea* L.) in India. *Integrated Pest Management Reviews*, 2(1): 1-15.

Ghewande, M.P., 1989a. Management of foliar diseases of groundnut (*Arachis hypogaea*) using plant extracts. *Inadian Journal of Agricultural Science*, 59: 133-134.

Ghewande, M.P., 1987. Virus diseases of groundnut and their management. In: *Symposium on Diseases of Oilseed and Pulse Crops and their Control* (Eds. S.C.Vyas and V.N.Shroff), Jawaharlal Nehru Krishi Vishwavidhyalaya, Indore, (M.P.) pp. 62-67.

Ghewande, M.P., 1989 b. Biological control of late leaf spot (*Phaeoisariopsis personata*) of groundnut (*Arachis hypogaea* L.) *Indian J. Agric. Sci.*, 59: 189-190.

Ghewande, M.P. and Desai, S., 1992. Integrated Management of Groundnut Diseases. In: *Advances in Plant Disease Management* (Eds. Vyas *et al.*) (In press).

Ghewande, M.P., 1985. Seed and seedling diseases of groundnut deserve attention for better productivity. *Seeds and Farms*, 11: 9-12 and 19.

Ghewande, M.P., Desai, S. Prem Narayan and Ingle, A.P., 1993. Integrated Management of foliar diseases of groundnut (*Arachis hypogaea* L.) in India. *Int. J. Pest Mgmt.*, 39: 375-378.

Ghewande, M.P., Desai, S., Prem Narayan and Kamble, S.D., 1992. Some sources of resistance to early leaf spot of groundnut (*Arachis hypogaea* L.) in India. *Trop. Agric.* (Trinidad), 69: 284-286.

Ghewande, M.P., 1983a. Effect of cultural practices on the disease (budnecrosis, collar rot, stem rot) incidence and yield of groundnut. *Indian Bot.Reptr*, 2: 176-177.

Ghewande, M.P., 1989a. Management of foliar diseases of groundnut (*Arachis hypogaea*) using plant extracts. *Indian J. Agric. Sci.*, 59: 133-134.

Ghewande, M.P., 1990a. Diseases of groundnut and their management. *J. Oilseeds Res.*, 7: 78-97.

Ghewande, M.P., Shukla, A.K. and Pande, R.N., 1986. Management of foliar disease of groundnut through agronomic practics. *Indian Bot. Reptr.*, 5: 179-181.

Ghewande, M.P., Desai, S., Prem Narayan and Ingle, A.P., 1991a. Evaluation of fungicides and plant products to manage foliar diseases of groundnut (*Arachis hypogaea* L.). *Proc. 4th Ann. Zonal Meet. Indian Phytopathology Soc.*, 1: 2.

Ghewande, M.P., Shukla, A.K., Pandey, R.N. and Misra, D.P., 1983. Losses in groundnut yields due to leafspots and rust at different intensity levels. *Indian J. Mycology and Pl. Path.*, 13: 125-127.

Ghorpade, S.A. and Thakur, S.G., 1989. Assessment of need based monocrotophos application for the control of major pest of groundnut. *Pesticides*, 11: 19-20.

Gorbet, D. W., Knauft, D.A. and Shokes, F.M., 1990. Response of peanut genotype with differential level of leafspot resistance to fungicides treatment. *Crop Science*, 30: 529-533.

Harvey Spurr and Bailey, 1983. Biological control of peanut leaf spot. *Proceedings of American Peanut Research and Education Society (APRES)*, 15: 93.

Huang Bingchao, 1991. Review for groundnut pests and diseases and IPM implementation in Guangdong Province of South China. Poster Abstract: Second International groundnut work shop, 25-29 November, 1991. Interational Research Institute for the Smi-Arid tropics, Patancheru, Adhra Pradesh,India. ICRISAT, 1980-1993. Annual Report, Internatioanl Crops Research Institute for Semi-Arid Crops, Patancheru, Andhra Pradesh, India.

Jayaraj, S. and Rabindra, R.J., 1985. Management of red hairy caterpillar *Amsacta* spp. (Lepidoptera:Arctiidae) on groundnut. In *Pest and Diseases Managemet–Oilseeds, Pulses, Millets and Cotton* (ed.S.Jayaraj) pp. 38-43.

Jayaraj, S. and Rabindra, R.J., 1985a. Integrated management of *Spodoptera litura* with special reference to NPV. In: *Pest and Diseases Managemet–Oilseeds, Pulses, Millets and Cotton* (Ed. S. Jayaraj), pp. 30-37.

Joshi, B.G., Rama Prasad, G., and Satiyanaryana, S.V.V., 1978. Relative hydroxide as antifeedant against tobacco caterpillar *Spodoptera litura* F. in the nursery. *Indian J. Agric. Sci.*, 48: 19-22.

Kennedy, F.J.S., Rajamanickam, K. and Raveendran, T.S., 1990. Effect of intercropping on insect pests of groundnut and their natural enemies. *Journal of Biological Control*, 4(1): 63-64.

Khan, M.I. and Raodeo, A.K., 1978. Importance of larval parasites in the control of *Stomopteryx subsecivella* Zeller. *Journal of Maharashtra Agricultural University* 3: 261-263.

Khan, M.I. and Raodeo, A.K., 1987. Importance of larval parasites in the control of *Stomopteryx subecivella* Zeller. *J. Maharashtra Agric. Univ.*, 3: 261-263.

Kokalis-Burelle, N., Backman, P.A., Rodriguez-Kabana, R. and Ploper, L.D., 1992. Potential for biocontrol of early leaf spot of peanut using *Bacillus cereus* and chitin as foliar amendments. *Biol. Cont.*, 2: 321-328.

Kothai, K., 1974. Bionomics and chemical control of the groundnut leaf miner *Stomopteryx subsecivella* (Zeller) (lepidoptea: Gelechiidae). *Msc(Ag) Thesis*, Univrsity of Agricultural Sciences, Bangalore, India, pp. 78.

Koul, D., 1985. Adiradichtin inetraction with development of *Spodoptera litura* Fab. *Ind. J. of Expt. Biology*, 23: 160-63

Kuhn, C.W., Paguio, D.R. and Adams, D.B., 1978. Tolerance in peanuts to peanut mottle virus. *Pl. Dis. Reptr.*, 62: 365-368.

Kuhn, C.W. and Demski, J.W., 1984. Peanut mottle. In: *Compendium of Peanut Diseases* (Eds. D.M. Porter, D.H. Smith., R. Rodriguez-Kabana. The American Phytopathological Society, pp. 73.

Lastin, S.M., Seif El- Nast, H.I., El-Nagar, M.A.A. and Nofal, M.A., 1989. Biological control of *Aspergillus niger* the causal organism of peanut crown rot by *Trichoderma harizanum*. *Ann. Agric. Sci.*, 34: 795-803.

Lingappa, S., 1990. Insect pest management under intercrop system,pp 197-206. In: *Intercropping Principles and Practices* (Eds. M.M. Hosmani, B.M. Chittpur and S.M. Hiramth). University of Agricultural Science, Dharward.

Logiswaran, G., Madhava Rao, G. Vasudevan and Kannan, V., 1982. Influence of time of sowing and weather factors on the infestation of leaf miner *Aproaerema modicella* Deventer and yiekd in rainfed groundnut. *Madras Agricultural Journal*, 69: 359-363.

Logiswaran, G. and Mohanasundram, M., 1985. Effect of intercropping, spacing and mulching in the control of groundnut leafminer, *Aproaerema modicella* Deventer (Gelechiidae: Lepidoptera). *Madras Agricultural Journal*, 72: 695-700.

Mohammad, A., 1981. The groundnut leaf miner (*Stomopteryx subsecivella* Zeller) (Lepidoptera: Gelechiidae). A eview of world literature. Occasional paper 3, Groundnut Improvement Program, Interational Crops Research Institute for Semi- Arid Tropics (ICRISAST), Patancheru, A.P., India. pp. 21.

Mohansundram, M., 1986. Insect pest management in groundnut. In: *Pest and Disease Management-Oilseeds, Pulses, Millets and Cotton* (Ed. S. Jayaraj). Tamil Nadu Agril. Univ., Coimbatore.

Moraes, S.A. and Godoy, I.J., 1985. Evaluation of resistance to *Cercosporidium personatum* In *Arachis hypogaea* genotypes. *Summa Phytopathologica*, 11: 140-151.

Muthiah, C. and Hussain. H.J., 1991. Bioefficacy of insecticides against groundnut pests. *Groundnut News*, 3: 6.

Muthiah, C., Senthivel, T., Venkatkrishnan, J. and Sivram, M.R., 1991. Effect of intercropping on incidence of pest and disease in groundnut (*Arachis hypogaea*). *Indian J. Agric. Sci.*, 61: 152-153.

Nandagopal, V., Soni, V.C., Hall, David R. and Gedia, M.V., 1995. Effects of certain components of IPM on the damage and yields in groundnut. Rsearch presented in the National Seminar on "*IPM in Agriculture*" held at the college of Agriculture, Nagpur on 29 and 30th of December, 1995.

Nandagopal. V., 1992. First record of insect pest and predator of thrips and jassids in groundnut. *International Arachis Newsletter*, 11: 26.

Nandagopal, V., 1988. *Crossopalpus* sp. an efficicent predator of jassids of groundnut. *Int. Arachis Newsl.*, 4: 18-19.

Nandagopal, V., 1992. Studies on the Integrated Pest Management in Groundnuut in Saurashtra. *Ph.D.Thesis*, Saurashtra University, Rajkot, pp. 246.

Nandagopal, V., Jaroli, R.K., Ramkumar and Reddy, P.S., 1991). Neem products: Possible insecticide on the groundnut jassid *Balclutha hortensis* Lindb. *Int. Arachis Newsl.*, 8: 22.

Nandagopal, V. and Reddy, P.S., 1990. Report of sex pheromone in groundnut leaf miner. *Int. Arachis Newsl.*, 7: 25-26.

Nandagopal. V., 1992a. First record of insect pest and predators of thrips and jassids in groundnut. *Int. Arachis Newsl.*, 11: 26.

Nofal, M.A., Seif EL-Nasr, H.I., Diab, M.M., EL-Nagar, M.A.A., and EL-Said, S.A.I., 1990. Effect of the systemic fungicides benlate and vitavax–capltan on *Aspergillus* crown rot incidence of peanut plants. *Ann. Agric. Sci.*, 15: 407-415.

Oblisami, G., Ramamurthy, K. and Rangaswami, G., 1969. Studies on the pathology of some crop pests of South India. *Mysore J. Agric. Sci.*, 3: 86-98.

Papavizas, G.C. and Collins, D.J., 1990. Influence of *Glioccadium virens* on germination and infectivity of sclerotia of *Sclerotium rolfsii*. *Phytopathology*, 80: 627-630.

Patel, H.K., 1968. Neem seed suspension: A sure deterrent to hairy caterpillar (*Amsacta moorie*). *Madras Agricultural J.*, 55(11): 509-510.

Patel, H.R., Vaishna, M.V. and Dhruj, I.U., 1986. Efficacy of aldicarb sulfone and carbofuron flowable seed treatment on plant growth and against *Meliodogyne arenaria* on groundnut. *Pesticides*, 20: 29-31.

Pawar, C.S. and Srivastava, C.P., 1988. Interaction between sex pheromones of *Spodoptera litura* Fab. and *Heliothis armigera* (Hub.). *Int. Arachis Newsl.*, 4: 22.

Pawar, C.S., Sithanantham, S., Bhatnagar, V.S., Srivastava, C.P. and Reed, W., 1988. The development of sex pheromone trapping of *Heliothis armigera* at ICRISAT, India. *Trop. Pest Mgmt.*, 34: 39-43.

Prasada Rao, R.D.V.J., Chakrabarty, S.K. and Reddy, A.S., 1988. First report of peanut stripe virus from India. *Plant Dis.*, 72: 912.

Ramkrishna, A., 1991. Integrated weed management: An efficient method of groundnut weed control. Poster Astracts, Second International Groundnut Workshop 25-29 November, ICRISAT, Patancheru, Andhra Pradesh India, pp. 633.

Rao, M.R. and Willey, R.W., 1980. Preliminary studies of intercropping combinations based on pigeonpea or sorghum. *Experimental Agriculture*, 16: 1-11.

Ratna, A.S., Rao, A.S., Reddy, B.L., Nolt, D.V.R., Reddy, M. Vijayalakshmi and McDonald, D., 1991. Studies on transmission of Indian peanut clump virus diseases by *Polymyxa graminis*. Annals of *Applied Biology*, 118: 71-78.

Reddy, D.V.R., Wightman, J.A., Bashear, R.J., Highland, B., Black, M., Sreenivasan, P., Dwivedi, S.L., Demski, J.W., McDonald, D., Smith, Jr. J.W. and Smith, D.H., 1991. Budnecrosis: A disease of groundut caused by tomato spotted wilt virus. Information Bulletin no. 31. ICRISAT, Patancheru, A.P., India.

Reddy, P.S. 1988. Genetics, breeding and varieties. In *Groundnut*, (ed. P.S. Reddy) pp. 508-525.

Reddy, D.V.R., Amin, P.W., McDonald, D. and Ghanekar, A.M., 1983. Epidemiology and control of groundnut bud necrosis and other diseases of legume crops in India caused by tomato spotted wilt virus. In: *Plant Disease Eplidemiology* (eds. R.T.Plumb and J.M. Tresh) Oxford, U.K., Blackwell. pp. 93-102.

Reddy, D.D.R., Subrahmanyam, P., Shanker Reddy, G.H., Raja Reddy, C. and Siva Rao, D.V., 1984. A nematode disease of peanut caused by *Tylenchorhynchus brevilineatus*. *Plant Dis.*, 68: 528-529.

Reddy, D.V.R. and Wihgtman, J.A., 1989. Tomato spotted wilt virus: Thrips transmission in disease vector research Vol. 5 (ed. K.F.Harris) pp. 203-219.

Rodriguez-Kabana, R., Roberson, D.G., Wells, L., Weaver, C.F. and King, P.S., 1991. Cotton as a rotation crop for the management of *Meloidogyne arenaria* and *Sclerotium rolfsii* in peanut. *Suppl. J. Nematology*, 23: 652-657.

Sakhuja, P.K. and Sethi, C.L., 1985. Screening of groundnut germplasm for resistance to root-kot nematode *Meloidogyne javanica*. *Indian J. Nematology*, 15: 129-130.

Salako, E.A. and Adu, J.K., 1990. Effect of foliar fungicides on foliar disease and pod, haulm and plant crude protein yield of groundnut. *Trop. Pest Mgmt.*, 16: 380-383.

Saxena, R.C., 1989. Insecticides from neem. In: *Insecticides of Plant Origin*. (Eds. J.T. Arnasan, J.R. Philogene and P. Morand). American Chemical Society Symposium Series 387:110-135, ACS, Washington, DC.

Schemutterer, H., 1990. Properties and potential of natural pesticides from the neem tree *Azadirachta indica*. *Annual Rev. Ent.*, 35: 271-297.

Shekhawat, P.S., Patel, V.N. and Patel, J.G., 1985. Economical fungicides spray schedule for control of tikka and rust disease of rainfed groundnut. *Indian J. Mycology Pl. Path.* 17: 11-16.

Shanmugam, N. and Govindasamy, G.V., 1973. Control of Marcrophomina root rot of Groundnut. *Madras Agric. J.* 60: 500-503.

Shanower, T.G., Wightman, J.A., Gutierrez, A.P. and Ranga Rao, G.V., 1992. Larval parasitoids and pathogens of the groundnut leaf miner, *Aproaerema modicella* (Lep.: Gelechiidae) in India. *Entomophaga*, 37(3): 419-427.

Sharma, S.B., Subrahmanyan, P. and Starr, E., 1990. Plant parasitic nematodes associated with groundnut in Niger. *Trop. Pest Mgmt.*, 36: 71-72.

Sharma, R.C., 1982. Studies on Groundnut root rot caused by *Macrophomina phaseolina* (Tessi) Goid (=*Rhizoctina batatiola* (Taub) Butler. *Ph.D. Thesis*, IARI, New Delhi.

Sharma, S.B. and McDoald, D., 1990. Global status of nematode problem of groundnut, pigeopea, chickpea, sorghum and pearlmillet and suggestions for future work. *Crop Prot.*, 9: 453-458.

Sharma, S.B., 1988. Survey for nematode diseases of Groundnut in some coastal and Rayalasama districts of Andhra Pradesh, India. *Int. Arachis Newsl.*, 3: 11-13.

Shekhawat, P.S., Patel, V.N. and Patel, J.G., 1985. Economical fungicides spray schedule for control of tikka and rust disease of rainfed groundnut. *Indian J. Mycology Pl. Path.*, 17: 11-16.

Shew, B.B., Wynne, J.C. and Beute, M.K., 1987. Field, microplot, and green house evaluations of resistance to *Sclerotium rolfsii* in peanut. *Plant Dis.*,71: 188-191.

Shivanarayan, N., 1985. Bird pests of national importance. In: *Integtrated Pest and Disease Management* (Ed. S.Jayaraj) pp. 285-289.

Srinivasan, S.D.V., Siva Rao and Sundrarajan, M., 1986. Influence of crop mannagement practices on incidences of leaf miner (*Aproaerema modicella* dev.) in groundnut. *J. Agric. Res. Andhra Pradesh Agric. Univ.*, 33: 381-382.

Starr, J.L., 1984. Expression of resistance in peanut, *Arachis hypogaea* to *Pratylenchus brachyrus*: Impact on screening for resistance. *J. Nematl.*, 16: 404-406.

Subrahmanyan, P. and McDonald, D., 1983. Rust disease of groundnut.Information Bulletin no.13, Patanncheru, AP, India, International Crops Research Institute for the Semi-Arid Tropics.

Subrahmanyan, P., Rao, V.R., McDonald, D., Moss, J.P. and Gibbons, R.W., 1989. Origin of resistance to rust and late leaf spot in peanut (*Arachis hypogaea* L.). *Econ. Bot.*, 43: 444–455.

Subrahmanyan, P. and McDonald, D., 1987. Groundnut rust disease: Epidemiology and control. ICRISAT (International Crops Research Innstitute for the Semi-Arid Tropics.),Groundnut rust disease. Proceedings of Discussion Group Meeting.24-28, September, ICRISAT Center, Patancheru, A.P., India.

Subramanyam, P. and Hildebrand, G.L., 1993. Current research on the management of early leaf spot of groundnut at SADC/ICRISAT Groundnut Project. In: *Proceedings of the third ICRISAT Regional Groundnut Meeting for West Africa*, Summary Ouagadougou, Burkina Faso, 14-17 Sept. pp. 49-50.

Upadhyay, V.R, and Vyas, H.N., 1987. Integrated pest control of major pest of groundnut in Gujarat in relation to population level and Coccinellid predators population. *Proc. Natn. Symp. Integrated Pest Cont.* Prog pros., 15- 17 October, Trivandrum, Kerala.

Usman, K.M., Durairaj, P., Karunanithi, K. and Thangavelu, S., 1991. Effect of neem products on the control of foliar disease of groundnut. Poster Abstract, *Second International Groundnut Workshop*, 25-November. ICRISAT, Patancheru, AP,India.

Varaprasad, K.S. and Sharma, S.B., 1990. First report of the lesion nematode, *Pratylenchus bachyurus* on groundnut in India. *Indian J. Pl. Prot.*, 18: 140.

Venkatarao, A. and Alice, D., 1991. Neem extract checks plant diseases. *Capart Press Clippings*, 7: 23.

Vidhyasekaran, P., 1981. Control of rust and tikka disease of groundnut. *Indian Phytopathology*, 34: 20-23.

Vijendra Singh, V., Ghewade, M.P. and Reddy, P.S., 1993. Peanut Stripe Virus disease-Present status and its management. *International Journal Pest Management*, 39: 422-430.

Vyas, S.C., Prasad, K.V.V., Shastry, P.P. and Shukla, B.N., 1986. Field control of leaf spots and rust of groundnut by fungicides. *Indian Phytopath.*, 39: 447-448.

Wightman, J.A. and Amin, P.W., 1988. Groundnut pests and their control in the Semi-Arid Tropics. *Trop. Pest Mgmt.*, 34: 218-226.

Yadav, C.P.C., 1985. *White grubs and their control (Bulletin)*. Indian Council of Agricultural Research, Krishi Bhavan, New Delhi, pp. 15.

Yadav, D.N., Patel, R.R. and Patel, R.C., 1987. Natural enemies of groundnut leaf miner *Aproaerema modicella* Deventer (Lepidoptera : Gelechiidae) and their impact in its infestation at Anand (Gujarat). *Gujarat Agric. Univ. Res. J.*, 13: 13-16.